Collaborative Statistics

By:

Barbara Illowsky, Ph.D.

Susan Dean

Collaborative Statistics

By:
Barbara Illowsky, Ph.D.
Susan Dean

Online:
<http://cnx.org/content/col10522/1.35/ >

C O N N E X I O N S

Rice University, Houston, Texas

Table of Contents

Preface[1]

Welcome to *Collaborative Statistics*, presented by Connexions. The initial section below introduces you to Connexions. If you are familiar with Connexions, please skip to About "Collaborative Statistics." (Section : About Connexions)

About Connexions

Connexions Modular Content
Connexions (cnx.org[2]) is an online, **open access** educational resource dedicated to providing high quality learning materials free online, free in printable PDF format, and at low cost in bound volumes through print-on-demand publishing. The *Collaborative Statistics* textbook is one of many **collections** available to Connexions users. Each **collection** is composed of a number of re-usable learning **modules** written in the Connexions XML markup language. Each module may also be re-used (or 're-purposed') as part of other collections and may be used outside of Connexions. Including *Collaborative Statistics*, Connexions currently offers over 6500 modules and more than 350 collections.

The modules of *Collaborative Statistics* are derived from the original paper version of the textbook under the same title, *Collaborative Statistics*. Each module represents a self-contained concept from the original work. Together, the modules comprise the original textbook.

Re-use and Customization
The Creative Commons (CC) Attribution license[3] applies to all Connexions modules. Under this license, any module in Connexions may be used or modified for any purpose as long as proper attribution to the original author(s) is maintained. Connexions' authoring tools make re-use (or re-purposing) easy. Therefore, instructors anywhere are permitted to create customized versions of the *Collaborative Statistics* textbook by editing modules, deleting unneeded modules, and adding their own supplementary modules. Connexions' authoring tools keep track of these changes and maintain the CC license's required attribution to the original authors. This process creates a new collection that can be viewed online, downloaded as a single PDF file, or ordered in any quantity by instructors and students as a low-cost printed textbook. To start building custom collections, please visit the help page, "Create a Collection with Existing Modules"[4] . For a guide to authoring modules, please look at the help page, "Create a Module in Minutes"[5] .

Read the book online, print the PDF, or buy a copy of the book.
To browse the *Collaborative Statistics* textbook online, visit the collection home page at cnx.org/content/col10522/latest[6]. You will then have three options.

[1]This content is available online at <http://cnx.org/content/m16026/1.16/>.
[2]http://cnx.org/
[3]http://creativecommons.org/licenses/by/2.0/
[4]http://cnx.org/help/CreateCollection
[5]http://cnx.org/help/ModuleInMinutes
[6]*Collaborative Statistics* <http://cnx.org/content/col10522/latest/>

1. You may obtain a PDF of the entire textbook to print or view offline by clicking on the "Download PDF" link in the "Content Actions" box.
2. You may order a bound copy of the collection by clicking on the "Order Printed Copy" button.
3. You may view the collection modules online by clicking on the "Start ≫" link, which takes you to the first module in the collection. You can then navigate through the subsequent modules by using their "Next ≫" and "Previous ≫" links to move forward and backward in the collection. You can jump to any module in the collection by clicking on that module's title in the "Collection Contents" box on the left side of the window. If these contents are hidden, make them visible by clicking on "[show table of contents]".

Accessibility and Section 508 Compliance

- For information on general Connexions accessibility features, please visit http://cnx.org/content/m17212/latest/[7].
- For information on accessibility features specific to the Collaborative Statistics textbook, please visit http://cnx.org/content/m17211/latest/[8].

Version Change History and Errata

- For a list of modifications, updates, and corrections, please visit http://cnx.org/content/m17360/latest/[9].

Adoption and Usage

- The Collaborative Statistics collection has been adopted and customized by a number of professors and educators for use in their classes. For a list of known versions and adopters, please visit http://cnx.org/content/m18261/latest/[10].

About "Collaborative Statistics"

Collaborative Statistics was written by Barbara Illowsky and Susan Dean, faculty members at De Anza College in Cupertino, California. The textbook was developed over several years and has been used in regular and honors-level classroom settings and in distance learning classes. Courses using this textbook have been articulated by the University of California for transfer of credit. The textbook contains full materials for course offerings, including expository text, examples, labs, homework, and projects. A Teacher's Guide is currently available in print form and on the Connexions site at http://cnx.org/content/col10547/latest/[11], and supplemental course materials including additional problem sets and video lectures are available at http://cnx.org/content/col10586/latest/[12]. The on-line text for each of these collections collections will meet the Section 508 standards for accessibility.

An on-line course based on the textbook was also developed by Illowsky and Dean. It has won an award as the best on-line California community college course. The on-line course will be available at a later date as a collection in Connexions, and each lesson in the on-line course will be linked to the on-line textbook chapter. The on-line course will include, in addition to expository text and examples, videos of course lectures in captioned and non-captioned format.

The original preface to the book as written by professors Illowsky and Dean, now follows:

[7]"Accessibility Features of Connexions" <http://cnx.org/content/m17212/latest/>
[8]"Collaborative Statistics: Accessibility" <http://cnx.org/content/m17211/latest/>
[9]"Collaborative Statistics: Change History" <http://cnx.org/content/m17360/latest/>
[10]"Collaborative Statistics: Adoption and Usage" <http://cnx.org/content/m18261/latest/>
[11]*Collaborative Statistics Teacher's Guide* <http://cnx.org/content/col10547/latest/>
[12]*Collaborative Statistics: Supplemental Course Materials* <http://cnx.org/content/col10586/latest/>

This book is intended for introductory statistics courses being taken by students at two– and four–year colleges who are majoring in fields other than math or engineering. Intermediate algebra is the only pre-requisite. The book focuses on applications of statistical knowledge rather than the theory behind it. The text is named *Collaborative Statistics* because students learn best by **doing**. In fact, they learn best by working in small groups. The old saying "two heads are better than one" truly applies here.

Our emphasis in this text is on four main concepts:

- thinking statistically
- incorporating technology
- working collaboratively
- writing thoughtfully

These concepts are integral to our course. Students learn the best by actively participating, not by just watching and listening. Teaching should be highly interactive. Students need to be thoroughly engaged in the learning process in order to make sense of statistical concepts. *Collaborative Statistics* provides techniques for students to write across the curriculum, to collaborate with their peers, to think statistically, and to incorporate technology.

This book takes students step by step. The text is interactive. Therefore, students can immediately apply what they read. Once students have completed the process of problem solving, they can tackle interesting and challenging problems relevant to today's world. The problems require the students to apply their newly found skills. In addition, technology (TI-83 graphing calculators are highlighted) is incorporated throughout the text and the problems, as well as in the special group activities and projects. The book also contains labs that use real data and practices that lead students step by step through the problem solving process.

At De Anza, along with hundreds of other colleges across the country, the college audience involves a large number of ESL students as well as students from many disciplines. The ESL students, as well as the non-ESL students, have been especially appreciative of this text. They find it extremely readable and understandable. *Collaborative Statistics* has been used in classes that range from 20 to 120 students, and in regular, honor, and distance learning classes.

Susan Dean

Barbara Illowsky

Additional Resources[13]

Additional Resources Currently Available

The following section describes some additional resources for learners and educators. These modules and collections are all available on the Connexions website (http://cnx.org/[14]) and can be viewed online, downloaded, printed, or ordered as appropriate.

Glossary
This module contains the entire glossary for the Collaborative Statistics textbook collection (col10522) since its initial release on 15 July 2008. The glossary is located at http://cnx.org/content/m16129/latest/[15].

View or Download This Textbook Online
The complete contents of this book are available at no cost on the Connexions website at http://cnx.org/content/col10522/latest/[16]. Anybody can view this content free of charge either as an online e-book or a downloadable PDF file. A low-cost printed version of this textbook is also available here[17] .

Collaborative Statistics Teacher's Guide
A complementary Teacher's Guide for Collaborative statistics is available through Connexions at http://cnx.org/content/col10547/latest/[18]. The Teacher's Guide includes suggestions for presenting concepts found throughout the book as well as recommended homework assignments. A low-cost printed version of this textbook is also available here[19] .

Supplemental Materials
This companion to Collaborative Statistics provides a number of additional resources for use by students and instructors based on the award winning Elementary Statistics Sofia online course[20] , also by textbook

[13] This content is available online at <http://cnx.org/content/m18746/1.5/>.
[14] http://cnx.org/
[15] "Collaborative Statistics: Glossary" <http://cnx.org/content/m16129/latest/>
[16] *Collaborative Statistics* <http://cnx.org/content/col10522/latest/>
[17] http://my.qoop.com/store/7064943342106149/7781159220340
[18] *Collaborative Statistics Teacher's Guide* <http://cnx.org/content/col10547/latest/>
[19] http://my.qoop.com/store/7064943342106149/8791310589747
[20] http://sofia.fhda.edu/gallery/statistics/index.html

authors Barbara Illowsky and Susan Dean. This content is designed to complement the textbook by providing video tutorials, course management materials, and sample problem sets. The Supplemental Materials collection can be found at http://cnx.org/content/col10586/latest/[21].

Video Lectures

- Video Lecture 1: Sampling and Data[22]
- Video Lecture 2: Descriptive Statistics[23]
- Video Lecture 3: Probability Topics[24]
- Video Lecture 4: Discrete Distributions[25]
- Video Lecture 5: Continuous Random Variables[26]
- Video Lecture 6: The Normal Distribution[27]
- Video Lecture 7: The Central Limit Theorem[28]
- Video Lecture 8: Confidence Intervals[29]
- Video Lecture 9: Hypothesis Testing with a Single Mean[30]
- Video Lecture 10: Hypothesis Testing with Two Means[31]
- Video Lecture 11: The Chi-Square Distribution[32]
- Video Lecture 12: Linear Regression and Correlation[33]

Version History
This module contains a listing of changes, updates, and corrections made to the Collaborative Statistics textbook collection (col10522) since its initial release on 15 July 2008. The Version History is located at http://cnx.org/content/m17360/latest/[34].

Textbook Adoption and Usage
This module is designed to track the various derivations of the Collaborative Statistics textbook and its various companion resources, as well as keep track of educators who have adopted various versions for their courses. New adopters are encouraged to provide their contact information and describe how they will use this book for their courses. The goal is to provide a list that will allow educators using this book to collaborate, share ideas, and make suggestions for future development of this text. The Adoption and Usage module is located at http://cnx.org/content/m18261/latest/[35].

Additional Technologies
In order to provide the most flexible learning resources possible, we invite collaboration from all instructors wishing to create customized versions of this content for use with other technologies. For instance, you may be interested in creating a set of instructions similar to this collection's calculator notes. If you would like to contribute to this collection, please use the contact the authors with any ideas or materials you have created.

---

[21] *Collaborative Statistics: Supplemental Course Materials* <http://cnx.org/content/col10586/latest/>
[22] "Elementary Statistics: Video Lecture - Sampling and Data" <http://cnx.org/content/m17561/latest/>
[23] "Elementary Statistics: Video Lecture - Descriptive Statistics" <http://cnx.org/content/m17562/latest/>
[24] "Elementary Statistics: Video Lecture - Probability Topics" <http://cnx.org/content/m17563/latest/>
[25] "Elementary Statistics: Video Lecture - Discrete Distributions" <http://cnx.org/content/m17565/latest/>
[26] "Elementary Statistics: Video Lecture - Continuous Random Variables" <http://cnx.org/content/m17566/latest/>
[27] "Elementary Statistics: Video Lecture - The Normal Distribution" <http://cnx.org/content/m17567/latest/>
[28] "Elementary Statistics: Video Lecture - The Central Limit Theorem" <http://cnx.org/content/m17568/latest/>
[29] "Elementary Statistics: Video Lecture - Confidence Intervals" <http://cnx.org/content/m17569/latest/>
[30] "Elementary Statistics: Video Lecture - Hypothesis Testing with a Single Mean" <http://cnx.org/content/m17570/latest/>
[31] "Elementary Statistics: Video Lecture - Hypothesis Testing with Two Means" <http://cnx.org/content/m17577/latest/>
[32] "Elementary Statistics: Video Lecture - The Chi-Square Distribution" <http://cnx.org/content/m17571/latest/>
[33] "Elementary Statistics: Video Lecture - Linear Regression and Correlation" <http://cnx.org/content/m17572/latest/>
[34] "Collaborative Statistics: Change History" <http://cnx.org/content/m17360/latest/>
[35] "Collaborative Statistics: Adoption and Usage" <http://cnx.org/content/m18261/latest/>

Accessibility and Section 508 Compliance

- For information on general Connexions accessibility features, please visit http://cnx.org/content/m17212/latest/[36].
- For information on accessibility features specific to the Collaborative Statistics textbook, please visit http://cnx.org/content/m17211/latest/[37].

[36]"Accessibility Features of Connexions" <http://cnx.org/content/m17212/latest/>
[37]"Collaborative Statistics: Accessibility" <http://cnx.org/content/m17211/latest/>

Author Ackowledgements [38]

We wish to acknowledge the many people who have helped us and have encouraged us in this project. At De Anza, Donald Rossi and Rupinder Sekhon and their contagious enthusiasm started us on our path to this book. Inna Grushko and Diane Mathios painstakingly checked every practice and homework problem. Inna also wrote the glossary and offered invaluable suggestions.

Kathy Plum co-taught with us the first term we introduced the TI-85. Lenore Desilets, Charles Klein, Kathy Plum, Janice Hector, Vernon Paige, Carol Olmstead, and Donald Rossi of De Anza College, Ann Flanigan of Kapiolani Community College, Birgit Aquilonius of West Valley College, and Terri Teegarden of San Diego Mesa College, graciously volunteered to teach out of our early editions. Janice Hector and Lenore Desilets also contributed problems. Diane Mathios and Carol Olmstead contributed labs as well. In addition, Diane and Kathy have been our "sounding boards" for new ideas. In recent years, Lisa Markus, Vladimir Logvinenko, and Roberta Bloom have contributed valuable suggestions.

Jim Lucas and Valerie Hauber of De Anza's Office of Institutional Research, along with Mary Jo Kane of Health Services, provided us with a wealth of data.

We would also like to thank the thousands of students who have used this text. So many of them gave us permission to include their outstanding word problems as homework. They encouraged us to turn our note packet into this book, have offered suggestions and criticisms, and keep us going.

Finally, we owe much to Frank, Jeffrey, and Jessica Dean and to Dan, Rachel, Matthew, and Rebecca Illowsky, who encouraged us to continue with our work and who had to hear more than their share of "I'm sorry, I can't" and "Just a minute, I'm working."

[38] This content is available online at <http://cnx.org/content/m16308/1.6/>.

Student Welcome Letter[39]

Dear Student:

Have you heard others say, "You're taking statistics? That's the hardest course I ever took!" They say that, because they probably spent the entire course confused and struggling. They were probably lectured to and never had the chance to experience the subject. You will not have that problem. Let's find out why.

There is a Chinese Proverb that describes our feelings about the field of statistics:

I HEAR, AND I FORGET

I SEE, AND I REMEMBER

I DO, AND I UNDERSTAND

Statistics is a "do" field. In order to learn it, you must "do" it. We have structured this book so that you will have hands-on experiences. They will enable you to truly understand the concepts instead of merely going through the requirements for the course.

What makes this book different from other texts? First, we have eliminated the drudgery of tedious calculations. You might be using computers or graphing calculators so that you do not need to struggle with algebraic manipulations. Second, this course is taught as a collaborative activity. With others in your class, you will work toward the common goal of learning this material.

Here are some hints for success in your class:

- Work hard and work every night.
- Form a study group and learn together.
- Don't get discouraged - you can do it!
- As you solve problems, ask yourself, "Does this answer make sense?"
- Many statistics words have the same meaning as in everyday English.
- Go to your teacher for help as soon as you need it.
- Don't get behind.
- Read the newspaper and ask yourself, "Does this article make sense?"
- Draw pictures - they truly help!

Good luck and don't give up!

Sincerely,
Susan Dean and Barbara Illowsky

De Anza College
21250 Stevens Creek Blvd.
Cupertino, California 95014

[39]This content is available online at <http://cnx.org/content/m16305/1.5/>.

Chapter 1

Sampling and Data

1.1 Sampling and Data[1]

1.1.1 Student Learning Objectives

By the end of this chapter, the student should be able to:

- Recognize and differentiate between key terms.
- Apply various types of sampling methods to data collection.
- Create and interpret frequency tables.

1.1.2 Introduction

You are probably asking yourself the question, "When and where will I use statistics?". If you read any newspaper or watch television, or use the Internet, you will see statistical information. There are statistics about crime, sports, education, politics, and real estate. Typically, when you read a newspaper article or watch a news program on television, you are given sample information. With this information, you may make a decision about the correctness of a statement, claim, or "fact." Statistical methods can help you make the "best educated guess."

Since you will undoubtedly be given statistical information at some point in your life, you need to know some techniques to analyze the information thoughtfully. Think about buying a house or managing a budget. Think about your chosen profession. The fields of economics, business, psychology, education, biology, law, computer science, police science, and early childhood development require at least one course in statistics.

Included in this chapter are the basic ideas and words of probability and statistics. You will soon understand that statistics and probability work together. You will also learn how data are gathered and what "good" data are.

1.2 Statistics[2]

The science of **statistics** deals with the collection, analysis, interpretation, and presentation of **data**. We see and use data in our everyday lives. To be able to use data correctly is essential to many professions and in your own best self-interest.

[1] This content is available online at <http://cnx.org/content/m16008/1.8/>.
[2] This content is available online at <http://cnx.org/content/m16020/1.12/>.

1.2.1 Optional Collaborative Classroom Exercise

In your classroom, try this exercise. Have class members write down the average time (in hours, to the nearest half-hour) they sleep per night. Your instructor will record the data. Then create a simple graph (called a **dot plot**) of the data. A dot plot consists of a number line and dots (or points) positioned above the number line. For example, consider the following data:

5; 5.5; 6; 6; 6; 6.5; 6.5; 6.5; 6.5; 7; 7; 8; 8; 9

The dot plot for this data would be as follows:

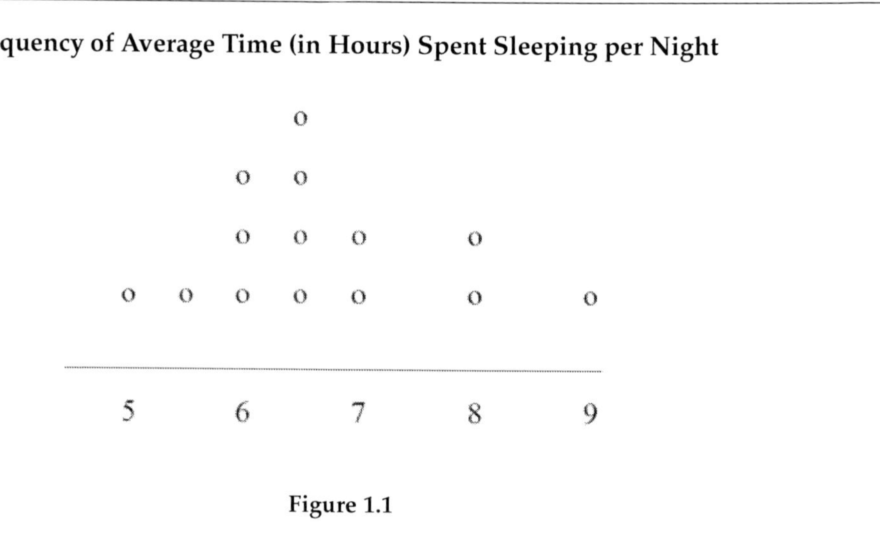

Figure 1.1

Does your dot plot look the same as or different from the example? Why? If you did the same example in an English class with the same number of students, do you think the results would be the same? Why or why not?

Where do your data appear to cluster? How could you interpret the clustering?

The questions above ask you to analyze and interpret your data. With this example, you have begun your study of statistics.

In this course, you will learn how to organize and summarize data. Organizing and summarizing data is called **descriptive statistics**. Two ways to summarize data are by graphing and by numbers (for example, finding an average). After you have studied probability and probability distributions, you will use formal methods for drawing conclusions from "good" data. The formal methods are called **inferential statistics**. Statistical inference uses probability to determine if conclusions drawn are reliable or not.

Effective interpretation of data (inference) is based on good procedures for producing data and thoughtful examination of the data. You will encounter what will seem to be too many mathematical formulas for interpreting data. The goal of statistics is not to perform numerous calculations using the formulas, but to gain an understanding of your data. The calculations can be done using a calculator or a computer. The understanding must come from you. If you can thoroughly grasp the basics of statistics, you can be more confident in the decisions you make in life.

1.3 Probability[3]

Probability is the mathematical tool used to study randomness. It deals with the chance of an event occurring. For example, if you toss a **fair** coin 4 times, the outcomes may not be 2 heads and 2 tails. However, if you toss the same coin 4,000 times, the outcomes will be close to 2,000 heads and 2,000 tails. The expected theoretical probability of heads in any one toss is $\frac{1}{2}$ or 0.5. Even though the outcomes of a few repetitions are uncertain, there is a regular pattern of outcomes when there are many repetitions. After reading about the English statistician Karl Pearson who tossed a coin 24,000 times with a result of 12,012 heads, one of the authors tossed a coin 2,000 times. The results were 996 heads. The fraction $\frac{996}{2000}$ is equal to 0.498 which is very close to 0.5, the expected probability.

The theory of probability began with the study of games of chance such as poker. Today, probability is used to predict the likelihood of an earthquake, of rain, or whether you will get a A in this course. Doctors use probability to determine the chance of a vaccination causing the disease the vaccination is supposed to prevent. A stockbroker uses probability to determine the rate of return on a client's investments. You might use probability to decide to buy a lottery ticket or not. In your study of statistics, you will use the power of mathematics through probability calculations to analyze and interpret your data.

1.4 Key Terms[4]

In statistics, we generally want to study a **population**. You can think of a population as an entire collection of persons, things, or objects under study. To study the larger population, we select a **sample**. The idea of **sampling** is to select a portion (or subset) of the larger population and study that portion (the sample) to gain information about the population. Data are the result of sampling from a population.

Because it takes a lot of time and money to examine an entire population, sampling is a very practical technique. If you wished to compute the overall grade point average at your school, it would make sense to select a sample of students who attend the school. The data collected from the sample would be the students' grade point averages. In presidential elections, opinion poll samples of 1,000 to 2,000 people are taken. The opinion poll is supposed to represent the views of the people in the entire country. Manufacturers of canned carbonated drinks take samples to determine if a 16 ounce can contains 16 ounces of carbonated drink.

From the sample data, we can calculate a statistic. A **statistic** is a number that is a property of the sample. For example, if we consider one math class to be a sample of the population of all math classes, then the average number of points earned by students in that one math class at the end of the term is an example of a statistic. The statistic is an estimate of a population parameter. A **parameter** is a number that is a property of the population. Since we considered all math classes to be the population, then the average number of points earned per student over all the math classes is an example of a parameter.

One of the main concerns in the field of statistics is how accurately a statistic estimates a parameter. The accuracy really depends on how well the sample represents the population. The sample must contain the characteristics of the population in order to be a **representative sample**. We are interested in both the sample statistic and the population parameter in inferential statistics. In a later chapter, we will use the sample statistic to test the validity of the established population parameter.

A **variable**, notated by capital letters like X and Y, is a characteristic of interest for each person or thing in a population. Variables may be **numerical** or **categorical**. **Numerical variables** take on values with equal units such as weight in pounds and time in hours. **Categorical variables** place the person or thing into a

[3]This content is available online at <http://cnx.org/content/m16015/1.9/>.

[4]This content is available online at <http://cnx.org/content/m16007/1.14/>.

category. If we let X equal the number of points earned by one math student at the end of a term, then X is a numerical variable. If we let Y be a person's party affiliation, then examples of Y include Republican, Democrat, and Independent. Y is a categorical variable. We could do some math with values of X (calculate the average number of points earned, for example), but it makes no sense to do math with values of Y (calculating an average party affiliation makes no sense).

Data are the actual values of the variable. They may be numbers or they may be words. Datum is a single value.

Two words that come up often in statistics are **average** and **proportion**. If you were to take three exams in your math classes and obtained scores of 86, 75, and 92, you calculate your average score by adding the three exam scores and dividing by three (your average score would be 84.3 to one decimal place). If, in your math class, there are 40 students and 22 are men and 18 are women, then the proportion of men students is $\frac{22}{40}$ and the proportion of women students is $\frac{18}{40}$. Average and proportion are discussed in more detail in later chapters.

Example 1.1
Define the key terms from the following study: We want to know the average amount of money first year college students spend at ABC College on school supplies that do not include books. We randomly survey 100 first year students at the college. Three of those students spent \$150, \$200, and \$225, respectively.

Solution
The **population** is all first year students attending ABC College this term.

The **sample** could be all students enrolled in one section of a beginning statistics course at ABC College (although this sample may not represent the entire population).

The **parameter** is the average amount of money spent (excluding books) by first year college students at ABC College this term.

The **statistic** is the average amount of money spent (excluding books) by first year college students in the sample.

The **variable** could be the amount of money spent (excluding books) by one first year student. Let X = the amount of money spent (excluding books) by one first year student attending ABC College.

The **data** are the dollar amounts spent by the first year students. Examples of the data are \$150, \$200, and \$225.

1.4.1 Optional Collaborative Classroom Exercise

Do the following exercise collaboratively with up to four people per group. Find a population, a sample, the parameter, the statistic, a variable, and data for the following study: You want to determine the average number of glasses of milk college students drink per day. Suppose yesterday, in your English class, you asked five students how many glasses of milk they drank the day before. The answers were 1, 0, 1, 3, and 4 glasses of milk.

1.5 Data[5]

Data may come from a population or from a sample. Small letters like x or y generally are used to represent data values. Most data can be put into the following categories:

- Qualitative
- Quantitative

Qualitative data are the result of categorizing or describing attributes of a population. Hair color, blood type, ethnic group, the car a person drives, and the street a person lives on are examples of qualitative data. Qualitative data are generally described by words or letters. For instance, hair color might be black, dark brown, light brown, blonde, gray, or red. Blood type might be AB+, O-, or B+. Qualitative data are not as widely used as quantitative data because many numerical techniques do not apply to the qualitative data. For example, it does not make sense to find an average hair color or blood type.

Quantitative data are always numbers and are usually the data of choice because there are many methods available for analyzing the data. Quantitative data are the result of **counting** or **measuring** attributes of a population. Amount of money, pulse rate, weight, number of people living in your town, and the number of students who take statistics are examples of quantitative data. Quantitative data may be either **discrete** or **continuous**.

All data that are the result of counting are called **quantitative discrete data**. These data take on only certain numerical values. If you count the number of phone calls you receive for each day of the week, you might get 0, 1, 2, 3, etc.

All data that are the result of measuring are **quantitative continuous data** assuming that we can measure accurately. Measuring angles in radians might result in the numbers $\frac{\pi}{6}$, $\frac{\pi}{3}$, $\frac{\pi}{2}$, π, $\frac{3\pi}{4}$, etc. If you and your friends carry backpacks with books in them to school, the numbers of books in the backpacks are discrete data and the weights of the backpacks are continuous data.

Example 1.2: Data Sample of Quantitative Discrete Data
The data are the number of books students carry in their backpacks. You sample five students. Two students carry 3 books, one student carries 4 books, one student carries 2 books, and one student carries 1 book. The numbers of books (3, 4, 2, and 1) are the quantitative discrete data.

Example 1.3: Data Sample of Quantitative Continuous Data
The data are the weights of the backpacks with the books in it. You sample the same five students. The weights (in pounds) of their backpacks are 6.2, 7, 6.8, 9.1, 4.3. Notice that backpacks carrying three books can have different weights. Weights are quantitative continuous data because weights are measured.

Example 1.4: Data Sample of Qualitative Data
The data are the colors of backpacks. Again, you sample the same five students. One student has a red backpack, two students have black backpacks, one student has a green backpack, and one student has a gray backpack. The colors red, black, black, green, and gray are qualitative data.

NOTE: You may collect data as numbers and report it categorically. For example, the quiz scores for each student are recorded throughout the term. At the end of the term, the quiz scores are reported as A, B, C, D, or F.

Example 1.5
Work collaboratively to determine the correct data type (quantitative or qualitative). Indicate whether quantitative data are continuous or discrete. Hint: Data that are discrete often start with the words "the number of."

[5] This content is available online at <http://cnx.org/content/m16005/1.12/>.

1. The number of pairs of shoes you own.
2. The type of car you drive.
3. Where you go on vacation.
4. The distance it is from your home to the nearest grocery store.
5. The number of classes you take per school year.
6. The tuition for your classes
7. The type of calculator you use.
8. Movie ratings.
9. Political party preferences.
10. Weight of sumo wrestlers.
11. Amount of money (in dollars) won playing poker.
12. Number of correct answers on a quiz.
13. Peoples' attitudes toward the government.
14. IQ scores. (This may cause some discussion.)

1.6 Sampling[6]

Gathering information about an entire population often costs too much or is virtually impossible. Instead, we use a sample of the population. **A sample should have the same characteristics as the population it is representing.** Most statisticians use various methods of random sampling in an attempt to achieve this goal. This section will describe a few of the most common methods.

There are several different methods of **random sampling**. In each form of random sampling, each member of a population initially has an equal chance of being selected for the sample. Each method has pros and cons. The easiest method to describe is called a **simple random sample**. Two simple random samples contain members equally representative of the entire population. In other words, each sample of the same size has an equal chance of being selected. For example, suppose Lisa wants to form a four-person study group (herself and three other people) from her pre-calculus class, which has 32 members including Lisa. To choose a simple random sample of size 3 from the other members of her class, Lisa could put all 32 names in a hat, shake the hat, close her eyes, and pick out 3 names. A more technological way is for Lisa to first list the last names of the members of her class together with a two-digit number as shown below.

[6]This content is available online at <http://cnx.org/content/m16014/1.13/>.

Class Roster

ID	Name
00	Anselmo
01	Bautista
02	Bayani
03	Cheng
04	Cuarismo
05	Cuningham
06	Fontecha
07	Hong
08	Hoobler
09	Jiao
10	Khan
11	King
12	Legeny
13	Lundquist
14	Macierz
15	Motogawa
16	Okimoto
17	Patel
18	Price
19	Quizon
20	Reyes
21	Roquero
22	Roth
23	Rowell
24	Salangsang
25	Slade
26	Stracher
27	Tallai
28	Tran
29	Wai
30	Wood

Table 1.1

Lisa can either use a table of random numbers (found in many statistics books as well as mathematical handbooks) or a calculator or computer to generate random numbers. For this example, suppose Lisa chooses to generate random numbers from a calculator. The numbers generated are:

.94360; .99832; .14669; .51470; .40581; .73381; .04399

Lisa reads two-digit groups until she has chosen three class members (that is, she reads .94360 as the groups 94, 43, 36, 60). Each random number may only contribute one class member. If she needed to, Lisa could have generated more random numbers.

The random numbers .94360 and .99832 do not contain appropriate two digit numbers. However the third random number, .14669, contains 14 (the fourth random number also contains 14), the fifth random number contains 05, and the seventh random number contains 04. The two-digit number 14 corresponds to Macierz, 05 corresponds to Cunningham, and 04 corresponds to Cuarismo. Besides herself, Lisa's group will consist of Marcierz, and Cunningham, and Cuarismo.

Sometimes, it is difficult or impossible to obtain a simple random sample because populations are too large. Then we choose other forms of sampling methods that involve a chance process for getting the sample. **Other well-known random sampling methods are the stratified sample, the cluster sample, and the systematic sample.**

To choose a **stratified sample**, divide the population into groups called strata and then take a sample from each stratum. For example, you could stratify (group) your college population by department and then choose a simple random sample from each stratum (each department) to get a stratified random sample. To choose a simple random sample from each department, number each member of the first department, number each member of the second department and do the same for the remaining departments. Then use simple random sampling to choose numbers from the first department and do the same for each of the remaining departments. Those numbers picked from the first department, picked from the second department and so on represent the members who make up the stratified sample.

To choose a **cluster sample**, divide the population into strata and then randomly select some of the strata. All the members from these strata are in the cluster sample. For example, if you randomly sample four departments from your stratified college population, the four departments make up the cluster sample. You could do this by numbering the different departments and then choose four different numbers using simple random sampling. All members of the four departments with those numbers are the cluster sample.

To choose a **systematic sample**, randomly select a starting point and take every nth piece of data from a listing of the population. For example, suppose you have to do a phone survey. Your phone book contains 20,000 residence listings. You must choose 400 names for the sample. Number the population 1 - 20,000 and then use a simple random sample to pick a number that represents the first name of the sample. Then choose every 50th name thereafter until you have a total of 400 names (you might have to go back to the of your phone list). Systematic sampling is frequently chosen because it is a simple method.

A type of sampling that is nonrandom is convenience sampling. **Convenience sampling** involves using results that are readily available. For example, a computer software store conducts a marketing study by interviewing potential customers who happen to be in the store browsing through the available software. The results of convenience sampling may be very good in some cases and highly biased (favors certain outcomes) in others.

Sampling data should be done very carefully. Collecting data carelessly can have devastating results. Surveys mailed to households and then returned may be very biased (for example, they may favor a certain group). It is better for the person conducting the survey to select the sample respondents.

When you analyze data, it is important to be aware of **sampling errors** and nonsampling errors. The actual process of sampling causes sampling errors. For example, the sample may not be large enough or representative of the population. Factors not related to the sampling process cause **nonsampling errors**. A defective counting device can cause a nonsampling error.

Example 1.6
Determine the type of sampling used (simple random, stratified, systematic, cluster, or convenience).

1. A soccer coach selects 6 players from a group of boys aged 8 to 10, 7 players from a group of boys aged 11 to 12, and 3 players from a group of boys aged 13 to 14 to form a recreational soccer team.
2. A pollster interviews all human resource personnel in five different high tech companies.
3. An engineering researcher interviews 50 women engineers and 50 men engineers.
4. A medical researcher interviews every third cancer patient from a list of cancer patients at a local hospital.
5. A high school counselor uses a computer to generate 50 random numbers and then picks students whose names correspond to the numbers.
6. A student interviews classmates in his algebra class to determine how many pairs of jeans a student owns, on the average.

Solution

1. stratified
2. cluster
3. stratified
4. systematic
5. simple random
6. convenience

If we were to examine two samples representing the same population, they would, more than likely, not be the same. Just as there is variation in data, there is variation in samples. As you become accustomed to sampling, the variability will seem natural.

Example 1.7
Suppose ABC College has 10,000 part-time students (the population). We are interested in the average amount of money a part-time student spends on books in the fall term. Asking all 10,000 students is an almost impossible task.

Suppose we take two different samples.

First, we use convenience sampling and survey 10 students from a first term organic chemistry class. Many of these students are taking first term calculus in addition to the organic chemistry class . The amount of money they spend is as follows:

$128; $87; $173; $116; $130; $204; $147; $189; $93; $153

The second sample is taken by using a list from the P.E. department of senior citizens who take P.E. classes and taking every 5th senior citizen on the list, for a total of 10 senior citizens. They spend:

$50; $40; $36; $15; $50; $100; $40; $53; $22; $22

Problem 1
Do you think that either of these samples is representative of (or is characteristic of) the entire 10,000 part-time student population?

Solution

No. The first sample probably consists of science-oriented students. Besides the chemistry course, some of them are taking first-term calculus. Books for these classes tend to be expensive. Most of these students are, more than likely, paying more than the average part-time student for their books. The second sample is a group of senior citizens who are, more than likely, taking courses for health and interest. The amount of money they spend on books is probably much less than the average part-time student. Both samples are biased. Also, in both cases, not all students have a chance to be in either sample.

Problem 2

Since these samples are not representative of the entire population, is it wise to use the results to describe the entire population?

Solution

No. Never use a sample that is not representative or does not have the characteristics of the population.

Now, suppose we take a third sample. We choose ten different part-time students from the disciplines of chemistry, math, English, psychology, sociology, history, nursing, physical education, art, and early childhood development. Each student is chosen using simple random sampling. Using a calculator, random numbers are generated and a student from a particular discipline is selected if he/she has a corresponding number. The students spend:

$180; $50; $150; $85; $260; $75; $180; $200; $200; $150

Problem 3

Do you think this sample is representative of the population?

Solution

Yes. It is chosen from different disciplines across the population.

Students often ask if it is "good enough" to take a sample, instead of surveying the entire population. If the survey is done well, the answer is yes.

1.6.1 Optional Collaborative Classroom Exercise

Exercise 1.6.1

As a class, determine whether or not the following samples are representative. If they are not, discuss the reasons.

1. To find the average GPA of all students in a university, use all honor students at the university as the sample.
2. To find out the most popular cereal among young people under the age of 10, stand outside a large supermarket for three hours and speak to every 20th child under age 10 who enters the supermarket.
3. To find the average annual income of all adults in the United States, sample U.S. congressmen. Create a cluster sample by considering each state as a stratum (group). By using simple random sampling, select states to be part of the cluster. Then survey every U.S. congressman in the cluster.

4. To determine the proportion of people taking public transportation to work, survey 20 people in New York City. Conduct the survey by sitting in Central Park on a bench and interviewing every person who sits next to you.
5. To determine the average cost of a two day stay in a hospital in Massachusetts, survey 100 hospitals across the state using simple random sampling.

1.7 Variation[7]

1.7.1 Variation in Data

Variation is present in any set of data. For example, 16-ounce cans of beverage may contain more or less than 16 ounces of liquid. In one study, eight 16 ounce cans were measured and produced the following amount (in ounces) of beverage:

15.8; 16.1; 15.2; 14.8; 15.8; 15.9; 16.0; 15.5

Measurements of the amount of beverage in a 16-ounce can may vary because different people make the measurements or because the exact amount, 16 ounces of liquid, was not put into the cans. Manufacturers regularly run tests to determine if the amount of beverage in a 16-ounce can falls within the desired range.

Be aware that as you take data, your data may vary somewhat from the data someone else is taking for the same purpose. This is completely natural. However, if two or more of you are taking the same data and get very different results, it is time for you and the others to reevaluate your data-taking methods and your accuracy.

1.7.2 Variation in Samples

It was mentioned previously that two or more **samples** from the same **population** and having the same characteristics as the population may be different from each other. Suppose Doreen and Jung both decide to study the average amount of time students sleep each night and use all students at their college as the population. Doreen uses systematic sampling and Jung uses cluster sampling. Doreen's sample will be different from Jung's sample even though both samples have the characteristics of the population. Even if Doreen and Jung used the same sampling method, in all likelihood their samples would be different. Neither would be wrong, however.

Think about what contributes to making Doreen's and Jung's samples different.

If Doreen and Jung took larger samples (i.e. the number of data values is increased), their sample results (the average amount of time a student sleeps) would be closer to the actual population average. But still, their samples would be, in all likelihood, different from each other. This **variability in samples** cannot be stressed enough.

1.7.2.1 Size of a Sample

The size of a sample (often called the number of observations) is important. The examples you have seen in this book so far have been small. Samples of only a few hundred observations, or even smaller, are sufficient for many purposes. In polling, samples that are from 1200 to 1500 observations are considered large enough and good enough if the survey is random and is well done. You will learn why when you study confidence intervals.

[7]This content is available online at <http://cnx.org/content/m16021/1.14/>.

1.7.2.2 Optional Collaborative Classroom Exercise

Exercise 1.7.1

Divide into groups of two, three, or four. Your instructor will give each group one 6-sided die. **Try this experiment twice.** Roll one fair die (6-sided) 20 times. Record the number of ones, twos, threes, fours, fives, and sixes you get below ("frequency" is the number of times a particular face of the die occurs):

First Experiment (20 rolls)

Face on Die	Frequency
1	
2	
3	
4	
5	
6	

Table 1.2

Second Experiment (20 rolls)

Face on Die	Frequency
1	
2	
3	
4	
5	
6	

Table 1.3

Did the two experiments have the same results? Probably not. If you did the experiment a third time, do you expect the results to be identical to the first or second experiment? (Answer yes or no.) Why or why not?

Which experiment had the correct results? They both did. The job of the statistician is to see through the variability and draw appropriate conclusions.

1.7.3 Critical Evaluation

We need to critically evaluate the statistical studies we read about and analyze before accepting the results of the study. Common problems to be aware of include

- Problems with Samples: A sample should be representative of the population. A sample that is not representative of the population is biased. Biased samples that are not representative of the population give results that are inaccurate and not valid.

- Self-Selected Samples: Responses only by people who choose to respond, such as call-in surveys are often unreliable.
- Sample Size Issues: Samples that are too small may be unreliable. Larger samples are better if possible. In some situations, small samples are unavoidable and can still be used to draw conclusions, even though larger samples are better. Examples: Crash testing cars, medical testing for rare conditions.
- Undue influence: Collecting data or asking questions in a way that influences the response.
- Non-response or refusal of subject to participate: The collected responses may no longer be representative of the population. Often, people with strong positive or negative opinions may answer surveys, which can affect the results.
- Causality: A relationship between two variables does not mean that one causes the other to occur. They may both be related (correlated) because of their relationship through a different variable.
- Self-Funded or Self-Interest Studies: A study performed by a person or organization in order to support their claim. Is the study impartial? Read the study carefully to evaluate the work. Do not automatically assume that the study is good but do not automatically assume the study is bad either. Evaluate it on its merits and the work done.
- Misleading Use of Data: Improperly displayed graphs, incomplete data, lack of context.
- Confounding: When the effects of multiple factors on a response cannot be separated. Confounding makes it difficult or impossible to draw valid conclusions about the effect of each factor.

1.8 Answers and Rounding Off[8]

A simple way to round off answers is to carry your final answer one more decimal place than was present in the original data. Round only the final answer. Do not round any intermediate results, if possible. If it becomes necessary to round intermediate results, carry them to at least twice as many decimal places as the final answer. For example, the average of the three quiz scores 4, 6, 9 is 6.3, rounded to the nearest tenth, because the data are whole numbers. Most answers will be rounded in this manner.

It is not necessary to reduce most fractions in this course. Especially in Probability Topics (Section 3.1), the chapter on probability, it is more helpful to leave an answer as an unreduced fraction.

1.9 Frequency[9]

Twenty students were asked how many hours they worked per day. Their responses, in hours, are listed below:

5; 6; 3; 3; 2; 4; 7; 5; 2; 3; 5; 6; 5; 4; 4; 3; 5; 2; 5; 3

Below is a frequency table listing the different data values in ascending order and their frequencies.

[8]This content is available online at <http://cnx.org/content/m16006/1.7/>.
[9]This content is available online at <http://cnx.org/content/m16012/1.15/>.

Frequency Table of Student Work Hours

DATA VALUE	FREQUENCY
2	3
3	5
4	3
5	6
6	2
7	1

Table 1.4

A **frequency** is the number of times a given datum occurs in a data set. According to the table above, there are three students who work 2 hours, five students who work 3 hours, etc. The total of the frequency column, 20, represents the total number of students included in the sample.

A **relative frequency** is the fraction of times an answer occurs. To find the relative frequencies, divide each frequency by the total number of students in the sample - in this case, 20. Relative frequencies can be written as fractions, percents, or decimals.

Frequency Table of Student Work Hours w/ Relative Frequency

DATA VALUE	FREQUENCY	RELATIVE FREQUENCY
2	3	$\frac{3}{20}$ or 0.15
3	5	$\frac{5}{20}$ or 0.25
4	3	$\frac{3}{20}$ or 0.15
5	6	$\frac{6}{20}$ or 0.30
6	2	$\frac{2}{20}$ or 0.10
7	1	$\frac{1}{20}$ or 0.05

Table 1.5

The sum of the relative frequency column is $\frac{20}{20}$, or 1.

Cumulative relative frequency is the accumulation of the previous relative frequencies. To find the cumulative relative frequencies, add all the previous relative frequencies to the relative frequency for the current row.

Frequency Table of Student Work Hours w/ Relative and Cumulative Relative Frequency

DATA VALUE	FREQUENCY	RELATIVE FREQUENCY	CUMULATIVE RELATIVE FREQUENCY

continued on next page

2	3	$\frac{3}{20}$ or 0.15	0.15
3	5	$\frac{5}{20}$ or 0.25	0.15 + 0.25 = 0.40
4	3	$\frac{3}{20}$ or 0.15	0.40 + 0.15 = 0.55
5	6	$\frac{6}{20}$ or 0.10	0.55 + 0.30 = 0.85
6	2	$\frac{2}{20}$ or 0.10	0.85 + 0.10 = 0.95
7	1	$\frac{1}{20}$ or 0.05	0.95 + 0.05 = 1.00

Table 1.6

The last entry of the cumulative relative frequency column is one, indicating that one hundred percent of the data has been accumulated.

> NOTE: Because of rounding, the relative frequency column may not always sum to one and the last entry in the cumulative relative frequency column may not be one. However, they each should be close to one.

The following table represents the heights, in inches, of a sample of 100 male semiprofessional soccer players.

Frequency Table of Soccer Player Height

HEIGHTS (INCHES)	FREQUENCY OF STUDENTS	RELATIVE FREQUENCY	CUMULATIVE RELATIVE FREQUENCY
59.95 - 61.95	5	$\frac{5}{100} = 0.05$	0.05
61.95 - 63.95	3	$\frac{3}{100} = 0.03$	0.05 + 0.03 = 0.08
63.95 - 65.95	15	$\frac{15}{100} = 0.15$	0.08 + 0.15 = 0.23
65.95 - 67.95	40	$\frac{40}{100} = 0.40$	0.23 + 0.40 = 0.63
67.95 - 69.95	17	$\frac{17}{100} = 0.17$	0.63 + 0.17 = 0.80
69.95 - 71.95	12	$\frac{12}{100} = 0.12$	0.80 + 0.12 = 0.92
71.95 - 73.95	7	$\frac{7}{100} = 0.07$	0.92 + 0.07 = 0.99
73.95 - 75.95	1	$\frac{1}{100} = 0.01$	0.99 + 0.01 = 1.00
	Total = 100	**Total = 1.00**	

Table 1.7

The data in this table has been **grouped** into the following intervals:

- 59.95 - 61.95 inches
- 61.95 - 63.95 inches
- 63.95 - 65.95 inches
- 65.95 - 67.95 inches
- 67.95 - 69.95 inches
- 69.95 - 71.95 inches
- 71.95 - 73.95 inches
- 73.95 - 75.95 inches

> NOTE: This example is used again in the Descriptive Statistics (Section 2.1) chapter, where the method used to compute the intervals will be explained.

In this sample, there are **5** players whose heights are between 59.95 - 61.95 inches, **3** players whose heights fall within the interval 61.95 - 63.95 inches, **15** players whose heights fall within the interval 63.95 - 65.95 inches, **40** players whose heights fall within the interval 65.95 - 67.95 inches, **17** players whose heights fall within the interval 67.95 - 69.95 inches, **12** players whose heights fall within the interval 69.95 - 71.95, 7 players whose height falls within the interval 71.95 - 73.95, and **1** player whose height falls within the interval 73.95 - 75.95. All heights fall between the endpoints of an interval and not at the endpoints.

Example 1.8

From the table, find the percentage of heights that are less than 65.95 inches.

Solution

If you look at the first, second, and third rows, the heights are all less than 65.95 inches. There are $5 + 3 + 15 = 23$ males whose heights are less than 65.95 inches. The percentage of heights less than 65.95 inches is then $\frac{23}{100}$ or 23%. This percentage is the cumulative relative frequency entry in the third row.

Example 1.9

From the table, find the percentage of heights that fall between 61.95 and 65.95 inches.

Solution

Add the relative frequencies in the second and third rows: $0.03 + 0.15 = 0.18$ or 18%.

Example 1.10

Use the table of heights of the 100 male semiprofessional soccer players. Fill in the blanks and check your answers.

1. The percentage of heights that are from 67.95 to 71.95 inches is:
2. The percentage of heights that are from 67.95 to 73.95 inches is:
3. The percentage of heights that are more than 65.95 inches is:
4. The number of players in the sample who are between 61.95 and 71.95 inches tall is:
5. What kind of data are the heights?
6. Describe how you could gather this data (the heights) so that the data are characteristic of all male semiprofessional soccer players.

Remember, you **count frequencies**. To find the relative frequency, divide the frequency by the total number of data values. To find the cumulative relative frequency, add all of the previous relative frequencies to the relative frequency for the current row.

1.9.1 Optional Collaborative Classroom Exercise

Exercise 1.9.1

In your class, have someone conduct a survey of the number of siblings (brothers and sisters) each student has. Create a frequency table. Add to it a relative frequency column and a cumulative relative frequency column. Answer the following questions:

1. What percentage of the students in your class have 0 siblings?
2. What percentage of the students have from 1 to 3 siblings?

3. What percentage of the students have fewer than 3 siblings?

Example 1.11

Nineteen people were asked how many miles, to the nearest mile they commute to work each day. The data are as follows:

2; 5; 7; 3; 2; 10; 18; 15; 20; 7; 10; 18; 5; 12; 13; 12; 4; 5; 10

The following table was produced:

Frequency of Commuting Distances

DATA	FREQUENCY	RELATIVE FREQUENCY	CUMULATIVE RELATIVE FREQUENCY
3	3	$\frac{3}{19}$	0.1579
4	1	$\frac{1}{19}$	0.2105
5	3	$\frac{3}{19}$	0.1579
7	2	$\frac{2}{19}$	0.2632
10	3	$\frac{4}{19}$	0.4737
12	2	$\frac{2}{19}$	0.7895
13	1	$\frac{1}{19}$	0.8421
15	1	$\frac{1}{19}$	0.8948
18	1	$\frac{1}{19}$	0.9474
20	1	$\frac{1}{19}$	1.0000

Table 1.8

Problem *(Solution on p. 46.)*

1. Is the table correct? If it is not correct, what is wrong?
2. True or False: Three percent of the people surveyed commute 3 miles. If the statement is not correct, what should it be? If the table is incorrect, make the corrections.
3. What fraction of the people surveyed commute 5 or 7 miles?
4. What fraction of the people surveyed commute 12 miles or more? Less than 12 miles? Between 5 and 13 miles (does not include 5 and 13 miles)?

1.10 Summary[10]

Statistics

- Deals with the collection, analysis, interpretation, and presentation of data

Probability

- Mathematical tool used to study randomness

Key Terms

- Population
- Parameter
- Sample
- Statistic
- Variable
- Data

Types of Data

- Quantitative Data (a number)
    - Discrete (You count it.)
    - Continuous (You measure it.)
- Qualitative Data (a category, words)

Sampling

- **With Replacement**: A member of the population may be chosen more than once
- **Without Replacement**: A member of the population may be chosen only once

Random Sampling

- Each member of the population has an equal chance of being selected

Sampling Methods

- Random
    - Simple random sample
    - Stratified sample
    - Cluster sample
    - Systematic sample
- Not Random
    - Convenience sample

NOTE: Samples must be representative of the population from which they come. They must have the same characteristics. However, they may vary but still represent the same population.

Frequency (freq. or f)

- The number of times an answer occurs

[10]This content is available online at <http://cnx.org/content/m16023/1.8/>.

Relative Frequency (rel. freq. or RF)

- The proportion of times an answer occurs
- Can be interpreted as a fraction, decimal, or percent

Cumulative Relative Frequencies (cum. rel. freq. or cum RF)

- An accumulation of the previous relative frequencies

1.11 Practice: Sampling and Data[11]

1.11.1 Student Learning Outcomes

- The student will practice constructing frequency tables.
- The student will differentiate between key terms.
- The student will compare sampling techniques.

1.11.2 Given

Studies are often done by pharmaceutical companies to determine the effectiveness of a treatment program. Suppose that a new AIDS antibody drug is currently under study. It is given to patients once the AIDS symptoms have revealed themselves. Of interest is the average length of time in months patients live once starting the treatment. Two researchers each follow a different set of 40 AIDS patients from the start of treatment until their deaths. The following data (in months) are collected.

Researcher 1 3; 4; 11; 15; 16; 17; 22; 44; 37; 16; 14; 24; 25; 15; 26; 27; 33; 29; 35; 44; 13; 21; 22; 10; 12; 8; 40; 32; 26; 27; 31; 34; 29; 17; 8; 24; 18; 47; 33; 34

Researcher 2 3; 14; 11; 5; 16; 17; 28; 41; 31; 18; 14; 14; 26; 25; 21; 22; 31; 2; 35; 44; 23; 21; 21; 16; 12; 18; 41; 22; 16; 25; 33; 34; 29; 13; 18; 24; 23; 42; 33; 29

1.11.3 Organize the Data

Complete the tables below using the data provided.

Researcher 1

Survival Length (in months)	Frequency	Relative Frequency	Cumulative Rel. Frequency
0.5 - 6.5			
6.5 - 12.5			
12.5 - 18.5			
18.5 - 24.5			
24.5 - 30.5			
30.5 - 36.5			
36.5 - 42.5			
42.5 - 48.5			

Table 1.9

Researcher 2

Survival Length (in months)	Frequency	Relative Frequency	Cumulative Rel. Frequency
			continued on next page

[11]This content is available online at <http://cnx.org/content/m16016/1.12/>.

0.5 - 6.5			
6.5 - 12.5			
12.5 - 18.5			
18.5 - 24.5			
24.5 - 30.5			
30.5 - 36.5			
36.5 - 42.5			
42.5 - 48.5			

Table 1.10

1.11.4 Key Terms

Define the key terms based upon the above example for Researcher 1.

Exercise 1.11.1
Population

Exercise 1.11.2
Sample

Exercise 1.11.3
Parameter

Exercise 1.11.4
Statistic

Exercise 1.11.5
Variable

Exercise 1.11.6
Data

1.11.5 Discussion Questions

Discuss the following questions and then answer in complete sentences.

Exercise 1.11.7
List two reasons why the data may differ.

Exercise 1.11.8
Can you tell if one researcher is correct and the other one is incorrect? Why?

Exercise 1.11.9
Would you expect the data to be identical? Why or why not?

Exercise 1.11.10
How could the researchers gather random data?

Exercise 1.11.11
Suppose that the first researcher conducted his survey by randomly choosing one state in the nation and then randomly picking 40 patients from that state. What sampling method would that researcher have used?

Exercise 1.11.12

Suppose that the second researcher conducted his survey by choosing 40 patients he knew. What sampling method would that researcher have used? What concerns would you have about this data set, based upon the data collection method?

1.12 Homework[12]

Exercise 1.12.1 *(Solution on p. 46.)*

For each item below:

i. Identify the type of data (quantitative - discrete, quantitative - continuous, or qualitative) that would be used to describe a response.
ii. Give an example of the data.

a. Number of tickets sold to a concert
b. Amount of body fat
c. Favorite baseball team
d. Time in line to buy groceries
e. Number of students enrolled at Evergreen Valley College
f. Most–watched television show
g. Brand of toothpaste
h. Distance to the closest movie theatre
i. Age of executives in Fortune 500 companies
j. Number of competing computer spreadsheet software packages

Exercise 1.12.2

Fifty part-time students were asked how many courses they were taking this term. The (incomplete) results are shown below:

Part-time Student Course Loads

# of Courses	Frequency	Relative Frequency	Cumulative Relative Frequency
1	30	0.6	
2	15		
3			

Table 1.11

a. Fill in the blanks in the table above.
b. What percent of students take exactly two courses?
c. What percent of students take one or two courses?

Exercise 1.12.3 *(Solution on p. 46.)*

Sixty adults with gum disease were asked the number of times per week they used to floss before their diagnoses. The (incomplete) results are shown below:

Flossing Frequency for Adults with Gum Disease

# Flossing per Week	Frequency	Relative Frequency	Cumulative Relative Freq.
0	27	0.4500	
1	18		
3			0.9333
6	3	0.0500	
7	1	0.0167	

[12]This content is available online at <http://cnx.org/content/m16010/1.16/>.

Table 1.12

a. Fill in the blanks in the table above.
b. What percent of adults flossed six times per week?
c. What percent flossed at most three times per week?

Exercise 1.12.4
A fitness center is interested in the average amount of time a client exercises in the center each week. Define the following in terms of the study. Give examples where appropriate.

a. Population
b. Sample
c. Parameter
d. Statistic
e. Variable
f. Data

Exercise 1.12.5 ***(Solution on p. 46.)***
Ski resorts are interested in the average age that children take their first ski and snowboard lessons. They need this information to optimally plan their ski classes. Define the following in terms of the study. Give examples where appropriate.

a. Population
b. Sample
c. Parameter
d. Statistic
e. Variable
f. Data

Exercise 1.12.6
A cardiologist is interested in the average recovery period for her patients who have had heart attacks. Define the following in terms of the study. Give examples where appropriate.

a. Population
b. Sample
c. Parameter
d. Statistic
e. Variable
f. Data

Exercise 1.12.7 ***(Solution on p. 46.)***
Insurance companies are interested in the average health costs each year for their clients, so that they can determine the costs of health insurance. Define the following in terms of the study. Give examples where appropriate.

a. Population
b. Sample
c. Parameter
d. Statistic
e. Variable
f. Data

Exercise 1.12.8

A politician is interested in the proportion of voters in his district that think he is doing a good job. Define the following in terms of the study. Give examples where appropriate.

- **a.** Population
- **b.** Sample
- **c.** Parameter
- **d.** Statistic
- **e.** Variable
- **f.** Data

Exercise 1.12.9 *(Solution on p. 47.)*

A marriage counselor is interested in the proportion the clients she counsels that stay married. Define the following in terms of the study. Give examples where appropriate.

- **a.** Population
- **b.** Sample
- **c.** Parameter
- **d.** Statistic
- **e.** Variable
- **f.** Data

Exercise 1.12.10

Political pollsters may be interested in the proportion of people that will vote for a particular cause. Define the following in terms of the study. Give examples where appropriate.

- **a.** Population
- **b.** Sample
- **c.** Parameter
- **d.** Statistic
- **e.** Variable
- **f.** Data

Exercise 1.12.11 *(Solution on p. 47.)*

A marketing company is interested in the proportion of people that will buy a particular product. Define the following in terms of the study. Give examples where appropriate.

- **a.** Population
- **b.** Sample
- **c.** Parameter
- **d.** Statistic
- **e.** Variable
- **f.** Data

Exercise 1.12.12

Airline companies are interested in the consistency of the number of babies on each flight, so that they have adequate safety equipment. Suppose an airline conducts a survey. Over Thanksgiving weekend, it surveys 6 flights from Boston to Salt Lake City to determine the number of babies on the flights. It determines the amount of safety equipment needed by the result of that study.

- **a.** Using complete sentences, list three things wrong with the way the survey was conducted.
- **b.** Using complete sentences, list three ways that you would improve the survey if it were to be repeated.

Exercise 1.12.13

Suppose you want to determine the average number of students per statistics class in your state. Describe a possible sampling method in 3 – 5 complete sentences. Make the description detailed.

Exercise 1.12.14

Suppose you want to determine the average number of cans of soda drunk each month by persons in their twenties. Describe a possible sampling method in 3 - 5 complete sentences. Make the description detailed.

Exercise 1.12.15

726 distance learning students at Long Beach City College in the 2004-2005 academic year were surveyed and asked the reasons they took a distance learning class. (*Source: Amit Schitai, Director of Instructional Technology and Distance Learning, LBCC*). The results of this survey are listed in the table below.

Reasons for Taking LBCC Distance Learning Courses

Convenience	87.6%
Unable to come to campus	85.1%
Taking on-campus courses in addition to my DL course	71.7%
Instructor has a good reputation	69.1%
To fulfill requirements for transfer	60.8%
To fulfill requirements for Associate Degree	53.6%
Thought DE would be more varied and interesting	53.2%
I like computer technology	52.1%
Had success with previous DL course	52.0%
On-campus sections were full	42.1%
To fulfill requirements for vocational certification	27.1%
Because of disability	20.5%

Table 1.13

Assume that the survey allowed students to choose from the responses listed in the table above.

a. Why can the percents add up to over 100%?
b. Does that necessarily imply a mistake in the report?
c. How do you think the question was worded to get responses that totaled over 100%?
d. How might the question be worded to get responses that totaled 100%?

Exercise 1.12.16

Nineteen immigrants to the U.S were asked how many years, to the nearest year, they have lived in the U.S. The data are as follows:

2; 5; 7; 2; 2; 10; 20; 15; 0; 7; 0; 20; 5; 12; 15; 12; 4; 5; 10

The following table was produced:

Frequency of Immigrant Survey Responses

Data	Frequency	Relative Frequency	Cumulative Relative Frequency
0	2	$\frac{2}{19}$	0.1053
2	3	$\frac{3}{19}$	0.2632
4	1	$\frac{1}{19}$	0.3158
5	3	$\frac{3}{19}$	0.1579
7	2	$\frac{2}{19}$	0.5789
10	2	$\frac{2}{19}$	0.6842
12	2	$\frac{2}{19}$	0.7895
15	1	$\frac{1}{19}$	0.8421
20	1	$\frac{1}{19}$	1.0000

Table 1.14

a. Fix the errors on the table. Also, explain how someone might have arrived at the incorrect number(s).
b. Explain what is wrong with this statement: "47 percent of the people surveyed have lived in the U.S. for 5 years."
c. Fix the statement above to make it correct.
d. What fraction of the people surveyed have lived in the U.S. 5 or 7 years?
e. What fraction of the people surveyed have lived in the U.S. at most 12 years?
f. What fraction of the people surveyed have lived in the U.S. fewer than 12 years?
g. What fraction of the people surveyed have lived in the U.S. from 5 to 20 years, inclusive?

Exercise 1.12.17

A "random survey" was conducted of 3274 people of the "microprocessor generation" (people born since 1971, the year the microprocessor was invented). It was reported that 48% of those individuals surveyed stated that if they had $2000 to spend, they would use it for computer equipment. Also, 66% of those surveyed considered themselves relatively savvy computer users. (*Source: San Jose Mercury News*)

a. Do you consider the sample size large enough for a study of this type? Why or why not?
b. Based on your "gut feeling," do you believe the percents accurately reflect the U.S. population for those individuals born since 1971? If not, do you think the percents of the population are actually higher or lower than the sample statistics? Why?

Additional information: The survey was reported by Intel Corporation of individuals who visited the Los Angeles Convention Center to see the Smithsonian Institure's road show called "America's Smithsonian."

c. With this additional information, do you feel that all demographic and ethnic groups were equally represented at the event? Why or why not?
d. With the additional information, comment on how accurately you think the sample statistics reflect the population parameters.

Exercise 1.12.18

a. List some practical difficulties involved in getting accurate results from a telephone survey.
b. List some practical difficulties involved in getting accurate results from a mailed survey.
c. With your classmates, brainstorm some ways to overcome these problems if you needed to conduct a phone or mail survey.

1.12.1 Try these multiple choice questions

The next four questions refer to the following: A Lake Tahoe Community College instructor is interested in the average number of days Lake Tahoe Community College math students are absent from class during a quarter.

Exercise 1.12.19 ***(Solution on p. 47.)***

What is the population she is interested in?

A. All Lake Tahoe Community College students
B. All Lake Tahoe Community College English students
C. All Lake Tahoe Community College students in her classes
D. All Lake Tahoe Community College math students

Exercise 1.12.20 ***(Solution on p. 47.)***

Consider the following:

X = **number of days a Lake Tahoe Community College math student is absent**

In this case, X is an example of a:

A. Variable
B. Population
C. Statistic
D. Data

Exercise 1.12.21 ***(Solution on p. 47.)***

The instructor takes her sample by gathering data on 5 randomly selected students from each Lake Tahoe Community College math class. The type of sampling she used is

A. Cluster sampling
B. Stratified sampling
C. Simple random sampling
D. Convenience sampling

Exercise 1.12.22 ***(Solution on p. 47.)***

The instructor's sample produces an average number of days absent of 3.5 days. This value is an example of a

A. Parameter
B. Data
C. Statistic
D. Variable

The next two questions refer to the following relative frequency table on hurricanes that have made direct hits on the U.S between 1851 and 2004. Hurricanes are given a strength category rating based on the minimum wind speed generated by the storm. (*http://www.nhc.noaa.gov/gifs/table5.gif*)

Frequency of Hurricane Direct Hits

Category	Number of Direct Hits	Relative Frequency	Cumulative Frequency
1	109	0.3993	0.3993
2	72	0.2637	0.6630
3	71	0.2601	
4	18		0.9890
5	3	0.0110	1.0000
	Total = 273		

Table 1.15

Exercise 1.12.23 *(Solution on p. 47.)*
What is the relative frequency of direct hits that were category 4 hurricanes?

A. 0.0768
B. 0.0659
C. 0.2601
D. Not enough information to calculate

Exercise 1.12.24 *(Solution on p. 47.)*
What is the relative frequency of direct hits that were AT MOST a category 3 storm?

A. 0.3480
B. 0.9231
C. 0.2601
D. 0.3370

The next three questions refer to the following: A study was done to determine the age, number of times per week and the duration (amount of time) of resident use of a local park in San Jose. The first house in the neighborhood around the park was selected randomly and then every 8th house in the neighborhood around the park was interviewed.

Exercise 1.12.25 *(Solution on p. 47.)*
'Number of times per week' is what type of data?

A. qualitative
B. quantitative - discrete
C. quantitative - continuous

Exercise 1.12.26 *(Solution on p. 47.)*
The sampling method was:

A. simple random
B. systematic
C. stratified
D. cluster

Exercise 1.12.27 *(Solution on p. 47.)*

'Duration (amount of time)' is what type of data?

A. qualitative
B. quantitative - discrete
C. quantitative - continuous

1.13 Lab 1: Data Collection[13]

Class Time:

Names:

1.13.1 Student Learning Outcomes

- The student will demonstrate the systematic sampling technique.
- The student will construct Relative Frequency Tables.
- The student will interpret results and their differences from different data groupings.

1.13.2 Movie Survey

Ask five classmates from a different class how many movies they saw last month at the theater. Do not include rented movies.

1. Record the data
2. In class, randomly pick one person. On the class list, mark that person's name. Move down four people's names on the class list. Mark that person's name. Continue doing this until you have marked 12 people's names. You may need to go back to the start of the list. For each marked name record below the five data values. You now have a total of 60 data values.
3. For each name marked, record the data:

______	______	______	______	______	______	______	______	______	______
______	______	______	______	______	______	______	______	______	______
______	______	______	______	______	______	______	______	______	______
______	______	______	______	______	______	______	______	______	______
______	______	______	______	______	______	______	______	______	______
______	______	______	______	______	______	______	______	______	______

Table 1.16

1.13.3 Order the Data

Complete the two relative frequency tables below using your class data.

---

[13]This content is available online at <http://cnx.org/content/m16004/1.11/>.

Frequency of Number of Movies Viewed

Number of Movies	Frequency	Relative Frequency	Cumulative Relative Frequency
0			
1			
2			
3			
4			
5			
6			
7+			

Table 1.17

Frequency of Number of Movies Viewed

Number of Movies	Frequency	Relative Frequency	Cumulative Relative Frequency
0-1			
2-3			
4-5			
6-7+			

Table 1.18

1. Using the tables, find the percent of data that is at most 2. Which table did you use and why?
2. Using the tables, find the percent of data that is at most 3. Which table did you use and why?
3. Using the tables, find the percent of data that is more than 2. Which table did you use and why?
4. Using the tables, find the percent of data that is more than 3. Which table did you use and why?

1.13.4 Discussion Questions

1. Is one of the tables above "more correct" than the other? Why or why not?
2. In general, why would someone group the data in different ways? Are there any advantages to either way of grouping the data?
3. Why did you switch between tables, if you did, when answering the question above?

1.14 Lab 2: Sampling Experiment[14]

Class Time:

Names:

1.14.1 Student Learning Outcomes

- The student will demonstrate the simple random, systematic, stratified, and cluster sampling techniques.
- The student will explain each of the details of each procedure used.

In this lab, you will be asked to pick several random samples. In each case, describe your procedure briefly, including how you might have used the random number generator, and then list the restaurants in the sample you obtained

NOTE: The following section contains restaurants stratified by city into columns and grouped horizontally by entree cost (clusters).

1.14.2 A Simple Random Sample

Pick a **simple random sample** of 15 restaurants.

1. Descibe the procedure:
2.

1. __________	6. __________	11. __________
2. __________	7. __________	12. __________
3. __________	8. __________	13. __________
4. __________	9. __________	14. __________
5. __________	10. __________	15. __________

Table 1.19

1.14.3 A Systematic Sample

Pick a **systematic sample** of 15 restaurants.

1. Descibe the procedure:
2.

1. __________	6. __________	11. __________
2. __________	7. __________	12. __________
3. __________	8. __________	13. __________
4. __________	9. __________	14. __________
5. __________	10. __________	15. __________

Table 1.20

[14]This content is available online at <http://cnx.org/content/m16013/1.12/>.

1.14.4 A Stratified Sample

Pick a **stratified sample**, by entree cost, of 20 restaurants with equal representation from each stratum.

1. Descibe the procedure:
2.

1. ________	6. ________	11. ________	16. ________
2. ________	7. ________	12. ________	17. ________
3. ________	8. ________	13. ________	18. ________
4. ________	9. ________	14. ________	19. ________
5. ________	10. ________	15. ________	20. ________

Table 1.21

1.14.5 A Stratified Sample

Pick a **stratified sample**, by city, of 21 restaurants with equal representation from each stratum.

1. Descibe the procedure:
2.

1. ________	6. ________	11. ________	16. ________
2. ________	7. ________	12. ________	17. ________
3. ________	8. ________	13. ________	18. ________
4. ________	9. ________	14. ________	19. ________
5. ________	10. ________	15. ________	20. ________
			21. ________

Table 1.22

1.14.6 A Cluster Sample

Pick a **cluster sample** of resturants from two cities. The number of restaurants will vary.

1. Descibe the procedure:
2.

1. ________	6. ________	11. ________	16. ________	21. ________
2. ________	7. ________	12. ________	17. ________	22. ________
3. ________	8. ________	13. ________	18. ________	23. ________
4. ________	9. ________	14. ________	19. ________	24. ________
5. ________	10. ________	15. ________	20. ________	25. ________

Table 1.23

1.14.7 Restaurants Stratified by City and Entree Cost

Restaurants Used in Sample

Entree Cost →	**Under $10**	**$10 to under $15**	**$15 to under $20**	**Over $20**
San Jose	El Abuelo Taq, Pasta Mia, Emma's Express, Bamboo Hut	Emperor's Guard, Creekside Inn	Agenda, Gervais, Miro's	Blake's, Eulipia, Hayes Mansion, Germania
Palo Alto	Senor Taco, Olive Garden, Taxi's	Ming's, P.A. Joe's, Stickney's	Scott's Seafood, Poolside Grill, Fish Market	Sundance Mine, Maddalena's, Spago's
Los Gatos	Mary's Patio, Mount Everest, Sweet Pea's, Andele Taqueria	Lindsey's, Willow Street	Toll House	Charter House, La Maison Du Cafe
Mountain View	Maharaja, New Ma's, Thai-Rific, Garden Fresh	Amber Indian, La Fiesta, Fiesta del Mar, Dawit	Austin's, Shiva's, Mazeh	Le Petit Bistro
Cupertino	Hobees, Hung Fu, Samrat, Panda Express	Santa Barb. Grill, Mand. Gourmet, Bombay Oven, Kathmandu West	Fontana's, Blue Pheasant	Hamasushi, Helios
Sunnyvale	Chekijababi, Taj India, Full Throttle, Tia Juana, Lemon Grass	Pacific Fresh, Charley Brown's, Cafe Cameroon, Faz, Aruba's	Lion & Compass, The Palace, Beau Sejour	
Santa Clara	Rangoli, Armadillo Willy's, Thai Pepper, Pasand	Arthur's, Katie's Cafe, Pedro's, La Galleria	Birk's, Truya Sushi, Valley Plaza	Lakeside, Mariani's

Table 1.24

NOTE: *The original lab was designed and contributed by Carol Olmstead.*

Solutions to Exercises in Chapter 1

Solution to Example 1.5 (p. 15)
Items 1, 5, 11, and 12 are quantitative discrete; items 4, 6, 10, and 14 are quantitative continuous; and items 2, 3, 7, 8, 9, and 13 are qualitative.
Solution to Example 1.10 (p. 26)

1. 29%
2. 36%
3. 77%
4. 87
5. quantitative continuous
6. get rosters from each team and choose a simple random sample from each

Solution to Example 1.11 (p. 27)

1. No. Frequency column sums to 18, not 19. Not all cumulative relative frequencies are correct.
2. False. Frequency for 3 miles should be 1; for 2 miles (left out), 2. Cumulative relative frequency column should read: 0.1052, 0.1579, 0.2105, 0.3684, 0.4737, 0.6316, 0.7368, 0.7895, 0.8421, 0.9474, 1.
3. $\frac{5}{19}$
4. $\frac{7}{19}$, $\frac{12}{19}$, $\frac{7}{19}$

Solutions to Homework

Solution to Exercise 1.12.1 (p. 33)

a. quantitative - discrete
b. quantitative - continuous
c. qualitative
d. quantitative - continuous
e. quantitative - discrete
f. qualitative
g. qualitative
h. quantitative - continuous
i. quantitative - continuous
j. quantitative - discrete

Solution to Exercise 1.12.3 (p. 33)

b. 5.00%
c. 93.33%

Solution to Exercise 1.12.5 (p. 34)

a. Children who take ski or snowboard lessons
b. A group of these children
c. The population average
d. The sample average
e. X = the age of one child who takes the first ski or snowboard lesson
f. A value for X, such as 3, 7, etc.

Solution to Exercise 1.12.7 (p. 34)

a. The clients of the insurance companies
b. A group of the clients

c. The average health costs of the clients
d. The average health costs of the sample
e. X = the health costs of one client
f. A value for X, such as 34, 9, 82, etc.

Solution to Exercise 1.12.9 (p. 35)

a. All the clients of the counselor
b. A group of the clients
c. The proportion of all her clients who stay married
d. The proportion of the sample who stay married
e. X = the number of couples who stay married
f. yes, no

Solution to Exercise 1.12.11 (p. 35)

a. All people (maybe in a certain geographic area, such as the United States)
b. A group of the people
c. The proportion of all people who will buy the product
d. The proportion of the sample who will buy the product
e. X = the number of people who will buy it
f. buy, not buy

Solution to Exercise 1.12.19 (p. 38)
D
Solution to Exercise 1.12.20 (p. 38)
A
Solution to Exercise 1.12.21 (p. 38)
B
Solution to Exercise 1.12.22 (p. 38)
C
Solution to Exercise 1.12.23 (p. 39)
B
Solution to Exercise 1.12.24 (p. 39)
B
Solution to Exercise 1.12.25 (p. 39)
B
Solution to Exercise 1.12.26 (p. 39)
B
Solution to Exercise 1.12.27 (p. 40)
C

Chapter 2

Descriptive Statistics

2.1 Descriptive Statistics[1]

2.1.1 Student Learning Objectives

By the end of this chapter, the student should be able to:

- Display data graphically and interpret graphs: stemplots, histograms and boxplots.
- Recognize, describe, and calculate the measures of location of data: quartiles and percentiles.
- Recognize, describe, and calculate the measures of the center of data: mean, median, and mode.
- Recognize, describe, and calculate the measures of the spread of data: variance, standard deviation, and range.

2.1.2 Introduction

Once you have collected data, what will you do with it? Data can be described and presented in many different formats. For example, suppose you are interested in buying a house in a particular area. You may have no clue about the house prices, so you might ask your real estate agent to give you a sample data set of prices. Looking at all the prices in the sample often is overwhelming. A better way might be to look at the median price and the variation of prices. The median and variation are just two ways that you will learn to describe data. Your agent might also provide you with a graph of the data.

In this chapter, you will study numerical and graphical ways to describe and display your data. This area of statistics is called **"Descriptive Statistics"**. You will learn to calculate, and even more importantly, to interpret these measurements and graphs.

2.2 Displaying Data[2]

A statistical graph is a tool that helps you learn about the shape or distribution of a sample. The graph can be a more effective way of presenting data than a mass of numbers because we can see where data clusters and where there are only a few data values. Newspapers and the Internet use graphs to show trends and to enable readers to compare facts and figures quickly.

Statisticians often graph data first in order to get a picture of the data. Then, more formal tools may be applied.

[1]This content is available online at <http://cnx.org/content/m16300/1.7/>.

[2]This content is available online at <http://cnx.org/content/m16297/1.7/>.

Some of the types of graphs that are used to summarize and organize data are the dot plot, the bar chart, the histogram, the stem-and-leaf plot, the frequency polygon (a type of broken line graph), pie charts, and the boxplot. In this chapter, we will briefly look at stem-and-leaf plots. Our emphasis will be on histograms and boxplots.

2.3 Stem and Leaf Graphs (Stemplots)[3]

One simple graph, the **stem-and-leaf graph** or **stemplot**, comes from the field of exploratory data analysis.It is a good choice when the data sets are small. To create the plot, divide each observation of data into a stem and a leaf. The leaf consists of **one digit**. For example, 23 has stem 2 and leaf 3. Four hundred thirty-two (432) has stem 43 and leaf 2. Five thousand four hundred thirty-two (5,432) has stem 543 and leaf 2. The decimal 9.3 has stem 9 and leaf 3. Write the stems in a vertical line from smallest the largest. Draw a vertical line to the right of the stems. Then write the leaves in increasing order next to their corresponding stem.

Example 2.1
For Susan Dean's spring pre-calculus class, scores for the first exam were as follows (smallest to largest):
33; 42; 49; 49; 53; 55; 55; 61; 63; 67; 68; 68; 69; 69; 72; 73; 74; 78; 80; 83; 88; 88; 88; 90; 92; 94; 94; 94; 94; 96; 100

Stem-and-Leaf Diagram

3	3
4	299
5	355
6	1378899
7	2348
8	03888
9	0244446
10	0

Table 2.1

The stemplot shows that most scores fell in the 60s, 70s, 80s, and 90s. Eight out of the 31 scores or approximately 26% of the scores were in the 90's or 100, a fairly high number of As.

The stemplot is a quick way to graph and gives an exact picture of the data. You want to look for an overall pattern and any outliers. An **outlier** is an observation of data that does not fit the rest of the data. It is sometimes called an **extreme value.** When you graph an outlier, it will appear not to fit the pattern of the graph. Some outliers are due to mistakes (for example, writing down 50 instead of 500) while others may indicate that something unusual is happening. It takes some background information to explain outliers. In the example above, there were no outliers.

Example 2.2
Create a stem plot using the data:

1.1; 1.5; 2.3; 2.5; 2.7; 3.2; 3.3; 3.3; 3.5; 3.8; 4.0; 4.2; 4.5; 4.5; 4.7; 4.8; 5.5; 5.6; 6.5; 6.7; 12.3

The data are the distance (in kilometers) from a home to the nearest supermarket.

[3] This content is available online at <http://cnx.org/content/m16849/1.7/>.

Problem *(Solution on p. 92.)*

1. Are there any outliers?
2. Do the data seem to have any concentration of values?

HINT: The leaves are to the right of the decimal.

NOTE: This book contains instructions for constructing a **histogram** and a **box plot** for the TI-83+ and TI-84 calculators. You can find additional instructions for using these calculators on the Texas Instruments (TI) website[4] .

2.4 Histograms[5]

For most of the work you do in this book, you will use a histogram to display the data. One advantage of a histogram is that it can readily display large data sets. A rule of thumb is to use a histogram when the data set consists of 100 values or more.

A **histogram** consists of contiguous boxes. It has both a horizontal axis and a vertical axis. The horizontal axis is labeled with what the data represents (for instance, distance from your home to school). The vertical axis is labeled either "frequency" or "relative frequency". The graph will have the same shape with either label. **Frequency** is commonly used when the data set is small and **relative frequency** is used when the data set is large or when we want to compare several distributions. The histogram (like the stemplot) can give you the shape of the data, the center, and the spread of the data. (The next section tells you how to calculate the center and the spread.)

The relative frequency is equal to the frequency for an observed value of the data divided by the total number of data values in the sample. (In the chapter on Sampling and Data (Section 1.1), we defined frequency as the number of times an answer occurs.) If:

- f = frequency
- n = total number of data values (or the sum of the individual frequencies), and
- RF = relative frequency,

then:

$$RF = \frac{f}{n} \tag{2.1}$$

For example, if 3 students in Mr. Ahab's English class of 40 students received an A, then,

$f = 3$, $n = 40$, and $RF = \frac{f}{n} = \frac{3}{40} = 0.075$

Seven and a half percent of the students received an A.

To construct a histogram, first decide how many **bars** or **intervals**, also called classes, represent the data. Many histograms consist of from 5 to 15 bars or classes for clarity. Choose a starting point for the first interval to be less than the smallest data value. A **convenient starting point** is a lower value carried out to one more decimal place than the value with the most decimal places. For example, if the value with the most decimal places is 6.1 and this is the smallest value, a convenient starting point is 6.05 (6.1 - 0.05 = 6.05). We say that 6.05 has more precision. If the value with the most decimal places is 2.23 and the lowest value

[4] http://education.ti.com/educationportal/sites/US/sectionHome/support.html
[5] This content is available online at <http://cnx.org/content/m16298/1.11/>.

is 1.5, a convenient starting point is 1.495 (1.5 - 0.005 = 1.495). If the value with the most decimal places is 3.234 and the lowest value is 1.0, a convenient starting point is 0.9995 (1.0 - .0005 = 0.9995). If all the data happen to be integers and the smallest value is 2, then a convenient starting point is 1.5 (2 - 0.5 = 1.5). Also, when the starting point and other boundaries are carried to one additional decimal place, no data value will fall on a boundary.

Example 2.3
The following data are the heights (in inches to the nearest half inch) of 100 male semiprofessional soccer players. The heights are **continuous** data since height is measured.

60; 60.5; 61; 61; 61.5

63.5; 63.5; 63.5

64; 64; 64; 64; 64; 64; 64; 64.5; 64.5; 64.5; 64.5; 64.5; 64.5; 64.5; 64.5

66; 66; 66; 66; 66; 66; 66; 66; 66; 66; 66.5; 66.5; 66.5; 66.5; 66.5; 66.5; 66.5; 66.5; 66.5; 66.5; 66.5; 67; 67; 67; 67; 67; 67; 67; 67; 67; 67; 67; 67; 67.5; 67.5; 67.5; 67.5; 67.5; 67.5; 67.5

68; 68; 69; 69; 69; 69; 69; 69; 69; 69; 69; 69; 69.5; 69.5; 69.5; 69.5; 69.5

70; 70; 70; 70; 70; 70; 70.5; 70.5; 70.5; 71; 71; 71

72; 72; 72; 72.5; 72.5; 73; 73.5

74

The smallest data value is 60. Since the data with the most decimal places has one decimal (for instance, 61.5), we want our starting point to have two decimal places. Since the numbers 0.5, 0.05, 0.005, etc. are convenient numbers, use 0.05 and subtract it from 60, the smallest value, for the convenient starting point.

60 - 0.05 = 59.95 which is more precise than, say, 61.5 by one decimal place. The starting point is, then, 59.95.

The largest value is 74. 74+ 0.05 = 74.05 is the ending value.

Next, calculate the width of each bar or class interval. To calculate this width, subtract the starting point from the ending value and divide by the number of bars (you must choose the number of bars you desire). Suppose you choose 8 bars.

$$\frac{74.05 - 59.95}{8} = 1.76 \tag{2.2}$$

NOTE: We will round up to 2 and make each bar or class interval 2 units wide. Rounding up to 2 is one way to prevent a value from falling on a boundary. For this example, using 1.76 as the width would also work.

The boundaries are:

- 59.95
- 59.95 + 2 = 61.95
- 61.95 + 2 = 63.95
- 63.95 + 2 = 65.95
- 65.95 + 2 = 67.95

- 67.95 + 2 = 69.95
- 69.95 + 2 = 71.95
- 71.95 + 2 = 73.95
- 73.95 + 2 = 75.95

The heights 60 through 61.5 inches are in the interval 59.95 - 61.95. The heights that are 63.5 are in the interval 61.95 - 63.95. The heights that are 64 through 64.5 are in the interval 63.95 - 65.95. The heights 66 through 67.5 are in the interval 65.95 - 67.95. The heights 68 through 69.5 are in the interval 67.95 - 69.95. The heights 70 through 71 are in the interval 69.95 - 71.95. The heights 72 through 73.5 are in the interval 71.95 - 73.95. The height 74 is in the interval 73.95 - 75.95.

The following histogram displays the heights on the x-axis and relative frequency on the y-axis.

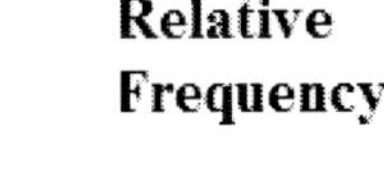

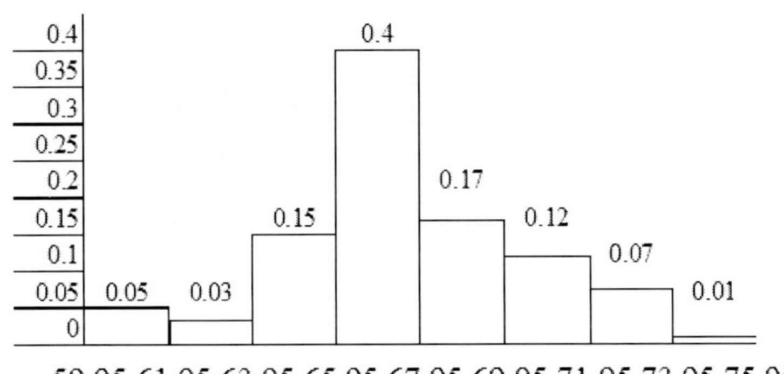

59.95 61.95 63.95 65.95 67.95 69.95 71.95 73.95 75.95

Heights

Example 2.4

The following data are the number of books bought by 50 part-time college students at ABC College. The number of books is discrete data since books are counted.

1; 1; 1; 1; 1; 1; 1; 1; 1; 1; 1

2; 2; 2; 2; 2; 2; 2; 2; 2; 2

3; 3; 3; 3; 3; 3; 3; 3; 3; 3; 3; 3; 3; 3; 3; 3

4; 4; 4; 4; 4; 4

5; 5; 5; 5; 5

6; 6

Eleven students buy 1 book. Ten students buy 2 books. Sixteen students buy 3 books. Six students buy 4 books. Five students buy 5 books. Two students buy 6 books.

Because the data are integers, subtract 0.5 from 1, the smallest data value and add 0.5 to 6, the largest data value. Then the starting point is 0.5 and the ending value is 6.5.

Problem *(Solution on p. 92.)*

Next, calculate the width of each bar or class interval. If the data are discrete and there are not too many different values, a width that places the data values in the middle of the bar or class interval is the most convenient. Since the data consist of the numbers 1, 2, 3, 4, 5, 6 and the starting point is 0.5, a width of one places the 1 in the middle of the interval from 0.5 to 1.5, the 2 in the middle of the interval from 1.5 to 2.5, the 3 in the middle of the interval from 2.5 to 3.5, the 4 in the middle of the interval from _______ to _______, the 5 in the middle of the interval from _______ to _______, and the _______ in the middle of the interval from _______ to _______ .

Calculate the number of bars as follows:

$$\frac{6.5 - 0.5}{bars} = 1 \tag{2.3}$$

where 1 is the width of a bar. Therefore, $bars = 6$.

The following histogram displays the number of books on the x-axis and the frequency on the y-axis.

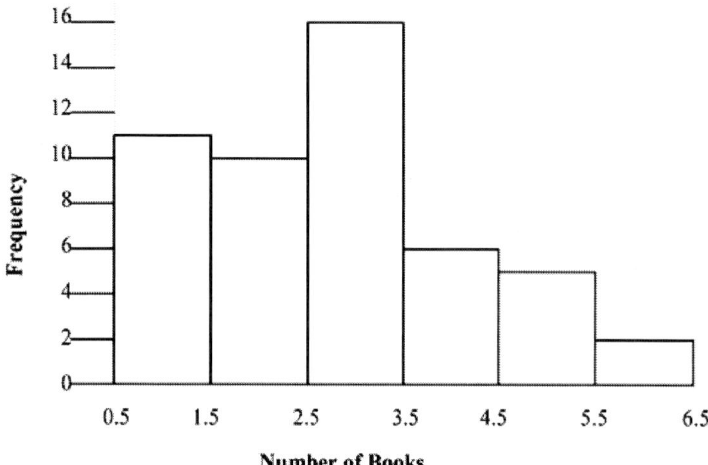

2.4.1 Optional Collaborative Exercise

Count the money (bills and change) in your pocket or purse. Your instructor will record the amounts. As a class, construct a histogram displaying the data. Discuss how many intervals you think is appropriate. You may want to experiment with the number of intervals. Discuss, also, the shape of the histogram.

Record the data, in dollars (for example, 1.25 dollars).

Construct a histogram.

2.5 Box Plots[6]

Box plots or **box-whisker plots** give a good graphical image of the concentration of the data. They also show how far from most of the data the extreme values are. The box plot is constructed from five values: the smallest value, the first quartile, the median, the third quartile, and the largest value. The median, the first quartile, and the third quartile will be discussed here, and then again in the section on measuring data in this chapter. We use these values to compare how close other data values are to them.

The **median**, a number, is a way of measuring the "center" of the data. You can think of the median as the "middle value," although it does not actually have to be one of the observed values. It is a number that separates ordered data into halves. Half the values are the same number or smaller than the median and half the values are the same number or larger. For example, consider the following data:

1; 11.5; 6; 7.2; 4; 8; 9; 10; 6.8; 8.3; 2; 2; 10; 1

Ordered from smallest to largest:

1; 1; 2; 2; 4; 6; **6.8**; **7.2**; 8; 8.3; 9; 10; 10; 11.5

The median is between the 7th value, 6.8, and the 8th value 7.2. To find the median, add the two values together and divide by 2.

$$\frac{6.8+7.2}{2}=7 \qquad (2.4)$$

The median is 7. Half of the values are smaller than 7 and half of the values are larger than 7.

Quartiles are numbers that separate the data into quarters. Quartiles may or may not be part of the data. To find the quartiles, first find the median or second quartile. The first quartile is the middle value of the lower half of the data and the third quartile is the middle value of the upper half of the data. To get the idea, consider the same data set shown above:

1; 1; 2; 2; 4; 6; 6.8; 7.2; 8; 8.3; 9; 10; 10; 11.5

The median or **second quartile** is 7. The lower half of the data is 1, 1, 2, 2, 4, 6, 6.8. The middle value of the lower half is 2.

1; 1; 2; **2**; 4; 6; 6.8

The number 2, which is part of the data, is the **first quartile**. One-fourth of the values are the same or less than 2 and three-fourths of the values are more than 2.

The upper half of the data is 7.2, 8, 8.3, 9, 10, 10, 11.5. The middle value of the upper half is 9.

7.2; 8; 8.3; **9**; 10; 10; 11.5

The number 9, which is part of the data, is the **third quartile**. Three-fourths of the values are less than 9 and one-fourth of the values are more than 9.

To construct a box plot, use a horizontal number line and a rectangular box. The smallest and largest data values label the endpoints of the axis. The first quartile marks one end of the box and the third quartile marks the other end of the box. **The middle fifty percent of the data fall inside the box.** The "whiskers" extend from the ends of the box to the smallest and largest data values. The box plot gives a good quick picture of the data.

[6]This content is available online at <http://cnx.org/content/m16296/1.8/>.

Consider the following data:

1; 1; 2; 2; 4; 6; 6.8 ; 7.2; 8; 8.3; 9; 10; 10; 11.5

The first quartile is 2, the median is 7, and the third quartile is 9. The smallest value is 1 and the largest value is 11.5. The box plot is constructed as follows (see calculator instructions in the back of this book or on the TI web site[7]):

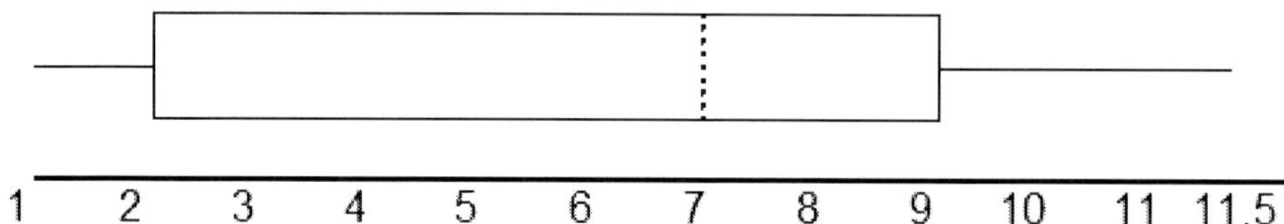

The two whiskers extend from the first quartile to the smallest value and from the third quartile to the largest value. The median is shown with a dashed line.

Example 2.5
The following data are the heights of 40 students in a statistics class.

59; 60; 61; 62; 62; 63; 63; 64; 64; 64; 65; 65; 65; 65; 65; 65; 65; 65; 65; 66; 66; 67; 67; 68; 68; 69; 70; 70; 70; 70; 70; 71; 71; 72; 72; 73; 74; 74; 75; 77

Construct a box plot with the following properties:

- Smallest value = 59
- Largest value = 77
- Q1: First quartile = 64.5
- Q2: Second quartile or median= 66
- Q3: Third quartile = 70

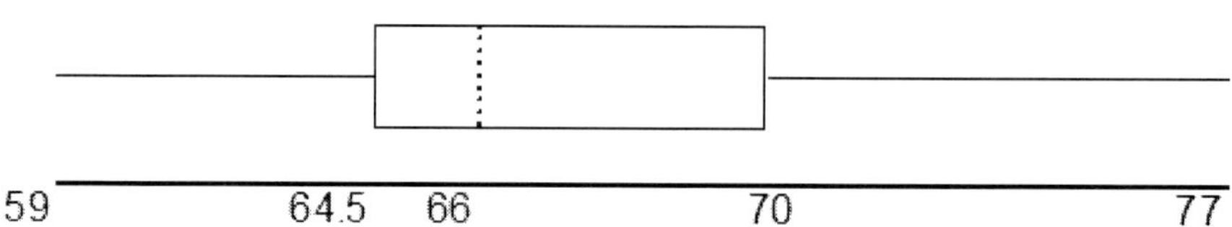

a. Each quarter has 25% of the data.
b. The spreads of the four quarters are 64.5 - 59 = 5.5 (first quarter), 66 - 64.5 = 1.5 (second quarter), 70 - 66 = 4 (3rd quarter), and 77 - 70 = 7 (fourth quarter). So, the second quarter has the smallest spread and the fourth quarter has the largest spread.
c. Interquartile Range: $IQR = Q3 - Q1 = 70 - 64.5 = 5.5$.
d. The interval 59 through 65 has more than 25% of the data so it has more data in it than the interval 66 through 70 which has 25% of the data.

[7]http://education.ti.com/educationportal/sites/US/sectionHome/support.html

For some sets of data, some of the largest value, smallest value, first quartile, median, and third quartile may be the same. For instance, you might have a data set in which the median and the third quartile are the same. In this case, the diagram would not have a dotted line inside the box displaying the median. The right side of the box would display both the third quartile and the median. For example, if the smallest value and the first quartile were both 1, the median and the third quartile were both 5, and the largest value was 7, the box plot would look as follows:

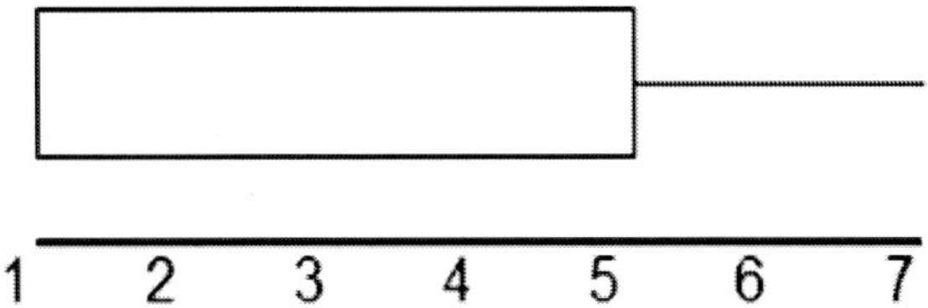

Example 2.6
Test scores for a college statistics class held during the day are:

99; 56; 78; 55.5; 32; 90; 80; 81; 56; 59; 45; 77; 84.5; 84; 70; 72; 68; 32; 79; 90

Test scores for a college statistics class held during the evening are:

98; 78; 68; 83; 81; 89; 88; 76; 65; 45; 98; 90; 80; 84.5; 85; 79; 78; 98; 90; 79; 81; 25.5

Problem *(Solution on p. 92.)*

- What are the smallest and largest data values for each data set?
- What is the median, the first quartile, and the third quartile for each data set?
- Create a boxplot for each set of data.
- Which boxplot has the widest spread for the middle 50% of the data (the data between the first and third quartiles)? What does this mean for that set of data in comparison to the other set of data?
- For each data set, what percent of the data is between the smallest value and the first quartile? (Answer: 25%) the first quartile and the median? (Answer: 25%) the median and the third quartile? the third quartile and the largest value? What percent of the data is between the first quartile and the largest value? (Answer: 75%)

The first data set (the top box plot) has the widest spread for the middle 50% of the data. $IQR = Q3 - Q1$ is $82.5 - 56 = 26.5$ for the first data set and $89 - 78 = 11$ for the second data set. So, the first set of data has its middle 50% of scores more spread out.

25% of the data is between M and $Q3$ and 25% is between $Q3$ and $Xmax$.

2.6 Measures of the Location of the Data[8]

The common measures of location are **quartiles** and **percentiles** (%iles). Quartiles are special percentiles. The first quartile, Q_1 is the same as the 25th percentile (25th %ile) and the third quartile, Q_3, is the same as the 75th percentile (75th %ile). The median, M, is called both the second quartile and the 50th percentile (50th %ile).

[8]This content is available online at <http://cnx.org/content/m16314/1.10/>.

To calculate quartiles and percentiles, the data must be ordered from smallest to largest. Recall that quartiles divide ordered data into quarters. Percentiles divide ordered data into hundredths. To score in the 90th percentile of an exam does not mean, necessarily, that you received 90% on a test. It means that your score was higher than 90% of the people who took the test and lower than the scores of the remaining 10% of the people who took the test. Percentiles are useful for comparing values. For this reason, universities and colleges use percentiles extensively.

The **interquartile range** is a number that indicates the spread of the middle half or the middle 50% of the data. It is the difference between the third quartile (Q_3) and the first quartile (Q_1).

$$IQR = Q_3 - Q_1 \tag{2.5}$$

The IQR can help to determine potential **outliers.** **A value is suspected to be a potential outlier if it is more than** *(1.5)(IQR)* **below the first quartile or more than** *(1.5)(IQR)* **above the third quartile**. Potential outliers always need further investigation.

Example 2.7
For the following 13 real estate prices, calculate the IQR and determine if any prices are outliers. Prices are in dollars. (*Source: San Jose Mercury News*)

389,950; 230,500; 158,000; 479,000; 639,000; 114,950; 5,500,000; 387,000; 659,000; 529,000; 575,000; 488,800; 1,095,000

Solution
Order the data from smallest to largest.

114,950; 158,000; 230,500; 387,000; 389,950; 479,000; 488,800; 529,000; 575,000; 639,000; 659,000; 1,095,000; 5,500,000

$M = 488,800$

$Q_1 = \frac{230500+387000}{2} = 308750$

$Q_3 = \frac{639000+659000}{2} = 649000$

$IQR = 649000 - 308750 = 340250$

$(1.5)(IQR) = (1.5)(340250) = 510375$

$Q_1 - (1.5)(IQR) = 308750 - 510375 = -201625$

$Q_3 + (1.5)(IQR) = 649000 + 510375 = 1159375$

No house price is less than -201625. However, 5,500,000 is more than 1,159,375. Therefore, 5,500,000 is a potential **outlier**.

Example 2.8
For the two data sets in the test scores example (p. 57), find the following:

a. The interquartile range. Compare the two interquartile ranges.
b. Any outliers in either set.
c. The 30th percentile and the 80th percentile for each set. How much data falls below the 30th percentile? Above the 80th percentile?

Example 2.9: Finding Quartiles and Percentiles Using a Table
Fifty statistics students were asked how much sleep they get per school night (rounded to the nearest hour). The results were (student data):

AMOUNT OF SLEEP-PER SCHOOL NIGHT (HOURS)	FREQUENCY	RELATIVE FREQUENCY	CUMULATIVE RELATIVE FREQUENCY
4	2	0.04	0.04
5	5	0.10	0.14
6	7	0.14	0.28
7	12	0.24	0.52
8	14	0.28	0.80
9	7	0.14	0.94
10	3	0.06	1.00

Table 2.2

Find the 28th percentile: Notice the 0.28 in the "cumulative relative frequency" column. 28% of 50 data values = 14. There are 14 values less than the 28th %ile. They include the two 4s, the five 5s, and the seven 6s. The 28th %ile is between the last 6 and the first 7. **The 28th %ile is 6.5.**

Find the median: Look again at the "cumulative relative frequency " column and find 0.52. The median is the 50th %ile or the second quartile. 50% of 50 = 25. There are 25 values less than the median. They include the two 4s, the five 5s, the seven 6s, and eleven of the 7s. The median or 50th %ile is between the 25th (7) and 26th (7) values. **The median is 7.**

Find the third quartile: The third quartile is the same as the 75th percentile. You can "eyeball" this answer. If you look at the "cumulative relative frequency" column, you find 0.52 and 0.80. When you have all the 4s, 5s, 6s and 7s, you have 52% of the data. When you include all the 8s, you have 80% of the data. **The 75th %ile, then, must be an 8** . Another way to look at the problem is to find 75% of 50 (= 37.5) and round up to 38. The third quartile, Q_3, is the 38th value which is an 8. You can check this answer by counting the values. (There are 37 values below the third quartile and 12 values above.)

Example 2.10
Using the table:

1. Find the 80th percentile.
2. Find the 90th percentile.
3. Find the first quartile. What is another name for the first quartile?
4. Construct a box plot of the data.

Collaborative Classroom Exercise: Your instructor or a member of the class will ask everyone in class how many sweaters they own. Answer the following questions.

1. How many students were surveyed?
2. What kind of sampling did you do?

3. Find the mean and standard deviation.
4. Find the mode.
5. Construct 2 different histograms. For each, starting value = _____ ending value = ____.
6. Find the median, first quartile, and third quartile.
7. Construct a box plot.
8. Construct a table of the data to find the following:
   - The 10th percentile
   - The 70th percentile
   - The percent of students who own less than 4 sweaters

2.7 Measures of the Center of the Data[9]

The "center" of a data set is also a way of describing location. The two most widely used measures of the "center" of the data are the **mean** (average) and the **median**. To calculate the **mean weight** of 50 people, add the 50 weights together and divide by 50. To find the **median weight** of the 50 people, order the data and find the number that splits the data into two equal parts (previously discussed under box plots in this chapter). The median is generally a better measure of the center when there are extreme values or outliers because it is not affected by the precise numerical values of the outliers. The mean is the most common measure of the center.

The mean can also be calculated by multiplying each distinct value by its frequency and then dividing the sum by the total number of data values. The letter used to represent the sample mean is an x with a bar over it (pronounced "x bar"): $\overline{x}$.

The Greek letter μ (pronounced "mew") represents the population mean. If you take a truly random sample, the sample mean is a good estimate of the population mean.

To see that both ways of calculating the mean are the same, consider the sample:

1; 1; 1; 2; 2; 3; 4; 4; 4; 4; 4

$$\overline{x} = \frac{1+1+1+2+2+3+4+4+4+4+4}{11} = 2.7 \tag{2.6}$$

$$\overline{x} = \frac{3\times1+2\times2+1\times3+5\times4}{11} = 2.7 \tag{2.7}$$

In the second example, the frequencies are 3, 2, 1, and 5.

You can quickly find the location of the median by using the expression $\frac{n+1}{2}$.

The letter n is the total number of data values in the sample. If n is an odd number, the median is the middle value of the ordered data (ordered smallest to largest). If n is an even number, the median is equal to the two middle values added together and divided by 2 after the data has been ordered. For example, if the total number of data values is 97, then $\frac{n+1}{2}= \frac{97+1}{2}= 49$. The median is the 49th value in the ordered data. If the total number of data values is 100, then $\frac{n+1}{2}= \frac{100+1}{2}= 50.5$. The median occurs midway between the 50th and 51st values. The location of the median and the median itself are **not** the same. The upper case letter M is often used to represent the median. The next example illustrates the location of the median and the median itself.

[9]This content is available online at <http://cnx.org/content/m17102/1.8/>.

Example 2.11
AIDS data indicating the number of months an AIDS patient lives after taking a new antibody drug are as follows (smallest to largest):

3; 4; 8; 8; 10; 11; 12; 13; 14; 15; 15; 16; 16; 17; 17; 18; 21; 22; 22; 24; 24; 25; 26; 26; 27; 27; 29; 29; 31; 32; 33; 33; 34; 34; 35; 37; 40; 44; 44; 47

Calculate the mean and the median.

Solution
The calculation for the mean is:

$$\overline{x} = \frac{[3+4+(8)(2)+10+11+12+13+14+(15)(2)+(16)(2)+...+35+37+40+(44)(2)+47]}{40} = 23.6$$

To find the median, **M**, first use the formula for the location. The location is:

$$\frac{n+1}{2} = \frac{40+1}{2} = 20.5$$

Starting at the smallest value, the median is located between the 20th and 21st values (the two 24s):

3; 4; 8; 8; 10; 11; 12; 13; 14; 15; 15; 16; 16; 17; 17; 18; 21; 22; 22; 24; 24; 25; 26; 26; 27; 27; 29; 29; 31; 32; 33; 33; 34; 34; 35; 37; 40; 44; 44; 47

$$M = \frac{24+24}{2} = 24$$

The median is 24.

Example 2.12
Suppose that, in a small town of 50 people, one person earns $5,000,000 per year and the other 49 each earn $30,000. Which is the better measure of the "center," the mean or the median?

Solution
$$\overline{x} = \frac{5000000+49\times30000}{50} = 129400$$

$$M = 30000$$

(There are 49 people who earn $30,000 and one person who earns $5,000,000.)

The median is a better measure of the "center" than the mean because 49 of the values are 30,000 and one is 5,000,000. The 5,000,000 is an outlier. The 30,000 gives us a better sense of the middle of the data.

Another measure of the center is the mode. The **mode** is the most frequent value. If a data set has two values that occur the same number of times, then the set is bimodal.

Example 2.13: Statistics exam scores for 20 students are as follows
Statistics exam scores for 20 students are as follows:

50 ; 53 ; 59 ; 59 ; 63 ; 63 ; 72 ; 72 ; 72 ; 72 ; 72 ; 76 ; 78 ; 81 ; 83 ; 84 ; 84 ; 84 ; 90 ; 93

Problem
Find the mode.

Solution
The most frequent score is 72, which occurs five times. Mode = 72.

Example 2.14
Five real estate exam scores are 430, 430, 480, 480, 495. The data set is bimodal because the scores 430 and 480 each occur twice.

When is the mode the best measure of the "center"? Consider a weight loss program that advertises an average weight loss of six pounds the first week of the program. The mode might indicate that most people lose two pounds the first week, making the program less appealing.

Statistical software will easily calculate the mean, the median, and the mode. Some graphing calculators can also make these calculations. In the real world, people make these calculations using software.

2.7.1 The Law of Large Numbers and the Mean

The Law of Large Numbers says that if you take samples of larger and larger size from any population, then the mean $\overline{x}$ of the sample gets closer and closer to μ. This is discussed in more detail in the section **The Central Limit Theorem** of this course.

NOTE: The formula for the mean is located in the Summary of Formulas (Section 2.10) section course.

2.8 Skewness and the Mean, Median, and Mode[10]

Consider the following data set:

4 ; 5 ; 6 ; 6 ; 6 ; 7 ; 7 ; 7 ; 7 ; 7 ; 7 ; 8 ; 8 ; 8 ; 9 ; 10

This data produces the histogram shown below. Each interval has width one and each value is located in the middle of an interval.

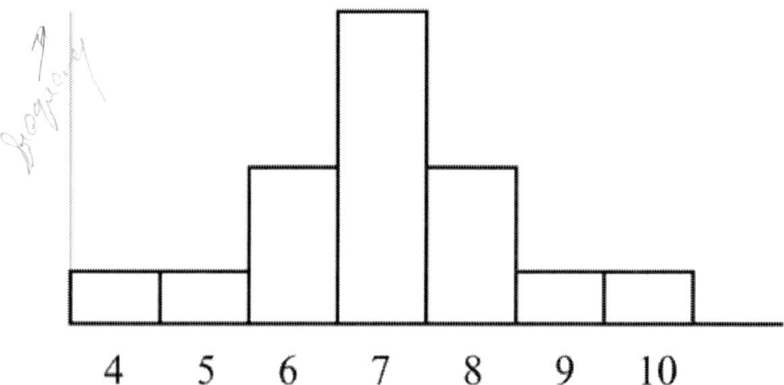

The histogram displays a **symmetrical** distribution of data. A distribution is symmetrical if a vertical line can be drawn at some point in the histogram such that the shape to the left and the right of the vertical

[10]This content is available online at <http://cnx.org/content/m17104/1.5/>.

line are mirror images of each other. The mean, the median, and the mode are each 7 for these data. **In a perfectly symmetrical distribution, the mean, the median, and the mode are the same.**

The histogram for the data:

4 ; 5 ; 6 ; 6 ; 6 ; 7 ; 7 ; 7 ; 7 ; 7 ; 7 ; 8

is not symmetrical. The right-hand side seems "chopped off" compared to the left side. The shape distribution is called **skewed to the left** because it is pulled out to the left.

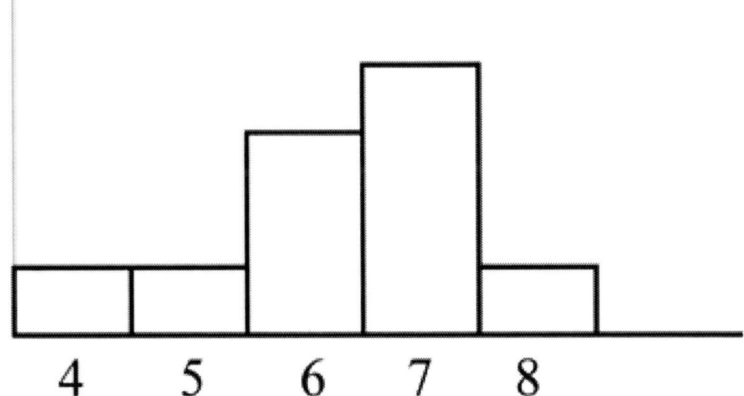

The mean is 6.3, the median is 6.5, and the mode is 7. **Notice that the mean is less than the median and they are both less than the mode.** The mean and the median both reflect the skewing but the mean more so.

The histogram for the data:

6 ; 7 ; 7 ; 7 ; 7 ; 7 ; 7 ; 8 ; 8 ; 8 ; 9 ; 10

is also not symmetrical. It is **skewed to the right**.

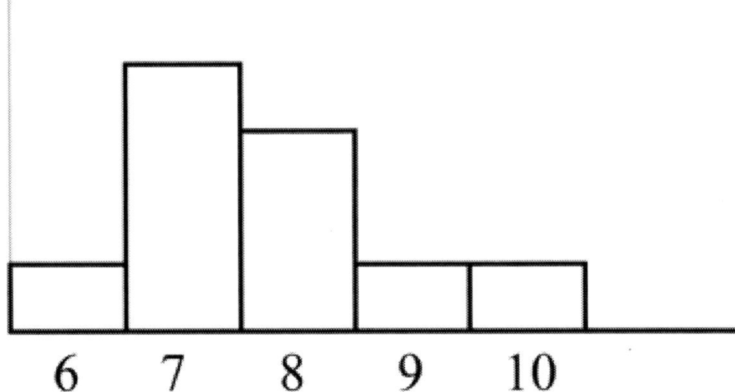

The mean is 7.7, the median is 7.5, and the mode is 7. **Notice that the mean is the largest statistic, while the mode is the smallest**. Again, the mean reflects the skewing the most.

To summarize, generally if the distribution of data is skewed to the left, the mean is less than the median, which is less than the mode. If the distribution of data is skewed to the right, the mode is less than the median, which is less than the mean.

Skewness and symmetry become important when we discuss probability distributions in later chapters.

2.9 Measures of the Spread of the Data[11]

The most common measure of spread is the standard deviation. The **standard deviation** is a number that measures how far data values are from their mean. For example, if the mean of a set of data containing 7 is 5 and the **standard deviation** is 2, then the value 7 is one (1) standard deviation from its mean because 5 + (1)(2) = 7.

The number line may help you understand standard deviation. If we were to put 5 and 7 on a number line, 7 is to the right of 5. We say, then, that 7 is **one** standard deviation to the **right** of 5. If 1 were also part of the data set, then 1 is **two** standard deviations to the **left** of 5 because 5 +(-2)(2) = 1.

1=5+(-2)(2) ; 7=5+(1)(2)

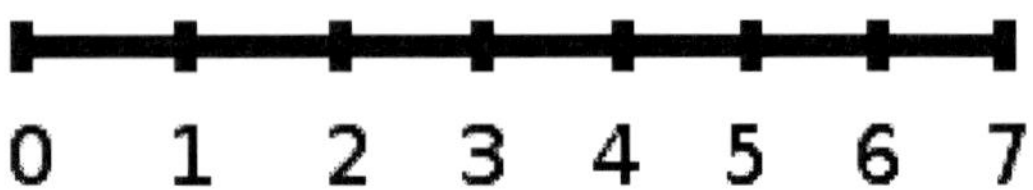

Formula: value = $\overline{x}$ + (#ofSTDEVs)(s)

Generally, a value = mean + (#ofSTDEVs)(standard deviation), where #ofSTDEVs = the number of standard deviations.

If x is a value and $\overline{x}$ is the sample mean, then $x - \overline{x}$ is called a deviation. In a data set, there are as many deviations as there are data values. Deviations are used to calculate the sample standard deviation.

Calculation of the Sample Standard Deviation
To calculate the standard deviation, calculate the variance first. The **variance** is the average of the squares of the deviations. The standard deviation is the square root of the variance. You can think of the standard deviation as a special average of the deviations (the $x - \overline{x}$ values). The lower case letter s represents the sample standard deviation and the Greek letter σ (sigma) represents the population standard deviation. We use s^2 to represent the sample variance and σ^2 to represent the population variance. If the sample has the same characteristics as the population, then s should be a good estimate of σ.

> NOTE: In practice, use either a calculator or computer software to calculate the standard deviation. However, please study the following step-by-step example.

Example 2.15
In a fifth grade class, the teacher was interested in the average age and the standard deviation of the ages of her students. What follows are the ages of her students to the nearest half year:

9 ; 9.5 ; 9.5 ; 10 ; 10 ; 10 ; 10 ; 10.5 ; 10.5 ; 10.5 ; 10.5 ; 11 ; 11 ; 11 ; 11 ; 11 ; 11 ; 11.5 ; 11.5 ; 11.5

$$\overline{x} = \frac{9 + 9.5 \times 2 + 10 \times 4 + 10.5 \times 4 + 11 \times 6 + 11.5 \times 3}{20} = 10.525 \quad (2.8)$$

[11]This content is available online at <http://cnx.org/content/m17103/1.8/>.

The average age is 10.53 years, rounded to 2 places.

The variance may be calculated by using a table. Then the standard deviation is calculated by taking the square root of the variance. We will explain the parts of the table after calculating s.

Data	**Freq.**	**Deviations**	*Deviations*^{2}	**(Freq.)**(*Deviations*^{2}**)**
x	f	$(x-\overline{x})$	$(x-\overline{x})^2$	$(f)\,(x-\overline{x})^2$
9	1	$9-10.525=-1.525$	$(-1.525)^2=2.325625$	$1\times 2.325625=2.325625$
9.5	2	$9.5-10.525=-1.025$	$(-1.025)^2=1.050625$	$2\times 1.050625=2.101250$
10	4	$10-10.525=-0.525$	$(-0.525)^2=0.275625$	$4\times .275625=1.1025$
10.5	4	$10.5-10.525=-0.025$	$(-0.025)^2=0.000625$	$4\times .000625=.0025$
11	6	$11-10.525=0.475$	$(0.475)^2=0.225625$	$6\times .225625=1.35375$
11.5	3	$11.5-10.525=0.975$	$(0.975)^2=0.950625$	$3\times .950625=2.851875$

Table 2.3

The sample variance, s^2, is equal to the sum of the last column (9.7375) divided by the total number of data values minus one (20 - 1):

$s^2=\frac{9.7375}{20-1}=0.5125$

The sample standard deviation, s, is equal to the square root of the sample variance:

$s=\sqrt{0.5125}=.0715891$ Rounded to two decimal places, $s=0.72$

Typically, you do the calculation for the standard deviation on your calculator or computer. The intermediate results are not rounded. This is done for accuracy.

Problem 1

Verify the mean and standard deviation calculated above on your calculator or computer. Find the median and mode.

Solution

- Median = 10.5
- Mode = 11

Problem 2

Find the value that is 1 standard deviation above the mean. Find $(\overline{x}+1s)$.

Solution

$(\overline{x}+1s)=10.53+(1)\,(0.72)=11.25$

Problem 3

Find the value that is two standard deviations below the mean. Find $(\overline{x}-2s)$.

Solution

$(\overline{x}-2s)=10.53-(2)\,(0.72)=9.09$

Problem 4
Find the values that are 1.5 standard deviations **from** (below and above) the mean.

Solution

- $(\bar{x} - 1.5s) = 10.53 - (1.5)(0.72) = 9.45$
- $(\bar{x} + 1.5s) = 10.53 + (1.5)(0.72) = 11.61$

Explanation of the table: The deviations show how spread out the data are about the mean. The value 11.5 is farther from the mean than 11. The deviations 0.975 and 0.475 indicate that. **If you add the deviations, the sum is always zero**. (For this example, there are 20 deviations.) So you cannot simply add the deviations to get the spread of the data. By squaring the deviations, you make them positive numbers. The variance, then, is the average squared deviation. It is small if the values are close to the mean and large if the values are far from the mean.

The variance is a squared measure and does not have the same units as the data. Taking the square root solves the problem. The standard deviation measures the spread in the same units as the data.

For the sample variance, we divide by the total number of data values minus one $(n-1)$. Why not divide by n? The answer has to do with the population variance. **The sample variance is an estimate of the population variance.** By dividing by $(n-1)$, we get a better estimate of the population variance.

Your concentration should be on what the standard deviation does, not on the arithmetic. The standard deviation is a number which measures how far the data are spread from the mean. Let a calculator or computer do the arithmetic.

The sample standard deviation, s , is either zero or larger than zero. When $s = 0$, there is no spread. When s is a lot larger than zero, the data values are very spread out about the mean. Outliers can make s very large.

The standard deviation, when first presented, can seem unclear. By graphing your data, you can get a better "feel" for the deviations and the standard deviation. You will find that in symmetrical distributions, the standard deviation can be very helpful but in skewed distributions, the standard deviation may not be much help. The reason is that the two sides of a skewed distribution have different spreads. In a skewed distribution, it is better to look at the first quartile, the median, the third quartile, the smallest value, and the largest value. Because numbers can be confusing, **always graph your data**.

NOTE: The formula for the standard deviation is at the end of the chapter.

Example 2.16
Use the following data (first exam scores) from Susan Dean's spring pre-calculus class:

33; 42; 49; 49; 53; 55; 55; 61; 63; 67; 68; 68; 69; 69; 72; 73; 74; 78; 80; 83; 88; 88; 88; 90; 92; 94; 94; 94; 94; 96; 100

a. Create a chart containing the data, frequencies, relative frequencies, and cumulative relative frequencies to three decimal places.
b. Calculate the following to one decimal place using a TI-83+ or TI-84 calculator:

 i. The sample mean
 ii. The sample standard deviation
 iii. The median
 iv. The first quartile

v. The third quartile
vi. IQR

c. Construct a box plot and a histogram on the same set of axes. Make comments about the box plot, the histogram, and the chart.

Solution

a.

Data	Frequency	Relative Frequency	Cumulative Relative Frequency
33	1	0.032	0.032
42	1	0.032	0.064
49	2	0.065	0.129
53	1	0.032	0.161
55	2	0.065	0.226
61	1	0.032	0.258
63	1	0.032	0.29
67	1	0.032	0.322
68	2	0.065	0.387
69	2	0.065	0.452
72	1	0.032	0.484
73	1	0.032	0.516
74	1	0.032	0.548
78	1	0.032	0.580
80	1	0.032	0.612
83	1	0.032	0.644
88	3	0.097	0.741
90	1	0.032	0.773
92	1	0.032	0.805
94	4	0.129	0.934
96	1	0.032	0.966
100	1	0.032	**0.998** (Why isn't this value 1?)

Table 2.4

b. i. The sample mean = 73.5
ii. The sample standard deviation = 17.9
iii. The median = 73
iv. The first quartile = 61
v. The third quartile = 90
vi. IQR = 90 - 61 = 29

c. The x-axis goes from 32.5 to 100.5; y-axis goes from -2.4 to 15 for the histogram; number of intervals is 5 for the histogram so the width of an interval is (100.5 - 32.5) divided by 5 which is equal to 13.6. Endpoints of the intervals: starting point is 32.5, 32.5+13.6 =

46.1, 46.1+13.6 = 59.7, 59.7+13.6 = 73.3, 73.3+13.6 = 86.9, 86.9+13.6 = 100.5 = the ending value; No data values fall on an interval boundary.

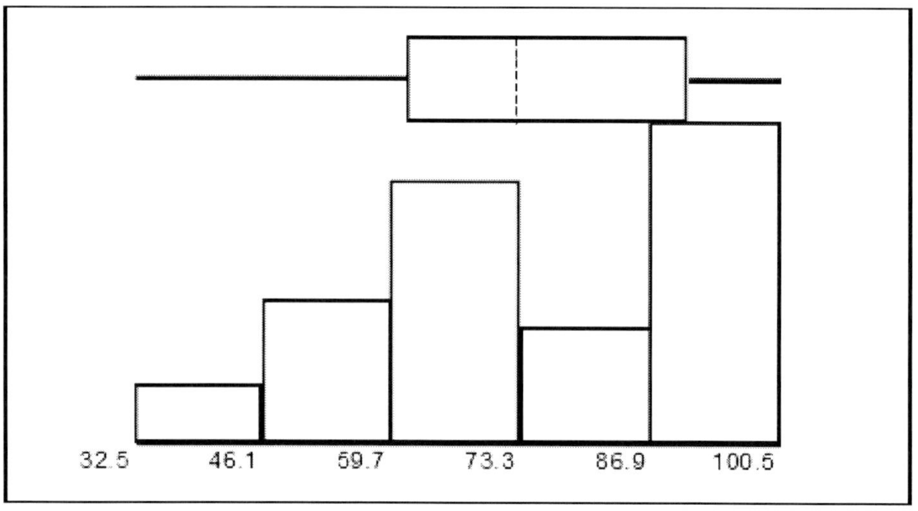

Figure 2.1

The long left whisker in the box plot is reflected in the left side of the histogram. The spread of the exam scores in the lower 50% is greater (73 - 33 = 40) than the spread in the upper 50% (100 - 73 = 27). The histogram, box plot, and chart all reflect this. There are a substantial number of A and B grades (80s, 90s, and 100). The histogram clearly shows this. The box plot shows us that the middle 50% of the exam scores (IQR = 29) are Ds, Cs, and Bs. The box plot also shows us that the lower 25% of the exam scores are Ds and Fs.

Example 2.17

Two students, John and Ali, from different high schools, wanted to find out who had the highest G.P.A. when compared to his school. Which student had the highest G.P.A. when compared to his school?

Student	GPA	School Mean GPA	School Standard Deviation
John	2.85	3.0	0.7
Ali	77	80	10

Table 2.5

Solution

Use the formula **value = mean + (#ofSTDEVs)(stdev)** and solve for #ofSTDEVs for each student (stdev = standard deviation):

$\#ofSTDEVs = \frac{value - mean}{stdev}$:

For John, $\#ofSTDEVs = \frac{2.85 - 3.0}{0.7} = -0.21$

For Ali, $\#ofSTDEVs = \frac{77 - 80}{10} = -0.3$

John has the better G.P.A. when compared to his school because his G.P.A. is 0.21 standard deviations **below his** mean while Ali's G.P.A. is 0.3 standard deviations **below his** mean.

2.10 Summary of Formulas[12]

Commonly Used Symbols

- The symbol Σ means to add or to find the sum.
- n = the number of data values in a sample
- N = the number of people, things, etc. in the population
- $\overline{x}$ = the sample mean
- s = the sample standard deviation
- μ = the population mean
- σ = the population standard deviation
- f = frequency
- x = numerical value

Commonly Used Expressions

- $x * f$ = A value multiplied by its respective frequency
- $\sum x$ = The sum of the values
- $\sum x * f$ = The sum of values multiplied by their respective frequencies
- $(x - \overline{x})$ or $(x - \mu)$ = Deviations from the mean (how far a value is from the mean)
- $(x - \overline{x})^2$ or $(x - \mu)^2$ = Deviations squared
- $f\,(x - \overline{x})^2$ or $f\,(x - \mu)^2$ = The deviations squared and multiplied by their frequencies

Mean Formulas:

- $\overline{x} = \frac{\sum x}{n}$ or $\overline{x} = \frac{\sum f \cdot x}{n}$
- $\mu = \frac{\sum x}{N}$ or $\mu = \frac{\sum f \cdot x}{N}$

Standard Deviation Formulas:

- $s = \sqrt{\frac{\Sigma (x - \overline{x})^2}{n-1}}$ or $s = \sqrt{\frac{\Sigma f \cdot (x - \overline{x})^2}{n-1}}$
- $\sigma = \sqrt{\frac{\Sigma (x - \overline{\mu})^2}{N}}$ or $\sigma = \sqrt{\frac{\Sigma f \cdot (x - \overline{\mu})^2}{N}}$

Formulas Relating a Value, the Mean, and the Standard Deviation:

- value = mean + (#ofSTDEVs)(standard deviation), where #ofSTDEVs = the number of standard deviations
- $x = \overline{x}$+ (#ofSTDEVs)(s)
- $x = \mu$ + (#ofSTDEVs)(σ)

[12]This content is available online at <http://cnx.org/content/m16310/1.8/>.

2.11 Practice 1: Center of the Data[13]

2.11.1 Student Learning Outcomes

- The student will calculate and interpret the center, spread, and location of the data.
- The student will construct and interpret histograms an box plots.

2.11.2 Given

Sixty-five randomly selected car salespersons were asked the number of cars they generally sell in one week. Fourteen people answered that they generally sell three cars; nineteen generally sell four cars; twelve generally sell five cars; nine generally sell six cars; eleven generally sell seven cars.

2.11.3 Complete the Table

Data Value (# cars)	Frequency	Relative Frequency	Cumulative Relative Frequency

Table 2.6

2.11.4 Discussion Questions

Exercise 2.11.1 *(Solution on p. 93.)*
What does the frequency column sum to? Why?

Exercise 2.11.2 *(Solution on p. 93.)*
What does the relative frequency column sum to? Why?

Exercise 2.11.3
What is the difference between relative frequency and frequency for each data value?

Exercise 2.11.4
What is the difference between cumulative relative frequency and relative frequency for each data value?

2.11.5 Enter the Data

Enter your data into your calculator or computer.

[13]This content is available online at <http://cnx.org/content/m16312/1.12/>.

2.11.6 Construct a Histogram

Determine appropriate minimum and maximum x and y values and the scaling. Sketch the histogram below. Label the horizontal and vertical axes with words. Include numerical scaling.

2.11.7 Data Statistics

Calculate the following values:

Exercise 2.11.5 *(Solution on p. 94.)*
Sample mean = $\overline{x}$ =

Exercise 2.11.6 *(Solution on p. 94.)*
Sample standard deviation = s_x =

Exercise 2.11.7 *(Solution on p. 94.)*
Sample size = n =

2.11.8 Calculations

Use the table in section 2.11.3 to calculate the following values:

Exercise 2.11.8 *(Solution on p. 94.)*
Median =

Exercise 2.11.9 *(Solution on p. 94.)*
Mode =

Exercise 2.11.10 *(Solution on p. 94.)*
First quartile =

Exercise 2.11.11 *(Solution on p. 94.)*
Second quartile = median = 50th percentile =

Exercise 2.11.12 *(Solution on p. 94.)*
Third quartile =

Exercise 2.11.13 *(Solution on p. 94.)*
Interquartile range (IQR) = _____ - _____ = _____

Exercise 2.11.14 *(Solution on p. 94.)*
10th percentile =

Exercise 2.11.15 *(Solution on p. 94.)*
70th percentile =

Exercise 2.11.16 *(Solution on p. 94.)*

Find the value that is 3 standard deviations:

- **a.** Above the mean
- **b.** Below the mean

2.11.9 Box Plot

Construct a box plot below. Use a ruler to measure and scale accurately.

2.11.10 Interpretation

Looking at your box plot, does it appear that the data are concentrated together, spread out evenly, or concentrated in some areas, but not in others? How can you tell?

2.12 Practice 2: Spread of the Data[14]

2.12.1 Student Learning Objectives

- The student will calculate measures of the center of the data.
- The student will calculate the spread of the data.

2.12.2 Given

The population parameters below describe the full-time equivalent number of students (FTES) each year at Lake Tahoe Community College from 1976-77 through 2004-2005. (*Source: Graphically Speaking by Bill King, LTCC Institutional Research, December 2005*).

Use these values to answer the following questions:

- $\mu = 1000$ FTES
- Median - 1014 FTES
- $\sigma = 474$ FTES
- First quartile = 528.5 FTES
- Third quartile = 1447.5 FTES
- $n = 29$ years

2.12.3 Calculate the Values

Exercise 2.12.1 ***(Solution on p. 94.)***

A sample of 11 years is taken. About how many are expected to have a FTES of 1014 or above? Explain how you determined your answer.

Exercise 2.12.2 ***(Solution on p. 94.)***

75% of all years have a FTES:

a. At or below:
b. At or above:

Exercise 2.12.3 ***(Solution on p. 94.)***

The population standard deviation =

Exercise 2.12.4 ***(Solution on p. 94.)***

What percent of the FTES were from 528.5 to 1447.5? How do you know?

Exercise 2.12.5 ***(Solution on p. 94.)***

What is the IQR? What does the IQR represent?

Exercise 2.12.6 ***(Solution on p. 94.)***

How many standard deviations away from the mean is the median?

[14]This content is available online at <http://cnx.org/content/m17105/1.10/>.

2.13 Homework[15]

Exercise 2.13.1 *(Solution on p. 94.)*

Twenty-five randomly selected students were asked the number of movies they watched the previous week. The results are as follows:

# of movies	Frequency	Relative Frequency	Cumulative Relative Frequency
0	5		
1	9		
2	6		
3	4		
4	1		

Table 2.7

a. Find the sample mean $\overline{x}$
b. Find the sample standard deviation, s
c. Construct a histogram of the data.
d. Complete the columns of the chart.
e. Find the first quartile.
f. Find the median.
g. Find the third quartile.
h. Construct a box plot of the data.
i. What percent of the students saw fewer than three movies?
j. Find the 40th percentile.
k. Find the 90th percentile.

Exercise 2.13.2

The median age for U.S. blacks currently is 30.1 years; for U.S. whites it is 36.6 years. (Source: U.S. Census)

a. Based upon this information, give two reasons why the black median age could be lower than the white median age.
b. Does the lower median age for blacks necessarily mean that blacks die younger than whites? Why or why not?
c. How might it be possible for blacks and whites to die at approximately the same age, but for the median age for whites to be higher?

Exercise 2.13.3 *(Solution on p. 95.)*

Forty randomly selected students were asked the number of pairs of sneakers they owned. Let X = the number of pairs of sneakers owned. The results are as follows:

[15]This content is available online at <http://cnx.org/content/m16801/1.12/>.

X	Frequency	Relative Frequency	Cumulative Relative Frequency
1	2		
2	5		
3	8		
4	12		
5	12		
7	1		

Table 2.8

a. Find the sample mean $\overline{x}$
b. Find the sample standard deviation, s
c. Construct a histogram of the data.
d. Complete the columns of the chart.
e. Find the first quartile.
f. Find the median.
g. Find the third quartile.
h. Construct a box plot of the data.
i. What percent of the students owned at least five pairs?
j. Find the 40th percentile.
k. Find the 90th percentile.

Exercise 2.13.4

600 adult Americans were asked by telephone poll, What do you think constitutes a middle-class income? The results are below. Also, include left endpoint, but not the right endpoint. (*Source: Time magazine; survey by Yankelovich Partners, Inc.*)

NOTE: "Not sure" answers were omitted from the results.

Salary ($)	Relative Frequency
< 20,000	0.02
20,000 - 25,000	0.09
25,000 - 30,000	0.19
30,000 - 40,000	0.26
40,000 - 50,000	0.18
50,000 - 75,000	0.17
75,000 - 99,999	0.02
100,000+	0.01

Table 2.9

a. What percent of the survey answered "not sure" ?
b. What percent think that middle-class is from $25,000 - $50,000 ?
c. Construct a histogram of the data

1. Should all bars have the same width, based on the data? Why or why not?
2. How should the <20,000 and the 100,000+ intervals be handled? Why?

d. Find the 40th and 80th percentiles

Exercise 2.13.5 *(Solution on p. 95.)*

Following are the published weights (in pounds) of all of the team members of the San Francisco 49ers from a previous year (Source: San Jose Mercury News).

177; 205; 210; 210; 232; 205; 185; 185; 178; 210; 206; 212; 184; 174; 185; 242; 188; 212; 215; 247; 241; 223; 220; 260; 245; 259; 278; 270; 280; 295; 275; 285; 290; 272; 273; 280; 285; 286; 200; 215; 185; 230; 250; 241; 190; 260; 250; 302; 265; 290; 276; 228; 265

a. Organize the data from smallest to largest value.
b. Find the median.
c. Find the first quartile.
d. Find the third quartile.
e. Construct a box plot of the data.
f. The middle 50% of the weights are from _______ to _______.
g. If our population were all professional football players, would the above data be a sample of weights or the population of weights? Why?
h. If our population were the San Francisco 49ers, would the above data be a sample of weights or the population of weights? Why?
i. Assume the population was the San Francisco 49ers. Find:

i. the population mean, μ.
ii. the population standard deviation, σ.
iii. the weight that is 2 standard deviations below the mean.
iv. When Steve Young, quarterback, played football, he weighed 205 pounds. How many standard deviations above or below the mean was he?

j. That same year, the average weight for the Dallas Cowboys was 240.08 pounds with a standard deviation of 44.38 pounds. Emmit Smith weighed in at 209 pounds. With respect to his team, who was lighter, Smith or Young? How did you determine your answer?

Exercise 2.13.6

An elementary school class ran 1 mile in an average of 11 minutes with a standard deviation of 3 minutes. Rachel, a student in the class, ran 1 mile in 8 minutes. A junior high school class ran 1 mile in an average of 9 minutes, with a standard deviation of 2 minutes. Kenji, a student in the class, ran 1 mile in 8.5 minutes. A high school class ran 1 mile in an average of 7 minutes with a standard deviation of 4 minutes. Nedda, a student in the class, ran 1 mile in 8 minutes.

a. Why is Kenji considered a better runner than Nedda, even though Nedda ran faster than he?
b. Who is the fastest runner with respect to his or her class? Explain why.

Exercise 2.13.7

In a survey of 20 year olds in China, Germany and America, people were asked the number of foreign countries they had visited in their lifetime. The following box plots display the results.

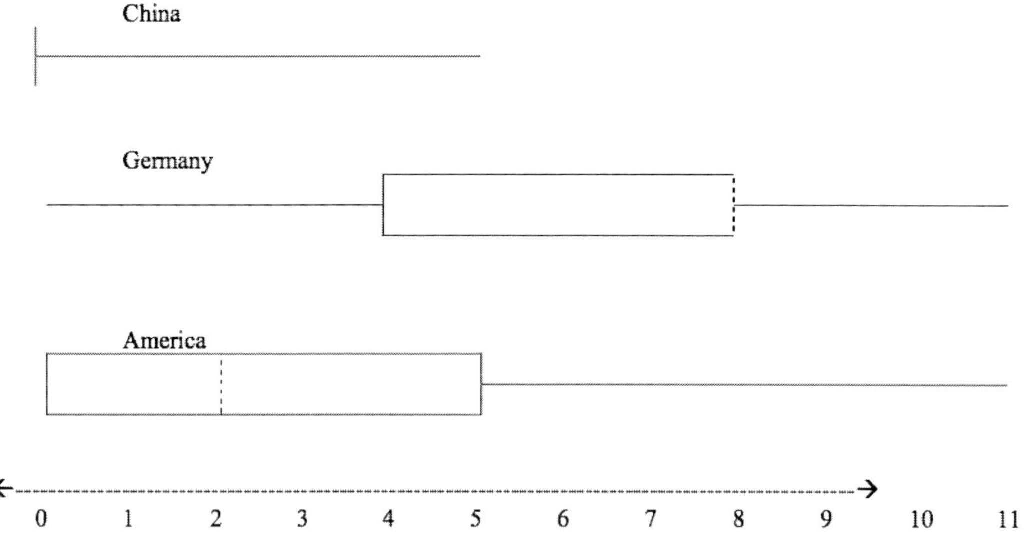

a. In complete sentences, describe what the shape of each box plot implies about the distribution of the data collected.
b. Explain how it is possible that more Americans than Germans surveyed have been to over eight foreign countries.
c. Compare the three box plots. What do they imply about the foreign travel of twenty year old residents of the three countries when compared to each other?

Exercise 2.13.8
Twelve teachers attended a seminar on mathematical problem solving. Their attitudes were measured before and after the seminar. A positive number change attitude indicates that a teacher's attitude toward math became more positive. The twelve change scores are as follows:

3; 8; -1; 2; 0; 5; -3; 1; -1; 6; 5; -2

a. What is the average change score?
b. What is the standard deviation for this population?
c. What is the median change score?
d. Find the change score that is 2.2 standard deviations below the mean.

Exercise 2.13.9 *(Solution on p. 95.)*
Three students were applying to the same graduate school. They came from schools with different grading systems. Which student had the best G.P.A. when compared to his school? Explain how you determined your answer.

Student	G.P.A.	School Ave. G.P.A.	School Standard Deviation
Thuy	2.7	3.2	0.8
Vichet	87	75	20
Kamala	8.6	8	0.4

Table 2.10

Exercise 2.13.10
Given the following box plot:

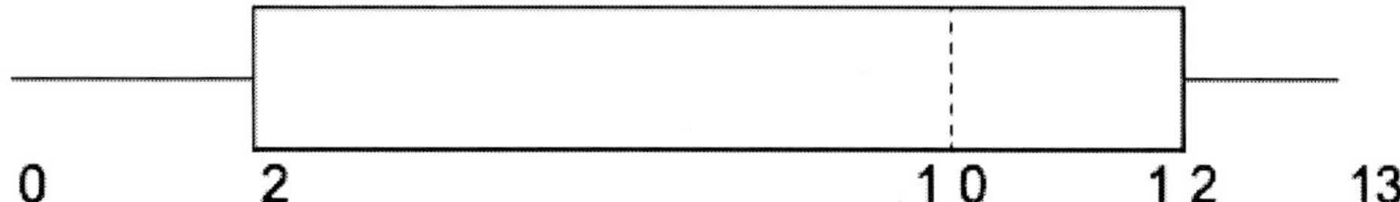

a. Which quarter has the smallest spread of data? What is that spread?
b. Which quarter has the largest spread of data? What is that spread?
c. Find the Inter Quartile Range (IQR).
d. Are there more data in the interval 5 - 10 or in the interval 10 - 13? How do you know this?
e. Which interval has the fewest data in it? How do you know this?

I. 0-2
II. 2-4
III. 10-12
IV. 12-13

Exercise 2.13.11
Given the following box plot:

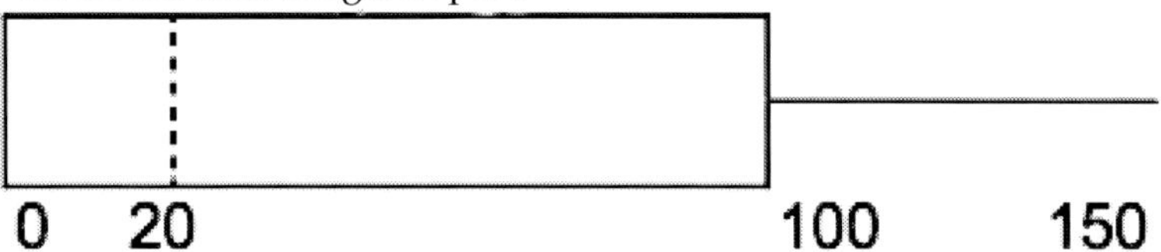

a. Think of an example (in words) where the data might fit into the above box plot. In 2-5 sentences, write down the example.
b. What does it mean to have the first and second quartiles so close together, while the second to fourth quartiles are far apart?

Exercise 2.13.12
Santa Clara County, CA, has approximately 27,873 Japanese-Americans. Their ages are as follows. (*Source: West magazine*)

Age Group	Percent of Community
0-17	18.9
18-24	8.0
25-34	22.8
35-44	15.0
45-54	13.1
55-64	11.9
65+	10.3

Table 2.11

a. Construct a histogram of the Japanese-American community in Santa Clara County, CA. The bars will **not** be the same width for this example. Why not?
b. What percent of the community is under age 35?
c. Which box plot most resembles the information above?

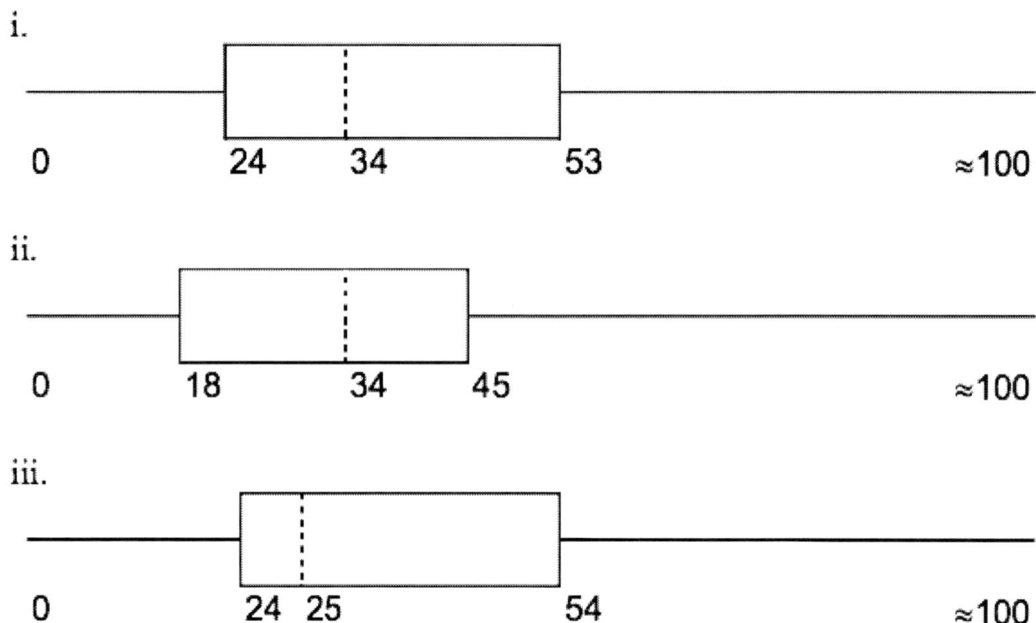

Exercise 2.13.13

Suppose that three book publishers were interested in the number of fiction paperbacks adult consumers purchase per month. Each publisher conducted a survey. In the survey, each asked adult consumers the number of fiction paperbacks they had purchased the previous month. The results are below.

Publisher A

# of books	Freq.	Rel. Freq.
0	10	
1	12	
2	16	
3	12	
4	8	
5	6	
6	2	
8	2	

Table 2.12

Publisher B

# of books	Freq.	Rel. Freq.
0	18	
1	24	
2	24	
3	22	
4	15	
5	10	
7	5	
9	1	

Table 2.13

Publisher C

# of books	Freq.	Rel. Freq.
0-1	20	
2-3	35	
4-5	12	
6-7	2	
8-9	1	

Table 2.14

a. Find the relative frequencies for each survey. Write them in the charts.
b. Using either a graphing calculator, computer, or by hand, use the frequency column to construct a histogram for each publisher's survey. For Publishers A and B, make bar widths of 1. For Publisher C, make bar widths of 2.
c. In complete sentences, give two reasons why the graphs for Publishers A and B are not identical.
d. Would you have expected the graph for Publisher C to look like the other two graphs? Why or why not?
e. Make new histograms for Publisher A and Publisher B. This time, make bar widths of 2.
f. Now, compare the graph for Publisher C to the new graphs for Publishers A and B. Are the graphs more similar or more different? Explain your answer.

Exercise 2.13.14

Often, cruise ships conduct all on-board transactions, with the exception of gambling, on a cashless basis. At the end of the cruise, guests pay one bill that covers all on-board transactions. Suppose that 60 single travelers and 70 couples were surveyed as to their on-board bills for a seven-day cruise from Los Angeles to the Mexican Riviera. Below is a summary of the bills for each group.

Singles

Amount($)	Frequency	Rel. Frequency
51-100	5	
101-150	10	
151-200	15	
201-250	15	
251-300	10	
301-350	5	

Table 2.15

Couples

Amount($)	Frequency	Rel. Frequency
100-150	5	
201-250	5	
251-300	5	
301-350	5	
351-400	10	
401-450	10	
451-500	10	
501-550	10	
551-600	5	
601-650	5	

Table 2.16

a. Fill in the relative frequency for each group.
b. Construct a histogram for the Singles group. Scale the x-axis by $50. widths. Use relative frequency on the y-axis.
c. Construct a histogram for the Couples group. Scale the x-axis by $50. Use relative frequency on the y-axis.
d. Compare the two graphs:

- **i.** List two similarities between the graphs.
- **ii.** List two differences between the graphs.
- **iii.** Overall, are the graphs more similar or different?

e. Construct a new graph for the Couples by hand. Since each couple is paying for two individuals, instead of scaling the x-axis by $50, scale it by $100. Use relative frequency on the y-axis.
f. Compare the graph for the Singles with the new graph for the Couples:

- **i.** List two similarities between the graphs.
- **ii.** Overall, are the graphs more similar or different?

i. By scaling the Couples graph differently, how did it change the way you compared it to the Singles?
j. Based on the graphs, do you think that individuals spend the same amount, more or less, as singles as they do person by person in a couple? Explain why in one or two complete sentences.

Exercise 2.13.15 *(Solution on p. 95.)*

Refer to the following histograms and box plot. Determine which of the following are true and which are false. Explain your solution to each part in complete sentences.

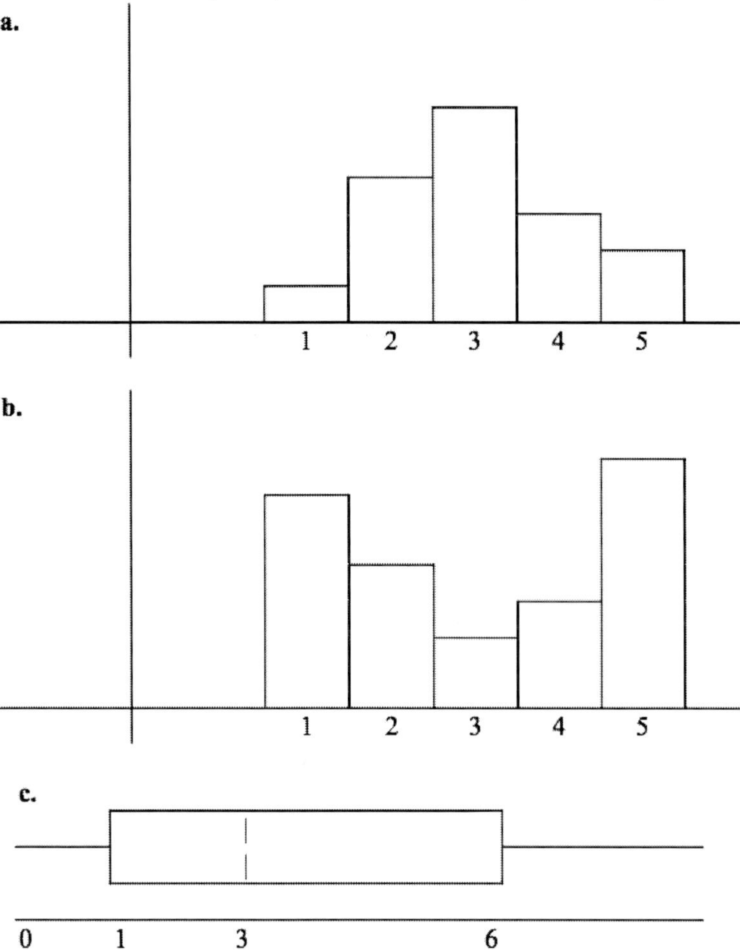

a. The medians for all three graphs are the same.
b. We cannot determine if any of the means for the three graphs is different.
c. The standard deviation for (b) is larger than the standard deviation for (a).
d. We cannot determine if any of the third quartiles for the three graphs is different.

Exercise 2.13.16

Refer to the following box plots.

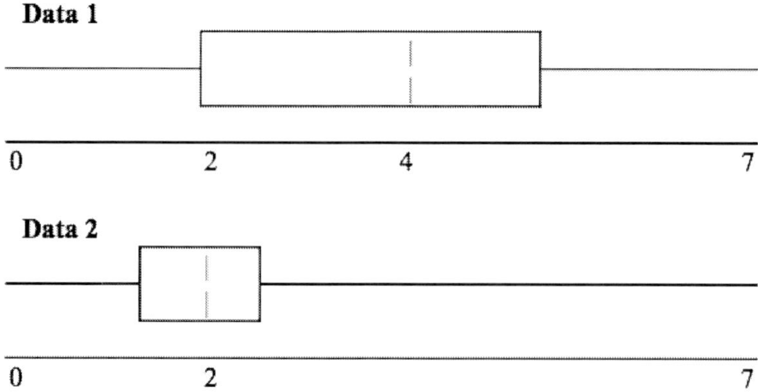

a. In complete sentences, explain why each statement is false.

- **i.** **Data 1** has more data values above 2 than **Data 2** has above 2.
- **ii.** The data sets cannot have the same mode.
- **iii.** For **Data 1**, there are more data values below 4 than there are above 4.

b. For which group, Data 1 or Data 2, is the value of "7" more likely to be an outlier? Explain why in complete sentences

Exercise 2.13.17 ***(Solution on p. 96.)***

In a recent issue of the *IEEE Spectrum*, 84 engineering conferences were announced. Four conferences lasted two days. Thirty-six lasted three days. Eighteen lasted four days. Nineteen lasted five days. Four lasted six days. One lasted seven days. One lasted eight days. One lasted nine days. Let X = the length (in days) of an engineering conference.

a. Organize the data in a chart.
b. Find the median, the first quartile, and the third quartile.
c. Find the 65th percentile.
d. Find the 10th percentile.
e. Construct a box plot of the data.
f. The middle 50% of the conferences last from _______ days to _______ days.
g. Calculate the sample mean of days of engineering conferences.
h. Calculate the sample standard deviation of days of engineering conferences.
i. Find the mode.
j. If you were planning an engineering conference, which would you choose as the length of the conference: mean; median; or mode? Explain why you made that choice.
k. Give two reasons why you think that 3 - 5 days seem to be popular lengths of engineering conferences.

Exercise 2.13.18

A survey of enrollment at 35 community colleges across the United States yielded the following figures (*source: Microsoft Bookshelf*):

6414; 1550; 2109; 9350; 21828; 4300; 5944; 5722; 2825; 2044; 5481; 5200; 5853; 2750; 10012; 6357; 27000; 9414; 7681; 3200; 17500; 9200; 7380; 18314; 6557; 13713; 17768; 7493; 2771; 2861; 1263; 7285; 28165; 5080; 11622

a. Organize the data into a chart with five intervals of equal width. Label the two columns "Enrollment" and "Frequency."
b. Construct a histogram of the data.

c. If you were to build a new community college, which piece of information would be more valuable: the mode or the average size?
d. Calculate the sample average.
e. Calculate the sample standard deviation.
f. A school with an enrollment of 8000 would be how many standard deviations away from the mean?

Exercise 2.13.19 *(Solution on p. 96.)*

The median age of the U.S. population in 1980 was 30.0 years. In 1991, the median age was 33.1 years. (*Source: Bureau of the Census*)

a. What does it mean for the median age to rise?
b. Give two reasons why the median age could rise.
c. For the median age to rise, is the actual number of children less in 1991 than it was in 1980? Why or why not?

Exercise 2.13.20

A survey was conducted of 130 purchasers of new BMW 3 series cars, 130 purchasers of new BMW 5 series cars, and 130 purchasers of new BMW 7 series cars. In it, people were asked the age they were when they purchased their car. The following box plots display the results.

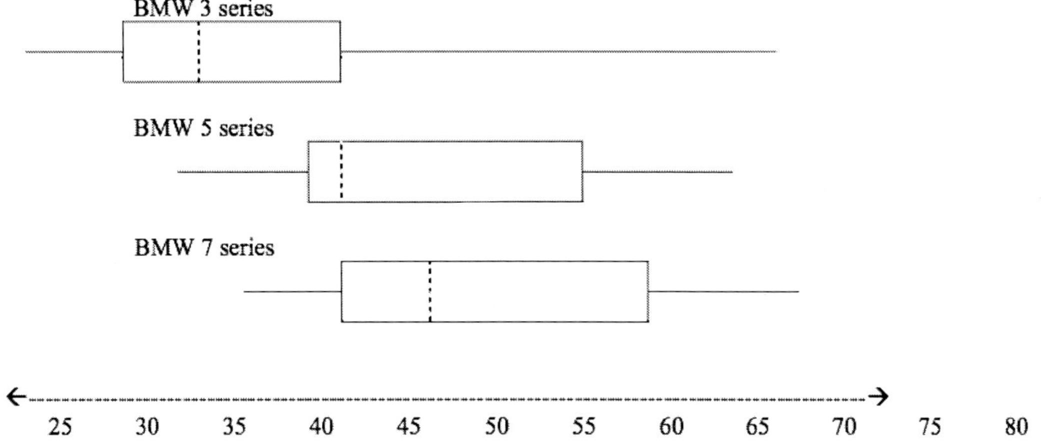

a. In complete sentences, describe what the shape of each box plot implies about the distribution of the data collected for that car series.
b. Which group is most likely to have an outlier? Explain how you determined that.
c. Compare the three box plots. What do they imply about the age of purchasing a BMW from the series when compared to each other?
d. Look at the BMW 5 series. Which quarter has the smallest spread of data? What is that spread?
e. Look at the BMW 5 series. Which quarter has the largest spread of data? What is that spread?
f. Look at the BMW 5 series. Find the Inter Quartile Range (IQR).
g. Look at the BMW 5 series. Are there more data in the interval 31-38 or in the interval 45-55? How do you know this?
h. Look at the BMW 5 series. Which interval has the fewest data in it? How do you know this?
   i. 31-35
   ii. 38-41

iii. 41-64

Exercise 2.13.21 *(Solution on p. 96.)*

The following box plot shows the U.S. population for 1990, the latest available year. (Source: Bureau of the Census, 1990 Census)

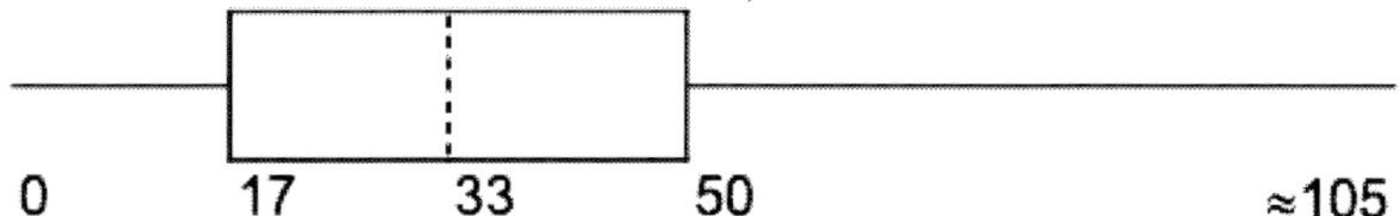

a. Are there fewer or more children (age 17 and under) than senior citizens (age 65 and over)? How do you know?
b. 12.6% are age 65 and over. Approximately what percent of the population are of working age adults (above age 17 to age 65)?

Exercise 2.13.22

Javier and Ercilia are supervisors at a shopping mall. Each was given the task of estimating the mean distance that shoppers live from the mall. They each randomly surveyed 100 shoppers. The samples yielded the following information:

	Javier	Ercilla
$\overline{x}$	6.0 miles	6.0 miles
s	4.0 miles	7.0 miles

Table 2.17

a. How can you determine which survey was correct ?
b. Explain what the difference in the results of the surveys implies about the data.
c. If the two histograms depict the distribution of values for each supervisor, which one depicts Ercilia's sample? How do you know?

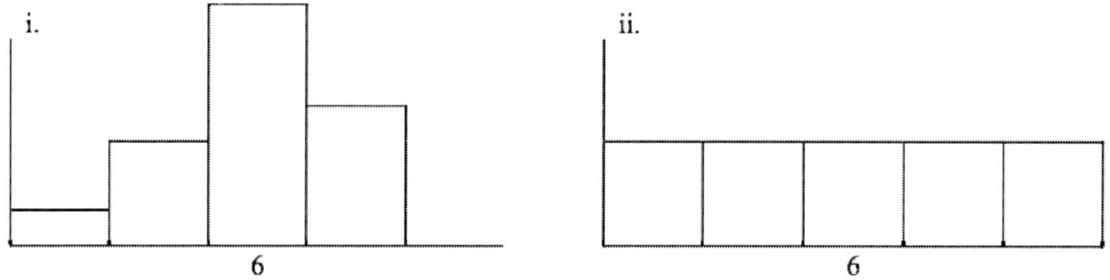

Figure 2.2

d. If the two box plots depict the distribution of values for each supervisor, which one depicts Ercilia's sample? How do you know?

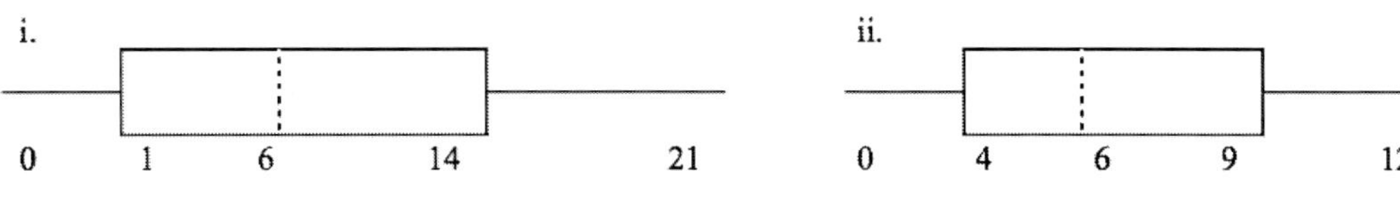

Figure 2.3

Exercise 2.13.23 *(Solution on p. 96.)*

Student grades on a chemistry exam were:

77, 78, 76, 81, 86, 51, 79, 82, 84, 99

a. Construct a stem-and-leaf plot of the data.
b. Are there any potential outliers? If so, which scores are they? Why do you consider them outliers?

2.13.1 Try these multiple choice questions.

The next three questions refer to the following information. We are interested in the number of years students in a particular elementary statistics class have lived in California. The information in the following table is from the entire section.

Number of years	Frequency
7	1
14	3
15	1
18	1
19	4
20	3
22	1
23	1
26	1
40	2
42	2
	Total = 20

Table 2.18

Exercise 2.13.24 *(Solution on p. 96.)*

What is the IQR?

A. 8

B. 11
C. 15
D. 35

Exercise 2.13.25 *(Solution on p. 96.)*
What is the mode?

A. 19
B. 19.5
C. 14 and 20
D. 22.65

Exercise 2.13.26 *(Solution on p. 96.)*
Is this a sample or the entire population?

A. sample
B. entire population
C. neither

The next two questions refer to the following table. X = the number of days per week that 100 clients use a particular exercise facility.

X	Frequency
0	3
1	12
2	33
3	28
4	11
5	9
6	4

Table 2.19

Exercise 2.13.27 *(Solution on p. 96.)*
The 80th percentile is:

A. 5
B. 80
C. 3
D. 4

Exercise 2.13.28 *(Solution on p. 96.)*
The number that is 1.5 standard deviations BELOW the mean is approximately:

A. 0.7
B. 4.8
C. -2.8
D. Cannot be determined

The next two questions refer to the following histogram. Suppose one hundred eleven people who shopped in a special T-shirt store were asked the number of T-shirts they own costing more than $19 each.

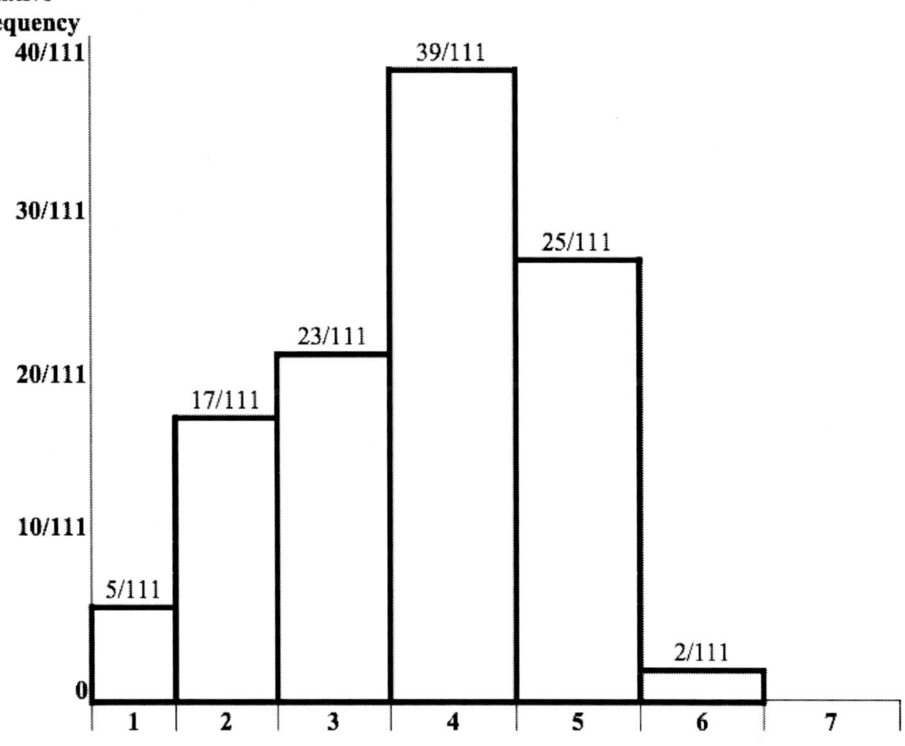

Exercise 2.13.29 *(Solution on p. 96.)*

The percent of people that own at most three (3) T-shirts costing more than $19 each is approximately:

- **A.** 21
- **B.** 59
- **C.** 41
- **D.** Cannot be determined

Exercise 2.13.30 *(Solution on p. 96.)*

If the data were collected by asking the first 111 people who entered the store, then the type of sampling is:

- **A.** cluster
- **B.** simple random
- **C.** stratified
- **D.** convenience

2.14 Lab: Descriptive Statistics[16]

Class Time:

Names:

2.14.1 Student Learning Objectives

- The student will construct a histogram and a box plot.
- The student will calculate univariate statistics.
- The student will examine the graphs to interpret what the data implies.

2.14.2 Collect the Data

Record the number of pairs of shoes you own:

1. Randomly survey 30 classmates. Record their values.

Survey Results

_____	_____	_____	_____	_____
_____	_____	_____	_____	_____
_____	_____	_____	_____	_____
_____	_____	_____	_____	_____
_____	_____	_____	_____	_____
_____	_____	_____	_____	_____

Table 2.20

2. Construct a histogram. Make 5-6 intervals. Sketch the graph using a ruler and pencil. Scale the axes.

[16]This content is available online at <http://cnx.org/content/m16299/1.12/>.

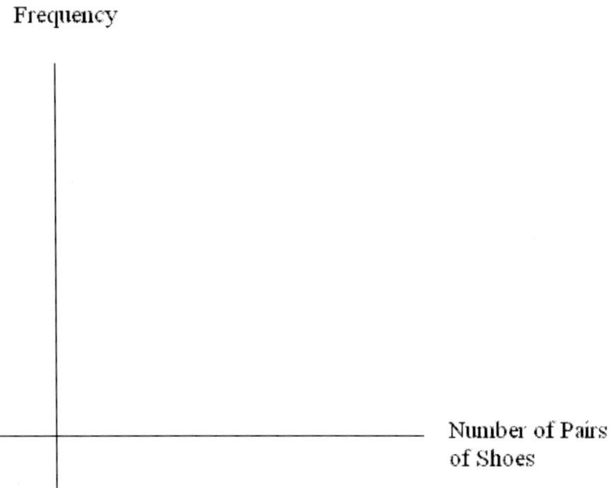

Figure 2.4

3. Calculate the following:
   - $\overline{x}$ =
   - s =
4. Are the data discrete or continuous? How do you know?
5. Describe the shape of the histogram. Use complete sentences.
6. Are there any potential outliers? Which value(s) is (are) it (they)? Use a formula to check the end values to determine if they are potential outliers.

2.14.3 Analyze the Data

1. Determine the following:
   - Minimum value =
   - Median =
   - Maximum value =
   - First quartile =
   - Third quartile =
   - IQR =
2. Construct a box plot of data
3. What does the shape of the box plot imply about the concentration of data? Use complete sentences.
4. Using the box plot, how can you determine if there are potential outliers?
5. How does the standard deviation help you to determine concentration of the data and whether or not there are potential outliers?
6. What does the IQR represent in this problem?
7. Show your work to find the value that is 1.5 standard deviations:
   a. Above the mean:
   b. Below the mean:

Solutions to Exercises in Chapter 2

Solution to Example 2.2 (p. 51)

The value 12.3 may be an outlier. Values appear to concentrate at 3 and 4 miles.

Stem	Leaf
1	1 5
2	3 5 7
3	3 3 3 5 8
4	0 2 5 5 7 8
5	5 6 6
6	5 7
7	
8	
9	
10	
11	
12	3

Table 2.21

Solution to Example 2.4 (p. 54)

- 3.5 to 4.5
- 4.5 to 5.5
- 6
- 5.5 to 6.5

Solution to Example 2.6 (p. 57)

First Data Set

- $Xmin = 32$
- $Q1 = 56$
- $M = 74.5$
- $Q3 = 82.5$
- $Xmax = 99$

Second Data Set

- $Xmin = 25.5$
- $Q1 = 78$
- $M = 81$
- $Q3 = 89$
- $Xmax = 98$

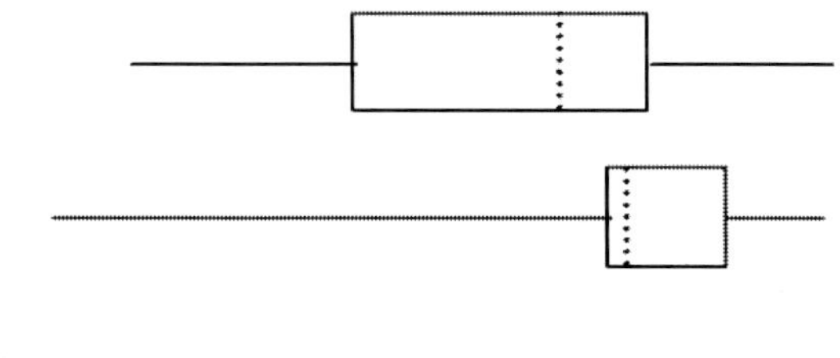

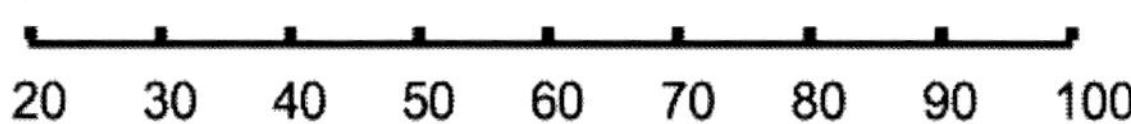

Solution to Example 2.8 (p. 58)

For the IQRs, see the answer to the test scores example (First Data Set, p. 92 Second Data Set, p. 92 p. 475). The first data set has the larger IQR, so the scores between $Q3$ and $Q1$ (middle 50%) for the first data set are more spread out and not clustered about the median.

First Data Set

- $\left(\frac{3}{2}\right) \cdot (IQR) = \left(\frac{3}{2}\right) \cdot (26.5) = 39.75$
- $Xmax - Q3 = 99 - 82.5 = 16.5$
- $Q1 - Xmin = 56 - 32 = 24$

$\left(\frac{3}{2}\right) \cdot (IQR) = 39.75$ is larger than 16.5 and larger than 24, so the first set has no outliers.

Second Data Set

- $\left(\frac{3}{2}\right) \cdot (IQR) = \left(\frac{3}{2}\right) \cdot (11) = 16.5$
- $Xmax - Q3 = 98 - 89 = 9$
- $Q1 - Xmin = 78 - 25.5 = 52.5$

$\left(\frac{3}{2}\right) \cdot (IQR) = 16.5$ is larger than 9 but smaller than 52.5, so for the second set 45 and 25.5 are outliers.

To find the percentiles, create a frequency, relative frequency, and cumulative relative frequency chart (see "Frequency" from the Sampling and Data Chapter (Section 1.9)). Get the percentiles from that chart.

First Data Set

- $30th\ \%ile\ (between\ the\ 6th\ and\ 7th\ values) = \frac{(56 + 59)}{2} = 57.5$
- $80th\ \%ile\ (between\ the\ 16th\ and\ 17th\ values) = \frac{(84 + 84.5)}{2} = 84.25$

Second Data Set

- $30th\ \%ile\ (7th\ value) = 78$
- $80th\ \%ile\ (18th\ value) = 90$

30% of the data falls below the 30th %ile, and 20% falls above the 80th %ile.

Solution to Example 2.10 (p. 59)

1. $\frac{(8+9)}{2} = 8.5$
2. 9
3. 6
4. First Quartile = 25th %ile

Solutions to Practice 1: Center of the Data

Solution to Exercise 2.11.1 (p. 71)

65

Solution to Exercise 2.11.2 (p. 71)
1
Solution to Exercise 2.11.5 (p. 72)
4.75
Solution to Exercise 2.11.6 (p. 72)
1.39
Solution to Exercise 2.11.7 (p. 72)
65
Solution to Exercise 2.11.8 (p. 72)
4
Solution to Exercise 2.11.9 (p. 72)
4
Solution to Exercise 2.11.10 (p. 72)
4
Solution to Exercise 2.11.11 (p. 72)
4
Solution to Exercise 2.11.12 (p. 72)
6
Solution to Exercise 2.11.13 (p. 72)
$6 - 4 = 2$
Solution to Exercise 2.11.14 (p. 72)
3
Solution to Exercise 2.11.15 (p. 72)
6
Solution to Exercise 2.11.16 (p. 73)

- **a.** 8.93
- **b.** 0.58

Solutions to Practice 2: Spread of the Data

Solution to Exercise 2.12.1 (p. 74)
6
Solution to Exercise 2.12.2 (p. 74)

- **a.** 1447.5
- **b.** 528.5

Solution to Exercise 2.12.3 (p. 74)
474 FTES
Solution to Exercise 2.12.4 (p. 74)
50%
Solution to Exercise 2.12.5 (p. 74)
919
Solution to Exercise 2.12.6 (p. 74)
0.03

Solutions to Homework

Solution to Exercise 2.13.1 (p. 75)

- **a.** 1.48
- **b.** 1.12

- **e.** 1
- **f.** 1
- **g.** 2

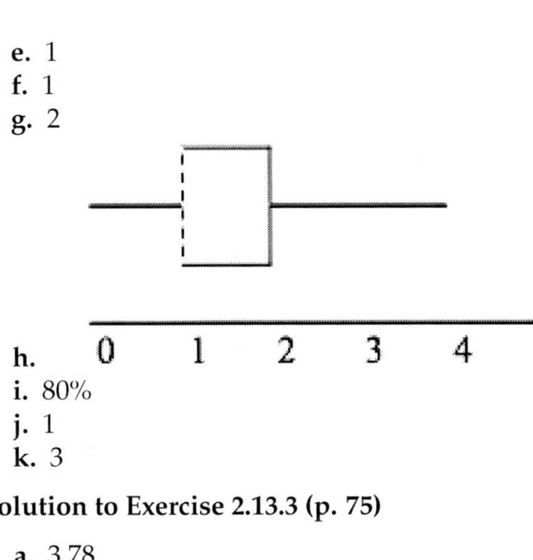

- **h.**
- **i.** 80%
- **j.** 1
- **k.** 3

Solution to Exercise 2.13.3 (p. 75)

- **a.** 3.78
- **b.** 1.29
- **e.** 3
- **f.** 4
- **g.** 5

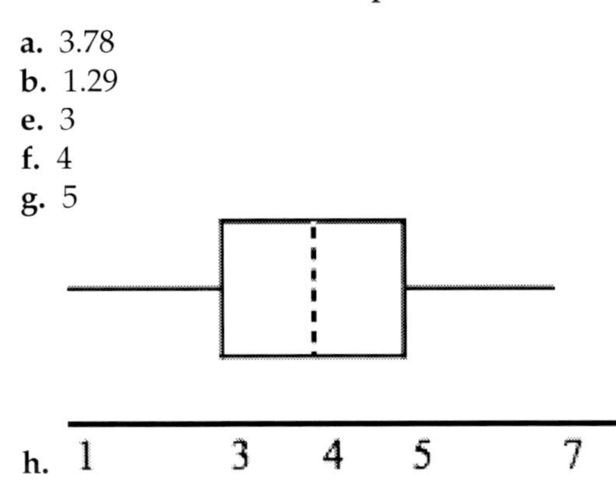

- **h.**
- **i.** 32.5%
- **j.** 4
- **k.** 5

Solution to Exercise 2.13.5 (p. 77)

- **b.** 241
- **c.** 205.5
- **d.** 272.5

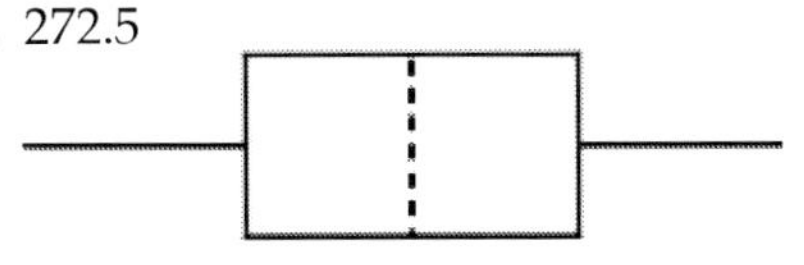

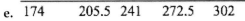

- **e.** 174 205.5 241 272.5 302
- **f.** 205.5, 272.5
- **g.** sample
- **h.** population
- **i.**
  - **i.** 236.34
  - **ii.** 37.50
  - **iii.** 161.34
  - **iv.** 0.84 std. dev. below the mean
- **j.** Young

Solution to Exercise 2.13.9 (p. 78)
Kamala

Solution to Exercise 2.13.15 (p. 83)

- **a.** True
- **b.** True
- **c.** True
- **d.** False

Solution to Exercise 2.13.17 (p. 84)

- **b.** 4,3,5
- **c.** 4
- **d.** 3

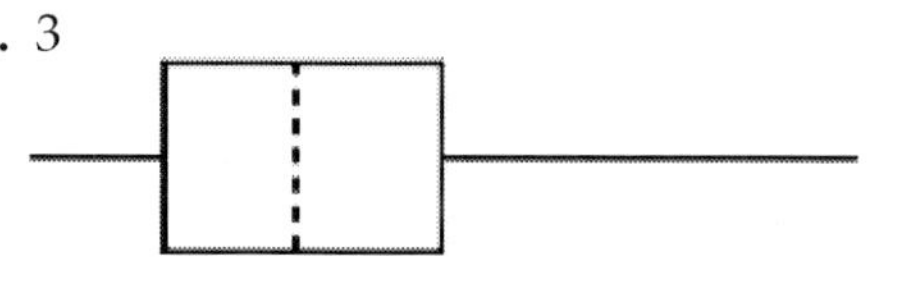

- **e.** 2 3 4 5 9
- **f.** 3,5
- **g.** 3.94
- **h.** 1.28
- **i.** 3
- **j.** mode

Solution to Exercise 2.13.19 (p. 85)

- **c.** Maybe

Solution to Exercise 2.13.21 (p. 86)

- **a.** more children
- **b.** 62.4%

Solution to Exercise 2.13.23 (p. 87)

- **b.** 51,99

Solution to Exercise 2.13.24 (p. 87)
A
Solution to Exercise 2.13.25 (p. 88)
A
Solution to Exercise 2.13.26 (p. 88)
B
Solution to Exercise 2.13.27 (p. 88)
D
Solution to Exercise 2.13.28 (p. 88)
A
Solution to Exercise 2.13.29 (p. 89)
C
Solution to Exercise 2.13.30 (p. 89)
D

Chapter 3

Probability Topics

3.1 Probability Topics[1]

3.1.1 Student Learning Objectives

By the end of this chapter, the student should be able to:

- Understand and use the terminology of probability.
- Determine whether two events are mutually exclusive and whether two events are independent.
- Calculate probabilities using the Addition Rules and Multiplication Rules.
- Construct and interpret Contingency Tables.
- Construct and interpret Venn Diagrams (optional).
- Construct and interpret Tree Diagrams (optional).

3.1.2 Introduction

It is often necessary to "guess" about the outcome of an event in order to make a decision. Politicians study polls to guess their likelihood of winning an election. Teachers choose a particular course of study based on what they think students can comprehend. Doctors choose the treatments needed for various diseases based on their assessment of likely results. You may have visited a casino where people play games chosen because of the belief that the likelihood of winning is good. You may have chosen your course of study based on the probable availability of jobs.

You have, more than likely, used probability. In fact, you probably have an intuitive sense of probability. Probability deals with the chance of an event occurring. Whenever you weigh the odds of whether or not to do your homework or to study for an exam, you are using probability. In this chapter, you will learn to solve probability problems using a systematic approach.

3.1.3 Optional Collaborative Classroom Exercise

Your instructor will survey your class. Count the number of students in the class today.

- Raise your hand if you have any change in your pocket or purse. Record the number of raised hands.
- Raise your hand if you rode a bus within the past month. Record the number of raised hands.
- Raise your hand if you answered "yes" to BOTH of the first two questions. Record the number of raised hands.

[1]This content is available online at <http://cnx.org/content/m16838/1.10/>.

Use the class data as estimates of the following probabilities. *P(change)* means the probability that a randomly chosen person in your class has change in his/her pocket or purse. *P(bus)* means the probability that a randomly chosen person in your class rode a bus within the last month and so on. Discuss your answers.

- Find *P(change)*.
- Find *P(bus)*.
- Find *P(change and bus)* Find the probability that a randomly chosen student in your class has change in his/her pocket or purse and rode a bus within the last month.
- Find *P(change | bus)* Find the probability that a randomly chosen student has change given that he/she rode a bus within the last month. Count all the students that rode a bus. From the group of students who rode a bus, count those who have change. The probability is equal to those who have change and rode a bus divided by those who rode a bus.

3.2 Terminology[2]

Probability measures the uncertainty that is associated with the outcomes of a particular experiment or activity. An **experiment** is a planned operation carried out under controlled conditions. If the result is not predetermined, then the experiment is said to be a **chance** experiment. Flipping one fair coin is an example of an experiment.

The result of an experiment is called an **outcome**. A **sample space** is a set of all possible outcomes. Three ways to represent a sample space are to list the possible outcomes, to create a tree diagram, or to create a Venn diagram. The uppercase letter S is used to denote the sample space. For example, if you flip one fair coin, $S = \{H, T\}$ where H = heads and T = tails are the outcomes.

An **event** is any combination of outcomes. Upper case letters like A and B represent events. For example, if the experiment is to flip one fair coin, event A might be getting at most one head. The probability of an event A is written $P(A)$.

The **probability** of any outcome is the **long-term relative frequency** of that outcome. For example, if you flip one fair coin from 20 to 2,000 times, the relative frequency of heads approaches 0.5 (the probability of heads). Probabilities are between 0 and 1, **inclusive** (includes 0 and 1 and all numbers between these values). $P(A) = 0$ means the event A can never happen. $P(A) = 1$ means the event A always happens.

To calculate the **probability of an event** A, count the outcomes for event A and divide by the total outcomes in the sample space. For example, if you toss a fair dime and a fair nickel, the sample space is $\{HH, TH, HT, TT\}$ where T = tails and H = heads. The sample space has four outcomes. If A denotes the probability of getting one head, then there are two outcomes $\{HT, TH\}$in the event. Thus, $P(A) = \frac{2}{4}$.

Equally likely means that each outcome of an experiment occurs with equal probability. For example, if you toss a fair, six-sided die, each face (1, 2, 3, 4, 5, or 6) is as likely to occur as any other face.

An outcome is in the event *A OR B* if the outcome is in A or is in B or is in both A and B. For example, let $A = \{1, 2, 3, 4, 5\}$ and $B = \{4, 5, 6, 7, 8\}$. $A\ OR\ B = \{1, 2, 3, 4, 5, 6, 7, 8\}$. Notice that 4 and 5 are NOT listed twice.

An outcome is in the event *A AND B* if the outcome is in both A and B at the same time. For example, let A and B be $\{1, 2, 3, 4, 5\}$ and $\{4, 5, 6, 7, 8\}$, respectively. Then $A\ AND\ B = \{4,5\}$.

The **complement** of event A is denoted A' (read "A prime"). A' consists of all outcomes that are **NOT** in A. Notice that $P(A) + P(A') = 1$. For example, let $S = \{1, 2, 3, 4, 5, 6\}$ and let $A = \{1, 2, 3, 4\}$. Then, $A' = \{5, 6\}$. $P(A) = \frac{4}{6}$, $P(A') = \frac{2}{6}$, *and* $P(A) + P(A') = \frac{4}{6} + \frac{2}{6} = 1$

[2]This content is available online at <http://cnx.org/content/m16845/1.8/>.

The **conditional probability** of A given B is written $P(A|B)$. $P(A|B)$ is the probability that event A will occur given that the event B has already occurred. **A conditional reduces the sample space**. We calculate the probability of A from the reduced sample space B. The formula to calculate $P(A|B)$ is

$$P(A|B) = \frac{P(AANDB)}{P(B)}$$

where $P(B)$ is greater than 0.

For example, suppose we toss one fair, six-sided die. The sample space $S = \{1, 2, 3, 4, 5, 6\}$. Let A = face is 2 or 3 and B = face is even (2, 4, 6). To calculate $P(A|B)$, we count the number of outcomes 2 or 3 in the sample space $B = \{2, 4, 6\}$. Then we divide that by the number of outcomes in B (and not S).

We get the same result by using the formula. Remember that S has 6 outcomes.

$$P(A|B) = \frac{P(A \text{ and } B)}{P(B)} = \frac{(\text{the number of outcomes that are 2 or 3 and even in S}) / 6}{(\text{the number of outcomes that are even in S}) / 6} = \frac{1/6}{3/6} = \frac{1}{3}$$

3.3 Independent and Mutually Exclusive Events[3]

Independent and mutually exclusive do **not** mean the same thing.

3.3.1 Independent Events

Two events are independent if the following are true:

- $P(A|B) = P(A)$
- $P(B|A) = P(B)$
- $P(A \text{ AND } B) = P(A) \cdot P(B)$

If A and B are **independent**, then the chance of A occurring does not affect the chance of B occurring and vice versa. For example, two roles of a fair die are independent events. The outcome of the first roll does not change the probability for the outcome of the second roll. To show two events are independent, you must show **only one** of the above conditions. If two events are NOT independent, then we say that they are **dependent**.

Sampling may be done **with replacement** or **without replacement**.

- **With replacement**: If each member of a population is replaced after it is picked, then that member has the possibility of being chosen more than once. When sampling is done with replacement, then events are considered to be independent, meaning the result of the first pick will not change the probabilities for the second pick.
- **Without replacement:**: When sampling is done without replacement, then each member of a population may be chosen only once. In this case, the probabilities for the second pick are affected by the result of the first pick. The events are considered to be dependent or not independent.

If it is not known whether A and B are independent or dependent, **assume they are dependent until you can show otherwise**.

[3]This content is available online at <http://cnx.org/content/m16837/1.10/>.

3.3.2 Mutually Exclusive Events

A and B are **mutually exclusive** events if they cannot occur at the same time. This means that A and B do not share any outcomes and $P(A\ AND\ B) = 0$.

For example, suppose the sample space $S = \{1, 2, 3, 4, 5, 6, 7, 8, 9, 10\}$. Let $A = \{1, 2, 3, 4, 5\}$, $B = \{4, 5, 6, 7, 8\}$, and $C = \{7, 9\}$. $A\ AND\ B = \{4,5\}$. $P(A\ AND\ B) = \frac{2}{10}$and is not equal to zero. Therefore, A and B are not mutually exclusive. A and C do not have any numbers in common so $P(A\ AND\ C) = 0$. Therefore, A and C are mutually exclusive.

If it is not known whether A and B are mutually exclusive, **assume they are not until you can show otherwise**.

The following examples illustrate these definitions and terms.

Example 3.1
Flip two fair coins. (This is an experiment.)

The sample space is $\{HH, HT, TH, TT\}$ where T = tails and H = heads. The outcomes are HH, HT, TH, and TT. The outcomes HT and TH are different. The HT means that the first coin showed heads and the second coin showed tails. The TH means that the first coin showed tails and the second coin showed heads.

- Let A = the event of getting **at most one tail**. (At most one tail means 0 or 1 tail.) Then A can be written as $\{HH, HT, TH\}$. The outcome HH shows 0 tails. HT and TH each show 1 tail.
- Let B = the event of getting all tails. B can be written as $\{TT\}$. B is the **complement** of A. So, $B = A'$. Also, $P(A) + P(B) = P(A) + P(A') = 1$.
- The probabilities for A and for B are $P(A) = \frac{3}{4}$ and $P(B) = \frac{1}{4}$.
- Let C = the event of getting all heads. $C = \{HH\}$. Since $B = \{TT\}$, $P(B\ AND\ C) = 0$. B and C are mutually exclusive. (B and C have no members in common because you cannot have all tails and all heads at the same time.)
- Let D = event of getting **more than one** tail. $D = \{TT\}$. $P(D) = \frac{1}{4}$.
- Let E = event of getting a head on the first roll. (This implies you can get either a head or tail on the second roll.) $E = \{HT, HH\}$. $P(E) = \frac{2}{4}$.
- Find the probability of getting **at least one** (1 or 2) tail in two flips. Let F = event of getting at least one tail in two flips. $F = \{HT, TH, TT\}$. $P(F) = \frac{3}{4}$

Example 3.2
Roll one fair 6-sided die. The sample space is $\{1, 2, 3, 4, 5, 6\}$. Let event A = a face is odd. Then $A = \{1, 3, 5\}$. Let event B = a face is even. Then $B = \{2, 4, 6\}$.

- Find the complement of A, A'. The complement of A, A', is B because A and B together make up the sample space. $P(A) + P(B) = P(A) + P(A') = 1$. Also, $P(A) = \frac{3}{6}$ and $P(B) = \frac{3}{6}$
- Let event C = odd faces larger than 2. Then $C = \{3,5\}$. Let event D = all even faces smaller than 5. Then $D = \{2,4\}$. $P(C\ and\ D) = 0$ because you cannot have an odd and even face at the same time. Therefore, C and D are mutually exclusive events.
- Let event E = all faces less than 5. $E = \{1,2,3,4\}$.

  Problem *(Solution on p. 132.)*
  Are C and E mutually exclusive events? (Answer yes or no.) Why or why not?

- *Find* $P(C|A)$. This is a conditional. Recall that the event C is $\{3, 5\}$ and event A is $\{1, 3, 5\}$. To find $P(C|A)$, find the probability of C using the sample space A. You have reduced the sample space from the original sample space $\{1, 2, 3, 4, 5, 6\}$ to $\{1, 3, 5\}$. So, $P(C|A) = \frac{2}{3}$

Example 3.3
Let event G = taking a math class. Let event H = taking a science class. Then, G *AND* H = taking a math class and a science class. Suppose $P(G) = 0.6$, $P(H) = 0.5$, and $P(G \text{ AND } H) = 0.3$. Are G and H independent?

If G and H are independent, then you must show **ONE** of the following:

- $P(G \mid H) = P(G)$
- $P(H \mid G) = P(H)$
- $P(G \text{ AND } H) = P(G) \cdot P(H)$

NOTE: **The choice you make depends on the information you have.** You could choose any of the methods here because you have the necessary information.

Problem 1
Show that $P(G \mid H) = P(G)$.

Solution
$P(G \mid H) = \frac{P(G \text{ AND } H)}{P(H)} = \frac{0.3}{0.5} = 0.6 = P(G)$

Problem 2
Show $P(G \text{ AND } H) = P(G) \cdot P(H)$.

Solution
$P(G) \cdot P(H) = 0.6 \cdot 0.5 = 0.3 = P(G \text{ AND } H)$

Since G and H are independent, then, knowing that a person is taking a science class does not change the chance that he/she is taking math. If the two events had not been independent (that is, they are dependent) then knowing that a person is taking a science class would change the chance he/she is taking math. For practice, show that $P(H \mid G) = P(H)$ to show that G and H are independent events.

Example 3.4
In a box there are 3 red cards and 5 blue cards. The red cards are marked with the numbers 1, 2, and 3, and the blue cards are marked with the numbers 1, 2, 3, 4, and 5. The cards are well-shuffled. You reach into the box (you cannot see into it) and draw one card.

Let R = red card is drawn, B = blue card is drawn, E = even-numbered card is drawn.

The sample space S = *R1, R2, R3, B1, B2, B3, B4, B5*. S has 8 outcomes.

- $P(R) = \frac{3}{8}$. $P(B) = \frac{5}{8}$. $P(R \text{ AND } B) = 0$. (You cannot draw one card that is both red and blue.)
- $P(E) = \frac{3}{8}$. (There are 3 even-numbered cards, $R2$, $B2$, and $B4$.)
- $P(E \mid B) = \frac{2}{5}$. (There are 5 blue cards: $B1$, $B2$, $B3$, $B4$, and $B5$. Out of the blue cards, there are 2 even cards: $B2$ and $B4$.)
- $P(B \mid E) = \frac{2}{3}$. (There are 3 even-numbered cards: $R2$, $B2$, and $B4$. Out of the even-numbered cards, 2 are blue: $B2$ and $B4$.)
- The events R and B are mutually exclusive because $P(R \text{ AND } B) = 0$.
- Let G = card with a number greater than 3. $G = \{B4, B5\}$. $P(G) = \frac{2}{8}$. Let H = blue card numbered between 1 and 4, inclusive. $H = \{B1, B2, B3, B4\}$. $P(G \mid H) = \frac{1}{4}$. (The only card in H that has a number greater than 3 is $B4$.) Since $\frac{2}{8} = \frac{1}{4}$, $P(G) = P(G \mid H)$ which means that G and H are independent.

3.4 Two Basic Rules of Probability[4]

3.4.1 The Multiplication Rule

If A and B are two events defined on a **sample space**, then: $P(A \text{ AND } B) = P(B) \cdot P(A \mid B)$.

This rule may also be written as : $P(A|B) = \frac{P(A \text{ AND } B)}{P(B)}$

(The probability of A given B equals the probability of A and B divided by the probability of B.)

If A and B are **independent**, then $P(A \mid B) = P(A)$. Then $P(A \text{ AND } B) = P(A \mid B)\, P(B)$ becomes $P(A \text{ AND } B) = P(A)\, P(B)$.

3.4.2 The Addition Rule

If A and B are defined on a sample space, then: $P(A \text{ OR } B) = P(A) + P(B) - P(A \text{ AND } B)$.

If A and B are **mutually exclusive**, then $P(A \text{ AND } B) = 0$. Then $P(A \text{ OR } B) = P(A) + P(B) - P(A \text{ AND } B)$ becomes $P(A \text{ OR } B) = P(A) + P(B)$.

Example 3.5

Klaus is trying to choose where to go on vacation. His two choices are: A = New Zealand and B = Alaska

- Klaus can only afford one vacation. The probability that he chooses A is $P(A) = 0.6$ and the probability that he chooses B is $P(B) = 0.35$.
- $P(A \text{ and } B) = 0$ because Klaus can only afford to take one vacation
- Therefore, the probability that he chooses either New Zealand or Alaska is $P(A \text{ OR } B) = P(A) + P(B) = 0.6 + 0.35 = 0.95$. Note that the probability that he does not choose to go anywhere on vacation must be 0.05.

Example 3.6

Carlos plays college soccer. He makes a goal 65% of the time he shoots. Carlos is going to attempt two goals in a row in the next game.

A = the event Carlos is successful on his first attempt. $P(A) = 0.65$. B = the event Carlos is successful on his second attempt. $P(B) = 0.65$. Carlos tends to shoot in streaks. The probability that he makes the second goal **GIVEN** that he made the first goal is 0.90.

Problem 1

What is the probability that he makes both goals?

Solution

The problem is asking you to find $P(A \text{ AND } B) = P(B \text{ AND } A)$. Since $P(B \mid A) = 0.90$:

$$P(B \text{ AND } A) = P(B \mid A)\, P(A) = 0.90 * 0.65 = 0.585 \quad (3.1)$$

Carlos makes the first and second goals with probability 0.585.

Problem 2

What is the probability that Carlos makes either the first goal or the second goal?

[4]This content is available online at <http://cnx.org/content/m16847/1.7/>.

Solution

The problem is asking you to find *P(A OR B)*.

$$P(A\ OR\ B) = P(A) + P(B) - P(A\ AND\ B) = 0.65 + 0.65 - 0.585 = 0.715 \quad (3.2)$$

Carlos makes either the first goal or the second goal with probability 0.715.

Problem 3

Are *A* and *B* independent?

Solution

No, they are not, because *P(B AND A)* = 0.585.

$$P(B) \cdot P(A) = (0.65) \cdot (0.65) = 0.423 \quad (3.3)$$

$$0.423 \neq 0.585 = P(B\ AND\ A) \quad (3.4)$$

So, *P(B AND A)* is **not** equal to *P(B)* · *P(A)*.

Problem 4

Are *A* and *B* mutually exclusive?

Solution

No, they are not because *P(A and B)* = 0.585.

To be mutually exclusive, *P(A AND B)* must equal 0.

Example 3.7

A community swim team has **150** members. **Seventy-five** of the members are advanced swimmers. **Forty-seven** of the members are intermediate swimmers. The remainder are novice swimmers. **Forty** of the advanced swimmers practice 4 times a week. **Thirty** of the intermediate swimmers practice 4 times a week. **Ten** of the novice swimmers practice 4 times a week. Suppose one member of the swim team is randomly chosen. Answer the questions (Verify the answers):

Problem 1

What is the probability that the member is a novice swimmer?

Solution

$\frac{28}{150}$

Problem 2

What is the probability that the member practices 4 times a week?

Solution

$\frac{80}{150}$

Problem 3

What is the probability that the member is an advanced swimmer and practices 4 times a week?

Solution
$\frac{40}{150}$

Problem 4
What is the probability that a member is an advanced swimmer and an intermediate swimmer? Are being an advanced swimmer and an intermediate swimmer mutually exclusive? Why or why not?

Solution
P(advanced AND intermediate) = 0, so these are mutually exclusive events. A swimmer cannot be an advanced swimmer and an intermediate swimmer at the same time.

Problem 5
Are being a novice swimmer and practicing 4 times a week independent events? Why or why not?

Solution
No, these are not independent events.

$$P(novice\ AND\ practices\ 4\ times\ per\ week) = 0.0667 \tag{3.5}$$

$$P(novice) \cdot P(practices\ 4\ times\ per\ week) = 0.0996 \tag{3.6}$$

$$0.0667 \neq 0.0996 \tag{3.7}$$

Example 3.8
Studies show that, if she lives to be 90, about 1 woman in 7 (approximately 14.3%) will develop breast cancer. Suppose that of those women who develop breast cancer, a test is negative 2% of the time. Also suppose that in the general population of women, the test for breast cancer is believed to be negative about 85% of the time. Let B = woman develops breast cancer and let N = tests negative.

Problem 1
What is the probability that a woman develops breast cancer? What is the probability that woman tests negative?

Solution
$P(B) = 0.143$; $P(N) = 0.85$

Problem 2
Given that a woman has breast cancer, what is the probability that she tests negative?

Solution
$P(N \mid B) = 0.02$

Problem 3
What is the probability that a woman has breast cancer AND tests negative?

Solution
$P(B\ AND\ N) = P(B) \cdot P(N \mid B) = (0.143) \cdot (0.02) = 0.0029$

Problem 4
What is the probability that a woman has breast cancer or tests negative?

Solution
$P(B\ OR\ N) = P(B) + P(N) - P(B\ AND\ N) = 0.143 + 0.85 - 0.0029 = 0.9901$

Problem 5
Are having breast cancer and testing negative independent events?

Solution
No. $P(N) = 0.85$; $P(N \mid B) = 0.02$. So, $P(N \mid B)$ does not equal $P(N)$

Problem 6
Are having breast cancer and testing negative mutually exclusive?

Solution
No. $P(B\ AND\ N) = 0.0020$. For B and N to be mutually exclusive, $P(B\ AND\ N)$ must be 0.

3.5 Contingency Tables[5]

A **contingency table** provides a different way of calculating probabilities. The table helps in determining conditional probabilities quite easily. The table displays sample values in relation to two different variables that may be dependent or contingent on one another. Later on, we will use contingency tables again, but in another manner.

Example 3.9
Suppose a study of speeding violations and drivers who use car phones produced the following fictional data:

	Speeding violation in the last year	**No speeding violation in the last year**	**Total**
Car phone user	25	280	305
Not a car phone user	45	405	450
Total	70	685	755

Table 3.1

The total number of people in the sample is 755. The row totals are 305 and 450. The column totals are 70 and 685. Notice that 305 + 450 = 755 and 70 + 685 = 755.

Calculate the following probabilities using the table

[5]This content is available online at <http://cnx.org/content/m16835/1.8/>.

Problem 1

P(person is a car phone user) =

Solution

$\frac{\textit{number of car phone users}}{\textit{total number in study}} = \frac{305}{755}$

Problem 2

P(person had no violation in the last year) =

Solution

$\frac{\textit{number that had no violation}}{\textit{total number in study}} = \frac{685}{755}$

Problem 3

P(person had no violation in the last year AND was a car phone user) =

Solution

$\frac{280}{755}$

Problem 4

P(person is a car phone user OR person had no violation in the last year) =

Solution

$\left(\frac{305}{755} + \frac{685}{755}\right) - \frac{280}{755} = \frac{710}{755}$

Problem 5

P(person is a car phone user GIVEN person had a violation in the last year) =

Solution

$\frac{25}{70}$ (The sample space is reduced to the number of persons who had a violation.)

Problem 6

P(person had no violation last year GIVEN person was not a car phone user) =

Solution

$\frac{405}{450}$ (The sample space is reduced to the number of persons who were not car phone users.)

Example 3.10

The following table shows a random sample of 100 hikers and the areas of hiking preferred:

Hiking Area Preference

Sex	The Coastline	Near Lakes and Streams	On Mountain Peaks	Total
Female	18	16	___	45
Male	___	___	14	55
Total	___	41	___	___

Table 3.2

Problem 1 *(Solution on p. 132.)*

Complete the table.

Problem 2 *(Solution on p. 132.)*

Are the events "being female" and "preferring the coastline" independent events?

Let F = being female and let C = preferring the coastline.

a. *P(F AND C)* =
b. $P(F) \cdot P(C) =$

Are these two numbers the same? If they are, then F and C are independent. If they are not, then F and C are not independent.

Problem 3 *(Solution on p. 132.)*

Find the probability that a person is male given that the person prefers hiking near lakes and streams. Let M = being male and let L = prefers hiking near lakes and streams.

a. What word tells you this is a conditional?
b. Fill in the blanks and calculate the probability: *P(___|___)* = ___.
c. Is the sample space for this problem all 100 hikers? If not, what is it?

Problem 4 *(Solution on p. 132.)*

Find the probability that a person is female or prefers hiking on mountain peaks. Let F = being female and let P = prefers mountain peaks.

a. *P(F)* =
b. *P(P)* =
c. *P(F AND P)* =
d. Therefore, *P(F OR P)* =

Example 3.11

Muddy Mouse lives in a cage with 3 doors. If Muddy goes out the first door, the probability that he gets caught by Alissa the cat is $\frac{1}{5}$ and the probability he is not caught is $\frac{4}{5}$. If he goes out the second door, the probability he gets caught by Alissa is $\frac{1}{4}$ and the probability he is not caught is $\frac{3}{4}$. The probability that Alissa catches Muddy coming out of the third door is $\frac{1}{2}$ and the probability she does not catch Muddy is $\frac{1}{2}$. It is equally likely that Muddy will choose any of the three doors so the probability of choosing each door is $\frac{1}{3}$.

Door Choice

Caught or Not	Door One	Door Two	Door Three	Total
Caught	$\frac{1}{15}$	$\frac{1}{12}$	$\frac{1}{6}$	___
Not Caught	$\frac{4}{15}$	$\frac{3}{12}$	$\frac{1}{6}$	___
Total	___	___	___	1

Table 3.3

- The first entry $\frac{1}{15} = \left(\frac{1}{5}\right)\left(\frac{1}{3}\right)$ is *P(Door One AND Caught)*.

- The entry $\frac{4}{15} = \left(\frac{4}{5}\right)\left(\frac{1}{3}\right)$ is *P(Door One AND Not Caught)*.

Verify the remaining entries.

Problem 1 ***(Solution on p. 132.)***

Complete the probability contingency table. Calculate the entries for the totals. Verify that the lower-right corner entry is 1.

Problem 2

What is the probability that Alissa does not catch Muddy?

Solution

$\frac{41}{60}$

Problem 3

What is the probability that Muddy chooses Door One **OR** Door Two given that Muddy is caught by Alissa?

Solution

$\frac{9}{19}$

NOTE: You could also do this problem by using a probability tree. See the Tree Diagrams (Optional) (Section 3.7) section of this chapter for examples.

3.6 Venn Diagrams (optional)[6]

A **Venn diagram** is a picture that represents the outcomes of an experiment. It generally consists of a box that represents the sample space S together with circles or ovals. The circles or ovals represent events.

Example 3.12

Suppose an experiment has the outcomes 1, 2, 3, ... , 12 where each outcome has an equal chance of occurring. Let event $A = \{1, 2, 3, 4, 5, 6\}$ and event $A = \{6, 7, 8, 9\}$. Then $A\ AND\ B = \{6\}$ and $A\ OR\ B = \{1, 2, 3, 4, 5, 6, 7, 8, 9\}$. The Venn diagram is as follows:

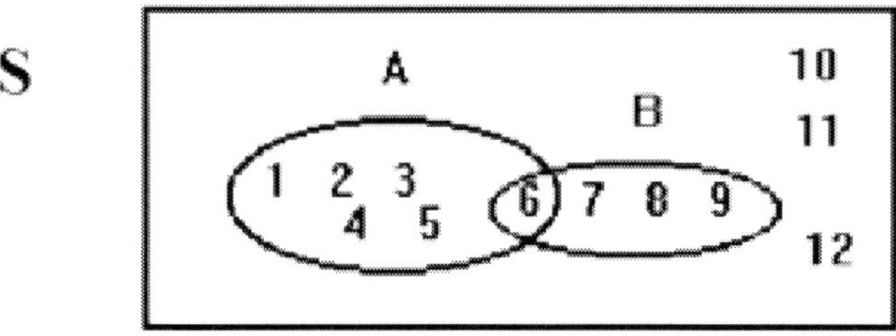

Example 3.13

Flip 2 fair coins. Let A = tails on the first coin. Let B = tails on the second coin. Then $A = \{TT, TH\}$ and $B = \{TT, HT\}$. Therefore, $A\ AND\ B = \{TT\}$. $A\ OR\ B = \{TH, TT, HT\}$.

[6]This content is available online at <http://cnx.org/content/m16848/1.9/>.

The sample space when you flip two fair coins is $S = \{HH, HT, TH, TT\}$. The outcome HH is in neither A nor B. The Venn diagram is as follows:

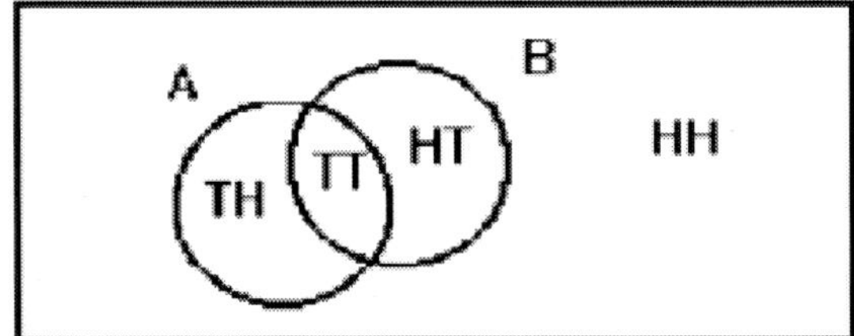

Example 3.14
Forty percent of the students at a local college belong to a club and **50%** work part time. **Five percent** of the students work part time and belong to a club. Draw a Venn diagram showing the relationships. Let C = student belongs to a club and PT = student works part time.

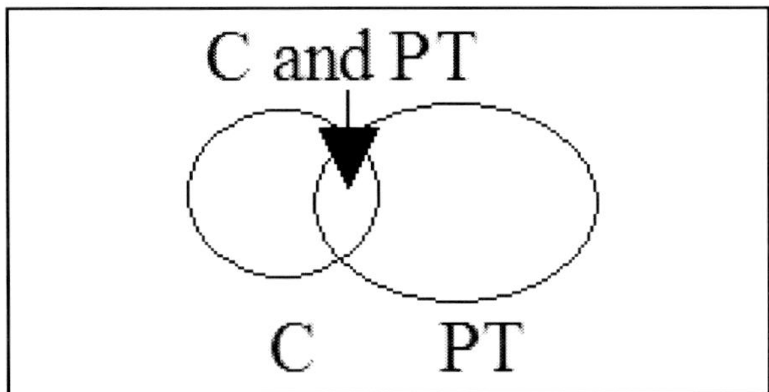

- The probability that a students belongs to a club is $P(C) = 0.40$.
- The probability that a student works part time is $P(PT) = 0.50$.
- The probability that a student belongs to a club AND works part time is $P(C\ AND\ PT) = 0.05$.
- The probability that a student belongs to a club **given** that the student works part time is:

$$P(C \mid PT) = \frac{P(C\ AND\ PT)}{P(PT)} = \frac{0.05}{0.50} = 0.1 \quad (3.8)$$

- The probability that a student belongs to a club **OR** works part time is:

$$P(C\ OR\ PT) = P(C) + P(PT) - P(C\ AND\ PT) = 0.40 + 0.50 - 0.05 = 0.85 \quad (3.9)$$

3.7 Tree Diagrams (optional)[7]

A **tree diagram** is a special type of graph used to determine the outcomes of an experiment. It consists of "branches" that are labeled with either frequencies or probabilities. Tree diagrams can make some probability problems easier to visualize and solve. The following example illustrates how to use a tree diagram.

[7]This content is available online at <http://cnx.org/content/m16846/1.10/>.

Example 3.15
In an urn, there are 11 balls. Three balls are red (R) and 8 balls are blue (B). Draw two balls, one at a time, **with replacement**. "With replacement" means that you put the first ball back in the urn before you select the second ball. The tree diagram using frequencies that show all the possible outcomes follows.

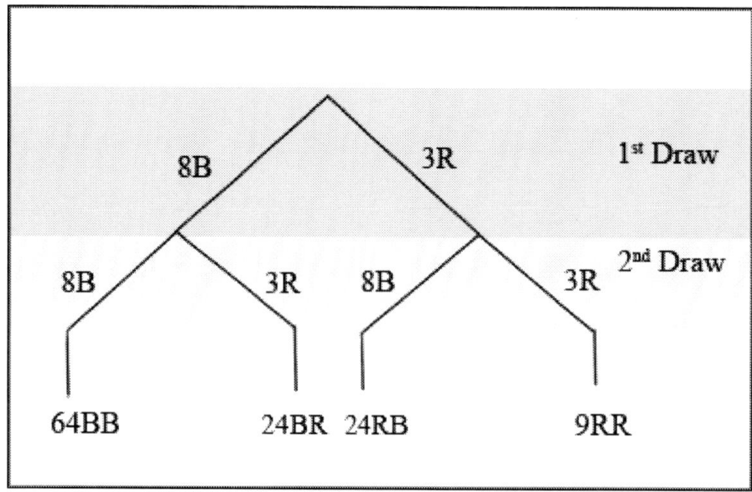

Figure 3.1: $Total = 64 + 24 + 24 + 9 = 121$

The first set of branches represents the first draw. The second set of branches represents the second draw. Each of the outcomes is distinct. In fact, we can list each red ball as $R1$, $R2$, and $R3$ and each blue ball as $B1$, $B2$, $B3$, $B4$, $B5$, $B6$, $B7$, and $B8$. Then the 9 RR outcomes can be written as:

$R1R1$; $R1R2$; $R1R3$; $R2R1$; $R2R2$; $R2R3$; $R3R1$; $R3R2$; $R3R3$

The other outcomes are similar.

There are a total of 11 balls in the urn. Draw two balls, one at a time, and with replacement. There are $11 \cdot 11 = 121$ outcomes, the size of the **sample space**.

Problem 1 *(Solution on p. 132.)*

List the 24 BR outcomes: $B1R1$, $B1R2$, $B1R3$, ...

Problem 2

Using the tree diagram, calculate *P(RR)*.

Solution

$P(RR) = \frac{3}{11} \cdot \frac{3}{11} = \frac{9}{121}$

Problem 3

Using the tree diagram, calculate *P(RB OR BR)*.

Solution

$P(RB\ OR\ BR) = \frac{3}{11} \cdot \frac{8}{11} + \frac{8}{11} \cdot \frac{3}{11} = \frac{48}{121}$

Problem 4

Using the tree diagram, calculate *P(R on 1st draw AND B on 2nd draw)*.

Solution

P(R on 1st draw AND B on 2nd draw) $= P(RB) = \frac{3}{11} \cdot \frac{8}{11} = \frac{24}{121}$

Problem 5

Using the tree diagram, calculate *P(R on 2nd draw given B on 1st draw)*.

Solution

P(R on 2nd draw given B on 1st draw) $= P(R \text{ on 2nd} \mid B \text{ on 1st}) = \frac{24}{88} = \frac{3}{11}$

This problem is a conditional. The sample space has been reduced to those outcomes that already have a blue on the first draw. There are $24 + 64 = 88$ possible outcomes (24 *BR* and 64 *BB*). Twenty-four of the 88 possible outcomes are *BR*. $\frac{24}{88} = \frac{3}{11}$.

Problem 6 ***(Solution on p. 132.)***

Using the tree diagram, calculate *P(BB)*.

Problem 7 ***(Solution on p. 133.)***

Using the tree diagram, calculate *P(B on the 2nd draw given R on the first draw)*.

Example 3.16

An urn has 3 red marbles and 8 blue marbles in it. Draw two marbles, one at a time, this time without replacement from the urn. **"Without replacement"** means that you do not put the first ball back before you select the second ball. Below is a tree diagram. The branches are labeled with probabilities instead of frequencies. The numbers at the ends of the branches are calculated by multiplying the numbers on the two corresponding branches, for example, $\frac{3}{11} \cdot \frac{2}{10} = \frac{6}{110}$.

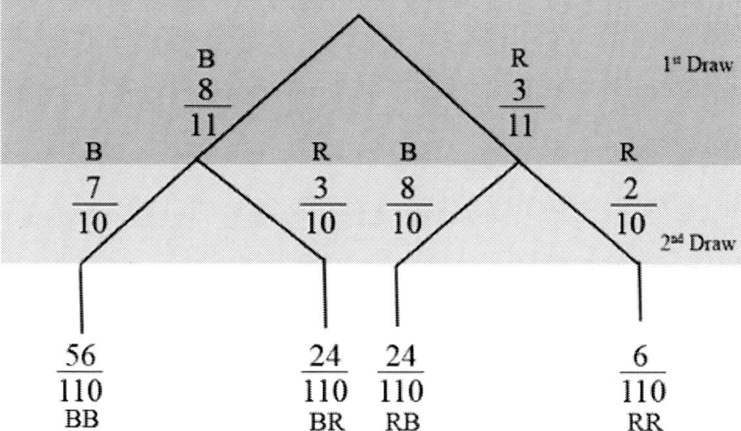

Figure 3.2: *Total* $= \frac{56 + 24 + 24 + 6}{110} = \frac{110}{110} = 1$

NOTE: If you draw a red on the first draw from the 3 red possibilities, there are 2 red left to draw on the second draw. You do not put back or replace the first ball after you have drawn it. You draw **without replacement**, so that on the second draw there are 10 marbles left in the urn.

Calculate the following probabilities using the tree diagram.

Problem 1

$P(RR) =$

Solution

$P(RR) = \frac{3}{11} \cdot \frac{2}{10} = \frac{6}{110}$

Problem 2 *(Solution on p. 133.)*

Fill in the blanks:

$P(RB\ OR\ BR) = \frac{3}{11} \cdot \frac{8}{10} + (\_\_)(\_\_) = \frac{48}{110}$

Problem 3 *(Solution on p. 133.)*

$P(R\ on\ 2d \mid B\ on\ 1st) =$

Problem 4 *(Solution on p. 133.)*

Fill in the blanks:

$P(R\ on\ 1st\ and\ B\ on\ 2nd) = P(RB) = (\_\_)(\_\_) = \frac{24}{110}$

Problem 5 *(Solution on p. 133.)*

$P(BB) =$

Problem 6

$P(B\ on\ 2nd \mid R\ on\ 1st) =$

Solution

There are $6 + 24$ outcomes that have R on the first draw (6 RR and 24 RB). The 6 and the 24 are frequencies. They are also the numerators of the fractions $\frac{6}{110}$ and $\frac{24}{110}$. The sample space is no longer 110 but $6 + 24 = 30$. Twenty-four of the 30 outcomes have B on the second draw. The probability is then $\frac{24}{30}$. Did you get this answer?

If we are using probabilities, we can label the tree in the following general way.

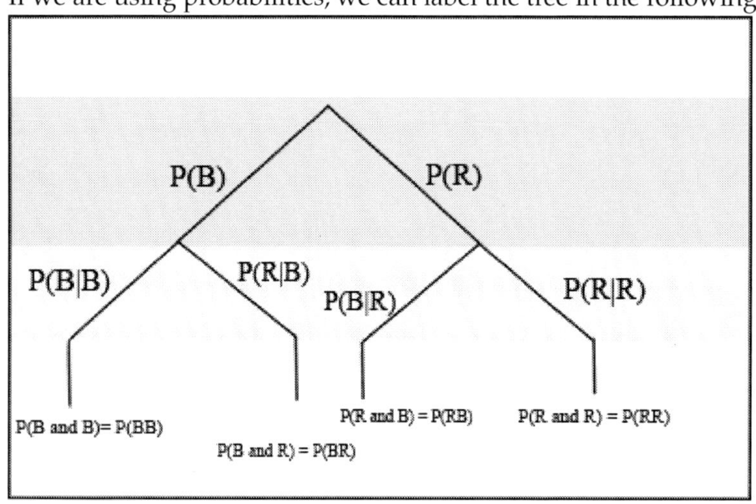

- *P(R | R)* here means *P(R on 2nd | R on 1st)*
- *P(B | R)* here means *P(B on 2nd | R on 1st)*
- *P(R | B)* here means *P(R on 2nd | B on 1st)*
- *P(B | B)* here means *P(B on 2nd | B on 1st)*

3.8 Summary of Formulas[8]

Formula 3.1: Compliment
If A and A' are complements then $P(A) + P(A') = 1$

Formula 3.2: Addition Rule
$P(A\ OR\ B) = P(A) + P(B) - P(A\ AND\ B)$

Formula 3.3: Mutually Exclusive
If A and B are mutually exclusive then $P(A\ AND\ B) = 0$; so $P(A\ OR\ B) = P(A) + P(B)$.

Formula 3.4: Multiplication Rule

- $P(A\ AND\ B) = P(B)P(A \mid B)$
- $P(A\ AND\ B) = P(A)P(B \mid A)$

Formula 3.5: Independence
If A and B are independent then:

- $P(A \mid B) = P(A)$
- $P(B \mid A) = P(B)$
- $P(A\ AND\ B) = P(A)P(B)$

[8]This content is available online at <http://cnx.org/content/m16843/1.4/>.

3.9 Practice 1: Contingency Tables[9]

3.9.1 Student Learning Objectives

- The student will practice constructing and interpreting contingency tables.

3.9.2 Given

An article in the *New England Journal of Medicine* (*by Haiman, Stram, Wilkens, Pike, et al., 1/26/06*), reported about a study of smokers in California and Hawaii. In one part of the report, the self-reported ethnicity and smoking levels per day were given. Of the people smoking at most 10 cigarettes per day, there were 9886 African Americans, 2745 Native Hawaiians, 12,831 Latinos, 8378 Japanese Americans, and 7650 Whites. Of the people smoking 11-20 cigarettes per day, there were 6514 African Americans, 3062 Native Hawaiians, 4932 Latinos, 10,680 Japanese Americans, and 9877 Whites. Of the people smoking 21-30 cigarettes per day, there were 1671 African Americans, 1419 Native Hawaiians, 1406 Latinos, 4715 Japanese Americans, and 6062 Whites. Of the people smoking at least 31 cigarettes per day, there were 759 African Americans, 788 Native Hawaiians, 800 Latinos, 2305 Japanese Americans, and 3970 Whites.

3.9.3 Complete the Table

Complete the table below using the data provided.

Smoking Levels by Ethnicity

Smoking Level	African American	Native Hawaiian	Latino	Japanese Americans	White	TOTALS
1-10						
11-20						
21-30						
31+						
TOTALS						

Table 3.4

3.9.4 Analyze the Data

Suppose that one person from the study is randomly selected.

Exercise 3.9.1 ***(Solution on p. 133.)***

Find the probability that person smoked 11-20 cigarettes per day.

Exercise 3.9.2 ***(Solution on p. 133.)***

Find the probability that person was Latino.

[9]This content is available online at <http://cnx.org/content/m16839/1.9/>.

3.9.5 Discussion Questions

Exercise 3.9.3 *(Solution on p. 133.)*

In words, explain what it means to pick one person from the study and that person is "Japanese American **AND** smokes 21-30 cigarettes per day." Also, find the probability.

Exercise 3.9.4 *(Solution on p. 133.)*

In words, explain what it means to pick one person from the study and that person is "Japanese American **OR** smokes 21-30 cigarettes per day." Also, find the probability.

Exercise 3.9.5 *(Solution on p. 133.)*

In words, explain what it means to pick one person from the study and that person is "Japanese American **GIVEN** that person smokes 21-30 cigarettes per day." Also, find the probability.

Exercise 3.9.6

Prove that smoking level/day and ethnicity are dependent events.

3.10 Practice 2: Calculating Probabilities[10]

3.10.1 Student Learning Objectives

- Students will define basic probability terms.
- Students will practice calculating probabilities.
- Students will determine whether two events are mutually exclusive or whether two events are independent.

NOTE: Use probability rules to solve the problems below. Show your work.

3.10.2 Given

68% of Californians support the death penalty. A majority of all racial groups in California support the death penalty, except for black Californians, of whom 45% support the death penalty (*Source: San Jose Mercury News, 12/2005*). 6% of all Californians are black (*Source: U.S. Census Bureau*).

In this problem, let:

- C = Californians supporting the death penalty
- B = Black Californians

Suppose that one Californian is randomly selected.

3.10.3 Analyze the Data

Exercise 3.10.1 *(Solution on p. 133.)*

$P(C) =$

Exercise 3.10.2 *(Solution on p. 133.)*

$P(B) =$

Exercise 3.10.3 *(Solution on p. 133.)*

$P(C|B) =$

Exercise 3.10.4

In words, what is " $C \mid B$"?

Exercise 3.10.5 *(Solution on p. 133.)*

$P(B \text{ AND } C) =$

Exercise 3.10.6

In words, what is "B and C"?

Exercise 3.10.7 *(Solution on p. 133.)*

Are B and C independent events? Show why or why not.

Exercise 3.10.8 *(Solution on p. 133.)*

$P(B \text{ OR } C) =$

Exercise 3.10.9

In words, what is "B or C"?

Exercise 3.10.10 *(Solution on p. 133.)*

Are B and C mutually exclusive events? Show why or why not.

---

[10]This content is available online at <http://cnx.org/content/m16840/1.10/>.

3.11 Homework[11]

Exercise 3.11.1 *(Solution on p. 133.)*

Suppose that you have 8 cards. 5 are green and 3 are yellow. The 5 green cards are numbered 1, 2, 3, 4, and 5. The 3 yellow cards are numbered 1, 2, and 3. The cards are well shuffled. You randomly draw one card.

- G = card drawn is green
- E = card drawn is even-numbered

a. List the sample space.
b. $P(G) =$
c. $P(G|E) =$
d. $P(G \text{ AND } E) =$
e. $P(G \text{ OR } E) =$
f. Are G and E mutually exclusive? Justify your answer numerically.

Exercise 3.11.2

Refer to the previous problem. Suppose that this time you randomly draw two cards, one at a time, and **with replacement**.

- G_1 = *first card is green*
- G_2 = *second card is green*

a. Draw a tree diagram of the situation.
b. $P(G_1 \text{ AND } G_2) =$
c. $P(\text{at least one green}) =$
d. $P(G_2 \mid G_1) =$
e. Are G_2 and G_1 independent events? Explain why or why not.

Exercise 3.11.3 *(Solution on p. 134.)*

Refer to the previous problems. Suppose that this time you randomly draw two cards, one at a time, and **without replacement**.

- G_1= first card is green
- G_2= second card is green

a. Draw a tree diagram of the situation.
b>. $P(G_1 \text{ AND } G_2) =$
c. *P(at least one green) =*
d. $P(G_2|G_1) =$
e. Are G_2 and G_1 independent events? Explain why or why not.

Exercise 3.11.4

Roll two fair dice. Each die has 6 faces.

a. List the sample space.
b. Let A be the event that either a 3 or 4 is rolled first, followed by an even number. Find $P(A)$.

[11] This content is available online at <http://cnx.org/content/m16836/1.10/>.

c. Let B be the event that the sum of the two rolls is at most 7. Find $P(B)$.
d. In words, explain what "$P(A|B)$" represents. Find $P(A|B)$.
e. Are A and B mutually exclusive events? Explain your answer in 1 - 3 complete sentences, including numerical justification.
f. Are A and B independent events? Explain your answer in 1 - 3 complete sentences, including numerical justification.

Exercise 3.11.5 ***(Solution on p. 134.)***

A special deck of cards has 10 cards. Four are green, three are blue, and three are red. When a card is picked, the color of it is recorded. An experiment consists of first picking a card and then tossing a coin.

a. List the sample space.
b. Let A be the event that a blue card is picked first, followed by landing a head on the coin toss. Find *P(A)*.
c. Let B be the event that a red or green is picked, followed by landing a head on the coin toss. Are the events A and B mutually exclusive? Explain your answer in 1 - 3 complete sentences, including numerical justification.
d. Let C be the event that a red or blue is picked, followed by landing a head on the coin toss. Are the events A and C mutually exclusive? Explain your answer in 1 - 3 complete sentences, including numerical justification.

Exercise 3.11.6

An experiment consists of first rolling a die and then tossing a coin:

a. List the sample space.
b. Let A be the event that either a 3 or 4 is rolled first, followed by landing a head on the coin toss. Find *P(A)*.
c. Let B be the event that a number less than 2 is rolled, followed by landing a head on the coin toss. Are the events A and B mutually exclusive? Explain your answer in 1 - 3 complete sentences, including numerical justification.

Exercise 3.11.7 ***(Solution on p. 134.)***

An experiment consists of tossing a nickel, a dime and a quarter. Of interest is the side the coin lands on.

a. List the sample space.
b. Let A be the event that there are at least two tails. Find *P(A)*.
c. Let B be the event that the first and second tosses land on heads. Are the events A and B mutually exclusive? Explain your answer in 1 - 3 complete sentences, including justification.

Exercise 3.11.8

Consider the following scenario:

- Let $P(C) = 0.4$
- Let $P(D) = 0.5$
- Let $P(C \mid D) = 0.6$

a. Find *P(C AND D)*.
b. Are C and D mutually exclusive? Why or why not?
c. Are C and D independent events? Why or why not?

d. Find *P(C AND D)* .
e. Find *P(D | C)*.

Exercise 3.11.9 *(Solution on p. 134.)*
E and F mutually exclusive events. $P(E) = 0.4$; $P(F) = 0.5$. Find $P(E \mid F)$.

Exercise 3.11.10
J and K are independent events. $P(J \mid K) = 0.3$. Find $P(J)$.

Exercise 3.11.11 *(Solution on p. 134.)*
U and V are mutually exclusive events. $P(U) = 0.26$; $P(V) = 0.37$. Find:

a. *P(U AND V)* =
b. *P(U | V)* =
c. *P(U OR V)* =

Exercise 3.11.12
Q and R are independent events. $P(Q) = 0.4$; $P(Q \text{ AND } R) = 0.1$. Find $P(R)$.

Exercise 3.11.13 *(Solution on p. 134.)*
Y and Z are independent events.

a. Rewrite the basic Addition Rule *P(Y OR Z)* $= P(Y) + P(Z) - P(Y \text{ AND } Z)$ using the information that Y and Z are independent events.
b. Use the rewritten rule to find $P(Z)$ if $P(Y \text{ OR } Z) = 0.71$ and $P(Y) = 0.42$.

Exercise 3.11.14
G and H are mutually exclusive events. $P(G) = 0.5$; $P(H) = 0.3$

a. Explain why the following statement MUST be false: $P(H \mid G) = 0.4$.
b. Find: *P(H OR G)*.
c. Are G and H independent or dependent events? Explain in a complete sentence.

Exercise 3.11.15 *(Solution on p. 134.)*
The following are real data from Santa Clara County, CA. As of March 31, 2000, there was a total of 3059 documented cases of AIDS in the county. They were grouped into the following categories (*Source: Santa Clara County Public H.D.*):

	Homosexual/Bisexual	**IV Drug User***	**Heterosexual Contact**	**Other**	**Totals**
Female	0	70	136	49	____
Male	2146	463	60	135	____
Totals	____	____	____	____	____

Table 3.5: * includes homosexual/bisexual IV drug users

Suppose one of the persons with AIDS in Santa Clara County is randomly selected. Compute the following:

a. *P(person is female)* =
b. *P(person has a risk factor Heterosexual Contact)* =
c. *P(person is female OR has a risk factor of IV Drug User)* =
d. *P(person is female AND has a risk factor of Homosexual/Bisexual)* =
e. *P(person is male AND has a risk factor of IV Drug User)* =

f. *P(female GIVEN person got the disease from heterosexual contact) =*
g. Construct a Venn Diagram. Make one group females and the other group heterosexual contact.

Exercise 3.11.16

Solve these questions using probability rules. Do NOT use the contingency table above. 3059 cases of AIDS had been reported in Santa Clara County, CA, through March 31, 2000. Those cases will be our population. Of those cases, 6.4% obtained the disease through heterosexual contact and 7.4% are female. Out of the females with the disease, 53.3% got the disease from heterosexual contact.

a. *P(person is female) =*
b. *P(person obtained the disease through heterosexual contact) =*
c. *P(female GIVEN person got the disease from heterosexual contact) =*
d. Construct a Venn Diagram. Make one group females and the other group heterosexual contact. Fill in all values as probabilities.

Exercise 3.11.17 ***(Solution on p. 135.)***

The following table identifies a group of children by one of four hair colors, and by type of hair.

Hair Type	Brown	Blond	Black	Red	Totals
Wavy	20		15	3	43
Straight	80	15		12	
Totals		20			215

Table 3.6

a. Complete the table above.
b. What is the probability that a randomly selected child will have wavy hair?
c. What is the probability that a randomly selected child will have either brown or blond hair?
d. What is the probability that a randomly selected child will have wavy brown hair?
e. What is the probability that a randomly selected child will have red hair, given that he has straight hair?
f. If B is the event of a child having brown hair, find the probability of the complement of B.
g. In words, what does the complement of B represent?

Exercise 3.11.18

A previous year, the weights of the members of the **San Francisco 49ers** and the **Dallas Cowboys** were published in the *San Jose Mercury News.* The factual data are compiled into the following table.

Shirt#	$\leq$ 210	211-250	251-290	290$\leq$
1-33	21	5	0	0
34-66	6	18	7	4
66-99	6	12	22	5

Table 3.7

For the following, suppose that you randomly select one player from the 49ers or Cowboys.

a. Find the probability that his shirt number is from 1 to 33.
b. Find the probability that he weighs at most 210 pounds.
c. Find the probability that his shirt number is from 1 to 33 AND he weighs at most 210 pounds.
d. Find the probability that his shirt number is from 1 to 33 OR he weighs at most 210 pounds.
e. Find the probability that his shirt number is from 1 to 33 GIVEN that he weighs at most 210 pounds.
f. If having a shirt number from 1 to 33 and weighing at most 210 pounds were independent events, then what should be true about *P(Shirt# 1-33* | $\leq$ *210 pounds)*?

Exercise 3.11.19 ***(Solution on p. 135.)***

Approximately 249,000,000 people live in the United States. Of these people, 31,800,000 speak a language other than English at home. Of those who speak another language at home, over 50 percent speak Spanish. (*Source: U.S. Bureau of the Census, 1990 Census*)

Let: E = speak English at home; E' = speak another language at home; S = speak Spanish at home

Finish each probability statement by matching the correct answer.

Probability Statements	**Answers**
a. P(E′) =	i. 0.8723
b. P(E) =	ii. > 0.50
c. P(S) =	iii. 0.1277
d. P(S | E′) =	iv. > 0.0639

Table 3.8

Exercise 3.11.20

The probability that a male develops some form of cancer in his lifetime is 0.4567 (Source: American Cancer Society). The probability that a male has at least one false positive test result (meaning the test comes back for cancer when the man does not have it) is 0.51 (Source: USA Today). Some of the questions below do not have enough information for you to answer them. Write "not enough information" for those answers.

Let: C = a man develops cancer in his lifetime; P = man has at least one false positive

a. Construct a tree diagram of the situation.
b. $P(C) =$
c. $P(P|C) =$
d. $P(P|C') =$
e. If a test comes up positive, based upon numerical values, can you assume that man has cancer? Justify numerically and explain why or why not.

Exercise 3.11.21 ***(Solution on p. 135.)***

In 1994, the U.S. government held a lottery to issue 55,000 Green Cards (permits for non-citizens to work legally in the U.S.). Renate Deutsch, from Germany, was one of approximately 6.5 million people who entered this lottery. Let G = *won Green Card*.

a. What was Renate's chance of winning a Green Card? Write your answer as a probability statement.

b. In the summer of 1994, Renate received a letter stating she was one of 110,000 finalists chosen. Once the finalists were chosen, assuming that each finalist had an equal chance to win, what was Renate's chance of winning a Green Card? Let F = *was a finalist*. Write your answer as a conditional probability statement.

c. Are G and F independent or dependent events? Justify your answer numerically and also explain why.

d. Are G and F mutually exclusive events? Justify your answer numerically and also explain why.

NOTE: P.S. Amazingly, on 2/1/95, Renate learned that she would receive her Green Card – true story!

Exercise 3.11.22

Three professors at George Washington University did an experiment to determine if economists are more selfish than other people. They dropped 64 stamped, addressed envelopes with $10 cash in different classrooms on the George Washington campus. 44% were returned overall. From the economics classes 56% of the envelopes were returned. From the business, psychology, and history classes 31% were returned. (*Source: Wall Street Journal*)

Let: R = money returned; E = economics classes; O = other classes

a. Write a probability statement for the overall percent of money returned.

b. Write a probability statement for the percent of money returned out of the economics classes.

c. Write a probability statement for the percent of money returned out of the other classes.

d. Is money being returned independent of the class? Justify your answer numerically and explain it.

e. Based upon this study, do you think that economists are more selfish than other people? Explain why or why not. Include numbers to justify your answer.

Exercise 3.11.23 ***(Solution on p. 135.)***

The chart below gives the number of suicides estimated in the U.S. for a recent year by age, race (black and white), and sex. We are interested in possible relationships between age, race, and sex. We will let suicide victims be our population. (*Source: The National Center for Health Statistics, U.S. Dept. of Health and Human Services*)

Race and Sex	1 - 14	15 - 24	25 - 64	over 64	TOTALS
white, male	210	3360	13,610		22,050
white, female	80	580	3380		4930
black, male	10	460	1060		1670
black, female	0	40	270		330
all others					
TOTALS	310	4650	18,780		29,760

Table 3.9

NOTE: Do not include "all others" for parts (f), (g), and (i).

- **a.** Fill in the column for the suicides for individuals over age 64.
- **b.** Fill in the row for all other races.
- **c.** Find the probability that a randomly selected individual was a white male.
- **d.** Find the probability that a randomly selected individual was a black female.
- **e.** Find the probability that a randomly selected individual was black
- **f.** Comparing "Race and Sex" to "Age," which two groups are mutually exclusive? How do you know?
- **g.** Find the probability that a randomly selected individual was male.
- **h.** Out of the individuals over age 64, find the probability that a randomly selected individual was a black or white male.
- **i.** Are being male and committing suicide over age 64 independent events? How do you know?

The next two questions refer to the following: The percent of licensed U.S. drivers (from a recent year) that are female is 48.60. Of the females, 5.03% are age 19 and under; 81.36% are age 20 - 64; 13.61% are age 65 or over. Of the licensed U.S. male drivers, 5.04% are age 19 and under; 81.43% are age 20 - 64; 13.53% are age 65 or over. (Source: Federal Highway Administration, U.S. Dept. of Transportation)

Exercise 3.11.24

Complete the following:

- **a.** Construct a table or a tree diagram of the situation.
- **b.** *P(driver is female) =*
- **c.** *P(driver is age 65 or over | driver is female) =*
- **d.** *P(driver is age 65 or over AND female) =*
- **e.** In words, explain the difference between the probabilities in part (c) and part (d).
- **f.** *P(driver is age 65 or over) =*
- **g.** Are being age 65 or over and being female mutually exclusive events? How do you know

Exercise 3.11.25 *(Solution on p. 135.)*

Suppose that 10,000 U.S. licensed drivers are randomly selected.

- **a.** How many would you expect to be male?
- **b.** Using the table or tree diagram from the previous exercise, construct a contingency table of gender versus age group.
- **c.** Using the contingency table, find the probability that out of the age 20 - 64 group, a randomly selected driver is female.

Exercise 3.11.26

Approximately 86.5% of Americans commute to work by car, truck or van. Out of that group, 84.6% drive alone and 15.4% drive in a carpool. Approximately 3.9% walk to work and approximately 5.3% take public transportation. (*Source: Bureau of the Census, U.S. Dept. of Commerce. Disregard rounding approximations.*)

- **a.** Construct a table or a tree diagram of the situation. Include a branch for all other modes of transportation to work.
- **b.** Assuming that the walkers walk alone, what percent of all commuters travel alone to work?
- **c.** Suppose that 1000 workers are randomly selected. How many would you expect to travel alone to work?

d. Suppose that 1000 workers are randomly selected. How many would you expect to drive in a carpool?

Exercise 3.11.27

Explain what is wrong with the following statements. Use complete sentences.

a. If there's a 60% chance of rain on Saturday and a 70% chance of rain on Sunday, then there's a 130% chance of rain over the weekend.
b. The probability that a baseball player hits a home run is greater than the probability that he gets a successful hit.

3.11.1 Try these multiple choice questions.

The next two questions refer to the following probability tree diagram which shows tossing an unfair coin **FOLLOWED BY** drawing one bead from a cup containing 3 red (R), 4 yellow (Y) and 5 blue (B) beads. For the coin, $P(H) = \frac{2}{3}$ and $P(T) = \frac{1}{3}$ where H = *"heads"* and T = *"tails"*.

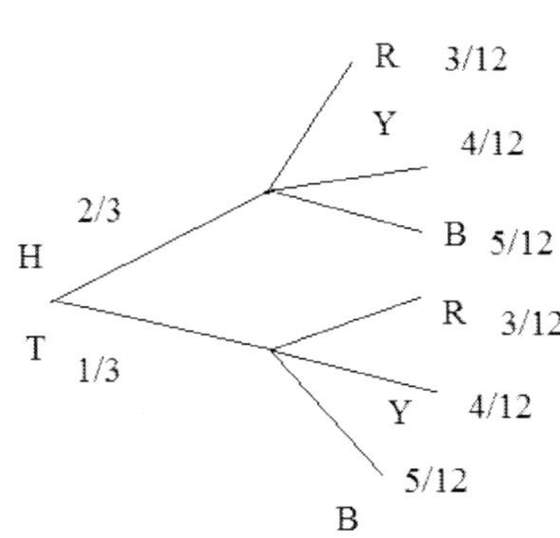

Figure 3.3

Exercise 3.11.28 *(Solution on p. 135.)*

Find *P(tossing a Head on the coin AND a Red bead)*

A. $\frac{2}{3}$
B. $\frac{5}{15}$
C. $\frac{6}{36}$
D. $\frac{5}{36}$

Exercise 3.11.29 *(Solution on p. 135.)*

Find *P(Blue bead)*.

A. $\frac{15}{36}$

B. $\frac{10}{36}$
C. $\frac{10}{12}$
D. $\frac{6}{36}$

The next three questions refer to the following table of data obtained from *www.baseball-almanac.com*[12] showing hit information for 4 well known baseball players.

NAME	Single	Double	Triple	Home Run	TOTAL HITS
Babe Ruth	1517	506	136	714	2873
Jackie Robinson	1054	273	54	137	1518
Ty Cobb	3603	174	295	114	4189
Hank Aaron	2294	624	98	755	3771
TOTAL	8471	1577	583	1720	12351

Table 3.10

Exercise 3.11.30 *(Solution on p. 135.)*

Find *P(hit was made by Babe Ruth)*.

A. $\frac{1518}{2873}$
B. $\frac{2873}{12351}$
C. $\frac{583}{12351}$
D. $\frac{4189}{12351}$

Exercise 3.11.31 *(Solution on p. 135.)*

Find *P(hit was made by Ty Cobb | The hit was a Home Run)*

A. $\frac{4189}{12351}$
B. $\frac{1141}{1720}$
C. $\frac{1720}{4189}$
D. $\frac{114}{12351}$

Exercise 3.11.32 *(Solution on p. 135.)*

Are *the hit being made by Hank Aaron* and *the hit being a double* independent events?

A. Yes, because *P(hit by Hank Aaron | hit is a double) = P(hit by Hank Aaron)*
B. No, because *P(hit by Hank Aaron | hit is a double)* ≠ *P(hit is a double)*
C. No, because *P(hit is by Hank Aaron | hit is a double)* ≠ *P(hit by Hank Aaron)*
D. Yes, because *P(hit is by Hank Aaron | hit is a double) = P(hit is a double)*

[12]http://cnx.org/content/m16836/latest/www.baseball-almanac.com

3.12 Review[13]

The first six exercises refer to the following study: In a survey of 100 stocks on NASDAQ, the average percent increase for the past year was 9% for NASDAQ stocks. Answer the following:

Exercise 3.12.1 *(Solution on p. 135.)*

The "average increase" for all NASDAQ stocks is the:

A. Population
B. Statistic
C. Parameter
D. Sample
E. Variable

Exercise 3.12.2 *(Solution on p. 136.)*

All of the NASDAQ stocks are the:

A. Population
B. Statistic
C. Parameter
D. Sample
E. Variable

Exercise 3.12.3 *(Solution on p. 136.)*

9% is the:

A. Population
B. Statistic
C. Parameter
D. Sample
E. Variable

Exercise 3.12.4 *(Solution on p. 136.)*

The 100 NASDAQ stocks in the survey are the:

A. Population
B. Statistic
C. Parameter
D. Sample
E. Variable

Exercise 3.12.5 *(Solution on p. 136.)*

The percent increase for one stock in the survey is the:

A. Population
B. Statistic
C. Parameter
D. Sample
E. Variable

---

[13]This content is available online at <http://cnx.org/content/m16842/1.9/>.

Exercise 3.12.6 *(Solution on p. 136.)*

Would the data collected be qualitative, quantitative – discrete, or quantitative – continuous?

The next two questions refer to the following study: Thirty people spent two weeks around Mardi Gras in New Orleans. Their two-week weight gain is below. (Note: a loss is shown by a negative weight gain.)

Weight Gain	Frequency
-2	3
-1	5
0	2
1	4
4	13
6	2
11	1

Table 3.11

Exercise 3.12.7 *(Solution on p. 136.)*

Calculate the following values:

- **a.** The average weight gain for the two weeks
- **b.** The standard deviation
- **c.** The first, second, and third quartiles

Exercise 3.12.8

Construct a histogram and a boxplot of the data.

3.13 Lab: Probability Topics[14]

Class time:

Names:

3.13.1 Student Learning Outcomes:

- The student will use theoretical and empirical methods to estimate probabilities.
- The student will appraise the differences between the two estimates.
- The student will demonstrate an understanding of long-term relative frequencies.

3.13.2 Do the Experiment:

Count out 40 mixed-color M&M's® which is approximately 1 small bag's worth (distance learning classes using the virtual lab would want to count out 25 M&M's®). Record the number of each color in the "Population" table. Use the information from this table to complete the theoretical probability questions. Next, put the M&M's in a cup. The experiment is to pick 2 M&M's, one at a time. Do **not** look at them as you pick them. The first time through, replace the first M&M before picking the second one. Record the results in the "With Replacement" column of the empirical table. Do this 24 times. The second time through, after picking the first M&M, do **not** replace it before picking the second one. Then, pick the second one. Record the results in the "Without Replacement" column section of the "Empirical Results" table. After you record the pick, put **both** M&M's back. Do this a total of 24 times, also. Use the data from the "Empirical Results" table to calculate the empirical probability questions. Leave your answers in unreduced fractional form. Do **not** multiply out any fractions.

Population

Color	Quantity
Yellow (Y)	
Green (G)	
Blue (BL)	
Brown (B)	
Orange (O)	
Red (R)	

Table 3.12

[14]This content is available online at <http://cnx.org/content/m16841/1.13/>.

Theoretical Probabilities

	With Replacement	Without Replacement
$P(2\ reds)$		
$P(R_1 B_2\ OR\ B_1 R_2)$		
$P(R_1\ AND\ G_2)$		
$P(G_2 \mid R_1)$		
$P(no\ yellows)$		
$P(doubles)$		
$P(no\ doubles)$		

Table 3.13: Note: G_2 = green on second pick; R_1 = red on first pick; B_1 = brown on first pick; B_2 = brown on second pick; doubles = both picks are the same colour.

Empirical Results

With Replacement	Without Replacement
(__ , __) (__ , __)	(__ , __) (__ , __)
(__ , __) (__ , __)	(__ , __) (__ , __)
(__ , __) (__ , __)	(__ , __) (__ , __)
(__ , __) (__ , __)	(__ , __) (__ , __)
(__ , __) (__ , __)	(__ , __) (__ , __)
(__ , __) (__ , __)	(__ , __) (__ , __)
(__ , __) (__ , __)	(__ , __) (__ , __)
(__ , __) (__ , __)	(__ , __) (__ , __)
(__ , __) (__ , __)	(__ , __) (__ , __)
(__ , __) (__ , __)	(__ , __) (__ , __)
(__ , __) (__ , __)	(__ , __) (__ , __)
(__ , __) (__ , __)	(__ , __) (__ , __)

Table 3.14

Empirical Probabilities

	With Replacement	Without Replacement
$P\,(2\ reds)$		
$P\,(R_1B_2\ OR\ B_1R_2)$		
$P\,(R_1\ AND\ G_2)$		
$P\,(G_2 \mid R_1)$		
$P\,(no\ yellows)$		
$P\,(doubles)$		
$P\,(no\ doubles)$		

Table 3.15: Note:

3.13.3 Discussion Questions

1. Why are the "With Replacement" and "Without Replacement" probabilities different?
2. Convert *P(no yellows)* to decimal format for both Theoretical "With Replacement" and for Empirical "With Replacement". Round to 4 decimal places.
   a. Theoretical "With Replacement": *P(no yellows)* =
   b. Empirical "With Replacement": *P(no yellows)* =
   c. Are the decimal values "close"? Did you expect them to be closer together or farther apart? Why?
3. If you increased the number of times you picked 2 M&M's to 240 times, why would empirical probability values change?
4. Would this change (see (3) above) cause the empirical probabilities and theoretical probabilities to be closer together or farther apart? How do you know?
5. Explain the differences in what $P\,(G_1\ AND\ R_2)$ and $P\,(R_1 \mid G_2)$ represent.

Solutions to Exercises in Chapter 3

Solution to Example 3.2 (p. 100)
No. $C = \{3, 5\}$ and $E = \{1, 2, 3, 4\}$. $P(C\ AND\ E) = \frac{1}{6}$. To be mutually exclusive, $P(C\ AND\ E)$ must be 0.
Solution to Example 3.10, Problem 1 (p. 107)

Hiking Area Preference

Sex	The Coastline	Near Lakes and Streams	On Mountain Peaks	Total
Female	18	16	**11**	45
Male	**16**	**25**	14	55
Total	**34**	41	**25**	**100**

Table 3.16

Solution to Example 3.10, Problem 2 (p. 107)

a. $P(F\ AND\ C) = \frac{18}{100} = 0.18$
b. $P(F) \cdot P(C) = \frac{45}{100} \cdot \frac{45}{100} = 0.45 \cdot 0.45 = 0.153$

$P(F\ AND\ C) \neq P(F) \cdot P(C)$, so the events F and C are not independent.
Solution to Example 3.10, Problem 3 (p. 107)

a. The word 'given' tells you that this is a conditional.
b. $P(M|L) = \frac{25}{41}$
c. No, the sample space for this problem is 41.

Solution to Example 3.10, Problem 4 (p. 107)

a. $P(F) = \frac{45}{100}$
b. $P(P) = \frac{25}{100}$
c. $P(F\ AND\ P) = \frac{11}{100}$
d. $P(F\ OR\ P) = \frac{45}{100} + \frac{25}{100} - \frac{11}{100} = \frac{59}{100}$

Solution to Example 3.11, Problem 1 (p. 108)

Door Choice

Caught or Not	Door One	Door Two	Door Three	Total
Caught	$\frac{1}{15}$	$\frac{1}{12}$	$\frac{1}{6}$	$\frac{19}{60}$
Not Caught	$\frac{4}{15}$	$\frac{3}{12}$	$\frac{1}{6}$	$\frac{31}{60}$
Total	$\frac{5}{15}$	$\frac{4}{12}$	$\frac{2}{6}$	1

Table 3.17

Solution to Example 3.15, Problem 1 (p. 110)
B1R1; B1R2; B1R3; B2R1; B2R2; B2R3; B3R1; B3R2; B3R3; B4R1; B4R2; B4R3; B5R1; B5R2; B5R3; B6R1; B6R2; B6R3; B7R1; B7R2; B7R3; B8R1; B8R2; B8R3

Solution to Example 3.15, Problem 6 (p. 111)
$P(BB) = \frac{64}{121}$
Solution to Example 3.15, Problem 7 (p. 111)
$P(B \text{ on 2nd draw} \mid R \text{ on 1st draw}) = \frac{8}{11}$

There are 9 + 24 outcomes that have *R* on the first draw (9 *RR* and 24 *RB*). The sample space is then 9 + 24 = 33. Twenty-four of the 33 outcomes have *B* on the second draw. The probability is then $\frac{24}{33}$.
Solution to Example 3.16, Problem 2 (p. 112)
$P(RB \text{ or } BR) = \frac{3}{11} \cdot \frac{8}{10} + \left(\frac{8}{11}\right)\left(\frac{3}{10}\right) = \frac{48}{110}$
Solution to Example 3.16, Problem 3 (p. 112)
$P(R \text{ on 2d} \mid B \text{ on 1st}) = \frac{3}{10}$
Solution to Example 3.16, Problem 4 (p. 112)
$P(R \text{ on 1st and } B \text{ on 2nd}) = P(RB) = \left(\frac{3}{11}\right)\left(\frac{8}{10}\right) = \frac{24}{110}$
Solution to Example 3.16, Problem 5 (p. 112)
$P(BB) = \frac{8}{11} \cdot \frac{7}{10}$

Solutions to Practice 1: Contingency Tables

Solution to Exercise 3.9.1 (p. 115)
$\frac{35{,}065}{100{,}450}$
Solution to Exercise 3.9.2 (p. 115)
$\frac{19{,}969}{100{,}450}$
Solution to Exercise 3.9.3 (p. 116)
$\frac{4{,}715}{100{,}450}$
Solution to Exercise 3.9.4 (p. 116)
$\frac{36{,}636}{100{,}450}$
Solution to Exercise 3.9.5 (p. 116)
$\frac{4715}{15{,}273}$

Solutions to Practice 2: Calculating Probabilities

Solution to Exercise 3.10.1 (p. 117)
0.68
Solution to Exercise 3.10.2 (p. 117)
0.06
Solution to Exercise 3.10.3 (p. 117)
0.45
Solution to Exercise 3.10.5 (p. 117)
0.027
Solution to Exercise 3.10.7 (p. 117)
No
Solution to Exercise 3.10.8 (p. 117)
0.713
Solution to Exercise 3.10.10 (p. 117)
No

Solutions to Homework

Solution to Exercise 3.11.1 (p. 118)

- **a.** $\{G1, G2, G3, G4, G5, Y1, Y2, Y3\}$
- **b.** $\frac{5}{8}$

c. $\frac{2}{3}$
d. $\frac{2}{8}$
e. $\frac{6}{8}$
f. No

Solution to Exercise 3.11.3 (p. 118)

b. $\left(\frac{5}{8}\right)\left(\frac{4}{7}\right)$
c. $\left(\frac{5}{8}\right)\left(\frac{3}{7}\right)+\left(\frac{3}{8}\right)\left(\frac{5}{7}\right)+\left(\frac{5}{8}\right)\left(\frac{4}{7}\right)$
d. $\frac{4}{7}$
e. No

Solution to Exercise 3.11.5 (p. 119)

a. $\{GH, GT, BH, BT, RH, RT\}$
b. $\frac{3}{20}$
c. Yes
d. No

Solution to Exercise 3.11.7 (p. 119)

a. $\{(HHH),(HHT),(HTH),(HTT),(THH),(THT),(TTH),(TTT)\}$
b. $\frac{4}{8}$
c. Yes

Solution to Exercise 3.11.9 (p. 120)
0
Solution to Exercise 3.11.11 (p. 120)

a. 0
b. 0
c. 0.63

Solution to Exercise 3.11.13 (p. 120)

b. 0.5

Solution to Exercise 3.11.15 (p. 120)
The completed contingency table is as follows:

	Homosexual/Bisexual	**IV Drug User***	**Heterosexual Contact**	**Other**	**Totals**
Female	0	70	136	49	**255**
Male	2146	463	60	135	**2804**
Totals	**2146**	**533**	**196**	**174**	**3059**

Table 3.18: * includes homosexual/bisexual IV drug users

a. $\frac{255}{3059}$
b. $\frac{196}{3059}$
c. $\frac{718}{3059}$
d. 0
e. $\frac{463}{3059}$

f. $\frac{136}{196}$

Solution to Exercise 3.11.17 (p. 121)

b. $\frac{43}{215}$
c. $\frac{120}{215}$
d. $\frac{20}{215}$
e. $\frac{12}{172}$
f. $\frac{115}{215}$

Solution to Exercise 3.11.19 (p. 122)

a. iii
b. i
c. iv
d. ii

Solution to Exercise 3.11.21 (p. 122)

a. $P(G) = 0.008$
b. 0.5
c. dependent
d. No

Solution to Exercise 3.11.23 (p. 123)

c. $\frac{22050}{29760}$
d. $\frac{330}{29760}$
e. $\frac{2000}{29760}$
f. $\frac{23720}{29760}$
g. $\frac{5010}{6020}$
h. Black females and ages 1-14
i. No

Solution to Exercise 3.11.25 (p. 124)

a. 5140
c. 0.49

Solution to Exercise 3.11.28 (p. 125)
C
Solution to Exercise 3.11.29 (p. 125)
A
Solution to Exercise 3.11.30 (p. 126)
B
Solution to Exercise 3.11.31 (p. 126)
B
Solution to Exercise 3.11.32 (p. 126)
C

Solutions to Review

Solution to Exercise 3.12.1 (p. 127)

C. Parameter

Solution to Exercise 3.12.2 (p. 127)

A. Population

Solution to Exercise 3.12.3 (p. 127)

B. Statistic

Solution to Exercise 3.12.4 (p. 127)

D. Sample

Solution to Exercise 3.12.5 (p. 127)

E. Variable

Solution to Exercise 3.12.6 (p. 128)
quantitative - continuous
Solution to Exercise 3.12.7 (p. 128)

a. 2.27
b. 3.04
c. -1, 4, 4

Chapter 4

Discrete Random Variables

4.1 Discrete Random Variables[1]

4.1.1 Student Learning Objectives

By the end of this chapter, the student should be able to:

- Recognize and understand discrete probability distribution functions, in general.
- Calculate and interpret expected values.
- Recognize the binomial probability distribution and apply it appropriately.
- Recognize the Poisson probability distribution and apply it appropriately (optional).
- Recognize the geometric probability distribution and apply it appropriately (optional).
- Recognize the hypergeometric probability distribution and apply it appropriately (optional).
- Classify discrete word problems by their distributions.

4.1.2 Introduction

A student takes a 10 question true-false quiz. Because the student had such a busy schedule, he or she could not study and randomly guesses at each answer. What is the probability of the student passing the test with at least a 70%?

Small companies might be interested in the number of long distance phone calls their employees make during the peak time of the day. Suppose the average is 20 calls. What is the probability that the employees make more than 20 long distance phone calls during the peak time?

These two examples illustrate two different types of probability problems involving discrete random variables. Recall that discrete data are data that you can count. A **random variable** describes the outcomes of a statistical experiment both in words. The values of a random variable can vary with each repetition of an experiment.

In this chapter, you will study probability problems involving discrete random distributions. You will also study long-term averages associated with them.

4.1.3 Random Variable Notation

Upper case letters like X or Y denote a random variable. Lower case letters like x or y denote the value of a random variable. If X **is a random variable, then** X **is defined in words.**

[1]This content is available online at <http://cnx.org/content/m16825/1.10/>.

For example, let X = the number of heads you get when you toss three fair coins. The sample space for the toss of three fair coins is TTT; THH; HTH; HHT; HTT; THT; TTH; HHH. Then, $x = 0, 1, 2, 3$. X is in words and x is a number. Notice that for this example, the x values are countable outcomes. Because you can count the possible values that X can take on and the outcomes are random (the x values 0, 1, 2, 3), X is a discrete random variable.

4.1.4 Optional Collaborative Classroom Activity

Toss a coin 10 times and record the number of heads. After all members of the class have completed the experiment (tossed a coin 10 times and counted the number of heads), fill in the chart using a heading like the one below. Let X = the number of heads in 10 tosses of the coin.

X	Frequency of X	Relative Frequency of X

Table 4.1

- Which value(s) of X occurred most frequently?
- If you tossed the coin 1,000 times, what values would X take on? Which value(s) of X do you think would occur most frequently?
- What does the relative frequency column sum to?

4.2 Probability Distribution Function (PDF) for a Discrete Random Variable[2]

A discrete **probability distribution function** has two characteristics:

- Each probability is between 0 and 1, inclusive.
- The sum of the probabilities is 1.

$P(X)$ is the notation used to represent a discrete **probability** distribution function.

Example 4.1

A child psychologist is interested in the number of times a newborn baby's crying wakes its mother after midnight. For a random sample of 50 mothers, the following information was obtained. Let X = the number of times a newborn wakes its mother after midnight. For this example, $x = 0, 1, 2, 3, 4, 5$.

$P(X = x)$ = probability that X takes on a value x.

[2] This content is available online at <http://cnx.org/content/m16831/1.11/>.

x	$P(X = x)$
0	$P(X=0) = \frac{2}{50}$
1	$P(X=1) = \frac{11}{50}$
2	$P(X=2) = \frac{23}{50}$
3	$P(X=3) = \frac{9}{50}$
4	$P(X=4) = \frac{4}{50}$
5	$P(X=5) = \frac{1}{50}$

Table 4.2

X takes on the values 0, 1, 2, 3, 4, 5. This is a discrete *PDF* because

1. Each $P(X = x)$ is between 0 and 1, inclusive.
2. The sum of the probabilities is 1, that is,

$$\frac{2}{50} + \frac{11}{50} + \frac{23}{50} + \frac{9}{50} + \frac{4}{50} + \frac{1}{50} = 1 \tag{4.1}$$

Example 4.2
Suppose Nancy has classes **3 days** a week. She attends classes 3 days a week **80%** of the time, **2 days 15%** of the time, **1 day 4%** of the time, and **no days 1%** of the time.

Problem 1 *(Solution on p. 182.)*
Let X = the number of days Nancy ____________________ .

Problem 2 *(Solution on p. 182.)*
X takes on what values?

Problem 3 *(Solution on p. 182.)*
Construct a probability distribution table (called a *PDF* table) like the one in the previous example. The table should have two columns labeled x and $P(X = x)$. What does the $P(X = x)$ column sum to?

4.3 Mean or Expected Value and Standard Deviation[3]

The **expected value** is often referred to as the **"long-term"average or mean** . This means that over the long term of doing an experiment over and over, you would **expect** this average.

The **mean** of a random variable X is μ. If we do an experiment many times (for instance, flip a fair coin, as Karl Pearson did, 24,000 times and let X = the number of heads) and record the value of X each time, the average gets closer and closer to μ as we keep repeating the experiment. This is known as the **Law of Large Numbers**.

> NOTE: To find the expected value or long term average, μ, simply multiply each value of the random variable by its probability and add the products.

[3]This content is available online at <http://cnx.org/content/m16828/1.12/>.

A Step-by-Step Example
A men's soccer team plays soccer 0, 1, or 2 days a week. The probability that they play 0 days is 0.2, the probability that they play 1 day is 0.5, and the probability that they play 2 days is 0.3. Find the long-term average, μ, or expected value of the days per week the men's soccer team plays soccer.

To do the problem, first let the random variable X = the number of days the men's soccer team plays soccer per week. X takes on the values 0, 1, 2. Construct a *PDF* table, adding a column $xP(X = x)$. In this column, you will multiply each x value by its probability.

Expected Value Table

x	P(X=x)	xP(X=x)
0	0.2	(0)(0.2) = 0
1	0.5	(1)(0.5) = 0.5
2	0.3	(2)(0.3) = 0.6

Table 4.4: This table is called an expected value table. The table helps you calculate the expected value or long-term average.

Add the last column to find the long term average or expected value: $(0)(0.2) + (1)(0.5) + (2)(0.3) = 0 + 0.5.\ 0.6 = 1.1$.

The expected value is 1.1. The men's soccer team would, on the average, expect to play soccer 1.1 days per week. The number 1.1 is the long term average or expected value if the men's soccer team plays soccer week after week after week. We say $\mu = 1.1$

Example 4.3
Find the expected value for the example about the number of times a newborn baby's crying wakes its mother after midnight. The expected value is the expected number of times a newborn wakes its mother after midnight.

x	P(X=x)	xP(X=x)
0	P(X=0) = $\frac{2}{50}$	$(0)\left(\frac{2}{50}\right) = 0$
1	P(X=1) = $\frac{11}{50}$	$(1)\left(\frac{11}{50}\right) = \frac{11}{50}$
2	P(X=2) = $\frac{23}{50}$	$(2)\left(\frac{23}{50}\right) = \frac{46}{50}$
3	P(X=3) = $\frac{9}{50}$	$(3)\left(\frac{9}{50}\right) = \frac{27}{50}$
4	P(X=4) = $\frac{4}{50}$	$(4)\left(\frac{4}{50}\right) = \frac{16}{50}$
5	P(X=5) = $\frac{1}{50}$	$(5)\left(\frac{1}{50}\right) = \frac{5}{50}$

Table 4.5: You expect a newborn to wake its mother after midnight 2.1 times, on the average.

Add the last column to find the expected value. μ = Expected Value = $\frac{105}{50} = 2.1$

Problem
Go back and calculate the expected value for the number of days Nancy attends classes a week. Construct the third column to do so.

Solution
2.74 days a week.

Example 4.4

Suppose you play a game of chance in which you choose 5 numbers from 0, 1, 2, 3, 4, 5, 6, 7, 8, 9. You may choose a number more than once. You pay \$2 to play and could profit \$100,000 if you match all 5 numbers in order (you get your \$2 back plus \$100,000). Over the long term, what is your **expected** profit of playing the game?

To do this problem, set up an expected value table for the amount of money you can profit.

Let X = the amount of money you profit. The values of x are not 0, 1, 2, 3, 4, 5, 6, 7, 8, 9. Since you are interested in your profit (or loss), the values of x are 100,000 dollars and -2 dollars.

To win, you must get all 5 numbers correct, in order. The probability of choosing one correct number is $\frac{1}{10}$ because there are 10 numbers. You may choose a number more than once. The probability of choosing all 5 numbers correctly and in order is:

$$\frac{1}{10}*\frac{1}{10}*\frac{1}{10}*\frac{1}{10}*\frac{1}{10}* \; = 1*10^{-5} = 0.00001 \tag{4.2}$$

Therefore, the probability of winning is 0.00001 and the probability of losing is

$$1 - 0.00001 = 0.99999 \tag{4.3}$$

The expected value table is as follows.

	x	$P(X{=}x)$	$xP(X{=}x)$
Loss	-2	0.99999	(-2)(0.99999)=-1.99998
Profit	100,000	0.00001	(100000)(0.00001)=1

Table 4.6: Add the last column. -1.99998 + 1 = -0.99998

Since -0.99998 is about -1, you would, on the average, expect to lose approximately one dollar for each game you play. However, each time you play, you either lose \$2 or profit \$100,000. The \$1 is the average or expected LOSS per game after playing this game over and over.

Example 4.5

Suppose you play a game with a biased coin. You play each game by tossing the coin once. *P(heads)* $= \frac{2}{3}$ and *P(tails)* $= \frac{1}{3}$. If you toss a head, you pay \$6. If you toss a tail, you win \$10. If you play this game many times, will you come out ahead?

Problem 1 *(Solution on p. 182.)*

Define a random variable X.

Problem 2 *(Solution on p. 182.)*

Complete the following expected value table.

	x	____	____
WIN	10	$\frac{1}{3}$	____
LOSE	____	____	$\frac{-12}{3}$

Table 4.7

Problem 3 *(Solution on p. 182.)*

What is the expected value, μ? Do you come out ahead?

Like data, probability distributions have standard deviations. To calculate the standard deviation (σ) of a probability distribution, find each deviation, square it, multiply it by its probability, add the products, and take the square root . To understand how to do the calculation, look at the table for the number of days per week a men's soccer team plays soccer. To find the standard deviation, add the entries in the column labeled $(x - \mu)^2 \cdot P(X = x)$ and take the square root.

x	$P(X=x)$	$xP(X=x)$	$(x - \mu)^2 P(X=x)$
0	0.2	(0)(0.2) = 0	$(0 - 1.1)^2 (.2) = 0.242$
1	0.5	(1)(0.5) = 0.5	$(1 - 1.1)^2 (.5) = 0.005$
2	0.3	(2)(0.3) = 0.6	$(2 - 1.1)^2 (.3) = 0.243$

Table 4.8

Add the last column in the table. $0.242 + 0.005 + 0.243 = 0.490$. The standard deviation is the square root of 0.49. $\sigma = \sqrt{0.49} = 0.7$

Generally for probability distributions, we use a calculator or a computer to calculate μ and σ to reduce roundoff error. For some probability distributions, there are short-cut formulas that calculate μ and σ.

4.4 Common Discrete Probability Distribution Functions[4]

Some of the more common discrete probability functions are binomial, geometric, hypergeometric, and Poisson. Most elementary courses do not cover the geometric, hypergeometric, and Poisson. Your instructor will let you know if he or she wishes to cover these distributions.

A probability distribution function is a pattern. You try to fit a probability problem into a **pattern** or distribution in order to perform the necessary calculations. These distributions are tools to make solving probability problems easier. Each distribution has its own special characteristics. Learning the characteristics enables you to distinguish among the different distributions.

4.5 Binomial[5]

The characteristics of a binomial experiment are:

1. There are a fixed number of trials. Think of trials as repetitions of an experiment. The letter n denotes the number of trials.
2. There are only 2 possible outcomes, called "success" and, "failure" for each trial. The letter p denotes the probability of a success on one trial and q denotes the probability of a failure on one trial. $p + q = 1$.
3. The n trials are independent and are repeated using identical conditions. Because the n trials are independent, the outcome of one trial does not affect the outcome of any other trial. Another way of saying this is that for each individual trial, the probability, p, of a success and probability, q, of a failure remain the same. For example, randomly guessing at a true - false statistics question has only two outcomes. If a success is guessing correctly, then a failure is guessing incorrectly. Suppose Joe always

[4]This content is available online at <http://cnx.org/content/m16821/1.5/>.

[5]This content is available online at <http://cnx.org/content/m16820/1.11/>.

guesses correctly on any statistics true - false question with probability $p = 0.6$. Then, $q = 0.4$ .This means that for every true - false statistics question Joe answers, his probability of success ($p = 0.6$) and his probability of failure ($q = 0.4$) remain the same.

The outcomes of a binomial experiment fit a **binomial probability distribution**. The random variable $X =$ the number of successes obtained in the n independent trials.

The mean, μ, and variance, σ^2, for the binomial probability distribution is $\mu = np$ and $\sigma^2 = npq$. The standard deviation, σ, is then $\sigma = \sqrt{npq}$.

Any experiment that has characteristics 2 and 3 is called a **Bernoulli Trial** (named after Jacob Bernoulli who, in the late 1600s, studied them extensively). A binomial experiment takes place when the number of successes are counted in one or more Bernoulli Trials.

Example 4.6
At ABC College, the withdrawal rate from an elementary physics course is 30% for any given term. This implies that, for any given term, 70% of the students stay in the class for the entire term. A "success" could be defined as an individual who withdrew. The random variable is $X =$ the number of students who withdraw from the elementary physics course per term.

Example 4.7
Suppose you play a game that you can only either win or lose. The probability that you win any game is 55% and the probability that you lose is 45%. If you play the game 20 times, what is the probability that you win 15 of the 20 games? Here, if you define $X =$ the number of wins, then X takes on the values $X = 0, 1, 2, 3, ..., 20$. The probability of a success is $p = 0.55$. The probability of a failure is $q = 0.45$. The number of trials is $n = 20$. The probability question can be stated mathematically as $P(X = 15)$.

Example 4.8
A fair coin is flipped 15 times. What is the probability of getting more than 10 heads? Let $X =$ the number of heads in 15 flips of the fair coin. X takes on the values $x = 0, 1, 2, 3, ..., 15$. Since the coin is fair, $p = 0.5$ and $q = 0.5$. The number of trials is $n = 15$. The probability question can be stated mathematically as $P(X > 10)$.

Example 4.9
Approximately 70% of statistics students do their homework in time for it to be collected and graded. In a statistics class of 50 students, what is the probability that at least 40 will do their homework on time?

Problem 1 *(Solution on p. 182.)*
This is a binomial problem because there is only a success or a __________, there are a definite number of trials, and the probability of a success is 0.70 for each trial.

Problem 2 *(Solution on p. 182.)*
If we are interested in the number of students who do their homework, then how do we define X?

Problem 3 *(Solution on p. 182.)*
What values does X take on?

Problem 4 *(Solution on p. 182.)*
What is a "failure", in words?

The probability of a success is $p = 0.70$. The number of trial is $n = 50$.

Problem 5 *(Solution on p. 182.)*
If $p + q = 1$, then what is q?

Problem 6 *(Solution on p. 182.)*

The words "at least" translate as what kind of inequality?

The probability question is $P(X \geq 40)$.

4.5.1 Notation for the Binomial: B = Binomial Probability Distribution Function

$X \sim B(n, p)$

Read this as "X is a random variable with a binomial distribution." The parameters are n and p. n = number of trials p = probability of a success on each trial

Example 4.10

It has been stated that about 41% of adult workers have a high school diploma but do not pursue any further education. If 20 adult workers are randomly selected, find the probability that at most 12 of them have a high school diploma but do not pursue any further education. How many adult workers do you expect to have a high school diploma but do not pursue any further education?

Let X = the number of workers who have a high school diploma but do not pursue any further education.

X takes on the values 0, 1, 2, ..., 20 where $n = 20$ and $p = 0.41$. $q = 1 - 0.41 = 0.59$. $X \sim B(20, 0.41)$

Find $P(X \leq 12)$. $P(X \leq 12) = 0.9738$. (calculator or computer)

Using the TI-83+ or the TI-84 calculators, the calculations are as follows. Go into 2nd DISTR. The syntax for the instructions are

To calculate (X **= value): binompdf(**n**,** p**, number)** If "number" is left out, the result is the binomial probability table.

To calculate $P(X \leq value)$**: binomcdf(**n**,** p**, number)** If "number" is left out, the result is the cumulative binomial probability table.

For this problem: After you are in 2nd DISTR, arrow down to A:binomcdf. Press ENTER. Enter 20,.41,12). The result is $P(X \leq 12) = 0.9738$**.**

NOTE: If you want to find $P(X = 12)$, use the pdf (0:binompdf). If you want to find $P(X > 12)$, use 1 - binomcdf(20,.41,12).

The probability at most 12 workers have a high school diploma but do not pursue any further education is 0.9738

The graph of $X \sim B(20, 0.41)$ is:

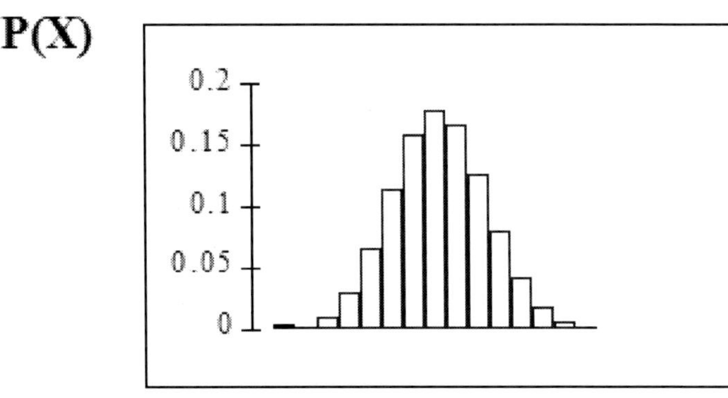

0 1 2 3 4 5............. .20

The y-axis contains the probability of X, where X = the number of workers who have only a high school diploma.

The number of adult workers that you expect to have a high school diploma but not pursue any further education is the mean, $\mu = np = (20)\,(0.41) = 8.2$.

The formula for the variance is $\sigma^2 = npq$. The standard deviation is $\sigma = \sqrt{npq}$. $\sigma = \sqrt{(20)\,(0.41)\,(0.59)} = 2.20$.

Example 4.11
The following example illustrates a problem that is **not** binomial. It violates the condition of independence. ABC College has a student advisory committee made up of 10 staff members and 6 students. The committee wishes to choose a chairperson and a recorder. What is the probability that the chairperson and recorder are both students? All names of the committee are put into a box and two names are drawn **without replacement**. The first name drawn determines the chairperson and the second name the recorder. There are two trials. However, the trials are not independent because the outcome of the first trial affects the outcome of the second trial. The probability of a student on the first draw is $\frac{6}{16}$. The probability of a student on the second draw is $\frac{5}{15}$, when the first draw produces a student. The probability is $\frac{6}{15}$ when the first draw produces a staff member. The probability of drawing a student's name changes for each of the trials and, therefore, violates the condition of independence.

4.6 Geometric (optional)[6]

The characteristics of a geometric experiment are:

1. There are one or more Bernoulli trials with all failures except the last one, which is a success. In other words, you keep repeating what you are doing until the first success. Then you stop. For example, you throw a dart at a bull's eye until you hit the bull's eye. The first time you hit the bull's eye is a "success" so you stop throwing the dart. It might take you 6 tries until you hit the bull's eye. You can think of the trials as failure, failure, failure, failure, failure, success. STOP.
2. In theory, the number of trials could go on forever. There must be at least one trial.
3. The probability,p, of a success and the probability, q, of a failure is the same for each trial. $p + q = 1$ and $q = 1 - p$. For example, the probability of rolling a 3 when you throw one fair die is $\frac{1}{6}$. This is true no matter how many times you roll the die. Suppose you want to know the probability of getting

[6]This content is available online at <http://cnx.org/content/m16822/1.13/>.

the first 3 on the fifth roll. On rolls 1, 2, 3, and 4, you do not get a face with a 3. The probability for each of rolls 1, 2, 3, and 4 is $q = \frac{5}{6}$, the probability of a failure. The probability of getting a 3 on the fifth roll is $\frac{5}{6} \cdot \frac{5}{6} \cdot \frac{5}{6} \cdot \frac{5}{6} \cdot \frac{1}{6} \cdot = 0.0804$

The outcomes of a geometric experiment fit a geometric probability distribution. The random variable X = the number of independent trials until the first success. The mean and variance are in the summary in this chapter.

Example 4.12
You play a game of chance that you can either win or lose (there are no other possibilities) **until** you lose. Your probability of losing is $p = 0.57$. What is the probability that it takes 5 games until you lose? Let X = the number of games you play until you lose (includes the losing game). Then X takes on the values 1, 2, 3, ... (could go on indefinitely). The probability question is $P(X = 5)$.

Example 4.13
A safety engineer feels that 35% of all industrial accidents in her plant are caused by failure of employees to follow instructions. She decides to look at the accident reports **until** she finds one that shows an accident caused by failure of employees to follow instructions. On the average, how many reports would the safety engineer **expect** to look at until she finds a report showing an accident caused by employee failure to follow instructions? What is the probability that the safety engineer will have to examine at least 3 reports until she finds a report showing an accident caused by employee failure to follow instructions?

Let X = the number of accidents the safety engineer must examine **until** she finds a report showing an accident caused by employee failure to follow instructions. X takes on the values 1, 2, 3, ... The first question asks you to find the **expected value** or the mean. The second question asks you to find $P(X \geq 3)$. ("At least" translates as a "greater than or equal to" symbol).

Example 4.14
Suppose that you are looking for a chemistry lab partner. The probability that someone agrees to be your lab partner is 0.55. Since you need a lab partner very soon, you ask every chemistry student you are acquainted with **until** one says that he/she will be your lab partner. What is the probability that the fourth person says yes?

This is a geometric problem because you may have a number of failures before you have the one success you desire. Also, the probability of a success stays the same each time you ask a chemistry student to be your lab partner. There is no definite number of trials (number of times you ask a chemistry student to be your partner).

Problem 1
Let X = the number of ____________ you must ask ____________ one says yes.

Solution
Let X = the number of **chemistry students** you must ask **until** one says yes.

Problem 2 *(Solution on p. 182.)*
What values does X take on?

Problem 3 *(Solution on p. 182.)*
What are p and q?

Problem 4 *(Solution on p. 183.)*
The probability question is P(_______).

4.6.1 Notation for the Geometric: G = Geometric Probability Distribution Function

$X \sim G(p)$

Read this as "X is a random variable with a geometric distribution." The parameter is p. p = the probability of a success for each trial.

Example 4.15

Assume that the probability of a defective computer component is 0.02. Find the probability that the first defect is caused by the 7th component tested. How many components do you expect to test until one is found to be defective?

Let X = the number of computer components tested until the first defect is found.

X takes on the values 1, 2, 3, ... where $p = 0.02$. $X \sim G(0.02)$

Find $P(X = 7)$. $P(X = 7) = 0.0177$. (calculator or computer)

TI-83+ and TI-84: **For a general discussion, see this example (binomial)**. The syntax is similar. The geometric parameter list is (p, number) If "number" is left out, the result is the geometric probability table. For this problem: **After you are in 2nd DISTR, arrow down to D:geometpdf. Press ENTER. Enter .02,7). The result is** $P(X = 7) = 0.0177$.

The probability that the 7th component is the first defect is 0.0177.

The graph of $X \sim G(0.02)$ is:

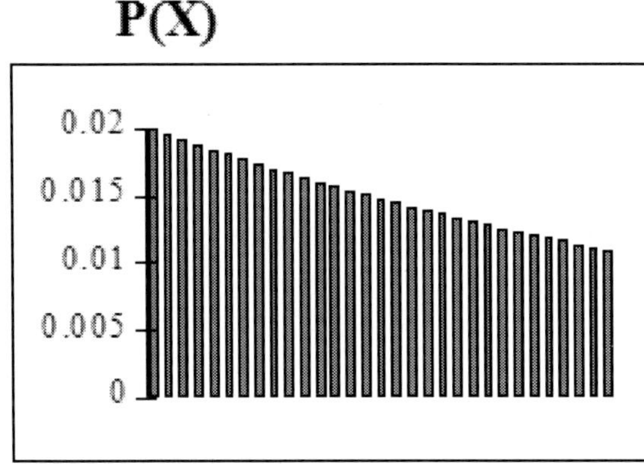

The y-axis contains the probability of X, where X = the number of computer components tested.

The number of components that you would expect to test until you find the first defective one is the mean, $\mu = 50$.

The formula for the mean is $\mu = \frac{1}{p} = \frac{1}{0.02} = 50$

The formula for the variance is $\sigma^2 = \frac{1}{p} \cdot \left(\frac{1}{p} - 1\right) = \frac{1}{0.02} \cdot \left(\frac{1}{0.02} - 1\right) = 2450$

The standard deviation is $\sigma = \sqrt{\frac{1}{p} \cdot \left(\frac{1}{p} - 1\right)} = \sqrt{\frac{1}{0.02} \cdot \left(\frac{1}{0.02} - 1\right)} = 49.5$

4.7 Hypergeometric (optional)[7]

The characteristics of a hypergeometric experiment are:

1. You take samples from **2** groups.
2. You are concerned with a group of interest, called the first group.
3. You sample **without replacement** from the combined groups. For example, you want to choose a softball team from a combined group of 11 men and 13 women. The team consists of 10 players.
4. Each pick is **not** independent, since sampling is without replacement. In the softball example, the probability of picking a women first is $\frac{13}{24}$. The probability of picking a man second is $\frac{11}{23}$ if a woman was picked first. It is $\frac{10}{23}$ if a man was picked first. The probability of the second pick depends on what happened in the first pick.
5. You are **not** dealing with Bernoulli Trials.

The outcomes of a hypergeometric experiment fit a **hypergeometric probability** distribution. The random variable X = the number of items from the group of interest. The mean and variance are given in the summary.

Example 4.16
A candy dish contains 100 jelly beans and 80 gumdrops. Fifty candies are picked at random. What is the probability that 35 of the 50 are gumdrops? The two groups are jelly beans and gumdrops. Since the probability question asks for the probability of picking gumdrops, the group of interest (first group) is gumdrops. The size of the group of interest (first group) is 80. The size of the second group is 100. The size of the sample is 50 (jelly beans or gumdrops). Let X = the number of gumdrops in the sample of 50. X takes on the values $x = 0, 1, 2, ..., 50$. The probability question is $P(X = 35)$.

Example 4.17
Suppose a shipment of 100 VCRs is known to have 10 defective VCRs. An inspector chooses 12 for inspection. He is interested in determining the probability that, among the 12, at most 2 are defective. The two groups are the 90 non-defective VCRs and the 10 defective VCRs. The group of interest (first group) is the defective group because the probability question asks for the probability of at most 2 defective VCRs. The size of the sample is 12 VCRs. (They may be non-defective or defective.) Let X = the number of defective VCRs in the sample of 12. X takes on the values 0, 1, 2, ..., 10. X may not take on the values 11 or 12. The sample size is 12, but there are only 10 defective VCRs. The inspector wants to know $P(X \leq 2)$ ("At most" means "less than or equal to").

Example 4.18
You are president of an on-campus special events organization. You need a committee of 7 to plan a special birthday party for the president of the college. Your organization consists of 18 women and 15 men. You are interested in the number of men on your committee. What is the probability that your committee has more than 4 men?

This is a hypergeometric problem because you are choosing your committee from two groups (men and women).

Problem 1 *(Solution on p. 183.)*
Are you choosing with or without replacement?

Problem 2 *(Solution on p. 183.)*
What is the group of interest?

Problem 3 *(Solution on p. 183.)*
How many are in the group of interest?

[7]This content is available online at <http://cnx.org/content/m16824/1.12/>.

Problem 4 *(Solution on p. 183.)*

How many are in the other group?

Problem 5 *(Solution on p. 183.)*

Let X = _________ on the committee. What values does X take on?

Problem 6 *(Solution on p. 183.)*

The probability question is *P(_______).*

4.7.1 Notation for the Hypergeometric: H = Hypergeometric Probability Distribution Function

$X \sim H\,(r,\ b,\ n)$

Read this as "X is a random variable with a hypergeometric distribution." The parameters are *r*, *b*, and *n*. *r* = the size of the group of interest (first group), *b* = the size of the second group, *n* = the size of the chosen sample

Example 4.19

A school site committee is to be chosen from 6 men and 5 women. If the committee consists of 4 members, what is the probability that 2 of them are men? How many men do you expect to be on the committee?

Let X = the number of men on the committee of 4. The men are the group of interest (first group).

X takes on the values 0, 1, 2, 3, 4, where $r = 6, b = 5$, and $n = 4$. $X \sim H(6,5,4)$

Find $P\,(X = 2)$. $P\,(X = 2) = 0.4545$(calculator or computer)

NOTE: Currently, the TI-83+ and TI-84 do not have hypergeometric probability functions. There are a number of computer packages, including Microsoft Excel, that do.

The probability that there are 2 men on the committee is about 0.45.

The graph of $X \sim H(6,5,4)$ is:

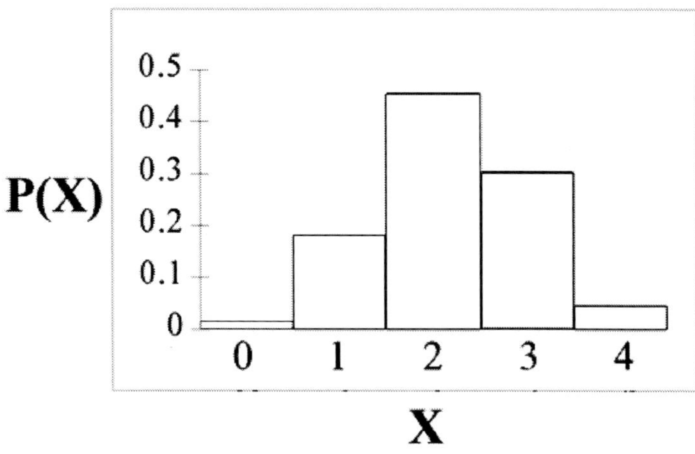

The *y*-axis contains the probability of X, where X = the number of men on the committee.

You would expect $m = 2.18$(about 2) men on the committee.

The formula for the mean is $\mu = \frac{n \cdot r}{r+b} = \frac{4 \cdot 6}{6+5} = 2.18$

The formula for the variance is fairly complex. You will find it in the Summary of the Discrete Probability Functions Chapter (Section 4.9).

4.8 Poisson[8]

Characteristics of a Poisson experiment are:

1. You are interested in the number of times something happens in a certain **interval**. For example, a book editor might be interested in the number of words spelled incorrectly in a particular book. It might be that, on the average, there are 5 words spelled incorrectly in 100 pages. The interval is the 100 pages.
2. The Poisson may be derived from the binomial if the probability of success is "small" (such as 0.01) and the number of trials is "large" (such as 1000). You will verify the relationship in the homework exercises. n is the number of trials and p is the probability of a "success."

The outcomes of a Poisson experiment fit a **Poisson probability distribution**. The random variable $X =$ the number of occurrences in the interval of interest. The mean and variance are given in the summary.

Example 4.20
The average number of loaves of bread put on a shelf in a bakery in a half-hour period is 12. What is the probability that the number of loaves put on the shelf in 5 minutes is 3? Of interest is the number of loaves of bread put on the shelf in 5 minutes. The time interval of interest is 5 minutes.

Let X = the number of loaves of bread put on the shelf in 5 minutes. If the average number of loaves put on the shelf in 30 minutes (half-hour) is 12, **then the average number of loaves put on the shelf in 5 minutes is**

$\left(\frac{5}{30}\right) \cdot 12 = 2$ loaves of bread

The probability question asks you to find $P(X = 3)$.

Example 4.21
A certain bank expects to receive 6 bad checks per day. What is the probability of the bank getting fewer than 5 bad checks on any given day? Of interest is the number of checks the bank receives in 1 day, so the time interval of interest is 1 day. Let X = the number of bad checks the bank receives in one day. If the bank expects to receive 6 bad checks per day then the average is 6 checks per day. The probability question asks for $P(X < 5)$.

Example 4.22
Your math instructor expects you to complete 2 pages of written math homework every day. What is the probability that you complete more than 2 pages a day?

This is a Poisson problem because your instructor is interested in knowing the number of pages of written math homework you complete in a day.

Problem 1 ***(Solution on p. 183.)***
What is the interval of interest?

Problem 2 ***(Solution on p. 183.)***
What is the average number of pages you should do in one day?

8This content is available online at <http://cnx.org/content/m16829/1.11/>.

Problem 3 *(Solution on p. 183.)*

Let X = ___________. What values does X take on?

Problem 4 *(Solution on p. 183.)*

The probability question is *P(______)*.

4.8.1 Notation for the Poisson: P = Poisson Probability Distribution Function

$X \sim P(\mu)$

Read this as "X is a random variable with a Poisson distribution." The parameter is μ (or λ). μ (or λ) = the mean for the interval of interest.

Example 4.23

Leah's answering machine receives about 6 telephone calls between 8 a.m. and 10 a.m. What is the probability that Leah receives more than 1 call **in the next 15 minutes?**

Let X = the number of calls Leah receives in 15 minutes. (The **interval of interest** is 15 minutes or $\frac{1}{4}$ hour.)

X takes on the values 0, 1, 2, 3, ...

If Leah receives, on the average, 6 telephone calls in 2 hours, and there are eight 15 minutes intervals in 2 hours, then Leah receives

$\frac{1}{8} \cdot 6 = 0.75$

calls in 15 minutes, on the average. So, μ = 0.75 for this problem.

$X \sim P(0.75)$

Find $P(X > 1)$. $P(X > 1) = 0.1734$ (calculator or computer)

TI-83+ and TI-84: For a general discussion, see **this example (Binomial)**. The syntax is similar. The Poisson parameter list is (μ for the interval of interest, number). **For this problem:**

Press 1- and then press 2nd DISTR. Arrow down to C:poissoncdf. Press ENTER. Enter .75,1). The result is $P(X > 1) = 0.1734$**. NOTE: The TI calculators use** λ **(lambda) for the mean.**

The probability that Leah receives more than 1 telephone call in the next fifteen minutes is about 0.1734.

The graph of $X \sim P(0.75)$ is:

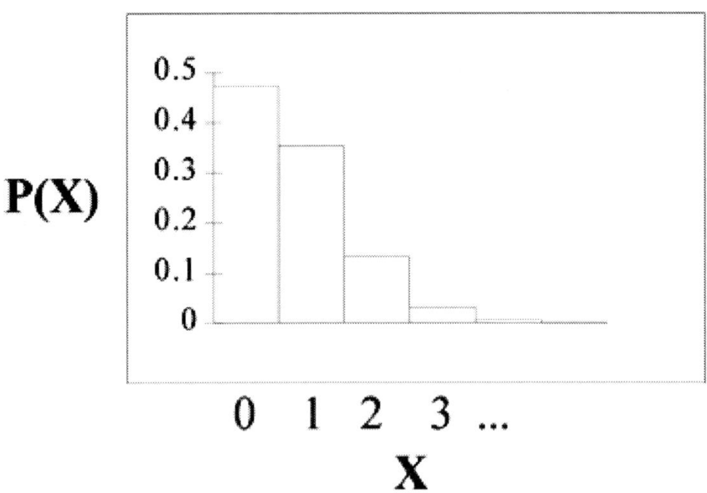

The y-axis contains the probability of X where X = the number of calls in 15 minutes.

4.9 Summary of Functions[9]

Formula 4.1: Binomial
$X{\sim}B\left(n,p\right)$

X = the number of successes in n independent trials

n = the number of independent trials

X takes on the values $x = 0,1, 2, 3, ...,n$

p = the probability of a success for any trial

q = the probability of a failure for any trial

$p+q=1 \quad q=1-p$

The mean is $\mu = np$. The standard deviation is $\sigma = \sqrt{npq}$.
Formula 4.2: Geometric
$X{\sim}G\left(p\right)$

X = the number of independent trials until the first success (count the failures and the first success)

X takes on the values x= 1, 2, 3, ...

p = the probability of a success for any trial

q = the probability of a failure for any trial

$p+q=1$

$q=1-p$

The mean is $\mu = \frac{1}{p}$

The standard deviation is $\sigma = \sqrt{\frac{1}{p}\left(\left(\frac{1}{p}\right)-1\right)}$
Formula 4.3: Hypergeometric
$X{\sim}H\left(r,b,n\right)$

X = the number of items from the group of interest that are in the chosen sample.

X may take on the values x= 0, 1, ..., up to the size of the group of interest. (The minimum value for X may be larger than 0 in some instances.)

r = the size of the group of interest (first group)

b= the size of the second group

n= the size of the chosen sample.

$n \le r+b$

The mean is: $\mu = \frac{nr}{r+b}$

[9]This content is available online at <http://cnx.org/content/m16833/1.8/>.

The standard deviation is: $\sigma = \sqrt{\frac{rbn(r+b+n)}{(r+b)^2(r+b-1)}}$

Formula 4.4: Poisson
$X \sim P(\mu)$

X = the number of occurrences in the interval of interest

X takes on the values $x = 0, 1, 2, 3, ...$

The mean μ is typically given. (λ is often used as the mean instead of μ.) When the Poisson is used to approximate the binomial, we use the binomial mean $\mu = np$. n is the binomial number of trials. p = the probability of a success for each trial. This formula is valid when n is "large" and p "small" (a general rule is that n should be greater than or equal to 20 and p should be less than or equal to 0.05). If n is large enough and p is small enough then the Poisson approximates the binomial very well. The standard deviation is $\sigma = \mu$.

4.10 Practice 1: Discrete Distribution[10]

4.10.1 Student Learning Objectives

- The student will investigate the properties of a discrete distribution.

4.10.2 Given:

A ballet instructor is interested in knowing what percent of each year's class will continue on to the next, so that she can plan what classes to offer. Over the years, she has established the following probability distribution.

- Let X = the number of years a student will study ballet with the teacher.
- Let $P(X = x)$ = the probability that a student will study ballet x years.

4.10.3 Organize the Data

Complete the table below using the data provided.

x	P(X=x)	x*P(X=x)
1	0.10	
2	0.05	
3	0.10	
4		
5	0.30	
6	0.20	
7	0.10	

Table 4.9

Exercise 4.10.1
In words, define the Random Variable X.

Exercise 4.10.2
$P(X = 4) =$

Exercise 4.10.3
$P(X < 4) =$

Exercise 4.10.4
On average, how many years would you expect a child to study ballet with this teacher?

4.10.4 Discussion Question

Exercise 4.10.5
What does the column "*P(X=x)*" sum to and why?

Exercise 4.10.6
What does the column "$x * P(X=x)$" sum to and why?

---

[10]This content is available online at <http://cnx.org/content/m16830/1.12/>.

4.11 Practice 2: Binomial Distribution[11]

4.11.1 Student Learning Outcomes

- The student will practice constructing Binomial Distributions.

4.11.2 Given

The Higher Education Research Institute at UCLA surveyed more than 263,000 incoming freshmen from 385 colleges. 36.7% of first-generation college students expected to work fulltime while in college. *(Source: Eric Hoover, The Chronicle of Higher Education, 2/3/2006).* Suppose that you randomly pick 8 college freshmen from the survey. You are interested in the number that expect to work full-time while in college.

4.11.3 Interpret the Data

Exercise 4.11.1 *(Solution on p. 183.)*
In words, define the random Variable X.

Exercise 4.11.2 *(Solution on p. 183.)*
$X\sim$___________

Exercise 4.11.3 *(Solution on p. 183.)*
What values does X take on?

Exercise 4.11.4
Construct the probability distribution function (PDF) for X.

x	$P(X=x)$

Table 4.10

Exercise 4.11.5 *(Solution on p. 183.)*
On average (u), how many would you expect to answer yes?

Exercise 4.11.6 *(Solution on p. 183.)*
What is the standard deviation (σ) ?

Exercise 4.11.7 *(Solution on p. 183.)*
What is the probability what at most 5 of the freshmen expect to work full-time?

[11]This content is available online at <http://cnx.org/content/m17107/1.11/>.

Exercise 4.11.8 *(Solution on p. 183.)*

What is the probability that at least 2 of the freshmen expect to work full-time?

Exercise 4.11.9

Construct a histogram or plot a line graph. Label the horizontal and vertical axes with words. Include numerical scaling.

4.12 Practice 3: Poisson Distribution[12]

4.12.1 Student Learning Objectives

- The student will investigate the properties of a Poisson distribution.

4.12.2 Given

On average, ten teens are killed in the U.S. in teen-driven autos per day (USA Today, 3/1/2005). As a result, states across the country are debating raising the driving age.

4.12.3 Interpret the Data

Exercise 4.12.1
In words, define the Random Variable X.

Exercise 4.12.2 *(Solution on p. 183.)*
$X \sim$_____________

Exercise 4.12.3 *(Solution on p. 183.)*
What values does X take on?

Exercise 4.12.4
For the given values of X, fill in the corresponding probabilities.

x	$P(X=x)$
0	
4	
8	
10	
11	
15	

Table 4.11

Exercise 4.12.5 *(Solution on p. 183.)*
Is it likely that there will be no teens killed in the U.S. in teen-driven autos on any given day? Numerically, why?

Exercise 4.12.6 *(Solution on p. 183.)*
Is it likely that there will be more than 20 teens killed in the U.S. in teen-driven autos on any given day? Numerically, why?

[12] This content is available online at <http://cnx.org/content/m17109/1.10/>.

4.13 Practice 4: Geometric Distribution[13]

4.13.1 Student Learning Objectives

- The student will investigate the properties of a geometric distribution.

4.13.2 Given:

Use the information from the Binomial Distribution Practice (Section 4.11). Suppose that you will randomly select one freshman from the study until you find one who expects to work full-time while in college. You are interested in the number of freshmen you must ask.

4.13.3 Interpret the Data

Exercise 4.13.1
In words, define the Random Variable X.

Exercise 4.13.2 *(Solution on p. 184.)*
$X \sim$

Exercise 4.13.3 *(Solution on p. 184.)*
What values does X take on?

Exercise 4.13.4
Construct the probability distribution function (PDF) for X. Stop at $X = 6$.

x	$P(X=x)$
0	
1	
2	
3	
4	
5	
6	

Table 4.12

Exercise 4.13.5 *(Solution on p. 184.)*
On average(μ), how many freshmen would you expect to have to ask until you found one who expects to work full-time while in college?

Exercise 4.13.6 *(Solution on p. 184.)*
What is the probability that you will need to ask fewer than 3 freshmen?

Exercise 4.13.7
Construct a histogram or plot a line graph. Label the horizontal and vertical axes with words. Include numerical scaling.

[13]This content is available online at <http://cnx.org/content/m17108/1.12/>.

4.14 Practice 5: Hypergeometric Distribution[14]

4.14.1 Student Learning Objectives

- The student will investigate the properties of a hypergeometric distribution.

4.14.2 Given

Suppose that a group of statistics students is divided into two groups: business majors and non-business majors. There are 16 business majors in the group and 7 non-business majors in the group. A random sample of 9 students is taken. We are interested in the number of business majors in the group.

4.14.3 Interpret the Data

Exercise 4.14.1
In words, define the Random Variable X.

Exercise 4.14.2 *(Solution on p. 184.)*
$X \sim$

Exercise 4.14.3 *(Solution on p. 184.)*
What values does X take on?

Exercise 4.14.4
Construct the probability distribution function (*PDF*) for X.

x	$P(X=x)$

Table 4.13

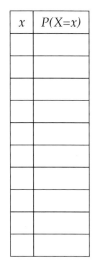

Exercise 4.14.5 *(Solution on p. 184.)*
On average(μ), how many would you expect to be business majors?

[14]This content is available online at <http://cnx.org/content/m17106/1.11/>.

4.15 Homework[15]

Exercise 4.15.1 *(Solution on p. 184.)*

1. Complete the PDF and answer the questions.

x	$P(X = x)$	$x \cdot P(X = x)$
0	0.3	
1	0.2	
2		
3	0.4	

Table 4.14

a. Find the probability that $X = 2$.
b. Find the expected value.

Exercise 4.15.2

Suppose that you are offered the following "deal." You roll a die. If you roll a 6, you win $10. If you roll a 4 or 5, you win $5. If you roll a 1, 2, or 3, you pay $6.

a. What are you ultimately interested in here (the value of the roll or the money you win)?
b. In words, define the Random Variable X.
c. List the values that X may take on.
d. Construct a PDF.
e. Over the long run of playing this game, what are your expected average winnings per game?
f. Based on numerical values, should you take the deal? Explain your decision in complete sentences.

Exercise 4.15.3 *(Solution on p. 184.)*

A venture capitalist, willing to invest $1,000,000, has three investments to choose from. The first investment, a software company, has a 10% chance of returning $5,000,000 profit, a 30% chance of returning $1,000,000 profit, and a 60% chance of losing the million dollars. The second company, a hardware company, has a 20% chance of returning $3,000,000 profit, a 40% chance of returning $1,000,000 profit, and a 40% chance of losing the million dollars. The third company, a biotech firm, has a 10% chance of returning $6,000,000 profit, a 70% of no profit or loss, and a 20% chance of losing the million dollars.

a. Construct a PDF for each investment.
b. Find the expected value for each investment.
c. Which is the safest investment? Why do you think so?
d. Which is the riskiest investment? Why do you think so?
e. Which investment has the highest expected return, on average?

Exercise 4.15.4

A theater group holds a fund-raiser. It sells 100 raffle tickets for $5 apiece. Suppose you purchase 4 tickets. The prize is 2 passes to a Broadway show, worth a total of $150.

---

[15]This content is available online at <http://cnx.org/content/m16823/1.12/>.

a. What are you interested in here?
b. In words, define the Random Variable X.
c. List the values that X may take on.
d. Construct a PDF.
e. If this fund-raiser is repeated often and you always purchase 4 tickets, what would be your expected average winnings per game?

Exercise 4.15.5 *(Solution on p. 184.)*

Suppose that 20,000 married adults in the United States were randomly surveyed as to the number of children they have. The results are compiled and are used as theoretical probabilities. Let X = the number of children

x	$P(X = x)$	$x \cdot P(X = x)$
0	0.10	
1	0.20	
2	0.30	
3		
4	0.10	
5	0.05	
6 (or more)	0.05	

Table 4.15

a. Find the probability that a married adult has 3 children.
b. In words, what does the expected value in this example represent?
c. Find the expected value.
d. Is it more likely that a married adult will have 2 – 3 children or 4 – 6 children? How do you know?

Exercise 4.15.6

Suppose that the PDF for the number of years it takes to earn a Bachelor of Science (B.S.) degree is given below.

x	$P(X = x)$
3	0.05
4	0.40
5	0.30
6	0.15
7	0.10

Table 4.16

a. In words, define the Random Variable X.
b. What does it mean that the values 0, 1, and 2 are not included for X on the PDF?
c. On average, how many years do you expect it to take for an individual to earn a B.S.?

4.15.1 For each problem:

a. In words, define the Random Variable X.
b. List the values hat X may take on.
c. Give the distribution of X. $X\sim$

Then, answer the questions specific to each individual problem.

Exercise 4.15.7 *(Solution on p. 184.)*
Six different colored dice are rolled. Of interest is the number of dice that show a "1."

d. On average, how many dice would you expect to show a "1"?
e. Find the probability that all six dice show a "1."
f. Is it more likely that 3 or that 4 dice will show a "1"? Use numbers to justify your answer numerically.

Exercise 4.15.8
According to a 2003 publication by Waits and Lewis *(source: http://nces.ed.gov/pubs2003/2003017.pdf*[16] *)*, by the end of 2002, 92% of U.S. public two-year colleges offered distance learning courses. Suppose you randomly pick 13 U.S. public two-year colleges. We are interested in the number that offer distance learning courses.

d. On average, how many schools would you expect to offer such courses?
e. Find the probability that at most 6 offer such courses.
f. Is it more likely that 0 or that 13 will offer such courses? Use numbers to justify your answer numerically and answer in a complete sentence.

Exercise 4.15.9 *(Solution on p. 184.)*
A school newspaper reporter decides to randomly survey 12 students to see if they will attend Tet festivities this year. Based on past years, she knows that 18% of students attend Tet festivities. We are interested in the number of students who will attend the festivities.

d. How many of the 12 students do we expect to attend the festivities?
e. Find the probability that at most 4 students will attend.
f. Find the probability that more than 2 students will attend.

Exercise 4.15.10
Suppose that about 85% of graduating students attend their graduation. A group of 22 graduating students is randomly chosen.

d. How many are expected to attend their graduation?
e. Find the probability that 17 or 18 attend.
f. Based on numerical values, would you be surprised if all 22 attended graduation? Justify your answer numerically.

Exercise 4.15.11 *(Solution on p. 185.)*
At The Fencing Center, 60% of the fencers use the foil as their main weapon. We randomly survey 25 fencers at The Fencing Center. We are interested in the numbers that do **not** use the foil as their main weapon.

d. How many are expected to **not** use the foil as their main weapon?
e. Find the probability that six do **not** use the foil as their main weapon.

[16] http://nces.ed.gov/pubs2003/2003017.pdf

f. Based on numerical values, would you be surprised if all 25 did **not** use foil as their main weapon? Justify your answer numerically.

Exercise 4.15.12

Approximately 8% of students at a local high school participate in after-school sports all four years of high school. A group of 60 seniors is randomly chosen. Of interest is the number that participated in after-school sports all four years of high school.

d. How many seniors are expected to have participated in after-school sports all four years of high school?
e. Based on numerical values, would you be surprised if none of the seniors participated in after-school sports all four years of high school? Justify your answer numerically.
f. Based upon numerical values, is it more likely that 4 or that 5 of the seniors participated in after-school sports all four years of high school? Justify your answer numerically.

Exercise 4.15.13 ***(Solution on p. 185.)***

The chance of having an extra fortune in a fortune cookie is about 3%. Given a bag of 144 fortune cookies, we are interested in the number of cookies with an extra fortune. Two distributions may be used to solve this problem. Use one distribution to solve the problem.

d. How many cookies do we expect to have an extra fortune?
e. Find the probability that none of the cookies have an extra fortune.
f. Find the probability that more than 3 have an extra fortune.
g. As n increases, what happens involving the probabilities using the two distributions? Explain in complete sentences.

Exercise 4.15.14

There are two games played for Chinese New Year and Vietnamese New Year. They are almost identical. In the Chinese version, fair dice with numbers 1, 2, 3, 4, 5, and 6 are used, along with a board with those numbers. In the Vietnamese version, fair dice with pictures of a gourd, fish, rooster, crab, crayfish, and deer are used. The board has those six objects on it, also. We will play with bets being $1. The player places a bet on a number or object. The "house" rolls three dice. If none of the dice show the number or object that was bet, the house keeps the $1 bet. If one of the dice shows the number or object bet (and the other two do not show it), the player gets back his $1 bet, plus $1 profit. If two of the dice show the number or object bet (and the third die does not show it), the player gets back his $1 bet, plus $2 profit. If all three dice show the number or object bet, the player gets back his $1 bet, plus $3 profit.

Let X = number of matches and Y= profit per game.

d. List the values that Y may take on. Then, construct one PDF table that includes both X & Y and their probabilities.
e. Calculate the average expected matches over the long run of playing this game for the player.
f. Calculate the average expected earnings over the long run of playing this game for the player.
g. Determine who has the advantage, the player or the house.

Exercise 4.15.15 ***(Solution on p. 185.)***

According to the South Carolina Department of Mental Health web site, for every 200 U.S. women, the average number who suffer from anorexia is one (*http://www.state.sc.us/dmh/anorexia/statistics.htm*[17]). Out of a randomly chosen group of 600 U.S. women:

[17] http://www.state.sc.us/dmh/anorexia/statistics.htm

d. How many are expected to suffer from anorexia?
e. Find the probability that no one suffers from anorexia.
f. Find the probability that more than four suffer from anorexia.

Exercise 4.15.16
The average number of children of middle-aged Japanese couples is 2.09 *(Source: The Yomiuri Shimbun, June 28, 2006).* Suppose that one middle-aged Japanese couple is randomly chosen.

d. Find the probability that they have no children.
e. Find the probability that they have fewer children than the Japanese average.
f. Find the probability that they have more children than the Japanese average .

Exercise 4.15.17 *(Solution on p. 185.)*
The average number of children per Spanish couples was 1.34 in 2005. Suppose that one Spanish couple is randomly chosen. *(Source: http://www.typicallyspanish.com/news/publish/article_4897.shtml*[18] *, June 16, 2006).*

d. Find the probability that they have no children.
e. Find the probability that they have fewer children than the Spanish average.
f. Find the probability that they have more children than the Spanish average .

Exercise 4.15.18
Fertile (female) cats produce an average of 3 litters per year. *(Source: The Humane Society of the United States).* Suppose that one fertile, female cat is randomly chosen. In one year, find the probability she produces:

d. No litters.
e. At least 2 litters.
f. Exactly 3 litters.

Exercise 4.15.19 *(Solution on p. 185.)*
A consumer looking to buy a used red Miata car will call dealerships until she finds a dealership that carries the car. She estimates the probability that any independent dealership will have the car will be 28%. We are interested in the number of dealerships she must call.

d. On average, how many dealerships would we expect her to have to call until she finds one that has the car?
e. Find the probability that she must call at most 4 dealerships.
f. Find the probability that she must call 3 or 4 dealerships.

Exercise 4.15.20
Suppose that the probability that an adult in America will watch the Super Bowl is 40%. Each person is considered independent. We are interested in the number of adults in America we must survey until we find one who will watch the Super Bowl.

d. How many adults in America do you expect to survey until you find one who will watch the Super Bowl?
e. Find the probability that you must ask 7 people.
f. Find the probability that you must ask 3 or 4 people.

---
[18] http://www.typicallyspanish.com/news/publish/article_4897.shtml

Exercise 4.15.21 *(Solution on p. 185.)*

A group of Martial Arts students is planning on participating in an upcoming demonstration. 6 are students of Tae Kwon Do; 7 are students of Shotokan Karate. Suppose that 8 students are randomly picked to be in the first demonstration. We are interested in the number of Shotokan Karate students in that first demonstration.

d. How many Shotokan Karate students do we expect to be in that first demonstration?
e. Find the probability that 4 students of Shotokan Karate are picked.
f. Find the probability that no more than 6 students of Shotokan Karate are picked.

Exercise 4.15.22

The chance of a IRS audit for a tax return with over $25,000 in income is about 2% per year. We are interested in the expected number of audits a person with that income has in a 20 year period. Assume each year is independent.

d. How many audits are expected in a 20 year period?
e. Find the probability that a person is not audited at all.
f. Find the probability that a person is audited more than twice.

Exercise 4.15.23 *(Solution on p. 185.)*

Refer to the previous problem. Suppose that 100 people with tax returns over $25,000 are randomly picked. We are interested in the number of people audited in 1 year. One way to solve this problem is by using the Binomial Distribution. Since n is large and p is small, another discrete distribution could be used to solve the following problems. Solve the following questions (d-f) using that distribution.

d. How many are expected to be audited?
e. Find the probability that no one was audited.
f. Find the probability that more than 2 were audited.

Exercise 4.15.24

Suppose that a technology task force is being formed to study technology awareness among instructors. Assume that 10 people will be randomly chosen to be on the committee from a group of 28 volunteers, 20 who are technically proficient and 8 who are not. We are interested in the number on the committee who are **not** technically proficient.

d. How many instructors do you expect on the committee who are **not** technically proficient?
e. Find the probability that at least 5 on the committee are not technically proficient.
f. Find the probability that at most 3 on the committee are not technically proficient.

Exercise 4.15.25 *(Solution on p. 186.)*

Refer back to Exercise 4.15.12. Solve this problem again, using a different, though still acceptable, distribution.

Exercise 4.15.26

Suppose that 9 Massachusetts athletes are scheduled to appear at a charity benefit. The 9 are randomly chosen from 8 volunteers from the Boston Celtics and 4 volunteers from the New England Patriots. We are interested in the number of Patriots picked.

d. Is it more likely that there will be 2 Patriots or 3 Patriots picked?
e. What is the probability that all of the volunteers will be from the Celtics
f. Is it more likely that more of the volunteers will be from the Patriots or from the Celtics? How do you know?

Exercise 4.15.27 *(Solution on p. 186.)*

On average, Pierre, an amateur chef, drops 3 pieces of egg shell into every 2 batters of cake he makes. Suppose that you buy one of his cakes.

d. On average, how many pieces of egg shell do you expect to be in the cake?
e. What is the probability that there will not be any pieces of egg shell in the cake?
f. Let's say that you buy one of Pierre's cakes each week for 6 weeks. What is the probability that there will not be any egg shell in any of the cakes?
g. Based upon the average given for Pierre, is it possible for there to be 7 pieces of shell in the cake? Why?

Exercise 4.15.28

It has been estimated that only about 30% of California residents have adequate earthquake supplies. Suppose we are interested in the number of California residents we must survey until we find a resident who does **not** have adequate earthquake supplies.

d. What is the probability that we must survey just 1 or 2 residents until we find a California resident who does not have adequate earthquake supplies?
e. What is the probability that we must survey at least 3 California residents until we find a California resident who does not have adequate earthquake supplies?
f. How many California residents do you expect to need to survey until you find a California resident who **does not** have adequate earthquake supplies?
g. How many California residents do you expect to need to survey until you find a California resident who **does** have adequate earthquake supplies?

Exercise 4.15.29 *(Solution on p. 186.)*

Refer to the above problem. Suppose you randomly survey 11 California residents. We are interested in the number who have adequate earthquake supplies.

d. What is the probability that at least 8 have adequate earthquake supplies?
e. Is it more likely that none or that all of the residents surveyed will have adequate earthquake supplies? Why?
f. How many residents do you expect will have adequate earthquake supplies?

The next 3 questions refer to the following: In one of its Spring catalogs, L.L. Bean® advertised footwear on 29 of its 192 catalog pages.

Exercise 4.15.30

Suppose we randomly survey 20 pages. We are interested in the number of pages that advertise footwear. Each page may be picked at most once.

d. How many pages do you expect to advertise footwear on them?
e. Is it probable that all 20 will advertise footwear on them? Why or why not?
f. What is the probability that less than 10 will advertise footwear on them?

Exercise 4.15.31 *(Solution on p. 186.)*

Suppose we randomly survey 20 pages. We are interested in the number of pages that advertise footwear. This time, each page may be picked more than once.

d. How many pages do you expect to advertise footwear on them?
e. Is it probable that all 20 will advertise footwear on them? Why or why not?
f. What is the probability that less than 10 will advertise footwear on them?

g. Suppose that a page may be picked more than once. We are interested in the number of pages that we must randomly survey until we find one that has footwear advertised on it. Define the random variable X and give its distribution.
h. Do you expect to survey more than 10 pages in order to find one that advertises footwear on it? Why?
i. What is the probability that you only need to survey at most 3 pages in order to find one that advertises footwear on it?
j. How many pages do you expect to need to survey in order to find one that advertises footwear?

Exercise 4.15.32
Suppose that you roll a fair die until each face has appeared at least once. It does not matter in what order the numbers appear. Find the expected number of rolls you must make until each face has appeared at least once.

4.15.2 Try these multiple choice problems.

For the next three problems: The probability that the San Jose Sharks will win any given game is 0.3694 based on their 13 year win history of 382 wins out of 1034 games played (as of a certain date). Their 2005 schedule for November contains 12 games. Let X= number of games won in November 2005

Exercise 4.15.33 *(Solution on p. 186.)*
The expected number of wins for the month of November 2005 is:

A. 1.67
B. 12
C. $\frac{382}{1043}$
D. 4.43

Exercise 4.15.34 *(Solution on p. 186.)*
What is the probability that the San Jose Sharks win 6 games in November?

A. 0.1476
B. 0.2336
C. 0.7664
D. 0.8903

Exercise 4.15.35 *(Solution on p. 186.)*
Find the probability that the San Jose Sharks win at least 5 games in November.

A. 0.3694
B. 0.5266
C. 0.4734
D. 0.2305

For the next three questions: The average number of times per week that Mrs. Plum's cats wake her up at night because they want to play is 10. We are interested in the number of times her cats wake her up each week.

Exercise 4.15.36 *(Solution on p. 186.)*
In words, the random variable X =

A. The number of times Mrs. Plum's cats wake her up each week

B. The number of times Mrs. Plum's cats wake her up each hour
C. The number of times Mrs. Plum's cats wake her up each night
D. The number of times Mrs. Plum's cats wake her up

Exercise 4.15.37 *(Solution on p. 186.)*

Find the probability that her cats will wake me up no more than 5 times next week.

A. 0.5000
B. 0.9329
C. 0.0378
D. 0.0671

4.16 Review[19]

The next two questions refer to the following:

A recent poll concerning credit cards found that 35 percent of respondents use a credit card that gives them a mile of air travel for every dollar they charge. Thirty percent of the respondents charge more than $2000 per month. Of those respondents who charge more than $2000, 80 percent use a credit card that gives them a mile of air travel for every dollar they charge.

Exercise 4.16.1 *(Solution on p. 186.)*

What is the probability that a randomly selected respondent expected to spend more than $2000 AND use a credit card that gives them a mile of air travel for every dollar they charge?

A. $(0.30)\,(0.35)$
B. $(0.80)\,(0.35)$
C. $(0.80)\,(0.30)$
D. (0.80)

Exercise 4.16.2 *(Solution on p. 186.)*

Based upon the above information, are using a credit card that gives a mile of air travel for each dollar spent AND charging more than $2000 per month independent events?

A. Yes
B. No, and they are not mutually exclusive either
C. No, but they are mutually exclusive
D. Not enough information given to determine the answer

Exercise 4.16.3 *(Solution on p. 186.)*

A sociologist wants to know the opinions of employed adult women about government funding for day care. She obtains a list of 520 members of a local business and professional women's club and mails a questionnaire to 100 of these women selected at random. 68 questionnaires are returned. What is the population in this study?

A. All employed adult women
B. All the members of a local business and professional women's club
C. The 100 women who received the questionnaire
D. All employed women with children

The next two questions refer to the following: An article from The San Jose Mercury News was concerned with the racial mix of the 1500 students at Prospect High School in Saratoga, CA. The table summarizes the results. (Male and female values are approximate.)

			Ethnic Group		
Gender	White	Asian	Hispanic	Black	American Indian
Male	400	168	115	35	16
Female	440	132	140	40	14

Table 4.17

[19] This content is available online at <http://cnx.org/content/m16832/1.9/>.

Exercise 4.16.4 *(Solution on p. 187.)*

Find the probability that a student is Asian or Male.

Exercise 4.16.5 *(Solution on p. 187.)*

Find the probability that a student is Black given that the student is Female.

Exercise 4.16.6 *(Solution on p. 187.)*

A sample of pounds lost, in a certain month, by individual members of a weight reducing clinic produced the following statistics:

- Mean = 5 lbs.
- Median = 4.5 lbs.
- Mode = 4 lbs.
- Standard deviation = 3.8 lbs.
- First quartile = 2 lbs.
- Third quartile = 8.5 lbs.

The correct statement is:

A. One fourth of the members lost exactly 2 pounds.
B. The middle fifty percent of the members lost from 2 to 8.5 lbs.
C. Most people lost 3.5 to 4.5 lbs.
D. All of the choices above are correct.

Exercise 4.16.7 *(Solution on p. 187.)*

What does it mean when a data set has a standard deviation equal to zero?

A. All values of the data appear with the same frequency.
B. The mean of the data is also zero.
C. All of the data have the same value.
D. There are no data to begin with.

Exercise 4.16.8 *(Solution on p. 187.)*

The statement that best describes the illustration below is:

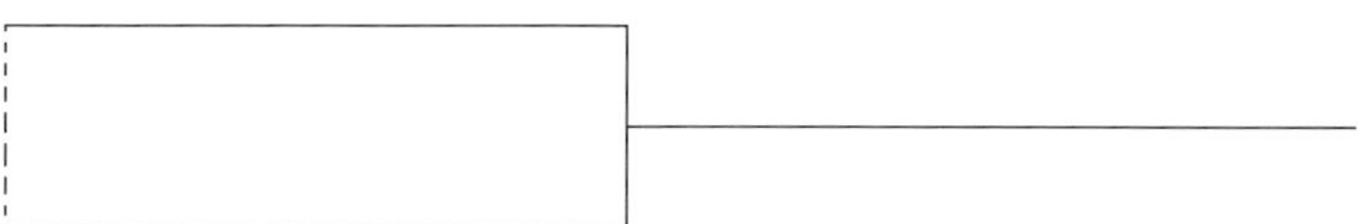

Figure 4.1

A. The mean is equal to the median.
B. There is no first quartile.
C. The lowest data value is the median.
D. The median equals $\frac{(Q1+Q3)}{2}$

Exercise 4.16.9 *(Solution on p. 187.)*

According to a recent article (San Jose Mercury News) the average number of babies born with significant hearing loss (deafness) is approximately 2 per 1000 babies in a healthy baby nursery. The number climbs to an average of 30 per 1000 babies in an intensive care nursery.

Suppose that 1000 babies from healthy nursery babies were surveyed. Find the probability that exactly 2 babies were born deaf.

Exercise 4.16.10 *(Solution on p. 187.)*

A "friend" offers you the following "deal." For a $10 fee, you may pick an envelope from a box containing 100 seemingly identical envelopes. However, each envelope contains a coupon for a free gift.

- 10 of the coupons are for a free gift worth $6.
- 80 of the coupons are for a free gift worth $8.
- 6 of the coupons are for a free gift worth $12.
- 4 of the coupons are for a free gift worth $40.

Based upon the financial gain or loss over the long run, should you play the game?

A. Yes, I expect to come out ahead in money.
B. No, I expect to come out behind in money.
C. It doesn't matter. I expect to break even.

The next four questions refer to the following: Recently, a nurse commented that when a patient calls the medical advice line claiming to have **the flu**, the chance that he/she truly has **the flu** (and not just a nasty cold) is only about 4%. Of the next 25 patients calling in claiming to have **the flu**, we are interested in how many actually have **the flu**.

Exercise 4.16.11 *(Solution on p. 187.)*

Define the Random Variable and list its possible values.

Exercise 4.16.12 *(Solution on p. 187.)*

State the distribution of X .

Exercise 4.16.13 *(Solution on p. 187.)*

Find the probability that at least 4 of the 25 patients actually have **the flu**.

Exercise 4.16.14 *(Solution on p. 187.)*

On average, for every 25 patients calling in, how many do you expect to have **the flu**?

The next two questions refer to the following: Different types of writing can sometimes be distinguished by the number of letters in the words used. A student interested in this fact wants to study the number of letters of words used by Tom Clancy in his novels. She opens a Clancy novel at random and records the number of letters of the first 250 words on the page.

Exercise 4.16.15 *(Solution on p. 187.)*

What kind of data was collected?

A. qualitative
B. quantitative - continuous
C. quantitative – discrete

Exercise 4.16.16 *(Solution on p. 187.)*

What is the population under study?

4.17 Lab 1: Discrete Distribution (Playing Card Experiment)[20]

Class Time:

Names:

4.17.1 Student Learning Outcomes:

- The student will compare empirical data and a theoretical distribution to determine if everyday experiment fits a discrete distribution.
- The student will demonstrate an understanding of long-term probabilities.

4.17.2 Supplies:

- One full deck of playing cards

4.17.3 Procedure

The experiment procedure is to pick one card from a deck of shuffled cards.

1. The theorectical probability of picking a diamond from a deck is: _________
2. Shuffle a deck of cards.
3. Pick one card from it.
4. Record whether it was a diamond or not a diamond.
5. Put the card back and reshuffle.
6. Do this a total of 10 times
7. Record the number of diamonds picked.
8. Let X = number of diamonds. Theoretically, $X \sim B$ (_____,_____)

4.17.4 Organize the Data

1. Record the number of diamonds picked for your class in the chart below. Then calculate the relative frequency.

[20]This content is available online at <http://cnx.org/content/m16827/1.10/>.

X	Frequency	Relative Frequency
0	__________	__________
1	__________	__________
2	__________	__________
3	__________	__________
4	__________	__________
5	__________	__________
6	__________	__________
7	__________	__________
8	__________	__________
9	__________	__________
10	__________	__________

Table 4.18

2. Calculate the following:
   a. $\overline{x} =$
   b. $s =$
3. Construct a histogram of the empirical data.

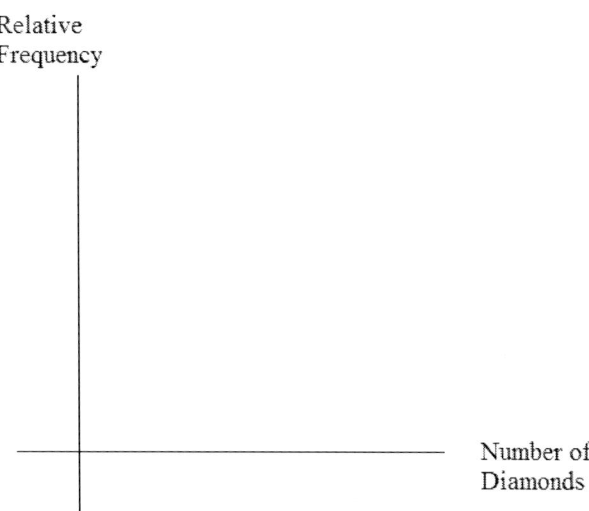

Figure 4.2

4.17.5 Theoretical Distribution

1. Build the theoretical PDF chart for X based on the distribution in the Procedure section above.

x	$P(X = x)$
0	
1	
2	
3	
4	
5	
6	
7	
8	
9	
10	

Table 4.19

2. Calculate the following:
   a. $\mu =$____________
   b. $\sigma =$____________
3. Constuct a histogram of the theoretical distribution.

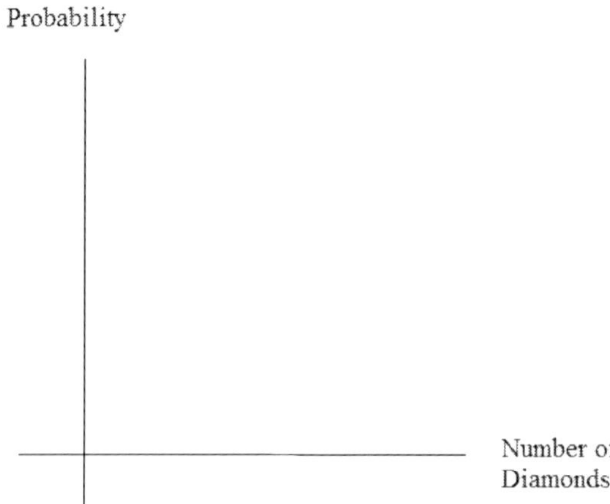

Figure 4.3

4.17.6 Using the Data

Calculate the following, rounding to 4 decimal places:

NOTE: RF = relative frequency

Use the table from the section titled "Theoretical Distribution" here:

- $P(X = 3) =$
- $P(1 < X < 4) =$
- $P(X \geq 8) =$

Use the data from the section titled "Organize the Data" here:

- $RF(X = 3) =$
- $RF(1 < X < 4) =$
- $RF(X \geq 8) =$

4.17.7 Discussion Questions

For questions 1. and 2., think about the shapes of the two graphs, the probabilities and the relative frequencies, the means, and the standard deviations.

1. Knowing that data vary, describe three similarities between the graphs and distributions of the theoretical and empirical distributions. Use complete sentences. (Note: These answers may vary and still be correct.)
2. Describe the three most significant differences between the graphs or distributions of the theoretical and empirical distributions. (Note: These answers may vary and still be correct.)
3. Using your answers from the two previous questions, does it appear that the data fit the theoretical distribution? In 1 - 3 complete sentences, explain why or why not.
4. Suppose that the experiment had been repeated 500 times. Which table (from "Organize the data" and "Theoretical Distributions") would you expect to change (and how would it change)? Why? Why wouldn't the other table change?

4.18 Lab 2: Discrete Distribution (Lucky Dice Experiment)[21]

Class Time:

Names:

4.18.1 Student Learning Outcomes:

- The student will compare empirical data and a theoretical distribution to determine if a Tet gambling game fits a discrete distribution.
- The student will demonstrate an understanding of long-term probabilities.

4.18.2 Supplies:

- 1 game "Lucky Dice" or 3 regular dice

NOTE: For a detailed game description, refer here. (The link goes to the beginning of Discrete Random Variables Homework. Please refer to Problem #14.)

NOTE: Round relative frequencies and probabilities to four decimal places.

4.18.3 The Procedure

1. The experiment procedure is to bet on one object. Then, roll 3 Lucky Dice and count the number of matches. The number of matches will decide your profit.
2. What is the theoretical probability of 1 die matching the object? __________
3. Choose one object to place a bet on. Roll the 3 Lucky Dice. Count the number of matches.
4. Let X = number of matches. Theoretically, $X \sim B$ (_______,_______)
5. Let Y = profit per game.

4.18.4 Organize the Data

In the chart below, fill in the Y value that corresponds to each X value. Next, record the number of matches picked for your class. Then, calculate the relative frequency.

1. Complete the table.

x	y	Frequency	Relative Frequency
0			
1			
2			
3			

Table 4.20

[21]This content is available online at <http://cnx.org/content/m16826/1.11/>.

2. Calculate the Following:
   a. $\overline{x} =$
   b. $s_x =$
   c. $\overline{y} =$
   d. $s_y =$
3. Explain what $\overline{x}$ represents.
4. Explain what $\overline{y}$ represents.
5. Based upon the experiment:
   a. What was the average profit per game?
   b. Did this represent an average win or loss per game?
   c. How do you know? Answer in complete sentences.
6. Construct a histogram of the empirical data

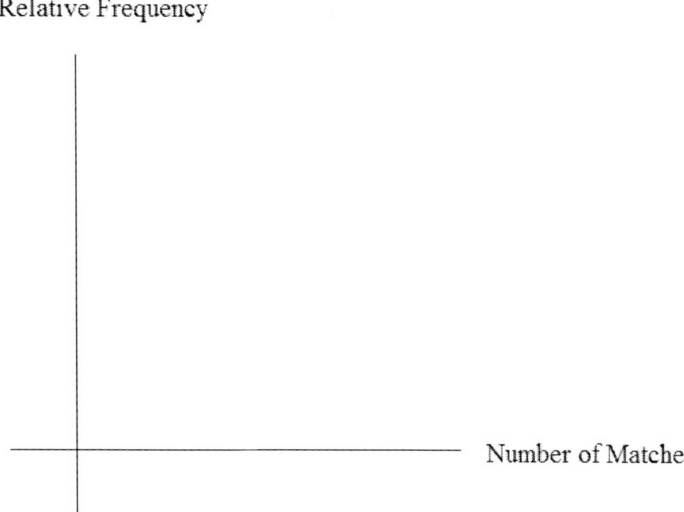

Figure 4.4

4.18.5 Theoretical Distribution

Build the theoretical PDF chart for X and Y based on the distribution from the section titled "The Procedure".

1.

x	y	$P(X = x) = P(Y = y)$	
0			
1			
2			
3			

Table 4.21

2. Calculate the following
   - **a.** $\mu_x =$
   - **b.** $\sigma_x =$
   - **c.** $\mu_y =$
3. Explain what μ_x represents.
4. Explain what μ_y represents.
5. Based upon theory:
   - **a.** What was the expected profit per game?
   - **b.** Did the expected profit represent an average win or loss per game?
   - **c.** How do you know? Answer in complete sentences.
6. Construct a histogram of the theoretical distribution.

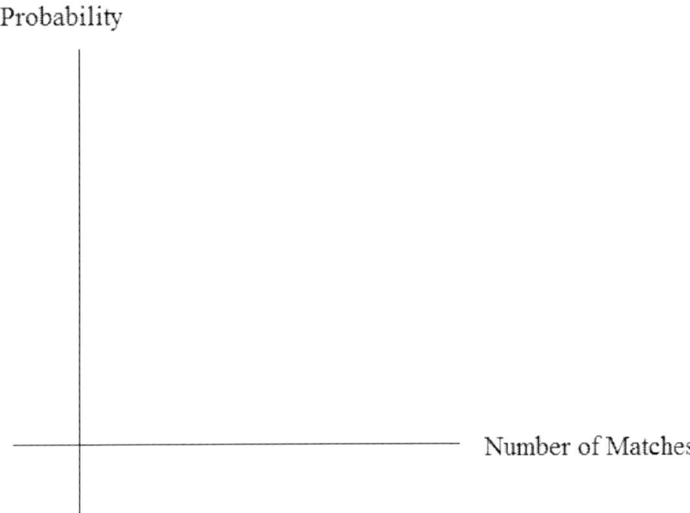

Figure 4.5

4.18.6 Use the Data

Calculate the following (rounded to 4 decimal places):

NOTE: *RF* = relative frequency

Use the data from the section titled "Theoretical Distribution" here:

1. $P(X = 3) =$__________
2. $P(0 < X < 3) =$__________
3. $P(X \geq 2) =$__________

Use the data from the section titled "Organize the Data" here:

1. $RF(X = 3) =$____________
2. $RF(0 < X < 3) =$____________
3. $RF(X \geq 2) =$ ____________

4.18.7 Discussion Question

For questions 1. and 2., consider the graphs, the probabilities and relative frequencies, the means and the standard deviations.

1. Knowing that data vary, describe three similarities between the graphs and distributions of the theoretical and empirical distributions. Use complete sentences. (Note: these answers may vary and still be correct.)
2. Describe the three most significant differences between the graphs or distributions of the theoretical and empirical distributions. (Note: these answers may vary and still be correct.)
3. Thinking about your answers to 1. and 2.,does it appear that the data fit the theoretical distribution? In 1 - 3 complete sentences, explain why or why not.
4. Suppose that the experiment had been repeated 500 times. Which table (from "Organize the Data" or "Theoretical Distribution") would you expect to change? Why? How might the table change?

Solutions to Exercises in Chapter 4

Solution to Example 4.2, Problem 1 (p. 139)
Let X = the number of days Nancy **attends class per week**.
Solution to Example 4.2, Problem 2 (p. 139)
0, 1, 2, and 3
Solution to Example 4.2, Problem 3 (p. 139)

x	$P(X = x)$
0	0.01
1	0.04
2	0.15
3	0.80

Table 4.22

Solution to Example 4.5, Problem 1 (p. 141)
X = amount of profit
Solution to Example 4.5, Problem 2 (p. 141)

	x	$P(X = x)$	$xP(X = x)$
WIN	10	$\frac{1}{3}$	$\frac{10}{3}$
LOSE	**-6**	$\frac{2}{3}$	$\frac{-12}{3}$

Table 4.23

Solution to Example 4.5, Problem 3 (p. 142)
Add the last column of the table. The expected value $\mu = \frac{-2}{3}$. You lose, on average, about 67 cents each time you play the game so you do not come out ahead.
Solution to Example 4.9, Problem 1 (p. 143)
failure
Solution to Example 4.9, Problem 2 (p. 143)
X = the number of statistics students who do their homework on time
Solution to Example 4.9, Problem 3 (p. 143)
$0, 1, 2, \ldots, 50$
Solution to Example 4.9, Problem 4 (p. 143)
Failure is a student who does not do his or her homework on time.
Solution to Example 4.9, Problem 5 (p. 143)
$q = 0.30$
Solution to Example 4.9, Problem 6 (p. 144)
greater than or equal to ($\geq$)
Solution to Example 4.14, Problem 2 (p. 146)
$1, 2, 3, \ldots$, (total number of chemistry students)
Solution to Example 4.14, Problem 3 (p. 146)

- $p = 0.55$
- $q = 0.45$

Solution to Example 4.14, Problem 4 (p. 146)
$P(X = 4)$
Solution to Example 4.18, Problem 1 (p. 148)
Without
Solution to Example 4.18, Problem 2 (p. 148)
The men
Solution to Example 4.18, Problem 3 (p. 148)
15 men
Solution to Example 4.18, Problem 4 (p. 149)
18 women
Solution to Example 4.18, Problem 5 (p. 149)
Let X = **the number of men** on the committee. $X = 0, 1, 2, \ldots, 7$.
Solution to Example 4.18, Problem 6 (p. 149)
$P(X>4)$
Solution to Example 4.22, Problem 1 (p. 150)
The 2 pages
Solution to Example 4.22, Problem 2 (p. 150)
2
Solution to Example 4.22, Problem 3 (p. 151)
Let X = **the number of pages of written math homework you do per day**.
Solution to Example 4.22, Problem 4 (p. 151)
$P(X > 2)$

Solutions to Practice 2: Binomial Distribution

Solution to Exercise 4.11.1 (p. 156)
X= the number that expect to work full-time.
Solution to Exercise 4.11.2 (p. 156)
B(8,0.367)
Solution to Exercise 4.11.3 (p. 156)
0,1,2,3,4,5,6,7,8
Solution to Exercise 4.11.5 (p. 156)
2.94
Solution to Exercise 4.11.6 (p. 156)
1.36
Solution to Exercise 4.11.7 (p. 156)
0.9677
Solution to Exercise 4.11.8 (p. 157)
0.8547

Solutions to Practice 3: Poisson Distribution

Solution to Exercise 4.12.2 (p. 158)
P(10)
Solution to Exercise 4.12.3 (p. 158)
0,1,2,3,4,...
Solution to Exercise 4.12.5 (p. 158)
No
Solution to Exercise 4.12.6 (p. 158)
No

Solutions to Practice 4: Geometric Distribution

Solution to Exercise 4.13.2 (p. 159)
G(0.367)
Solution to Exercise 4.13.3 (p. 159)
0,1,2,. . .
Solution to Exercise 4.13.5 (p. 159)
2.72
Solution to Exercise 4.13.6 (p. 159)
0.5993

Solutions to Practice 5: Hypergeometric Distribution

Solution to Exercise 4.14.2 (p. 161)
H(16,7,9)
Solution to Exercise 4.14.3 (p. 161)
2,3,4,5,6,7,8,9
Solution to Exercise 4.14.5 (p. 161)
6.26

Solutions to Homework

Solution to Exercise 4.15.1 (p. 162)

- **a.** 0.1
- **b.** 1.6

Solution to Exercise 4.15.3 (p. 162)

- **b.** $200,000;$600,000;$400,000
- **c.** third investment
- **d.** first investment
- **e.** second investment

Solution to Exercise 4.15.5 (p. 163)

- **a.** 0.2
- **c.** 2.35
- **d.** 2-3 children

Solution to Exercise 4.15.7 (p. 164)

- **a.** X = the number of dice that show a 1
- **b.** 0,1,2,3,4,5,6
- **c.** $X \sim B\left(6, \frac{1}{6}\right)$
- **d.** 1
- **e.** 0.00002
- **f.** 3 dice

Solution to Exercise 4.15.9 (p. 164)

- **a.** X = the number of students that will attend Tet.
- **b.** 0, 1, 2, 3, 4, 5, 6, 7, 8, 9, 10, 11, 12
- **c.** $X \sim B(12{,}0.18)$
- **d.** 2.16

- **e.** 0.9511
- **f.** 0.3702

Solution to Exercise 4.15.11 (p. 164)

- **a.** X = the number of fencers that do **not** use foil as their main weapon
- **b.** 0, 1, 2, 3,... 25
- **c.** $X \sim B(25,0.40)$
- **d.** 10
- **e.** 0.0442
- **f.** Yes

Solution to Exercise 4.15.13 (p. 165)

- **a.** X = the number of fortune cookies that have an extra fortune
- **b.** 0, 1, 2, 3,... 144
- **c.** $X \sim B(25,0.40)$ or $P(4.32)$
- **d.** 4.32
- **e.** 0.0124 or 0.0133
- **f.** 0.6300 or 0.6264

Solution to Exercise 4.15.15 (p. 165)

- **a.** X = the number of women that suffer from anorexia
- **b.** 0, 1, 2, 3,... 600 (can leave off 600)
- **c.** $X \sim P(3)$
- **d.** 3
- **e.** 0.0498
- **f.** 0.1847

Solution to Exercise 4.15.17 (p. 166)

- **a.** X = the number of children for a Spanish couple
- **b.** 0, 1, 2, 3,...
- **c.** $X \sim P(1.34)$
- **d.** 0.2618
- **e.** 0.6127
- **f.** 0.3873

Solution to Exercise 4.15.19 (p. 166)

- **a.** X = the number of dealers she calls until she finds one with a used red Miata
- **b.** 0, 1, 2, 3,...
- **c.** $X \sim G(0.28)$
- **d.** 3.57
- **e.** 0.7313
- **f.** 0.2497

Solution to Exercise 4.15.21 (p. 167)

- **d.** 4.31
- **e.** 0.4079
- **f.** 0.9953

Solution to Exercise 4.15.23 (p. 167)

- **d.** 2

e. 0.1353
f. 0.3233

Solution to Exercise 4.15.25 (p. 167)

a. X = the number of seniors that participated in after-school sports all 4 years of high school
b. 0, 1, 2, 3,... 60
c. $X \sim P\,(4.8)$
d. 4.8
e. Yes
f. 4

Solution to Exercise 4.15.27 (p. 168)

a. X = the number of shell pieces in one cake
b. 0, 1, 2, 3,...
c. $X \sim P\,(1.5)$
d. 1.5
e. 0.2231
f. 0.0001
g. Yes

Solution to Exercise 4.15.29 (p. 168)

d. 0.0043
e. none
f. 3.3

Solution to Exercise 4.15.31 (p. 168)

d. 3.02
e. No
f. 0.9997
h. 0.2291
i. 0.3881
j. 6.6207 pages

Solution to Exercise 4.15.33 (p. 169)
D: 4.43
Solution to Exercise 4.15.34 (p. 169)
A: 0.1476
Solution to Exercise 4.15.35 (p. 169)
C: 0.4734
Solution to Exercise 4.15.36 (p. 169)
A: The number of times Mrs. Plum's cats wake her up each week
Solution to Exercise 4.15.37 (p. 170)
D: 0.0671

Solutions to Review

Solution to Exercise 4.16.1 (p. 171)
C
Solution to Exercise 4.16.2 (p. 171)
B

Solution to Exercise 4.16.3 (p. 171)
A
Solution to Exercise 4.16.4 (p. 172)
0.5773
Solution to Exercise 4.16.5 (p. 172)
0.0522
Solution to Exercise 4.16.6 (p. 172)
B
Solution to Exercise 4.16.7 (p. 172)
C
Solution to Exercise 4.16.8 (p. 172)
C
Solution to Exercise 4.16.9 (p. 173)
0.2709
Solution to Exercise 4.16.10 (p. 173)
B
Solution to Exercise 4.16.11 (p. 173)
X = the number of patients calling in claiming to have **the flu**, who actually have **the flu**. $X = 0, 1, 2, ...25$
Solution to Exercise 4.16.12 (p. 173)
$B(25, 0.04)$
Solution to Exercise 4.16.13 (p. 173)
0.0165
Solution to Exercise 4.16.14 (p. 173)
1
Solution to Exercise 4.16.15 (p. 173)
C
Solution to Exercise 4.16.16 (p. 173)
All words used by Tom Clancy in his novels

Chapter 5

Continuous Random Variables

5.1 Continuous Random Variables[1]

5.1.1 Student Learning Objectives

By the end of this chapter, the student should be able to:

- Recognize and understand continuous probability density functions in general.
- Recognize the uniform probability distribution and apply it appropriately.
- Recognize the exponential probability distribution and apply it appropriately.

5.1.2 Introduction

Continuous random variables have many applications. Baseball batting averages, IQ scores, the length of time a long distance telephone call lasts, the amount of money a person carries, the length of time a computer chip lasts, and SAT scores are just a few. The field of reliability depends on a variety of continuous random variables.

This chapter gives an introduction to continuous random variables and the many continuous distributions. We will be studying these continuous distributions for several chapters.

The characteristics of continuous random variables are:

- The outcomes are measured, not counted.
- Geometrically, the probability of an outcome is equal to an area under a mathematical curve called the density curve, $f(x)$.
- Each individual value has zero probability of occurring. Instead we find the probability that the value is between two endpoints.

We will start with the two simplest continuous distributions, the **Uniform** and the **Exponential**.

> NOTE: The values of discrete and continuous random variables can be ambiguous. For example, if X is equal to the number of miles (to the nearest mile) you drive to work, then X is a discrete random variable. You count the miles. If X is the distance you drive to work, then you measure values of X and X is a continuous random variable. How the random variable is defined is very important.

[1]This content is available online at <http://cnx.org/content/m16808/1.9/>.

5.2 Continuous Probability Functions[2]

We begin by defining a continuous probability density function. We use the function notation $f(X)$. Intermediate algebra may have been your first formal introduction to functions. In the study of probability, the functions we study are special. We define the function $f(X)$ so that the area between it and the x-axis is equal to a probability. Since the maximum probability is one, the maximum area is also one.

For continuous probability distributions, PROBABILITY = AREA.

Example 5.1
Consider the function $f(X) = \frac{1}{20}$ for $0 \leq X \leq 20$. X = a real number. The graph of $f(X) = \frac{1}{20}$ is a horizontal line. However, since $0 \leq X \leq 20$, $f(X)$ is restricted to the portion between $X = 0$ and $X = 20$, inclusive .

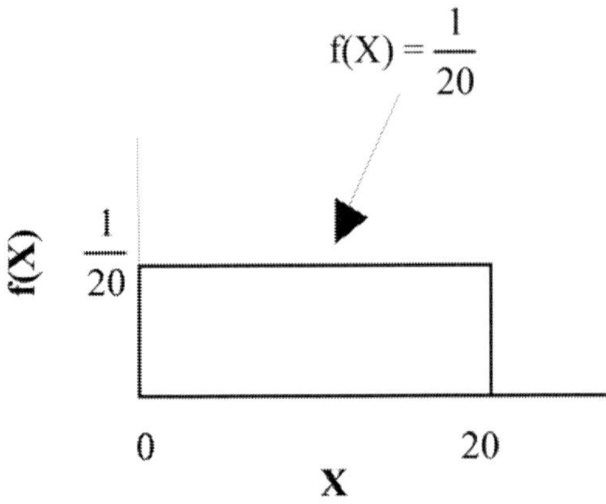

$f(X) = \frac{1}{20}$ **for** $0 \leq X \leq 20$.

The graph of $f(X) = \frac{1}{20}$ is a horizontal line segment when $0 \leq X \leq 20$.

The area between $f(X) = \frac{1}{20}$ where $0 \leq X \leq 20$. and the x-axis is the area of a rectangle with base = 20 and height = $\frac{1}{20}$.

$AREA = 20 \cdot \frac{1}{20} = 1$

This particular function, where we have restricted X so that the area between the function and the x-axis is 1, is an example of a continuous probability density function. It is used as a tool to calculate probabilities.

Suppose we want to find the area between $f(X) = \frac{1}{20}$ **and the x-axis where** $0 < X < 2$ **.**

f(X) 1/20

0 2 20

X

$AREA = (2 - 0) \cdot \frac{1}{20} = 0.1$

[2]This content is available online at <http://cnx.org/content/m16805/1.8/>.

$(2-0) = 2 =$ *base of a rectangle*

$\frac{1}{20}$ = the height.

The area corresponds to a probability. The probability that X is between 0 and 2 is 0.1, which can be written mathematically as $P(0<X<2) = P(X<2) = 0.1$.

Suppose we want to find the area between $f(X) = \frac{1}{20}$ and the x-axis where $4 < X < 15$.

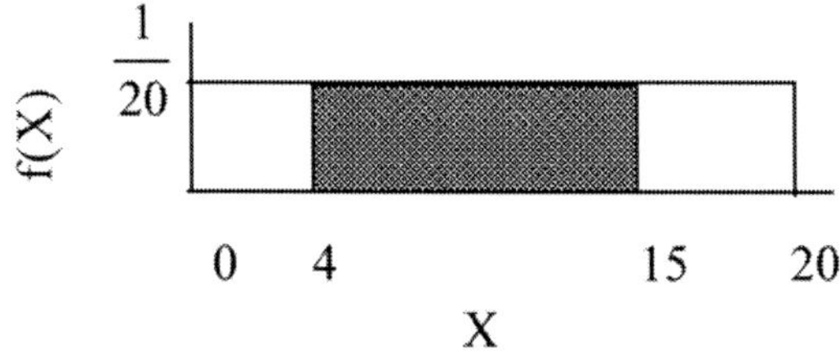

$AREA = (15-4) \cdot \frac{1}{20} = 0.55$

$(15-4) = 11 =$ *the base of a rectangle*

$\frac{1}{20}$ = the height.

The area corresponds to the probability $P(4 < X < 15) = 0.55$.

Suppose we want to find $P(X = 15)$. On an x-y graph, $X = 15$ is a vertical line. A vertical line has no width (or 0 width). Therefore, $P(X = 15) = (base)(height) = (0)\left(\frac{1}{20}\right) = 0$.

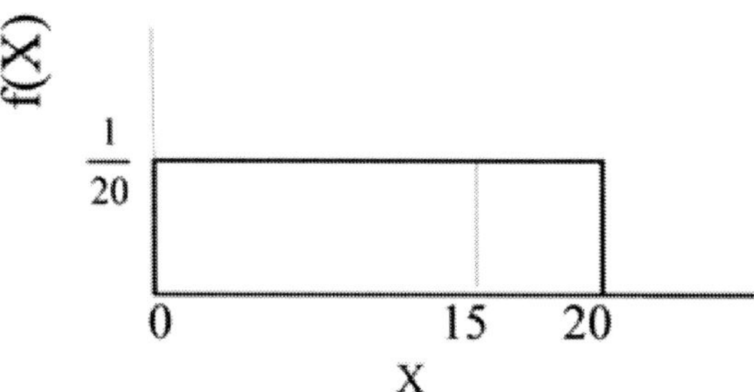

$P(X \leq x)$ (can be written as $P(X < x)$ for continuous distributions) is called the cumulative distribution function or *CDF*. Notice the "less than or equal to" symbol. We can use the *CDF* to calculate $P(X > x)$. The *CDF* gives "area to the left" and $P(X > x)$ gives "area to the right." We calculate $P(X > x)$ for continuous distributions as follows: $P(X > x) = 1 - P(X < x)$.

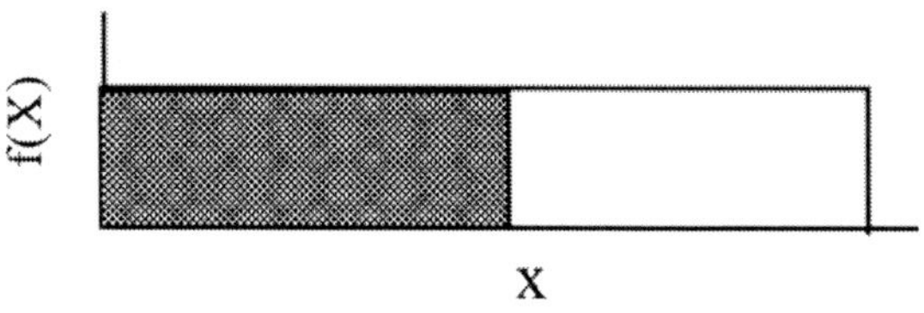

f(X)

X

P(X < x) P(X > x) = 1 – P(X < x)

Label the graph with $f(X)$ and X. Scale the x and y axes with the maximum x and y values. $f(X) = \frac{1}{20}, 0 \leq X \leq 20$.

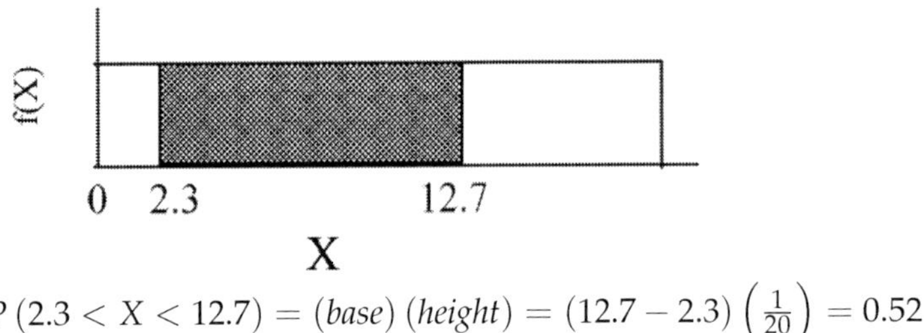

$P(2.3 < X < 12.7) = (base)(height) = (12.7 - 2.3)\left(\frac{1}{20}\right) = 0.52$

5.3 The Uniform Distribution[3]

Example 5.2
The previous problem is an example of the **uniform probability distribution**.

Illustrate the uniform distribution. The data that follows are 55 smiling times, in seconds, of an eight-week old baby.

10.4	19.6	18.8	13.9	17.8	16.8	21.6	17.9	12.5	11.1	4.9
12.8	14.8	22.8	20.0	15.9	16.3	13.4	17.1	14.5	19.0	22.8
1.3	0.7	8.9	11.9	10.9	7.3	5.9	3.7	17.9	19.2	9.8
5.8	6.9	2.6	5.8	21.7	11.8	3.4	2.1	4.5	6.3	10.7
8.9	9.4	9.4	7.6	10.0	3.3	6.7	7.8	11.6	13.8	18.6

Table 5.1

sample mean = 11.49 and sample standard deviation = 6.23

We will assume that the smiling times, in seconds, follow a uniform distribution between 0 and 23 seconds, inclusive. This means that any smiling time from 0 to and including 23 seconds is **equally likely**. The histogram that could be constructed from the sample is an empirical distribution that closely matches the theoretical uniform distribution.

Let X = length, in seconds, of an eight-week old baby's smile.

The notation for the uniform distribution is

$X \sim U(a,b)$ where a = the lowest value of X and b = the highest value of X.

The probability density function is $f(X) = \frac{1}{b-a}$ for $a \leq X \leq b$.

For this example, $X \sim U(0, 23)$ and $f(X) = \frac{1}{23-0}$ for $0 \leq X \leq 23$.

Formulas for the theoretical mean and standard deviation are

$\mu = \frac{a+b}{2}$ and $\sigma = \sqrt{\frac{(b-a)^2}{12}}$

For this problem, the theoretical mean and standard deviation are

[3] This content is available online at <http://cnx.org/content/m16819/1.14/>.

$\mu = \frac{0+23}{2} = 11.50$ seconds and $\sigma = \sqrt{\frac{(23-0)^2}{12}} = 6.64$ seconds

Notice that the theoretical mean and standard deviation are close to the sample mean and standard deviation.

Example 5.3
Problem 1
What is the probability that a randomly chosen eight-week old baby smiles between 2 and 18 seconds?

Solution
Find $P(2 < X < 18)$.

$P(2 < X < 18) = (base)\,(height) = (18 - 2) \cdot \frac{1}{23} = \frac{16}{23}$.

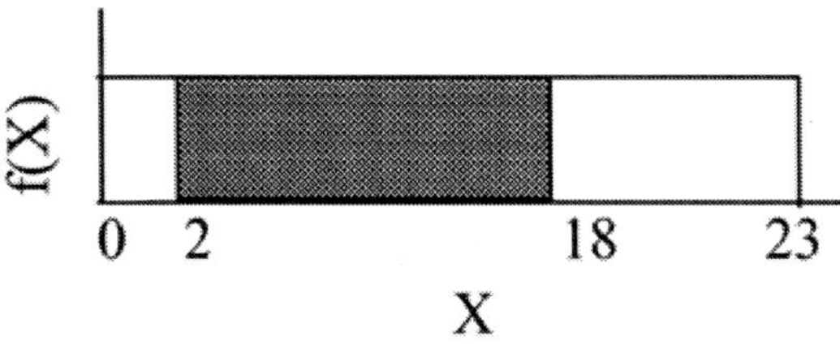

Problem 2
Find the 90th percentile for an eight week old baby's smiling time.

Solution
Ninety percent of the smiling times fall below the 90th percentile, k, so $P(X < k) = 0.90$

$P(X < k) = 0.90$

$(base)\,(height) = 0.90$

$(k - 0) \cdot \frac{1}{23} = 0.90$

$k = 23 \cdot 0.90 = 20.7$

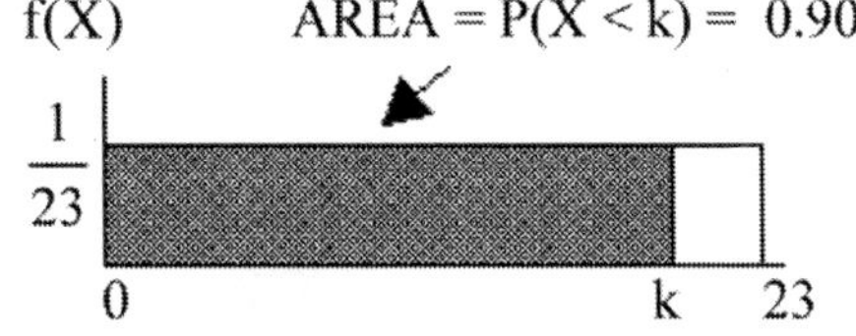

Problem 3
Find the probability that a random eight week old baby smiles more than 12 seconds **KNOWING** that the baby smiles **MORE THAN 8 SECONDS**.

Solution
Find $P(X > 12|X > 8)$ There are two ways to do the problem. **For the first way**, use the fact that this is a **conditional** and changes the sample space. The graph illustrates the new sample space. You already know the baby smiled more than 8 seconds.

Write a new $f(X)$**:** $f(X) = \frac{1}{23-8} = \frac{1}{15}$

for $8 < X < 23$

$P(X > 12|X > 8) = (23 - 12) \cdot \frac{1}{15} = \frac{11}{15}$

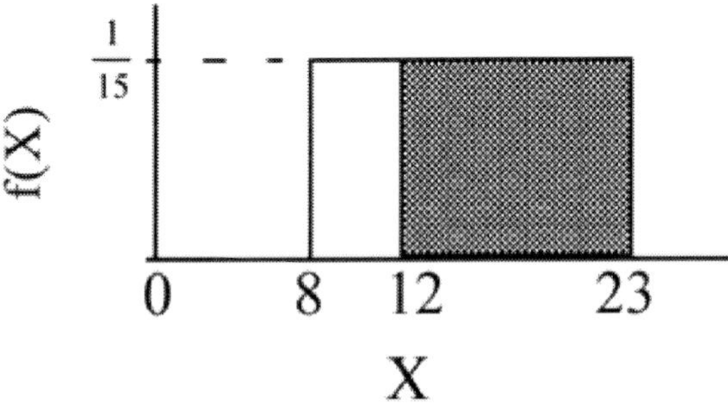

For the second way, use the conditional formula from **Probability Topics** with the original distribution $X \sim U(0, 23)$:

$P(A|B) = \frac{P(A\ AND\ B)}{P(B)}$ For this problem, A is $(X > 12)$ and B is $(X > 8)$.

So, $P(X > 12|X > 8) = \frac{(X>12\ AND\ X>8)}{P(X>8)} = \frac{P(X>12)}{P(X>8)} = \frac{\frac{11}{23}}{\frac{15}{23}} = 0.733$

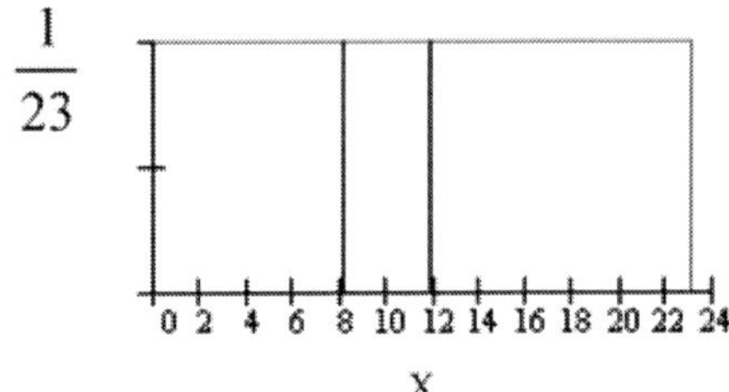

Example 5.4
Uniform: The amount of time, in minutes, that a person must wait for a bus is uniformly distributed between 0 and 15 minutes, inclusive.

Problem 1
What is the probability that a person waits fewer than 12.5 minutes?

Solution
Let X = the number of minutes a person must wait for a bus. $a = 0$ and $b = 15$. $X \sim U(0, 15)$. Write the probability density function. $f(X) = \frac{1}{15-0} = \frac{1}{15}$ for $0 \leq X \leq 15$.

Find $P(X < 12.5)$. Draw a graph.

$P(X < k) = (base)(height) = (12.5 - 0) \cdot \frac{1}{15} = 0.8333$

The probability a person waits less than 12.5 minutes is 0.8333.

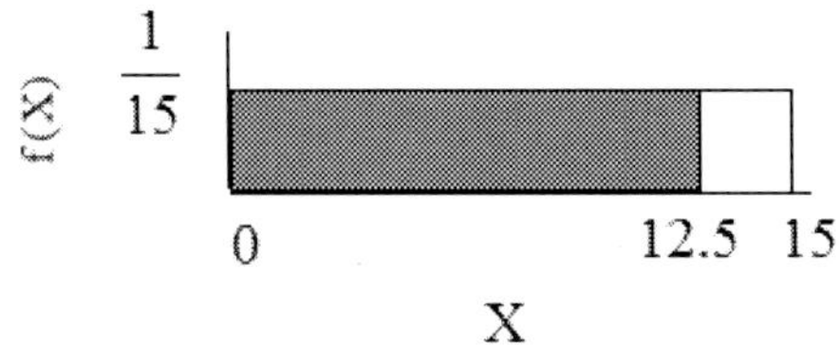

Problem 2

On the average, how long must a person wait?

Find the mean, μ, and the standard deviation, σ.

Solution

$\mu = \frac{a+b}{2} = \frac{15+0}{2} = 7.5$. On the average, a person must wait 7.5 minutes.

$\sigma = \sqrt{\frac{(b-a)^2}{12}} = \sqrt{\frac{(15-0)^2}{12}} = 4.3$. The Standard deviation is 4.3 minutes.

Problem 3

Ninety percent of the time, the time a person must wait falls below what value?

NOTE: This asks for the 90th percentile.

Solution

Find the 90th percentile. Draw a graph. Let k = the 90th percentile.

$P(X < k) = (base)(height) = (k - 0) \cdot \left(\frac{1}{15}\right)$

$0.90 = k \cdot \frac{1}{15}$

$k = (0.90)(15) = 13.5$

k is sometimes called a critical value.

The 90th percentile is 13.5 minutes. Ninety percent of the time, a person must wait at most 13.5 minutes.

AREA = $P(X < k) = 0.90$

f(X) 1/15

0 k

X

Example 5.5
Uniform: The average number of donuts a nine-year old child eats per month is uniformly distributed from 0.5 to 4 donuts, inclusive. Let X = the average number of donuts a nine-year old child eats per month. Then $X \sim U(0.5, 4)$.

Problem 1 *(Solution on p. 220.)*
The probability that a randomly selected nine-year old child eats an average of more than two donuts is _______.

Problem 2 *(Solution on p. 220.)*
Find the probability that a different nine-year old child eats an average of more than two donuts given that his or her amount is more than 1.5 donuts.

The second probability question has a **conditional** (refer to "Probability Topics (Section 3.1)"). You are asked to find the probability that a nine-year old eats an average of more than two donuts given that his/her amount is more than 1.5 donuts. Solve the problem two different ways (see the first example (Example 5.2)). You must reduce the sample space. **First way**: Since you already know the child eats more than 1.5 donuts, you are no longer starting at $a = 0.5$ donut. Your starting point is 1.5 donuts.

Write a new f(X):

$f(X) = \frac{1}{4-1.5} = \frac{2}{5}$ for $1.5 \leq X \leq 4$.

Find $P(X > 2|X > 1.5)$. Draw a graph.

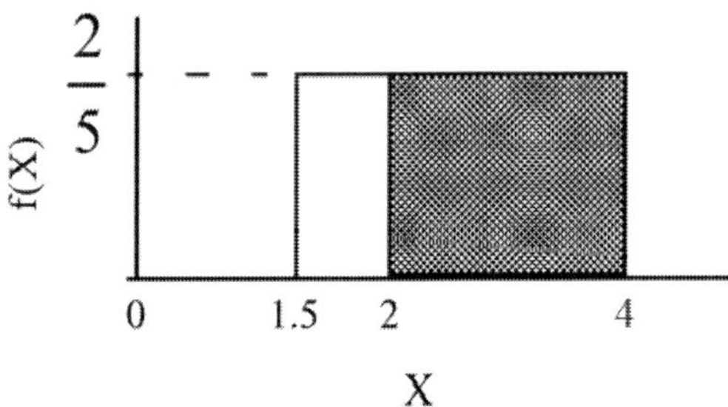

$P(X > 2|X > 1.5) = (base)(new\ height) = (4-2)(2/5) = ?$
The probability that a nine-year old child eats an average of more than 2 donuts when he/she has already eaten more than 1.5 donuts is $\frac{4}{5}$.

Second way: Draw the original graph for $X \sim U(0.5, 4)$. Use the conditional formula

$P(X > 2|X > 1.5) = \frac{P(X>2\ AND\ X>1.5)}{P(X>1.5)} = \frac{P(X>2)}{P(X>1.5)} = \frac{\frac{2}{3.5}}{\frac{2.5}{3.5}} = 0.8 = \frac{4}{5}$

NOTE: See "Summary of the Uniform and Exponential Probability Distributions (Section 5.5)" for a full summary.

5.4 The Exponential Distribution[4]

The **exponential** distribution is often concerned with the amount of time until some specific event occurs. For example, the amount of time (beginning now) until an earthquake occurs has an exponential distribution. Other examples include the length, in minutes, of long distance business telephone calls, and the amount of time, in months, a car battery lasts. It can be shown, too, that the amount of change that you have in your pocket or purse follows an exponential distribution.

Values for an exponential random variable occur in the following way. There are fewer large values and more small values. For example, the amount of money customers spend in one trip to the supermarket follows an exponential distribution. There are more people that spend less money and fewer people that spend large amounts of money.

The exponential distribution is widely used in the field of reliability. Reliability deals with the amount of time a product lasts.

Example 5.6

Illustrates the exponential distribution: Let X = amount of time (in minutes) a postal clerk spends with his/her customer. The time is known to have an exponential distribution with the average amount of time equal to 4 minutes.

X is a **continuous random variable** since time is measured. It is given that $\mu = 4$ minutes. To do any calculations, you must know m, the decay parameter.

$m = \frac{1}{\mu}$. Therefore, $m = \frac{1}{4} = 0.25$

The standard deviation, σ, is the same as the mean. $\mu = \sigma$

The distribution notation is $X \sim Exp\,(m)$. Therefore, $X \sim Exp\,(0.25)$.

The probability density function is $f\,(X) = m \cdot e^{-m \cdot x}$ The number e = 2.71828182846... It is a number that is used often in mathematics. Scientific calculators have the key "e^x." If you enter 1 for x, the calculator will display the value e.

The curve is:

$f\,(X) = 0.25 \cdot e^{-\,0.25 \cdot X}$ where X is at least 0 and $m = 0.25$.

For example, $f\,(5) = 0.25 \cdot e^{-\,0.25 \cdot 5} = 0.072$

The graph is as follows:

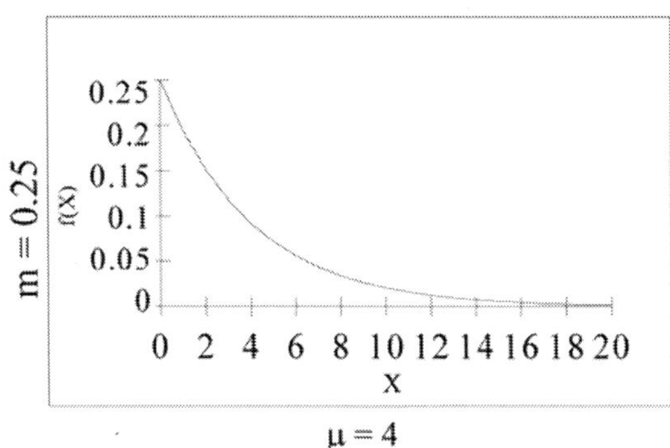

[4]This content is available online at <http://cnx.org/content/m16816/1.12/>.

Notice the graph is a declining curve. When X = 0,

$f(X) = 0.25 \cdot e^{-0.25 \cdot 0} = 0.25 \cdot 1 = 0.25 = m$

Example 5.7
Problem 1
Find the probability that a clerk spends four to five minutes with a randomly selected customer.

Solution
Find $P(4 < X < 5)$.

The **cumulative distribution function (CDF)** gives the area to the left.

$P(X < x) = 1 - e^{-m \cdot x}$

$P(X < 5) = 1 - e^{-0.25 \cdot 5} = 0.7135$ and $P(X < 4) = 1 - e^{-0.25 \cdot 4} = 0.6321$

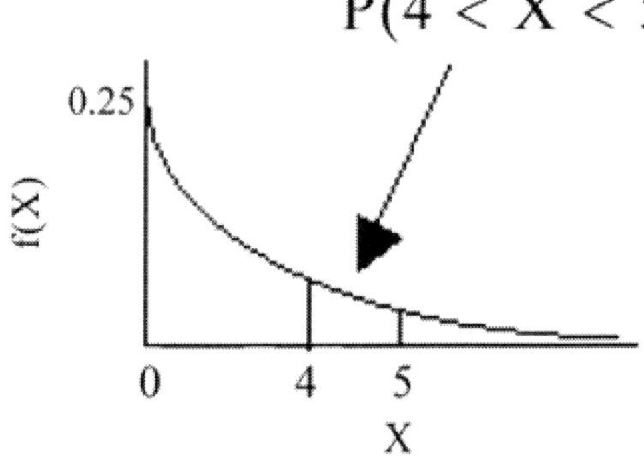

NOTE: You can do these calculations easily on a calculator.

The probability that a postal clerk spends four to five minutes with a randomly selected customer is

$P(4 < X < 5) = P(X < 5) - P(X < 4) = 0.7135 - 0.6321 = 0.0814$

NOTE: TI-83+ and TI-84: On the home screen, enter (1-e^(-.25*5))-(1-e^(-.25*4)) or enter e^(-.25*4)-e^(-.25*5).

Problem 2
Half of all customers are finished within how long? (Find the 50th percentile)

Solution
Find the 50th percentile.

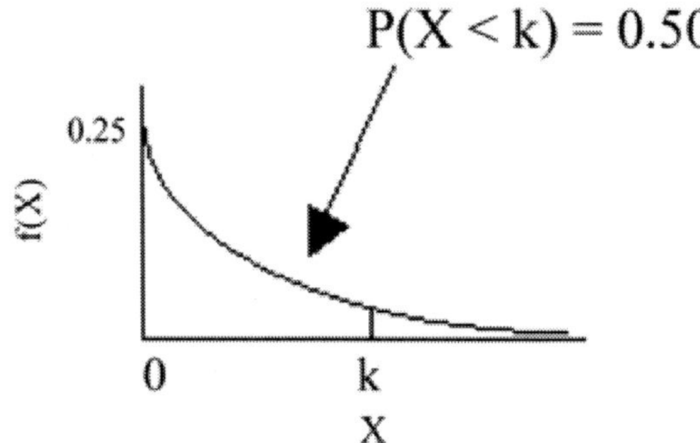

$P(X < k) = 0.50$, $k = 2.8$ minutes (calculator or computer)

Half of all customers are finished within 2.8 minutes.

You can also do the calculation as follows:

$P(X < k) = 0.50$ and $P(X < k) = 1 - e^{-0.25 \cdot k}$

Therefore, $0.50 = 1 - e^{-0.25 \cdot k}$ and $e^{-0.25 \cdot k} = 1 - 0.50 = 0.5$

Take natural logs: $ln\left(e^{-0.25 \cdot k}\right) = ln\,(0.50)$. So, $-0.25 \cdot k = ln\,(0.50)$

Solve for k: $k = \frac{ln(.50)}{-0.25} = 2.8$ minutes

NOTE: A formula for the percentile k is $k = \frac{LN(1 - AreaToTheLeft)}{-m}$ where LN is the natural log.

NOTE: TI-83+ and TI-84: On the home screen, enter LN(1-.50)/-.25. Press the (-) for the negative.

Problem 3
Which is larger, the mean or the median?

Solution
Is the mean or median larger?

From part b, the median or 50th percentile is 2.8 minutes. The theoretical mean is 4 minutes. The mean is larger.

5.4.1 Optional Collaborative Classroom Activity

Have each class member count the change he/she has in his/her pocket or purse. Your instructor will record the amounts in dollars and cents. Construct a histogram of the data taken by the class. Use 5 intervals. Draw a smooth curve through the bars. The graph should look approximately exponential. Then calculate the mean.

Let X = the amount of money a student in your class has in his/her pocket or purse.

The distribution for X is approximately exponential with mean, μ = _______ and m = _______. The standard deviation, σ = ________.

Draw the appropriate exponential graph. You should label the x and y axes, the decay rate, and the mean. Shade the area that represents the probability that one student has less than \$.40 in his/her pocket or purse. (Shade $P(X < 0.40)$).

Example 5.8
On the average, a certain computer part lasts 10 years. The length of time the computer part lasts is exponentially distributed.

Problem 1
What is the probability that a computer part lasts more than 7 years?

Solution
Let X = the amount of time (in years) a computer part lasts.

$\mu = 10$ so $m = \frac{1}{\mu} = \frac{1}{10} = 0.1$

Find $P(X > 7)$. Draw a graph.

$P(X > 7) = 1 - P(X < 7)$.

Since $P(X < x) = 1 - e^{-mx}$ then $P(X > x) = 1 - (1 - e^{-m \cdot x}) = e^{-m \cdot x}$

$P(X > 7) = 1 - e^{-0.1 \cdot 7} = 0.4966$. The probability that a computer part lasts more than 7 years is 0.4966.

NOTE: TI-83+ and TI-84: On the home screen, enter e^(-.1*7).

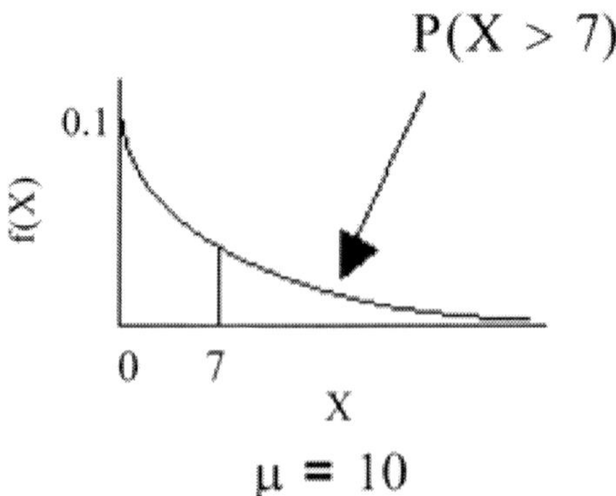

Problem 2
On the average, how long would 5 computer parts last if they are used one after another?

Solution
On the average, 1 computer part lasts 10 years. Therefore, 5 computer parts, if they are used one right after the other would last, on the average,

$(5)(10) = 50$ years.

Problem 3
Eighty percent of computer parts last at most how long?

Solution
Find the 80th percentile. Draw a graph. Let k = the 80th percentile.

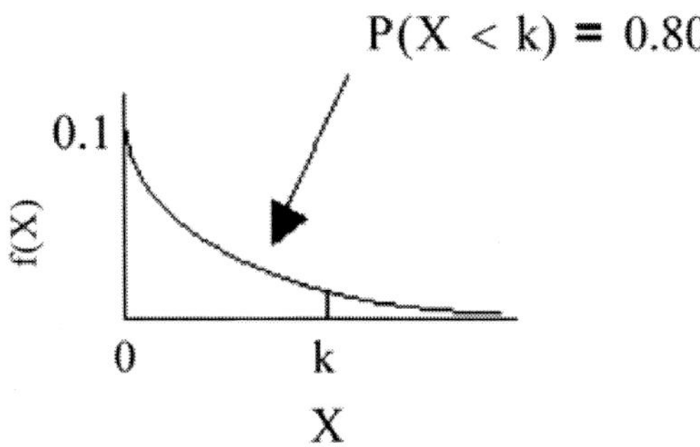

Solve for k: $k = \frac{ln(1-.80)}{-0.1} = 16.1$ years

Eighty percent of the computer parts last at most 16.1 years.

NOTE: TI-83+ and TI-84: On the home screen, enter LN(1 - .80)/-.1

Problem 4
What is the probability that a computer part lasts between 9 and 11 years?

Solution
Find $P(9 < X < 11)$. Draw a graph.

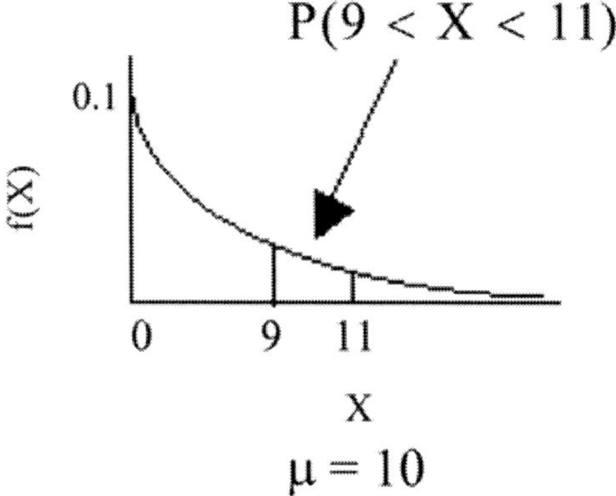

$P(9 < X < 11) = P(X < 11) - P(X < 9) = \left(1 - e^{-0.1 \cdot 11}\right) - \left(1 - e^{-0.1 \cdot 9}\right) = 0.6671 - 0.5934 =$ 0.0737. (calculator or computer)

The probability that a computer part lasts between 9 and 11 years is 0.0737.

NOTE: TI-83+ and TI-84: On the home screen, enter e^(-.1*9) - e^(-.1*11).

Example 5.9

Suppose that the length of a phone call, in minutes, is an exponential random variable with decay parameter = $\frac{1}{12}$. If another person arrives at a public telephone just before you, find the probability that you will have to wait more than 5 minutes. Let X = the length of a phone call, in minutes.

Problem ***(Solution on p. 220.)***

What is m, μ, and σ? The probability that you must wait more than 5 minutes is _______ .

NOTE: A summary for exponential distribution is available in "Summary of The Uniform and Exponential Probability Distributions (Section 5.5)".

5.5 Summary of the Uniform and Exponential Probability Distributions[5]

Formula 5.1: Uniform
X = a real number between a and b (in some instances, X can take on the values a and b). a = smallest X ; b = largest X

$X \sim U\,(a,b)$

The mean is $\mu = \frac{a+b}{2}$

The standard deviation is $\sigma = \sqrt{\frac{(b-a)^2}{12}}$

Probability density function: $f\,(X) = \frac{1}{b-a}$ for $a \leq X \leq b$

Area to the Left of x: $P\,(X < x) =$ *(base)(height)*

Area to the Right of x: $P\,(X > x) =$ *(base)(height)*

Area Between c and d: $P\,(c < X < d) = (base)\,(height) = (d-c)\,(height)$.

Formula 5.2: Exponential
$X \sim Exp\,(m)$

X = a real number, 0 or larger. m = the parameter that controls the rate of decay or decline

The mean and standard deviation **are the same.**

$\mu = \sigma = \frac{1}{m}$ and $m = \frac{1}{\mu} = \frac{1}{\sigma}$

The probability density function: $f\,(X) = m \cdot e^{-m \cdot X}$, $X \geq 0$

Area to the Left of x: $P\,(X < x) = 1 - e^{-m \cdot x}$

Area to the Right of x: $P\,(X > x) = e^{-m \cdot x}$

Area Between c and d: $P\,(c < X < d) = P\,(X < d) - P\,(X < c) = \left(1 - e^{-\,m \cdot d}\right) - (1 - e^{-\,m \cdot c}) = e^{-\,m \cdot c} - e^{-\,m \cdot d}$

Percentile, k: $k = \frac{LN(1\text{-}AreaToTheLeft)}{-m}$

[5]This content is available online at <http://cnx.org/content/m16813/1.10/>.

5.6 Practice 1: Uniform Distribution[6]

5.6.1 Student Learning Outcomes

- The student will explore the properties of data with a uniform distribution.

5.6.2 Given

The age of cars in the staff parking lot of a suburban college is uniformly distributed from six months (0.5 years) to 9.5 years.

5.6.3 Describe the Data

Exercise 5.6.1 *(Solution on p. 220.)*
What is being measured here?

Exercise 5.6.2 *(Solution on p. 220.)*
In words, define the Random Variable X.

Exercise 5.6.3 *(Solution on p. 220.)*
Are the data discrete or continuous?

Exercise 5.6.4 *(Solution on p. 220.)*
The interval of values for X is:

Exercise 5.6.5 *(Solution on p. 220.)*
The distribution for X is:

5.6.4 Probability Distribution

Exercise 5.6.6 *(Solution on p. 220.)*
Write the probability density function.

Exercise 5.6.7 *(Solution on p. 220.)*
Graph the probability distribution.

a. Sketch the graph of the probability distribution.

Figure 5.1

[6]This content is available online at <http://cnx.org/content/m16812/1.10/>.

b. Identify the following values:

i. Lowest value for X:
ii. Highest value for X:
iii. Height of the rectangle:
iv. Label for x-axis (words):
v. Label for y-axis (words):

5.6.5 Random Probability

Exercise 5.6.8 *(Solution on p. 220.)*

Find the probability that a randomly chosen car in the lot was less than 4 years old.

a. Sketch the graph. Shade the area of interest.

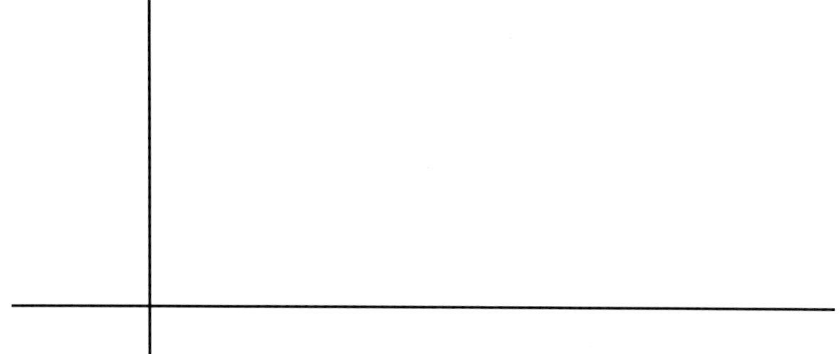

Figure 5.2

b. Find the probability. $P(X < 4) =$

Exercise 5.6.9 *(Solution on p. 220.)*

Out of just the cars less than 7.5 years old, find the probability that a randomly chosen car in the lot was less than 4 years old.

a. Sketch the graph. Shade the area of interest.

Figure 5.3

b. Find the probability. $P(X < 4 \mid X < 7.5) =$

Exercise 5.6.10: Discussion Question

What has changed in the previous two problems that made the solutions different?

5.6.6 Quartiles

Exercise 5.6.11 *(Solution on p. 220.)*

Find the average age of the cars in the lot.

Exercise 5.6.12 *(Solution on p. 220.)*

Find the third quartile of ages of cars in the lot. This means you will have to find the value such that $\frac{3}{4}$, or 75%, of the cars are at most (less than or equal to) that age.

a. Sketch the graph. Shade the area of interest.

Figure 5.4

b. Find the value k such that $P(X < k) = 0.75$.

c. The third quartile is:

5.7 Practice 2: Exponential Distribution[7]

5.7.1 Student Learning Outcomes

- The student will explore the properties of data with a exponential distribution.

5.7.2 Given

Carbon-14 is a radioactive element with a half-life of about 5730 years. Carbon-14 is said to decay exponentially. The decay rate is 0.000121 . We start with 1 gram of carbon-14. We are interested in the time (years) it takes to decay carbon-14.

5.7.3 Describe the Data

Exercise 5.7.1
What is being measured here?

Exercise 5.7.2 *(Solution on p. 221.)*
Are the data discrete or continuous?

Exercise 5.7.3 *(Solution on p. 221.)*
In words, define the Random Variable X.

Exercise 5.7.4 *(Solution on p. 221.)*
What is the decay rate (m)?

Exercise 5.7.5 *(Solution on p. 221.)*
The distribution for X is:

5.7.4 Probability

Exercise 5.7.6 *(Solution on p. 221.)*
Find the amount (percent of 1 gram) of carbon-14 lasting less than 5730 years. This means, find $P(X < 5730)$.

a. Sketch the graph. Shade the area of interest.

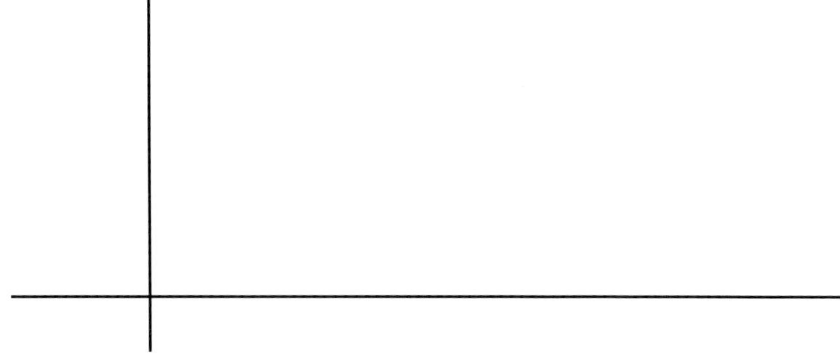

Figure 5.5

[7]This content is available online at <http://cnx.org/content/m16811/1.10/>.

b. Find the probability. $P(X < 5730) =$

Exercise 5.7.7 *(Solution on p. 221.)*

Find the percentage of carbon-14 lasting longer than 10,000 years.

a. Sketch the graph. Shade the area of interest.

Figure 5.6

b. Find the probability. $P(X > 10000) =$

Exercise 5.7.8 *(Solution on p. 221.)*

Thirty percent (30%) of carbon-14 will decay within how many years?

a. Sketch the graph. Shade the area of interest.

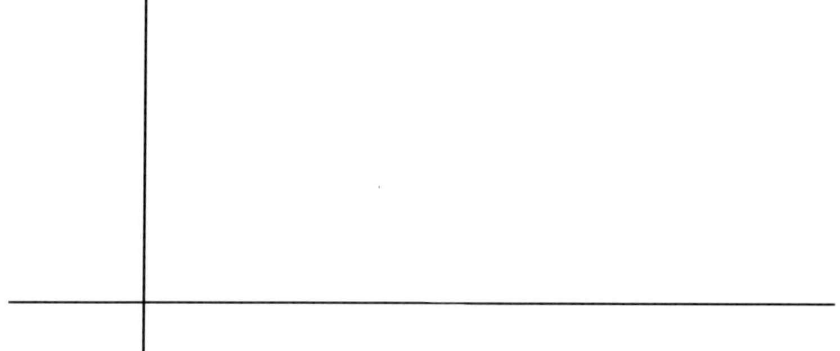

Figure 5.7

b. Find the value k such that $P(X < k) = 0.30$.

5.8 Homework[8]

For each probability and percentile problem, DRAW THE PICTURE!

Exercise 5.8.1

Consider the following experiment. You are one of 100 people enlisted to take part in a study to determine the percent of nurses in America with an R.N. (registered nurse) degree. You ask nurses if they have an R.N. degree. The nurses answer "yes" or "no." You then calculate the percentage of nurses with an R.N. degree. You give that percentage to your supervisor.

a. What part of the experiment will yield discrete data?
b. What part of the experiment will yield continuous data?

Exercise 5.8.2

When age is rounded to the nearest year, do the data stay continuous, or do they become discrete? Why?

Exercise 5.8.3 *(Solution on p. 221.)*

Births are approximately uniformly distributed between the 52 weeks of the year. They can be said to follow a Uniform Distribution from 1 – 53 (spread of 52 weeks).

a. $X\sim$
b. Graph the probability distribution.
c. $f(x)=$
d. $\mu=$
e. $\sigma=$
f. Find the probability that a person is born at the exact moment week 19 starts. That is, find $P(X=19)$.
g. $P(2 < X < 31) =$
h. Find the probability that a person is born after week 40.
i. $P(12 < X \mid X < 28)=$
j. Find the 70th percentile.
k. Find the minimum for the upper quarter.

Exercise 5.8.4

A random number generator picks a number from 1 to 9 in a uniform manner.

a. $X\sim$
b. Graph the probability distribution.
c. $f(x) =$
d. $\mu =$
e. $\sigma =$
f. $P(3.5 < X < 7.25) =$
g. $P(X > 5.67) =$
h. $P(X > 5 \mid X > 3) =$
i. Find the 90th percentile.

Exercise 5.8.5 *(Solution on p. 221.)*

The speed of cars passing through the intersection of Blossom Hill Road and the Almaden Expressway varies from 10 to 35 mph and is uniformly distributed. None of the cars travel over 35 mph through the intersection.

a. $X =$

---

[8]This content is available online at <http://cnx.org/content/m16807/1.10/>.

b. $X \sim$
c. Graph the probability distribution.
d. $f(x) =$
e. $\mu =$
f. $\sigma =$
g. What is the probability that the speed of a car is at most 30 mph?
h. What is the probability that the speed of a car is between 16 and 22 mph.
i. $P(20 < X < 53) =$ State this in a probability question (similar to **g** and **h**), draw the picture, and find the probability.
j. Find the 90th percentile. This means that 90% of the time, the speed is less than _____ mph while passing through the intersection per minute.
k. Find the 75th percentile. In a complete sentence, state what this means. (See **j**.)
l. Find the probability that the speed is more than 24 mph given (or knowing that) it is at least 15 mph.

Exercise 5.8.6

According to a study by Dr. John McDougall of his live-in weight loss program at St. Helena Hospital, the people who follow his program lose between 6 and 15 pounds a month until they approach trim body weight. Let's suppose that the weight loss is uniformly distributed. We are interested in the weight loss of a randomly selected individual following the program for one month. (Source: **The McDougall Program for Maximum Weight Loss** by John A. McDougall, M.D.)

a. $X =$
b. $X \sim$
c. Graph the probability distribution.
d. $f(x) =$
e. $\mu =$
f. $\sigma =$
g. Find the probability that the individual lost more than 10 pounds in a month.
h. Suppose it is known that the individual lost more than 10 pounds in a month. Find the probability that he lost less than 12 pounds in the month.
i. $P(7 < X < 13 \mid X > 9) =$ State this in a probability question (similar to g and h), draw the picture, and find the probability.

Exercise 5.8.7 *(Solution on p. 221.)*

A subway train on the Red Line arrives every 8 minutes during rush hour. We are interested in the length of time a commuter must wait for a train to arrive. The time follows a uniform distribution.

a. $X =$
b. $X \sim$
c. Graph the probability distribution.
d. $f(x) =$
e. $\mu =$
f. $\sigma =$
g. Find the probability that the commuter waits less than one minute.
h. Find the probability that the commuter waits between three and four minutes.
i. 60% of commuters wait more than how long for the train? State this in a probability question (similar to **g** and **h**), draw the picture, and find the probability.

Exercise 5.8.8

The age of a first grader on September 1 at Garden Elementary School is uniformly distributed from 5.8 to 6.8 years. We randomly select one first grader from the class.

a. $X =$
b. $X\sim$
c. Graph the probability distribution.
d. $f(x) =$
e. $\mu =$
f. $\sigma =$
g. Find the probability that she is over 6.5 years.
h. Find the probability that she is between 4 and 6 years.
i. Find the 70th percentile for the age of first graders on September 1 at Garden Elementary School.

Exercise 5.8.9 *(Solution on p. 222.)*

Let $X\sim\text{Exp}(0.1)$

a. decay rate=
b. $\mu =$
c. Graph the probability distribution function.
d. On the above graph, shade the area corresponding to $P(X < 6)$ and find the probability.
e. Sketch a new graph, shade the area corresponding to $P(3 < X < 6)$ and find the probability.
f. Sketch a new graph, shade the area corresponding to $P(X > 7)$ and find the probability.
g. Sketch a new graph, shade the area corresponding to the 40th percentile and find the value.
h. Find the average value of X.

Exercise 5.8.10

Suppose that the length of long distance phone calls, measured in minutes, is known to have an exponential distribution with the average length of a call equal to 8 minutes.

a. $X =$
b. Is X continuous or discrete?
c. $X\sim$
d. $\mu =$
e. $\sigma =$
f. Draw a graph of the probability distribution. Label the axes.
g. Find the probability that a phone call lasts less than 9 minutes.
h. Find the probability that a phone call lasts more than 9 minutes.
i. Find the probability that a phone call lasts between 7 and 9 minutes.
j. If 25 phone calls are made one after another, on average, what would you expect the total to be? Why?

Exercise 5.8.11 *(Solution on p. 222.)*

Suppose that the useful life of a particular car battery, measured in months, decays with parameter 0.025. We are interested in the life of the battery.

a. $X =$
b. Is X continuous or discrete?
c. $X\sim$
d. On average, how long would you expect 1 car battery to last?
e. On average, how long would you expect 9 car batteries to last, if they are used one after another?
f. Find the probability that a car battery lasts more than 36 months.

g. 70% of the batteries last at least how long?

Exercise 5.8.12

The percent of persons (ages 5 and older) in each state who speak a language at home other than English is approximately exponentially distributed with a mean of 9.848 . Suppose we randomly pick a state. (Source: Bureau of the Census, U.S. Dept. of Commerce)

a. $X =$
b. Is X continuous or discrete?
c. $X \sim$
d. $\mu =$
e. $\sigma =$
f. Draw a graph of the probability distribution. Label the axes.
g. Find the probability that the percent is less than 12.
h. Find the probability that the percent is between 8 and 14.
i. The percent of all individuals living in the United States who speak a language at home other than English is 13.8 .

  i. Why is this number different from 9.848%?
  ii. What would make this number higher than 9.848%?

Exercise 5.8.13 *(Solution on p. 222.)*

The time (in years) **after** reaching age 60 that it takes an individual to retire is approximately exponentially distributed with a mean of about 5 years. Suppose we randomly pick one retired individual. We are interested in the time after age 60 to retirement.

a. $X =$
b. Is X continuous or discrete?
c. $X \sim$
d. $\mu =$
e. $\sigma =$
f. Draw a graph of the probability distribution. Label the axes.
g. Find the probability that the person retired after age 70.
h. Do more people retire before age 65 or after age 65?
i. In a room of 1000 people over age 80, how many do you expect will NOT have retired yet?

Exercise 5.8.14

The cost of all maintenance for a car during its first year is approximately exponentially distributed with a mean of $150.

a. $X =$
b. $X \sim$
c. $\mu =$
d. $\sigma =$
e. Draw a graph of the probability distribution. Label the axes.
f. Find the probability that a car required over $300 for maintenance during its first year.

5.8.1 Try these multiple choice problems

The next three questions refer to the following information. The average lifetime of a certain new cell phone is 3 years. The manufacturer will replace any cell phone failing within 2 years of the date of purchase. The lifetime of these cell phones is known to follow an exponential distribution.

Exercise 5.8.15 *(Solution on p. 222.)*

The decay rate is

A. 0.3333
B. 0.5000
C. 2.0000
D. 3.0000

Exercise 5.8.16 *(Solution on p. 222.)*

What is the probability that a phone will fail within 2 years of the date of purchase?

A. 0.8647
B. 0.4866
C. 0.2212
d. 0.9997

Exercise 5.8.17 *(Solution on p. 222.)*

What is the median lifetime of these phones (in years)?

A. 0.1941
B. 1.3863
C. 2.0794
D. 5.5452

The next three questions refer to the following information. The Sky Train from the terminal to the rental car and long term parking center is supposed to arrive every 8 minutes. The waiting times for the train are known to follow a uniform distribution.

Exercise 5.8.18 *(Solution on p. 222.)*

What is the average waiting time (in minutes)?

A. 0.0000
B. 2.0000
C. 3.0000
D. 4.0000

Exercise 5.8.19 *(Solution on p. 222.)*

Find the 30th percentile for the waiting times (in minutes).

A. 2.0000
B. 2.4000
C. 2.750
D. 3.000

Exercise 5.8.20 *(Solution on p. 222.)*

The probability of waiting more than 7 minutes given a person has waited more than 4 minutes is?

A. 0.1250
B. 0.2500
C. 0.5000
D. 0.7500

5.9 Review[9]

Exercise 5.9.1 – Exercise 5.9.7 refer to the following study: A recent study of mothers of junior high school children in Santa Clara County reported that 76% of the mothers are employed in paid positions. Of those mothers who are employed, 64% work full-time (over 35 hours per week), and 36% work part-time. However, out of all of the mothers in the population, 49% work full-time. The population under study is made up of mothers of junior high school children in Santa Clara County.

Let E =employed, Let F =full-time employment

Exercise 5.9.1 *(Solution on p. 222.)*

a. Find the percent of all mothers in the population that NOT employed.
b. Find the percent of mothers in the population that are employed part-time.

Exercise 5.9.2 *(Solution on p. 222.)*
The type of employment is considered to be what type of data?

Exercise 5.9.3 *(Solution on p. 222.)*
In symbols, what does the 36% represent?

Exercise 5.9.4 *(Solution on p. 223.)*
Find the probability that a randomly selected person from the population will be employed OR work full-time.

Exercise 5.9.5 *(Solution on p. 223.)*
Based upon the above information, are being employed AND working part-time:

a. mutually exclusive events? Why or why not?
b. independent events? Why or why not?

Exercise 5.9.6 - Exercise 5.9.7 refer to the following: We randomly pick 10 mothers from the above population. We are interested in the number of the mothers that are employed. Let X =number of mothers that are employed.

Exercise 5.9.6 *(Solution on p. 223.)*
State the distribution for X.

Exercise 5.9.7 *(Solution on p. 223.)*
Find the probability that at least 6 are employed.

Exercise 5.9.8 *(Solution on p. 223.)*
We expect the Statistics Discussion Board to have, on average, 14 questions posted to it per week. We are interested in the number of questions posted to it per day.

a. Define X.
b. What are the values that the random variable may take on?
c. State the distribution for X.
d. Find the probability that from 10 to 14 (inclusive) questions are posted to the Listserv on a randomly picked day.

Exercise 5.9.9 *(Solution on p. 223.)*
A person invests $1000 in stock of a company that hopes to go public in 1 year.

- The probability that the person will lose all his money after 1 year (i.e. his stock will be worthless) is 35%.

[9]This content is available online at <http://cnx.org/content/m16810/1.10/>.

- The probability that the person's stock will still have a value of $1000 after 1 year (i.e. no profit and no loss) is 60%.
- The probability that the person's stock will increase in value by $10,000 after 1 year (i.e. will be worth $11,000) is 5%.

Find the expected PROFIT after 1 year.

Exercise 5.9.10 *(Solution on p. 223.)*

Rachel's piano cost $3000. The average cost for a piano is $4000 with a standard deviation of $2500. Becca's guitar cost $550. The average cost for a guitar is $500 with a standard deviation of $200. Matt's drums cost $600. The average cost for drums is $700 with a standard deviation of $100. Whose cost was lowest when compared to his or her own instrument? Justify your answer.

Exercise 5.9.11 *(Solution on p. 223.)*

For the following data, which of the measures of central tendency would be the LEAST useful: mean, median, mode? Explain why. Which would be the MOST useful? Explain why.

$4, 6, 6, 12, 18, 18, 18, 200$

Exercise 5.9.12 *(Solution on p. 223.)*

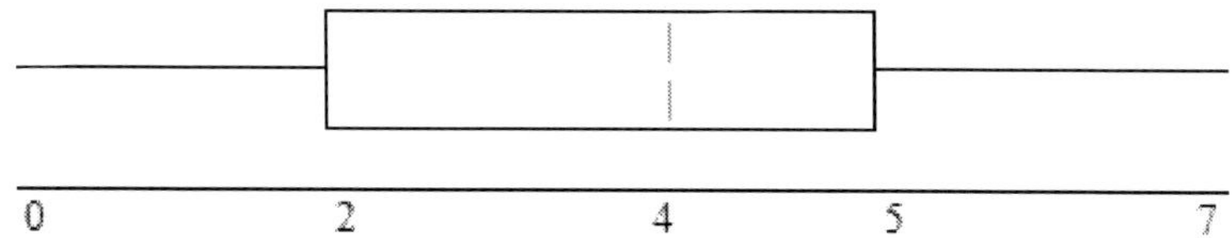

For each statement below, explain why each is either true or false.

a. 25% of the data are at most 5.
b. There is the same amount of data from 4 – 5 as there is from 5 – 7.
c. There are no data values of 3.
d. 50% of the data are 4.

Exercise 5.9.13 – Exercise 5.9.14 refer to the following: 64 faculty members were asked the number of cars they owned (including spouse and children's cars). The results are given in the following graph:

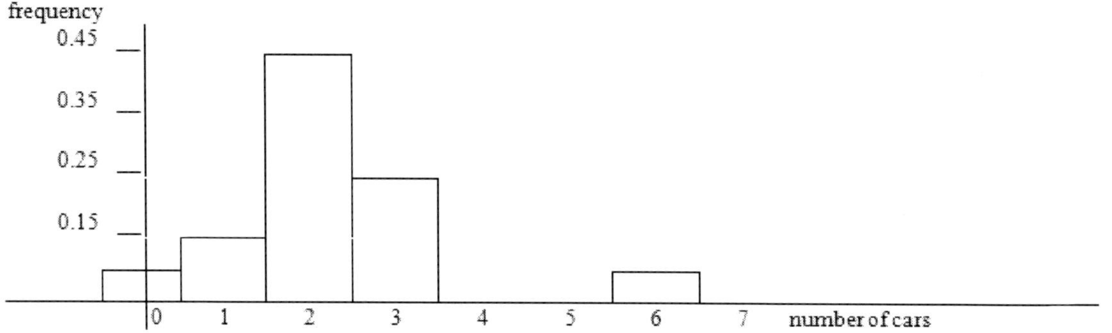

Exercise 5.9.13 *(Solution on p. 223.)*

Find the approximate number of responses that were "3."

Exercise 5.9.14 *(Solution on p. 223.)*

Find the first, second and third quartiles. Use them to construct a box plot of the data.

Exercise 5.9.15 – Exercise 5.9.16 refer to the following study done of the Girls soccer team "Snow Leopards":

Hair Style		Hair Color	
	blond	brown	black
ponytail	3	2	5
plain	2	2	1

Table 5.2

Suppose that one girl from the Snow Leopards is randomly selected.

Exercise 5.9.15 *(Solution on p. 223.)*

Find the probability that the girl has black hair GIVEN that she wears a ponytail.

Exercise 5.9.16 *(Solution on p. 223.)*

Find the probability that the girl wears her hair plain OR has brown hair.

Exercise 5.9.17 *(Solution on p. 223.)*

Find the probability that the girl has blond hair AND that she wears her hair plain.

5.10 Lab: Continuous Distribution[10]

Class Time:

Names:

5.10.1 Student Learning Outcomes:

- The student will compare and contrast empirical data from a random number generator with the Uniform Distribution.

5.10.2 Collect the Data

Use a random number generator to generate 50 values between 0 and 1 (inclusive). List them below. Round the numbers to 4 decimal places or set the calculator MODE to 4 places.

1. Complete the table:

Table 5.3

2. Calculate the following:

   a. $\overline{x} =$
   b. $s =$
   c. 40th percentile =
   d. 3rd quartile =
   e. Median =

5.10.3 Organize the Data

1. Construct a histogram of the empirical data. Make 8 bars.

[10]This content is available online at <http://cnx.org/content/m16803/1.11/>.

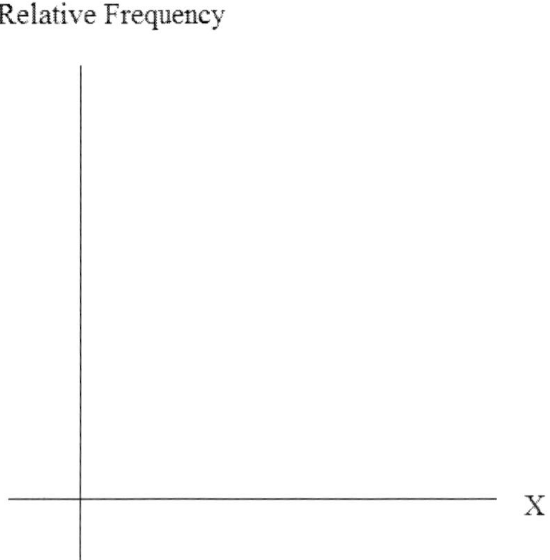

Figure 5.8

2. Construct a histogram of the empirical data. Make 5 bars.

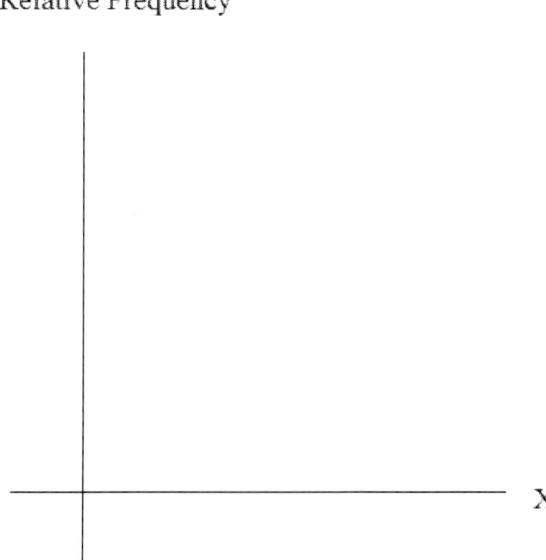

Figure 5.9

5.10.4 Describe the Data

1. Describe the shape of each graph. Use 2 – 3 complete sentences. (Keep it simple. Does the graph go straight across, does it have a V shape, does it have a hump in the middle or at either end, etc.? One way to help you determine a shape, is to roughly draw a smooth curve through the top of the bars.)
2. Describe how changing the number of bars might change the shape.

5.10.5 Theoretical Distribution

1. In words, $X =$
2. The theoretical distribution of X is $X \sim U(0,1)$. Use it for this part.
3. In theory, based upon the distribution $X \sim U(0,1)$, complete the following.
   a. $\mu =$
   b. $\sigma =$
   c. 40th percentile =
   d. 3rd quartile =
   e. median = __________
4. Are the empirical values (the data) in the section titled "Collect the Data" close to the corresponding theoretical values above? Why or why not?

5.10.6 Plot the Data

1. Construct a box plot of the data. Be sure to use a ruler to scale accurately and draw straight edges.
2. Do you notice any potential outliers? If so, which values are they? Either way, numerically justify your answer. (Recall that any DATA are less than Q1 – 1.5*IQR or more than Q3 + 1.5*IQR are potential outliers. IQR means interquartile range.)

5.10.7 Compare the Data

1. For each part below, use a complete sentence to comment on how the value obtained from the data compares to the theoretical value you expected from the distribution in the section titled "Theoretical Distribution."
   a. minimum value:
   b. 40th percentile:
   c. median:
   d. third quartile:
   e. maximum value:
   f. width of IQR:
   g. overall shape:
2. Based on your comments in the section titled "Collect the Data", how does the box plot fit or not fit what you would expect of the distribution in the section titled "Theoretical Distribution?"

5.10.8 Discussion Question

1. Suppose that the number of values generated was 500, not 50. How would that affect what you would expect the empirical data to be and the shape of its graph to look like?

Solutions to Exercises in Chapter 5

Solution to Example 5.5, Problem 1 (p. 196)
0.5714
Solution to Example 5.5, Problem 2 (p. 196)
$\frac{4}{5}$
Solution to Example 5.9 (p. 202)

- $m = \frac{1}{12}$
- $\mu = 12$
- $\sigma = 12$

$P(X > 5) = 0.6592$

Solutions to Practice 1: Uniform Distribution

Solution to Exercise 5.6.1 (p. 204)
The age of cars in the staff parking lot
Solution to Exercise 5.6.2 (p. 204)
X = The age (in years) of cars in the staff parking lot
Solution to Exercise 5.6.3 (p. 204)
Continuous
Solution to Exercise 5.6.4 (p. 204)
0.5 - 9.5
Solution to Exercise 5.6.5 (p. 204)
$X \sim U(0.5, 9.5)$
Solution to Exercise 5.6.6 (p. 204)
$f(x) = \frac{1}{9}$
Solution to Exercise 5.6.7 (p. 204)

b.i. 0.5
b.ii. 9.5
b.iii. Age of Cars
b.iv. $\frac{1}{9}$
b.v. $f(x)$

Solution to Exercise 5.6.8 (p. 205)

b.: $\frac{3.5}{9}$

Solution to Exercise 5.6.9 (p. 205)

b: $\frac{3.5}{7}$

Solution to Exercise 5.6.11 (p. 206)
$\mu = 5$
Solution to Exercise 5.6.12 (p. 206)

b. $k = 7.25$

Solutions to Practice 2: Exponential Distribution

Solution to Exercise 5.7.2 (p. 207)
Continuous
Solution to Exercise 5.7.3 (p. 207)
X = Time (years) to decay carbon-14
Solution to Exercise 5.7.4 (p. 207)
$m = 0.000121$
Solution to Exercise 5.7.5 (p. 207)
$X \sim \text{Exp}(0.000121)$
Solution to Exercise 5.7.6 (p. 207)

b. $P(X < 5730) = 0.5001$

Solution to Exercise 5.7.7 (p. 208)

b. $P(X > 10000) = 0.2982$

Solution to Exercise 5.7.8 (p. 208)

b. $k = 2947.73$

Solutions to Homework

Solution to Exercise 5.8.3 (p. 209)

a. $X \sim U(1, 53)$
c. $f(x) = \frac{1}{52}$ where $1 \leq x \leq 53$
d. 27
e. 15.01
f. 0
g. $\frac{29}{52}$
h. $\frac{13}{52}$
i. $\frac{16}{27}$
j. 37.4
k. 40

Solution to Exercise 5.8.5 (p. 209)

b. $X \sim U(10, 35)$
d. $f(x) = \frac{1}{25}$ where $10 \leq X \leq 35$
e. $\frac{45}{2}$
f. 7.22
g. $\frac{4}{5}$
h. $\frac{6}{25}$
i. $\frac{3}{5}$
j. 32.5
k. 28.75
l. $\frac{11}{20}$

Solution to Exercise 5.8.7 (p. 210)

b. $X \sim U(0, 8)$
d. $f(x) = \frac{1}{8}$ where $0 \leq X \leq 8$
e. 4

- **f.** 2.31
- **g.** $\frac{1}{8}$
- **h.** $\frac{1}{8}$
- **i.** 3.2

Solution to Exercise 5.8.9 (p. 211)

- **a.** 0.1
- **b.** 10
- **d.** 0.4512
- **e.** 0.1920
- **f.** 0.4966
- **g.** 5.11
- **h.** 10

Solution to Exercise 5.8.11 (p. 211)

- **c.** $X \sim Exp\,(0.025)$
- **d.** 40 months
- **e.** 360 months
- **f.** 0.4066
- **g.** 14.27

Solution to Exercise 5.8.13 (p. 212)

- **c.** $X \sim Exp\left(\frac{1}{5}\right)$
- **d.** 5
- **e.** 5
- **g.** 0.1353
- **h.** Before
- **i.** 18.3

Solution to Exercise 5.8.15 (p. 213)
A
Solution to Exercise 5.8.16 (p. 213)
B
Solution to Exercise 5.8.17 (p. 213)
C
Solution to Exercise 5.8.18 (p. 213)
D
Solution to Exercise 5.8.19 (p. 213)
B
Solution to Exercise 5.8.20 (p. 213)
B

Solutions to Review

Solution to Exercise 5.9.1 (p. 214)

- **a.** 24%
- **b.** 27%

Solution to Exercise 5.9.2 (p. 214)
Qualitative

Solution to Exercise 5.9.3 (p. 214)
$P(PT \mid E)$
Solution to Exercise 5.9.4 (p. 214)
0.7336
Solution to Exercise 5.9.5 (p. 214)

a. No,
b. No,

Solution to Exercise 5.9.6 (p. 214)
$B(10, 0.76)$
Solution to Exercise 5.9.7 (p. 214)
0.9330
Solution to Exercise 5.9.8 (p. 214)

a. X = the number of questions posted to the Statistics Listserv per day
b. $x = 0, 1, 2, ...$
c. $X \sim P(2)$
d. 0

Solution to Exercise 5.9.9 (p. 214)
\$150
Solution to Exercise 5.9.10 (p. 215)
Matt
Solution to Exercise 5.9.11 (p. 215)
Mean
Solution to Exercise 5.9.12 (p. 215)

a. False
b. True
c. False
d. False

Solution to Exercise 5.9.13 (p. 215)
16
Solution to Exercise 5.9.14 (p. 215)
2, 2, 3
Solution to Exercise 5.9.15 (p. 216)
$\frac{5}{10} = 0.5$
Solution to Exercise 5.9.16 (p. 216)
$\frac{7}{15}$
Solution to Exercise 5.9.17 (p. 216)
$\frac{2}{15}$

Chapter 6

The Normal Distribution

6.1 The Normal Distribution[1]

6.1.1 Student Learning Objectives

By the end of this chapter, the student should be able to:

- Recognize the normal probability distribution and apply it appropriately.
- Recognize the standard normal probability distribution and apply it appropriately.
- Compare normal probabilities by converting to the standard normal distribution.

6.1.2 Introduction

The normal, a continuous distribution, is the most important of all the distributions. It is widely used and even more widely abused. Its graph is bell-shaped. You see the bell curve in almost all disciplines. Some of these include psychology, business, economics, the sciences, nursing, and, of course, mathematics. Some of your instructors may use the normal distribution to help determine your grade. Most IQ scores are normally distributed. Often real estate prices fit a normal distribution. The normal distribution is extremely important but it cannot be applied to everything in the real world.

In this chapter, you will study the normal distribution, the standard normal, and applications associated with them.

6.1.3 Optional Collaborative Classroom Activity

Your instructor will record the heights of both men and women in your class, separately. Draw histograms of your data. Then draw a smooth curve through each histogram. Is each curve somewhat bell-shaped? Do you think that if you had recorded 200 data values for men and 200 for women that the curves would look bell-shaped? Calculate the mean for each data set. Write the means on the x-axis of the appropriate graph below the peak. Shade the approximate area that represents the probability that one randomly chosen male is taller than 72 inches. Shade the approximate area that represents the probability that one randomly chosen female is shorter than 60 inches. If the total area under each curve is one, does either probability appear to be more than 0.5?

[1]This content is available online at <http://cnx.org/content/m16979/1.9/>.

The normal distribution has two parameters (two numerical descriptive measures), the mean (μ) and the standard deviation (σ). If X is a quantity to be measured that has a normal distribution with mean (μ) and the standard deviation (σ), we designate this by writing

NORMAL:$X \sim N(\mu, \sigma)$

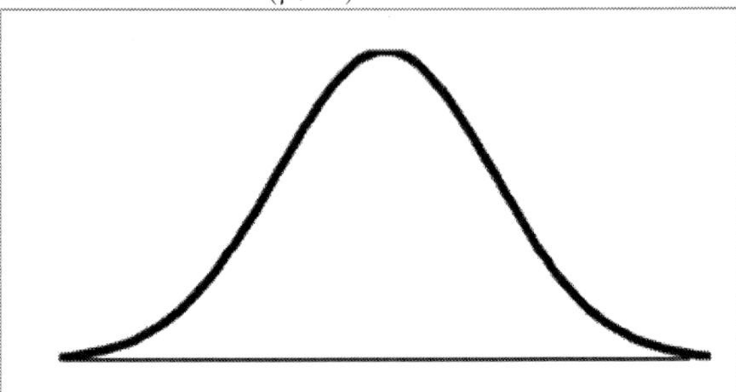

The probability density function is a rather complicated function. **Do not memorize it**. It is not necessary.

$$f(x) = \frac{1}{\sigma \cdot \sqrt{2 \cdot \pi}} \cdot e^{-\frac{1}{2} \cdot \left(\frac{x-\mu}{\sigma}\right)^2}$$

The cumulative distribution function is $P(X < x)$ It is calculated either by a calculator or a computer or it is looked up in a table

The curve is symmetrical about a vertical line drawn through the mean, μ. In theory, the mean is the same as the median since the graph is symmetric about μ. As the notation indicates, the normal distribution depends only on the mean and the standard deviation. Since the area under the curve must equal one, a change in the standard deviation, σ, causes a change in the shape of the curve; the curve becomes fatter or skinnier depending on σ. A change in μ causes the graph to shift to the left or right. This means there are an infinite number of normal probability distributions. One of special interest is called the standard **normal distribution**.

6.2 The Standard Normal Distribution[2]

The **standard normal distribution** is a normal distribution of **standardized values called z-scores. A z-score is measured in units of the standard deviation.** For example, if the mean of a normal distribution is 5 and the standard deviation is 2, the value 11 is 3 standard deviations above (or to the right of) the mean. The calculation is:

$$x = \mu + (z)\sigma = 5 + (3)(2) = 11 \qquad (6.1)$$

The z-score is 3.

The mean for the standard normal distribution is 0 and the standard deviation is 1. The transformation

$z = \frac{x-\mu}{\sigma}$ produces the distribution $Z \sim N(0,1)$. The value x comes from a normal distribution with mean μ and standard deviation σ.

[2]This content is available online at <http://cnx.org/content/m16986/1.7/>.

6.3 Z-scores[3]

If X is a normally distributed random variable and $X \sim N(\mu, \sigma)$, then the z-score is:

$$z = \frac{x - \mu}{\sigma} \tag{6.2}$$

The z-score tells you how many standard deviations that the value x is above (to the right of) or below (to the left of) the mean, μ. Values of x that are larger than the mean have positive z-scores and values of x that are smaller than the mean have negative z-scores. If x equals the mean, then x has a z-score of 0.

Example 6.1
Suppose $X \sim N(5, 6)$. This says that X is a normally distributed random variable with mean $\mu = 5$ and standard deviation $\sigma = 6$. Suppose $x = 17$. Then:

$$z = \frac{x - \mu}{\sigma} = \frac{17 - 5}{6} = 2 \tag{6.3}$$

This means that $x = 17$ is **2 standard deviations** (2σ) above or to the right of the mean $\mu = 5$. The standard deviation is $\sigma = 6$.

Notice that:

$$5 + 2 \cdot 6 = 17 \qquad \textit{(The pattern is } \mu + z\sigma = x.\textit{)} \tag{6.4}$$

Now suppose $x = 1$. Then:

$$z = \frac{x - \mu}{\sigma} = \frac{1 - 5}{6} = -0.67 \qquad \textit{(rounded to two decimal places)} \tag{6.5}$$

This means that $x = 1$ is 0.67 standard deviations (-0.67σ) below or to the left of the mean $\mu = 5$. Notice that:

$5 + (-0.67)(6)$ is approximately equal to 1 (This has the pattern $\mu + (-0.67)\sigma = 1$)

Summarizing, when z is positive, x is above or to the right of μ and when z is negative, x is to the left of or below μ.

Example 6.2
Some doctors believe that a person can lose 5 pounds, on the average, in a month by reducing his/her fat intake and by exercising consistently. Suppose weight loss has a normal distribution. Let X = the amount of weight lost (in pounds) by a person in a month. Use a standard deviation of 2 pounds. $X \sim N(5, 2)$. Fill in the blanks.

Problem 1 ***(Solution on p. 247.)***
Suppose a person **lost** 10 pounds in a month. The z-score when $x = 10$ pounds is $z = 2.5$ (verify). This z-score tells you that $x = 10$ is ________ standard deviations to the ________ (right or left) of the mean _____ (What is the mean?).

Problem 2 ***(Solution on p. 247.)***
Suppose a person **gained** 3 pounds (a negative weight loss). Then $z =$ __________. This z-score tells you that $x = -3$ is ________ standard deviations to the __________ (right or left) of the mean.

Suppose the random variables X and Y have the following normal distributions: $X \sim N(5, 6)$ and $Y \sim N(2, 1)$. If $x = 17$, then $z = 2$. (This was previously shown.) If $y = 4$, what is z?

$$z = \frac{y - \mu}{\sigma} = \frac{4 - 2}{1} = 2 \qquad \textit{where } \mu\textit{=2 and } \sigma\textit{=1.} \tag{6.6}$$

[3]This content is available online at <http://cnx.org/content/m16991/1.7/>.

The z-score for $y = 4$ is $z = 2$. This means that 4 is $z = 2$ standard deviations to the right of the mean. Therefore, $x = 17$ and $y = 4$ are both 2 (of **their**) standard deviations to the right of **their** respective means.

The z-score allows us to compare data that are scaled differently. To understand the concept, suppose $X \sim N(5, 6)$ represents weight gains for one group of people who are trying to gain weight in a 6 week period and $Y \sim N(2, 1)$ measures the same weight gain for a second group of people. A negative weight gain would be a weight loss. Since $x = 17$ and $y = 4$ are each 2 standard deviations to the right of their means, they represent the same weight gain **in relationship to their means**.

6.4 Areas to the Left and Right of x[4]

The arrow in the graph below points to the area to the left of x. This area is represented by the probability $P(X < x)$. Normal tables, computers, and calculators provide or calculate the probability $P(X < x)$.

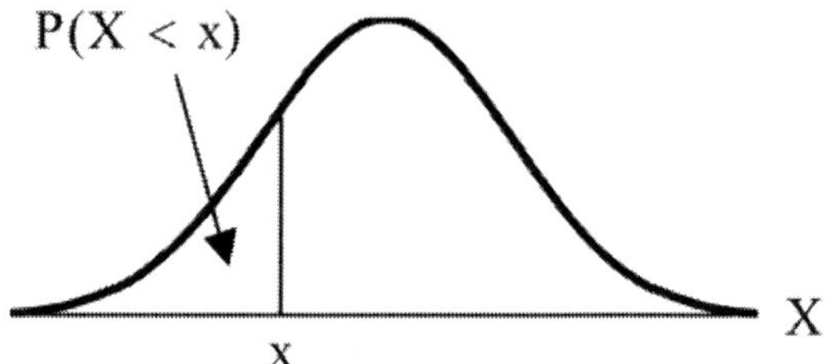

The area to the right is then $P(X > x) = 1 - P(X < x)$**.**

Remember, $P(X < x) =$ **Area to the left** of the vertical line through x.

$P(X > x) = 1 - P(X < x) =$. **Area to the right** of the vertical line through x

$P(X < x)$ is the same as $P(X \leq x)$ and $P(X > x)$ is the same as $P(X \geq x)$ for continuous distributions.

6.5 Calculations of Probabilities[5]

Probabilities are calculated by using technology. There are instructions in the chapter for the TI-83+ and TI-84 calculators.

NOTE: In the Table of Contents for **Collaborative Statistics**, entry **15. Tables** has a link to a table of normal probabilities. Use the probability tables if so desired, instead of a calculator.

Example 6.3
If the area to the left is 0.0228, then the area to the right is $1 - 0.0228 = 0.9772$.

Example 6.4
The final exam scores in a statistics class were normally distributed with a mean of 63 and a standard deviation of 5.

Problem 1
Find the probability that a randomly selected student scored more than 65 on the exam.

[4]This content is available online at <http://cnx.org/content/m16976/1.5/>.
[5]This content is available online at <http://cnx.org/content/m16977/1.9/>.

Solution
Let X = a score on the final exam. $X \sim N(63, 5)$, where $\mu = 63$ and $\sigma = 5$

Draw a graph.

Then, find $P(X > 65)$.

$P(X > 65) = 0.3446$ (calculator or computer)

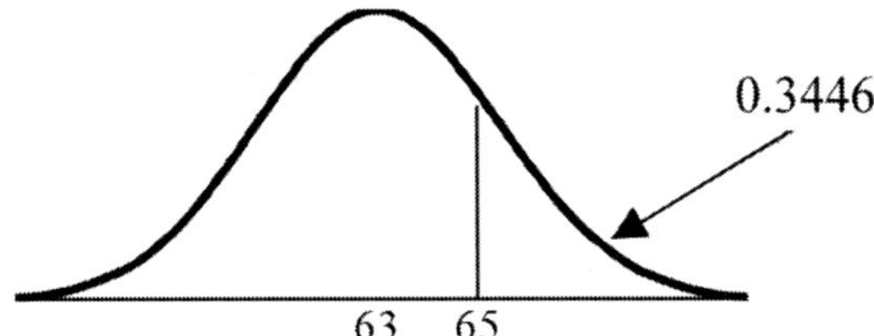

The probability that one student scores more than 65 is 0.3446.

Using the TI-83+ or the TI-84 calculators, the calculation is as follows. Go into `2nd DISTR`.

After pressing `2nd DISTR`, press `2:normalcdf`.

The syntax for the instructions are shown below.

normalcdf(lower value, upper value, mean, standard deviation) For this problem: normalcdf(65,1E99,63,5) = 0.3446. You get 1E99 (= 10^{99}) by pressing `1`, the `EE` key (a 2nd key) and then `99`. Or, you can enter `10^99` instead. The number 10^{99} is way out in the right tail of the normal curve. We are calculating the area between 65 and 10^{99}. In some instances, the lower number of the area might be -1E99 (= -10^{99}). The number -10^{99} is way out in the left tail of the normal curve.

HISTORICAL NOTE: The TI probability program calculates a z-score and then the probability from the z-score. Before technology, the z-score was looked up in a standard normal probability table (because the math involved is too cumbersome) to find the probability. In this example, a standard normal table with area to the left of the z-score was used. You calculate the z-score and look up the area to the left. The probability is the area to the right.

$z = \frac{65-63}{5} = 0.4$. Area to the left is 0.6554. $P(X > 65) = P(Z > 0.4) = 1 - 0.6554 = 0.3446$

Problem 2
Find the probability that a randomly selected student scored less than 85.

Solution
Draw a graph.

Then find $P(X < 85)$. Shade the graph. $P(X < 85) = 1$ (calculator or computer)

The probability that one student scores less than 85 is approximately 1 (or 100%).

The TI-instructions and answer are as follows:

normalcdf(0,85,63,5) = 1 (rounds to 1)

Problem 3
Find the 90th percentile (that is, find the score k that has 90 % of the scores below k and 10% of the scores above k).

Solution
Find the 90th percentile. For each problem or part of a problem, draw a new graph. Draw the x-axis. Shade the area that corresponds to the 90th percentile.

Let k = the 90th percentile. k is located on the x-axis. $P(X < k)$ is the area to the left of k. The 90th percentile k separates the exam scores into those that are the same or lower than k and those that are the same or higher. Ninety percent of the test scores are the same or lower than k and 10% are the same or higher. k is often called a **critical value**.

$k = 69.4$ (calculator or computer)

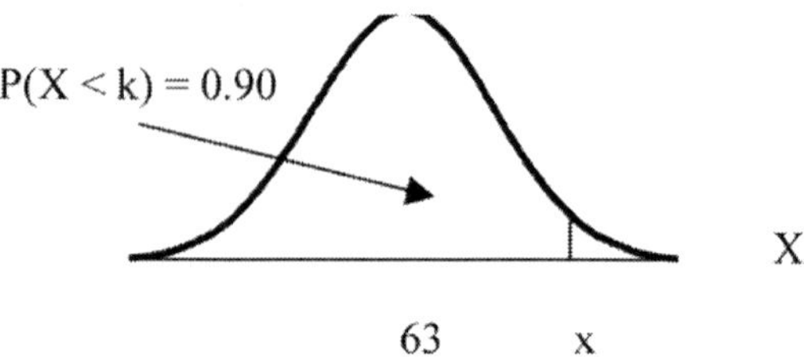

The 90th percentile is 69.4. This means that 90% of the test scores fall at or below 69.4 and 10% fall at or above. For the TI-83+ or TI-84 calculators, use `invNorm` in `2nd DISTR`. invNorm(area to the left, mean, standard deviation) For this problem, invNorm(.90,63,5) = 69.4

Problem 4
Find the 70th percentile (that is, find the score k such that 70% of scores are below k and 30% of the scores are above k).

Solution
Find the 70th percentile.

Draw a new graph and label it appropriately. $k = 65.6$

The 70th percentile is 65.6. This means that 70% of the test scores fall at or below 65.5 and 30% fall at or above.

invNorm(.70,63,5) = 65.6

Example 6.5
More and more households in the United States have at least one computer. The computer is used for office work at home, research, communication, personal finances, education, entertainment, and a myriad of other things. Suppose the average number of hours a household personal computer is used for entertainment is 2 hours per day. Assume the times for entertainment are normally distributed and the standard deviation for the times is half an hour.

Problem 1
Find the probability that a household personal computer is used between 1.8 and 2.75 hours per day.

Solution

Let X = the amount of time (in hours) a household personal computer is used for entertainment. $X \sim N(2, 0.5)$ where $\mu = 2$ and $\sigma = 0.5$.

Find $P(1.8 < X < 2.75)$.

The probability for which you are looking is the area **between** $x = 1.8$ and $x = 2.75$. $P(1.8 < X < 2.75) = 0.5886$

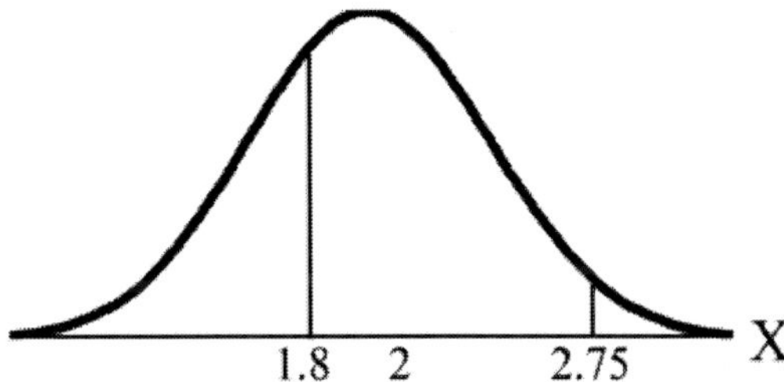

normalcdf(1.8,2.75,2,.5) = 0.5886

The probability that a household personal computer is used between 1.8 and 2.75 hours per day for entertainment is 0.5886.

Problem 2

Find the maximum number of hours per day that the bottom quartile of households use a personal computer for entertainment.

Solution

To find the maximum number of hours per day that the bottom quartile of households uses a personal computer for entertainment, **find the 25th percentile,** k, where $P(X < k) = 0.25$.

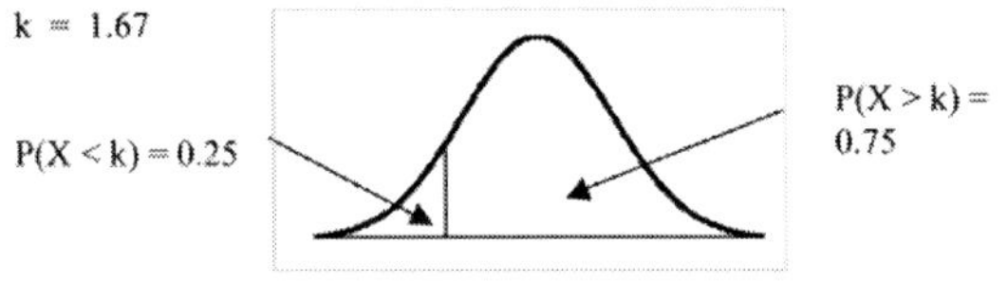

k = 1.67

P(X < k) = 0.25

P(X > k) = 0.75

invNorm(.25,2,.5) = 1.67

The maximum number of hours per day that the bottom quartile of households uses a personal computer for entertainment is 1.67 hours.

6.6 Summary of Formulas[6]

Formula 6.1: Normal Probability Distribution
$X \sim N(\mu, \sigma)$

μ = the mean σ = the standard deviation

Formula 6.2: Standard Normal Probability Distribution
$Z \sim N(0, 1)$

Z = a standardized value (z-score)

mean = 0 standard deviation = 1

Formula 6.3: Finding the kth Percentile
To find the **kth** percentile when the z-score is known: $k = \mu + (z)\sigma$

Formula 6.4: z-score
$z = \frac{x-\mu}{\sigma}$

Formula 6.5: Finding the area to the left
The area to the left: $P(X < x)$

Formula 6.6: Finding the area to the right
The area to the right: $P(X > x) = 1 - P(X < x)$

[6]This content is available online at <http://cnx.org/content/m16987/1.4/>.

6.7 Practice: The Normal Distribution[7]

6.7.1 Student Learning Outcomes

- The student will explore the properties of data with a normal distribution.

6.7.2 Given

The life of Sunshine CD players is normally distributed with a mean of 4.1 years and a standard deviation of 1.3 years. A CD player is guaranteed for 3 years. We are interested in the length of time a CD player lasts.

6.7.3 Normal Distribution

Exercise 6.7.1
Define the Random Variable X in words. $X =$

Exercise 6.7.2
$X\sim$

Exercise 6.7.3 *(Solution on p. 247.)*
Find the probability that a CD player will break down during the guarantee period.

a. Sketch the situation. Label and scale the axes. Shade the region corresponding to the probability.

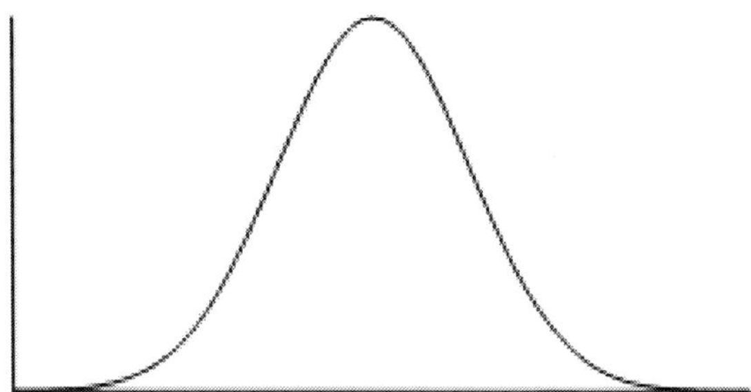

Figure 6.1

b. $P(0 < X <$ ________ $) =$ ________

Exercise 6.7.4 *(Solution on p. 247.)*
Find the probability that a CD player will last between 2.8 and 6 years.

a. Sketch the situation. Label and scale the axes. Shade the region corresponding to the probability.

[7]This content is available online at <http://cnx.org/content/m16983/1.9/>.

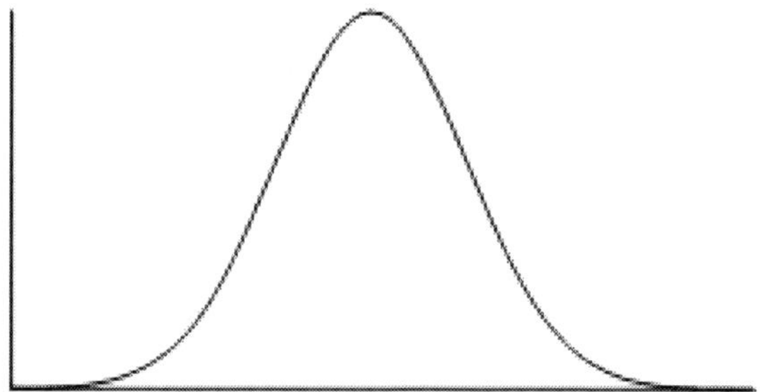

Figure 6.2

b. $P(\underline{\quad 2.8 \quad} < X < \underline{\quad 6 \quad}) = \underline{\qquad\qquad}$

Exercise 6.7.5 ***(Solution on p. 247.)***

Find the 70th percentile of the distribution for the time a CD player lasts.

a. Sketch the situation. Label and scale the axes. Shade the region corresponding to the lower 70%.

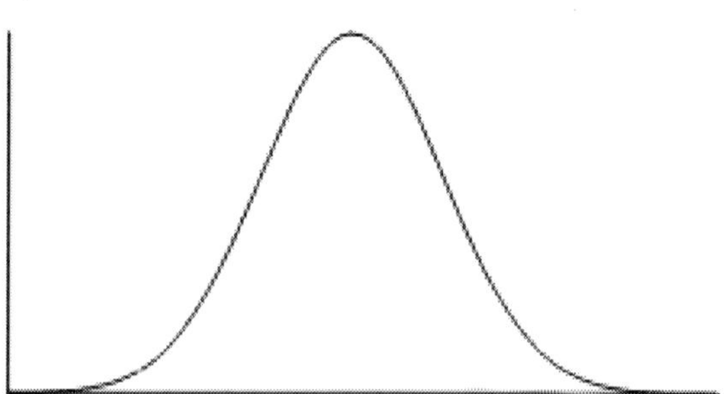

Figure 6.3

b. $P(X < k) = \underline{\qquad\qquad}$. Therefore, $k = \underline{\qquad\qquad}$.

6.8 Homework[8]

Exercise 6.8.1 *(Solution on p. 247.)*

According to a study done by De Anza students, the height for Asian adult males is normally distributed with an average of 66 inches and a standard deviation of 2.5 inches. Suppose one Asian adult male is randomly chosen. Let X =height of the individual.

a. $X\sim$_______(_______,_______)
b. Find the probability that the person is between 65 and 69 inches. Include a sketch of the graph and write a probability statement.
c. Would you expect to meet many Asian adult males over 72 inches? Explain why or why not, and justify your answer numerically.
d. The middle 40% of heights fall between what two values? Sketch the graph and write the probability statement.

Exercise 6.8.2

IQ is normally distributed with a mean of 100 and a standard deviation of 15. Suppose one individual is randomly chosen. Let X =IQ of an individual.

a. $X\sim$_______(_______,_______)
b. Find the probability that the person has an IQ greater than 120. Include a sketch of the graph and write a probability statement.
c. Mensa is an organization whose members have the top 2% of all IQs. Find the minimum IQ needed to qualify for the Mensa organization. Sketch the graph and write the probability statement.
d. The middle 50% of IQs fall between what two values? Sketch the graph and write the probability statement.

Exercise 6.8.3 *(Solution on p. 247.)*

The percent of fat calories that a person in America consumes each day is normally distributed with a mean of about 36 and a standard deviation of 10. Suppose that one individual is randomly chosen. Let X =percent of fat calories.

a. $X\sim$_______(_______,_______)
b. Find the probability that the percent of fat calories a person consumes is more than 40. Graph the situation. Shade in the area to be determined.
c. Find the maximum number for the lower quarter of percent of fat calories. Sketch the graph and write the probability statement.

Exercise 6.8.4

Suppose that the distance of fly balls hit to the outfield (in baseball) is normally distributed with a mean of 250 feet and a standard deviation of 50 feet.

a. If X = distance in feet for a fly ball, then $X\sim$_______(_______,_______)
b. If one fly ball is randomly chosen from this distribution, what is the probability that this ball traveled fewer than 220 feet? Sketch the graph. Scale the horizontal axis X. Shade the region corresponding to the probability. Find the probability.
c. Find the 80th percentile of the distribution of fly balls. Sketch the graph and write the probability statement.

[8] This content is available online at <http://cnx.org/content/m16978/1.12/>.

Exercise 6.8.5 *(Solution on p. 247.)*

In China, 4-year-olds average 3 hours a day unsupervised. Most of the unsupervised children live in rural areas, considered safe. Suppose that the standard deviation is 1.5 hours and the amount of time spent alone is normally distributed. We randomly survey one Chinese 4-year-old living in a rural area. We are interested in the amount of time the child spends alone per day. (Source: **San Jose Mercury News**)

a. In words, define the random variable X. $X =$
b. $X\sim$
c. Find the probability that the child spends less than 1 hour per day unsupervised. Sketch the graph and write the probability statement.
d. What percent of the children spend over 10 hours per day unsupervised?
e. 70% of the children spend at least how long per day unsupervised?

Exercise 6.8.6

In the 1992 presidential election, Alaska's 40 election districts averaged 1956.8 votes per district for President Clinton. The standard deviation was 572.3. (There are only 40 election districts in Alaska.) The distribution of the votes per district for President Clinton was bell-shaped. Let $X =$ number of votes for President Clinton for an election district. (Source: **The World Almanac and Book of Facts**)

a. State the approximate distribution of X. $X\sim$
b. Is 1956.8 a population mean or a sample mean? How do you know?
c. Find the probability that a randomly selected district had fewer than 1600 votes for President Clinton. Sketch the graph and write the probability statement.
d. Find the probability that a randomly selected district had between 1800 and 2000 votes for President Clinton.
e. Find the third quartile for votes for President Clinton.

Exercise 6.8.7 *(Solution on p. 247.)*

Suppose that the duration of a particular type of criminal trial is known to be normally distributed with a mean of 21 days and a standard deviation of 7 days.

a. In words, define the random variable X. $X =$
b. $X\sim$
c. If one of the trials is randomly chosen, find the probability that it lasted at least 24 days. Sketch the graph and write the probability statement.
d. 60% of all of these types of trials are completed within how many days?

Exercise 6.8.8

Terri Vogel, an amateur motorcycle racer, averages 129.71 seconds per 2.5 mile lap (in a 7 lap race) with a standard deviation of 2.28 seconds . The distribution of her race times is normally distributed. We are interested in one of her randomly selected laps. (Source: log book of Terri Vogel)

a. In words, define the random variable X. $X =$
b. $X\sim$
c. Find the percent of her laps that are completed in less than 130 seconds.
d. The fastest 3% of her laps are under _______ .
e. The middle 80% of her laps are from _______ seconds to _______ seconds.

Exercise 6.8.9 *(Solution on p. 247.)*

Thuy Dau, Ngoc Bui, Sam Su, and Lan Voung conducted a survey as to how long customers at Lucky claimed to wait in the checkout line until their turn. Let X =time in line. Below are the ordered real data (in minutes):

0.50	4.25	5	6	7.25
1.75	4.25	5.25	6	7.25
2	4.25	5.25	6.25	7.25
2.25	4.25	5.5	6.25	7.75
2.25	4.5	5.5	6.5	8
2.5	4.75	5.5	6.5	8.25
2.75	4.75	5.75	6.5	9.5
3.25	4.75	5.75	6.75	9.5
3.75	5	6	6.75	9.75
3.75	5	6	6.75	10.75

Table 6.1

a. Calculate the sample mean and the sample standard deviation.
b. Construct a histogram. Start the $x - axis$ at -0.375 and make bar widths of 2 minutes.
c. Draw a smooth curve through the midpoints of the tops of the bars.
d. In words, describe the shape of your histogram and smooth curve.
e. Let the sample mean approximate μ and the sample standard deviation approximate σ. The distribution of X can then be approximated by $X\sim$
f. Use the distribution in (e) to calculate the probability that a person will wait fewer than 6.1 minutes.
g. Determine the cumulative relative frequency for waiting less than 6.1 minutes.
h. Why aren't the answers to (f) and (g) exactly the same?
i. Why are the answers to (f) and (g) as close as they are?
j. If only 10 customers were surveyed instead of 50, do you think the answers to (f) and (g) would have been closer together or farther apart? Explain your conclusion.

Exercise 6.8.10

Suppose that Ricardo and Anita attend different colleges. Ricardo's GPA is the same as the average GPA at his school. Anita's GPA is 0.70 standard deviations above her school average. In complete sentences, explain why each of the following statements may be false.

a. Ricardo's actual GPA is lower than Anita's actual GPA.
b. Ricardo is not passing since his z-score is zero.
c. Anita is in the 70th percentile of students at her college.

Exercise 6.8.11 *(Solution on p. 248.)*

Below is the number of AIDS cases for Santa Clara County by year of diagnosis.

Year	# cases		Year	# cases		Year	# cases		Year	# cases
1983	10		1990	225		1997	170		2004	95
1984	26		1991	243		1998	137		2005	98
1985	60		1992	357		1999	137		2006	112
1986	76		1993	382		2000	121		2007	81
1987	134		1994	277		2001	119			
1988	151		1995	249		2002	131			
1989	175		1996	197		2003	116			

Table 6.2

a. Calculate the sample mean and the sample standard deviation for the number of AIDS cases (the data).
b. Construct a histogram of the data.
c. Draw a smooth curve through the midpoints of the tops of the bars of the histogram.
d. In words, describe the shape of your histogram and smooth curve.
e. Let the sample mean approximate μ and the sample standard deviation approximate σ. The distribution of X can then be approximated by $X\sim$
f. Use the distribution in (e) to calculate the probability that the number of AIDS cases is less than 150.
g. Determine the cumulative relative frequency that the number of AIDS cases is less than 150. Hint: Order the data and count the number of AIDS cases that are less than 150. Divide by the total number of AIDS cases.
h. Why aren't the answers to (f) and (g) exactly the same?

6.8.1 Try These Multiple Choice Questions

The questions below refer to the following: The patient recovery time from a particular surgical procedure is normally distributed with a mean of 5.3 days and a standard deviation of 2.1 days.

Exercise 6.8.12 *(Solution on p. 248.)*

What is the median recovery time?

A. 2.7
B. 5.3
C. 7.4
D. 2.1

Exercise 6.8.13 *(Solution on p. 248.)*

What is the z-score for a patient who takes 10 days to recover?

A. 1.5
B. 0.2
C. 2.2
D. 7.3

Exercise 6.8.14 *(Solution on p. 248.)*

What is the probability of spending more than 2 days in recovery?

A. 0.0580
B. 0.8447
C. 0.0553
D. 0.9420

Exercise 6.8.15 *(Solution on p. 248.)*

The 90th percentile for recovery times is?

A. 8.89
B. 7.07
C. 7.99
D. 4.32

The questions below refer to the following: The length of time to find a parking space at 9 A.M. follows a normal distribution with a mean of 5 minutes and a standard deviation of 2 minutes.

Exercise 6.8.16 *(Solution on p. 248.)*

Based upon the above information and numerically justified, would you be surprised if it took less than 1 minute to find a parking space?

A. Yes
B. No
C. Unable to determine

Exercise 6.8.17 *(Solution on p. 248.)*

Find the probability that it takes at least 8 minutes to find a parking space.

A. 0.0001
B. 0.9270
C. 0.1862
D. 0.0668

Exercise 6.8.18 *(Solution on p. 248.)*

Seventy percent of the time, it takes more than how many minutes to find a parking space?

A. 1.24
B. 2.41
C. 3.95
D. 6.05

Exercise 6.8.19 *(Solution on p. 248.)*

If the mean is significantly greater than the standard deviation, which of the following statements is true?

I . The data cannot follow the uniform distribution.
II . The data cannot follow the exponential distribution..
III . The data cannot follow the normal distribution.

A. I only
B. II only
C. III only
D. I, II, and III

6.9 Review[9]

The next two questions refer to: $X \sim U\,(3, 13)$

Exercise 6.9.1 *(Solution on p. 248.)*

Explain which of the following are false and which are true.

a: $f\,(x) = \frac{1}{10}, 3 \leq x \leq 13$
b: There is no mode.
c: The median is less than the mean.
d: $P\,(X > 10) = P\,(X \leq 6)$

Exercise 6.9.2 *(Solution on p. 248.)*

Calculate:

a: Mean
b: Median
c: 65th percentile.

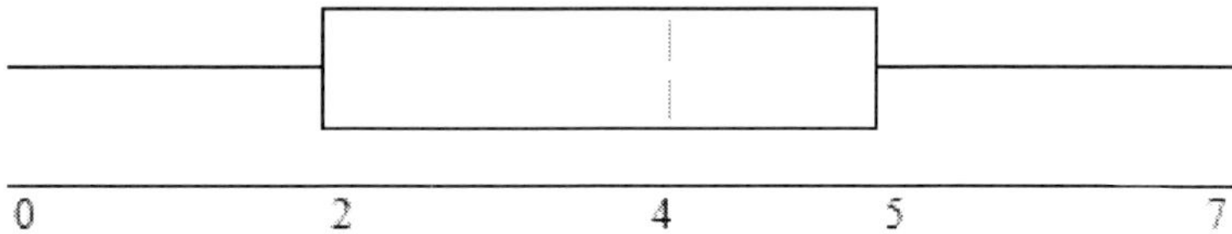

Exercise 6.9.3 *(Solution on p. 248.)*

Which of the following is true for the above box plot?

a: 25% of the data are at most 5.
b: There is about the same amount of data from 4 – 5 as there is from 5 – 7.
c: There are no data values of 3.
d: 50% of the data are 4.

Exercise 6.9.4 *(Solution on p. 248.)*

If $P\,(G \mid H) = P\,(G)$, then which of the following is correct?

A: G and H are mutually exclusive events.
B: $P\,(G) = P\,(H)$
C: Knowing that H has occurred will affect the chance that G will happen.
D: G and H are independent events.

Exercise 6.9.5 *(Solution on p. 248.)*

If $P\,(J) = 0.3$, $P\,(K) = 0.6$, and J and K are independent events, then explain which are correct and which are incorrect.

A: $P\,(JandK) = 0$
B: $P\,(JorK) = 0.9$
C: $P\,(JorK) = 0.72$
D: $P\,(J) \neq P\,(J \mid K)$

[9]This content is available online at <http://cnx.org/content/m16985/1.9/>.

Exercise 6.9.6 *(Solution on p. 249.)*

On average, 5 students from each high school class get full scholarships to 4-year colleges. Assume that most high school classes have about 500 students.

X = the number of students from a high school class that get full scholarships to 4-year school. Which of the following is the distribution of X?

A. P(5)
B. B(500,5)
C. Exp(1/5)
D. N(5, (0.01)(0.99)/500)

6.10 Lab 1: Normal Distribution (Lap Times)[10]

Class Time:

Names:

6.10.1 Student Learning Outcome:

- The student will compare and contrast empirical data and a theoretical distribution to determine if Terry Vogel's lap times fit a continuous distribution.

6.10.2 Directions:

Round the relative frequencies and probabilities to 4 decimal places. Carry all other decimal answers to 2 places.

6.10.3 Collect the Data

1. Use the data from Terri Vogel's Log Book (Section 14.3.1: Lap Times). Use a Stratified Sampling Method by Lap (Races 1 – 20) and a random number generator to pick 6 lap times from each stratum. Record the lap times below for Laps 2 – 7.

________	________	________	________	________	________
________	________	________	________	________	________
________	________	________	________	________	________
________	________	________	________	________	________
________	________	________	________	________	________
________	________	________	________	________	________

Table 6.3

2. Construct a histogram. Make 5 - 6 intervals. Sketch the graph using a ruler and pencil. Scale the axes.

[10]This content is available online at <http://cnx.org/content/m16981/1.16/>.

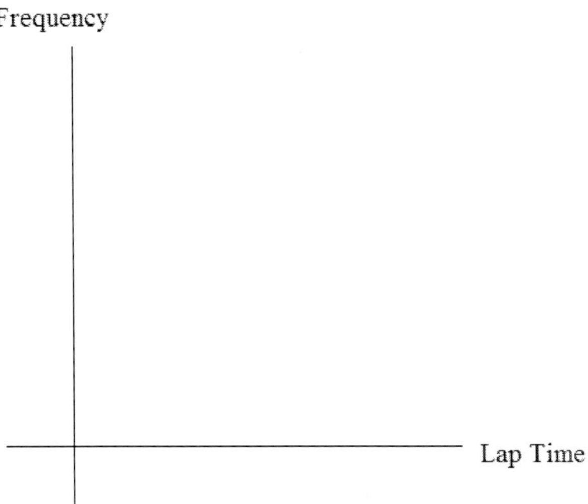

Figure 6.4

3. Calculate the following.
   a. $\overline{x} =$
   b. $s =$
4. Draw a smooth curve through the tops of the bars of the histogram. Use 1 – 2 complete sentences to describe the general shape of the curve. (Keep it simple. Does the graph go straight across, does it have a V-shape, does it have a hump in the middle or at either end, etc.?)

6.10.4 Analyze the Distribution

Using your sample mean, sample standard deviation, and histogram to help, what was the approximate theoretical distribution of the data?

- $X \sim$
- How does the histogram help you arrive at the approximate distribution?

6.10.5 Describe the Data

Use the Data from the section titled "Collect the Data" to complete the following statements.

- The IQR goes from __________ to __________.
- IQR = __________. (IQR=Q3-Q1)
- The 15th percentile is:
- The 85th percentile is:
- The median is:
- The empirical probability that a randomly chosen lap time is more than 130 seconds =
- Explain the meaning of the 85th percentile of this data.

6.10.6 Theoretical Distribution

Using the theoretical distribution from the section titled "Analyse the Distribution" complete the following statements:

- The IQR goes from __________ to __________.
- IQR =
- The 15th percentile is:
- The 85th percentile is:
- The median is:
- The probability that a randomly chosen lap time is more than 130 seconds =
- Explain the meaning the 85th percentile of this distribution.

6.10.7 Discussion Questions

- Do the data from the section titled "Collect the Data" give a close approximation to the theoretical distibution in the section titled "Analyze the Distribution"? In complete sentences and comparing the result in the sections titled "Describe the Data" and "Theoretical Distribution", explain why or why not.

6.11 Lab 2: Normal Distribution (Pinkie Length)[11]

Class Time:

Names:

6.11.1 Student Learning Outcomes:

- The student will compare empirical data and a theoretical distribution to determine if an everyday experiment fits a continuous distribution.

6.11.2 Collect the Data

Measure the length of your pinkie finger (in cm.)

1. Randomly survey 30 adults. Round to the nearest 0.5 cm.

______	______	______	______	______
______	______	______	______	______
______	______	______	______	______
______	______	______	______	______
______	______	______	______	______
______	______	______	______	______

Table 6.4

2. Construct a histogram. Make 5-6 intervals. Sketch the graph using a ruler and pencil. Scale the axes.

Frequency

Length of Finger

3. Calculate the Following

[11]This content is available online at <http://cnx.org/content/m16980/1.13/>.

a. $\overline{x} =$
b. $s =$

4. Draw a smooth curve through the top of the bars of the histogram. Use 1-2 complete sentences to describe the general shape of the curve. (Keep it simple. Does the graph go straight across, does it have a V-shape, does it have a hump in the middle or at either end, etc.?)

6.11.3 Analyze the Distribution

Using your sample mean, sample standard deviation, and histogram to help, what was the approximate theoretical distribution of the data from the section titled "Collect the Data"?

- $X \sim$
- How does the histogram help you arrive at the approximate distribution?

6.11.4 Describe the Data

Using the data in the section titled "Collect the Data" complete the following statements. (Hint: order the data)

REMEMBER: $(IQR = Q3 - Q1)$

- IQR =
- 15th percentile is:
- 85th percentile is:
- Median is:
- What is the empirical probability that a randomly chosen pinkie length is more than 6.5 cm?
- Explain the meaning the 85th percentile of this data.

6.11.5 Theoretical Distribution

Using the Theoretical Distribution in the section titled "Analyze the Distribution"

- IQR =
- 15th percentile is:
- 85th percentile is:
- Median is:
- What is the empirical probability that a randomly chosen pinkie length is more than 6.5 cm?
- Explain the meaning of the 85th percentile of this data.

6.11.6 Discussion Questions

- Do the data from the section entitled "Collect the Data" give a close approximation to the theoretical distribution in "Analyze the Distribution." In complete sentences and comparing the results in the sections titled "Describe the Data" and "Theoretical Distribution", explain why or why not.

Solutions to Exercises in Chapter 6

Solution to Example 6.2, Problem 1 (p. 227)
This z-score tells you that $x = 10$ is **2.5** standard deviations to the **right** of the mean **5**.
Solution to Example 6.2, Problem 2 (p. 227)
z = **-4**. This z-score tells you that $x = -3$ is **4** standard deviations to the **left** of the mean.

Solutions to Practice: The Normal Distribution

Solution to Exercise 6.7.3 (p. 233)

b. $3, 0.1979$

Solution to Exercise 6.7.4 (p. 233)

b. $2.8, 6, 0.7694$

Solution to Exercise 6.7.5 (p. 234)

b. $0.70, 4.78$years

Solutions to Homework

Solution to Exercise 6.8.1 (p. 235)

a. $N(66, 2.5)$
b. 0.5404
c. No
d. Between 64.7 and 67.3 inches

Solution to Exercise 6.8.3 (p. 235)

a. $N(36,10)$
b. 0.3446
c. 29.3

Solution to Exercise 6.8.5 (p. 236)

a. the time (in hours) a 4-year-old in China spends unsupervised per day
b. $N(3, 1.5)$
c. 0.0912
d. 0
e. 2.21 hours

Solution to Exercise 6.8.7 (p. 236)

a. The duration of a criminal trial
b. $N(21, 7)$
c. 0.3341
d. 22.77

Solution to Exercise 6.8.9 (p. 237)

a. The sample mean is 5.51 and the sample standard deviation is 2.15
e. $N(5.51, 2.15)$
f. 0.6081
g. 0.64

Solution to Exercise 6.8.11 (p. 237)

a. The sample mean is 155.16 and the sample standard deviation is 92.1605.
e. $N\left(155.16, 92.1605\right)$
f. 0.4315
g. 0.3408

Solution to Exercise 6.8.12 (p. 238)
B
Solution to Exercise 6.8.13 (p. 238)
C
Solution to Exercise 6.8.14 (p. 239)
D
Solution to Exercise 6.8.15 (p. 239)
C
Solution to Exercise 6.8.16 (p. 239)
C
Solution to Exercise 6.8.17 (p. 239)
D
Solution to Exercise 6.8.18 (p. 239)
C
Solution to Exercise 6.8.19 (p. 239)
B

Solutions to Review

Solution to Exercise 6.9.1 (p. 240)

a: True
b: True
c: False – the median and the mean are the same for this symmetric distribution
d: True

Solution to Exercise 6.9.2 (p. 240)

a: 8
b: 8
c: $P\left(X < k\right) = 0.65 = \left(k - 3\right) * \left(\frac{1}{10}\right)$. $k = 9.5$

Solution to Exercise 6.9.3 (p. 240)

a: False – $\frac{3}{4}$ of the data are at most 5
b: True – each quartile has 25% of the data
c: False – that is unknown
d: False – 50% of the data are 4 or less

Solution to Exercise 6.9.4 (p. 240)
D
Solution to Exercise 6.9.5 (p. 240)

A: False - J and K are independent so they are not mutually exclusive which would imply dependency (meaning P(J and K) is not 0).
B: False - see answer C.
C: True - P(J or K) = P(J) + P(K) - P(J and K) = P(J) + P(K) - P(J)P(K) = 0.3 + 0.6 - (0.3)(0.6) = 0.72. Note that P(J and K) = P(J)P(K) because J and K are independent.

D: False - J and K are independent so P(J) = P(J | K).

Solution to Exercise 6.9.6 (p. 241)
A

Chapter 7

The Central Limit Theorem

7.1 The Central Limit Theorem[1]

7.1.1 Student Learning Objectives

By the end of this chapter, the student should be able to:

- Recognize the Central Limit Theorem problems.
- Classify continuous word problems by their distributions.
- Apply and interpret the Central Limit Theorem for Averages.
- Apply and interpret the Central Limit Theorem for Sums.

7.1.2 Introduction

What does it mean to be average? Why are we so concerned with averages? Two reasons are that they give us a middle ground for comparison and they are easy to calculate. In this chapter, you will study averages and the Central Limit Theorem.

The Central Limit Theorem (CLT for short) is one of the most powerful and useful ideas in all of statistics. Both alternatives are concerned with drawing finite samples of size n from a population with a known mean, μ, and a known standard deviation, σ. The first alternative says that if we collect samples of size n and n is "large enough," calculate each sample's mean, and create a histogram of those means, then the resulting histogram will tend to have an approximate normal bell shape. The second alternative says that if we again collect samples of size n that are "large enough," calculate the sum of each sample and create a histogram, then the resulting histogram will again tend to have a normal bell-shape.

In either case, it does not matter what the distribution of the original population is, or whether you even need to know it. The important fact is that the sample means (averages) and the sums tend to follow the normal distribution. And, the rest you will learn in this chapter.

The size of the sample, n, that is required in order to be to be 'large enough' depends on the original population from which the samples are drawn. If the original population is far from normal then more observations are needed for the sample averages or the sample sums to be normal. **Sampling is done with replacement.**

[1]This content is available online at <http://cnx.org/content/m16953/1.12/>.

Optional Collaborative Classroom Activity

Do the following example in class: Suppose 8 of you roll 1 fair die 10 times, 7 of you roll 2 fair dice 10 times, 9 of you roll 5 fair dice 10 times, and 11 of you roll 10 fair dice 10 times. (The 8, 7, 9, and 11 were randomly chosen.)

Each time a person rolls more than one die, he/she calculates the **average** of the faces showing. For example, one person might roll 5 fair dice and get a 2, 2, 3, 4, 6 on one roll.

The average is $\frac{2+2+3+4+6}{5} = 3.4$. The 3.4 is one average when 5 fair dice are rolled. This same person would roll the 5 dice 9 more times and calculate 9 more averages for a total of 10 averages.

Your instructor will pass out the dice to several people as described above. Roll your dice 10 times. For each roll, record the faces and find the average. Round to the nearest 0.5.

Your instructor (and possibly you) will produce one graph (it might be a histogram) for 1 die, one graph for 2 dice, one graph for 5 dice, and one graph for 10 dice. Since the "average" when you roll one die, is just the face on the die, what distribution do these "averages" appear to be representing?

Draw the graph for the averages using 2 dice. Do the averages show any kind of pattern?

Draw the graph for the averages using 5 dice. Do you see any pattern emerging?

Finally, draw the graph for the averages using 10 dice. Do you see any pattern to the graph? What can you conclude as you increase the number of dice?

As the number of dice rolled increases from 1 to 2 to 5 to 10, the following is happening:

1. The average of the averages remains approximately the same.
2. The spread of the averages (the standard deviation of the averages) gets smaller.
3. The graph appears steeper and thinner.

You have just demonstrated the Central Limit Theorem (CLT).

The Central Limit Theorem tells you that as you increase the number of dice, **the sample means (averages) tend toward a normal distribution.**

7.2 The Central Limit Theorem for Sample Means (Averages)[2]

Suppose X is a random variable with a distribution that may be known or unknown (it can be any distribution). Using a subscript that matches the random variable, suppose:

a. μ_X = the mean of X
b. σ_X = the standard deviation of X

If you draw random samples of size n, then as n increases, the random variable $\overline{X}$ which consists of sample means, tends to be **normally distributed** and

$$\overline{X} \sim N\left(\mu_X, \frac{\sigma_X}{\sqrt{n}}\right)$$

The Central Limit Theorem for Sample Means (Averages) says that if you keep drawing larger and larger samples (like rolling 1, 2, 5, and, finally, 10 dice) and **calculating their means** the sample means (averages)

[2]This content is available online at <http://cnx.org/content/m16947/1.14/>.

form their own **normal distribution**. This distribution has the same mean as the original distribution and a variance that equals the original variance divided by n, the sample size. n is the number of values that are averaged together not the number of times the experiment is done.

The random variable $\overline{X}$ has a different z-score associated with it than the random variable X. $\overline{x}$ is the value of $\overline{X}$ in one sample.

$$z = \frac{\overline{x} - \mu_X}{\left(\frac{\sigma_X}{\sqrt{n}}\right)} \tag{7.1}$$

μ_X is both the average of X and of $\overline{X}$.

$\sigma_{\overline{X}} = \frac{\sigma_X}{\sqrt{n}}$ = standard deviation of $\overline{X}$ and is called the **standard error of the mean.**

Example 7.1

An unknown distribution has a mean of 90 and a standard deviation of 15. Samples of size $n = 25$ are drawn randomly from the population.

Problem 1

Find the probability that the **sample mean** is between 85 and 92.

Solution

Let X = one value from the original unknown population. The probability question asks you to find a probability for the **sample mean (or average)**.

Let $\overline{X}$ = the mean or average of a sample of size 25. Since $\mu_X = 90$, $\sigma_X = 15$, and $n = 25$;

then $\overline{X} \sim N\left(90, \frac{15}{\sqrt{25}}\right)$

Find $P\left(85 < \overline{X} < 92\right)$ Draw a graph.

$P\left(85 < \overline{X} < 92\right) = 0.6997$

The probability that the sample mean is between 85 and 92 is 0.6997.

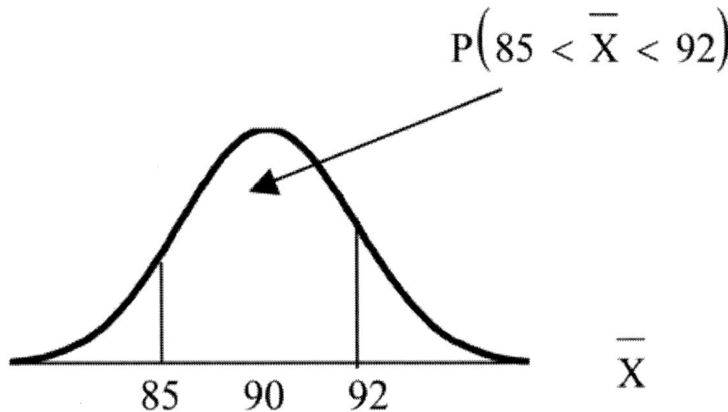

TI-83: `normalcdf`(lower value, upper value, mean for averages, `stdev` for averages)

`stdev` = standard deviation

The parameter list is abbreviated (lower, upper, μ, $\frac{\sigma}{\sqrt{n}}$)

$\texttt{normalcdf}\left(85, 92, 90, \frac{15}{\sqrt{25}}\right) = 0.6997$

Problem 2

Find the average value that is 2 standard deviations above the the mean of the averages.

Solution

To find the average value that is 2 standard deviations above the mean of the averages, use the formula

value = $\mu_X + (\#ofSTDEVs)\left(\frac{\sigma_X}{\sqrt{n}}\right)$

value = $90 + 2 \cdot \frac{15}{\sqrt{25}} = 96$

So, the average value that is 2 standard deviations above the mean of the averages is 96.

Example 7.2

The length of time, in hours, it takes an "over 40" group of people to play one soccer match is normally distributed with a **mean of 2 hours** and a **standard deviation of 0.5 hours**. A **sample of size** n **= 50** is drawn randomly from the population.

Problem 1

Find the probability that the **sample mean** is between 1.8 hours and 2.3 hours.

Solution

Let X = the time, in hours, it takes to play one soccer match.

The probability question asks you to find a probability for the **sample mean or average time, in hours**, it takes to play one soccer match.

Let $\overline{X}$ = the **average** time, in hours, it takes to play one soccer match.

Problem 2 ***(Solution on p. 284.)***

If μ_X = ________, σ_X = ________, and n = ________, then $\overline{X} \sim N($______, ______$)$ by the Central Limit Theorem for Averages of Sample Means.

Find $P\left(1.8 < \overline{X} < 2.3\right)$. Draw a graph.

$P\left(1.8 < \overline{X} < 2.3\right) = 0.9977$

$\texttt{normalcdf}\left(1.8, 2.3, 2, \frac{.5}{\sqrt{50}}\right) = 0.9977$

The probability that the sample mean is between 1.8 hours and 2.3 hours is ______.

7.3 The Central Limit Theorem for Sums[3]

Suppose X is a random variable with a distribution that may be **known or unknown** (it can be any distribution) and suppose:

[3]This content is available online at <http://cnx.org/content/m16948/1.10/>.

a. μ_X = the mean of X
b. σ_X = the standard deviation of X

If you draw random samples of size n, then as n increases, the random variable ΣX which consists of sums tends to be **normally distributed** and

$\Sigma X \sim N\left(n \cdot \mu_X, \sqrt{n} \cdot \sigma_X\right)$

The Central Limit Theorem for Sums says that if you keep drawing larger and larger samples and taking their sums, the sums form their own normal distribution. **The distribution has a mean equal to the original mean multiplied by the sample size and a standard deviation equal to the original standard deviation multiplied by the square root of the sample size.**

The random variable ΣX has the following z-score associated with it:

a. Σx is one sum.
b. $z = \frac{\Sigma x - n \cdot \mu_X}{\sqrt{n} \cdot \sigma_X}$

a. $n \cdot \mu_X$ = the mean of ΣX
b. $\sqrt{n} \cdot \sigma_X$ = standard deviation of ΣX

Example 7.3
An unknown distribution has a mean of 90 and a standard deviation of 15. A sample of size 80 is drawn randomly from the population.

Problem

a. Find the probability that the sum of the 80 values (or the total of the 80 values) is more than 7500.
b. Find the sum that is 1.5 standard deviations below the mean of the sums.

Solution
Let X = one value from the original unknown population. The probability question asks you to find a probability for **the sum (or total of) 80 values.**

ΣX = the sum or total of 80 values. Since $\mu_X = 90$, $\sigma_X = 15$, and $\sigma_X = 80$, then

$\Sigma X \sim N\left(80 \cdot 90, \sqrt{80} \cdot 15\right)$

a. mean of the sums = $n \cdot \mu_X = (80)(90) = 7200$
b. standard deviation of the sums = $\sqrt{n} \cdot \sigma_X = \sqrt{80} \cdot 15$
c. sum of 80 values = $\Sigma x = 7500$

Find $P(\Sigma X > 7500)$ Draw a graph.

$P(\Sigma X > 7500) = 0.0127$

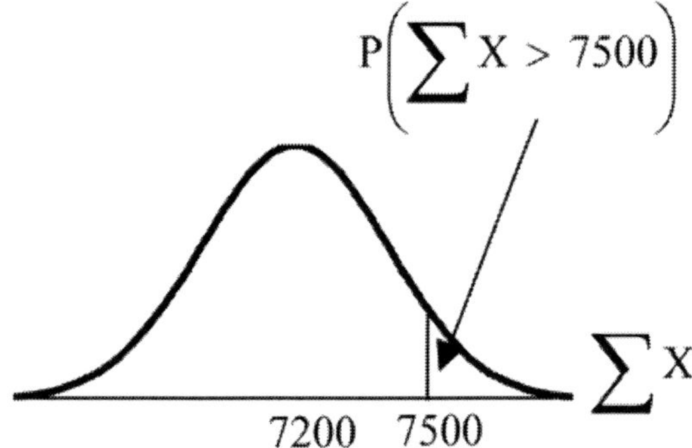

`normalcdf`(lower value, upper value, mean of sums, `stdev` of sums)

The parameter list is abbreviated (lower, upper, $n \cdot \mu_X$, $\sqrt{n} \cdot \sigma_X$)

`normalcdf`(7500,1E99, $80 \cdot 90$, $\sqrt{80} \cdot 15 = 0.0127$

Reminder: $1E99 = 10^{99}$. Press the EE key for E.

7.4 Using the Central Limit Theorem[4]

It is important for you to understand when to use the **CLT**. If you are being asked to find the probability of an average or mean, use the CLT for means or averages. If you are being asked to find the probability of a sum or total, use the CLT for sums. This also applies to percentiles for averages and sums.

NOTE: If you are being asked to find the probability of an **individual** value, do **not** use the CLT. **Use the distribution of its random variable.**

7.4.1 Law of Large Numbers

The Law of Large Numbers says that if you take samples of larger and larger size from any population, then the mean $\overline{x}$ of the sample gets closer and closer to μ. From the Central Limit Theorem, we know that as n gets larger and larger, the sample averages follow a normal distribution. The larger n gets, the smaller the standard deviation gets. (Remember that the standard deviation for $\overline{X}$ is $\frac{\sigma}{\sqrt{n}}$.) This means that the sample mean $\overline{x}$ must be close to the population mean μ. We can say that μ is the value that the sample averages approach as n gets larger. The Central Limit Theorem illustrates the Law of Large Numbers.

Example 7.4
A study involving stress is done on a college campus among the students. **The stress scores follow a uniform distribution** with the lowest stress score equal to 1 and the highest equal to 5. Using a sample of 75 students, find:

1. The probability that the **average stress score** for the 75 students is less than 2.
2. The 90th percentile for the **average stress score** for the 75 students.
3. The probability that the **total of the 75 stress scores** is less than 200.
4. The 90th percentile for the **total stress score** for the 75 students.

[4]This content is available online at <http://cnx.org/content/m16958/1.10/>.

Let X = one stress score.

Problems 1. and 2. ask you to find a probability or a percentile for an **average** or **mean**. Problems c and d ask you to find a probability or a percentile for a **total or sum**. The sample size, n, is equal to 75.

Since the individual stress scores follow a uniform distribution, $X \sim U(1,5)$ where $a = 1$ and $b = 5$ (See the chapter on Continuous Random Variables (Section 5.1)).

$\mu_X = \frac{a+b}{2} = \frac{1+5}{2} = 3$

$\sigma_X = \sqrt{\frac{(b-a)^2}{12}} = \sqrt{\frac{(5-1)^2}{12}} = 1.15$

For problems a and b, let $\overline{X}$ = the average stress score for the 75 students. Then,

$\overline{X} \sim N\left(3, \frac{1.15}{\sqrt{75}}\right)$ where $n = 75$.

Problem 1
Find $P\left(\overline{X} < 2\right)$. Draw the graph.

Solution
$P\left(\overline{X} < 2\right) = 0$

The probability that the average stress score is less than 2 is about 0.

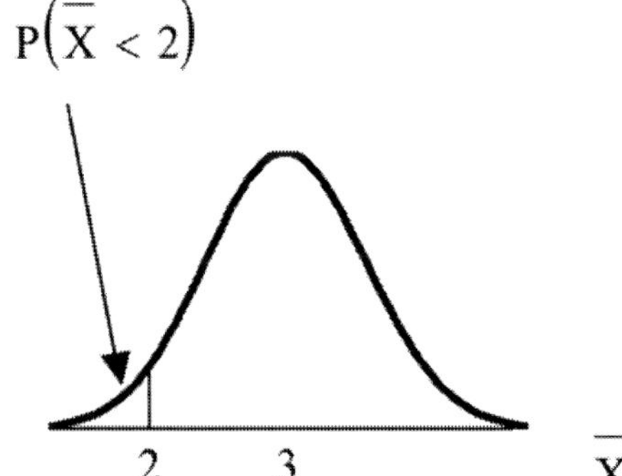

$\texttt{normalcdf}\left(1, 2, 3, \frac{1.15}{\sqrt{75}}\right) = 0$

REMINDER: The smallest stress score is 1. Therefore, the smallest average for 75 stress scores is 1.

Problem 2
Find the 90th percentile for the average of 75 stress scores. Draw a graph.

Solution
Let k = the 90th precentile.

Find k where $P\left(\overline{X} < k\right) = 0.90$.

$k = 3.2$

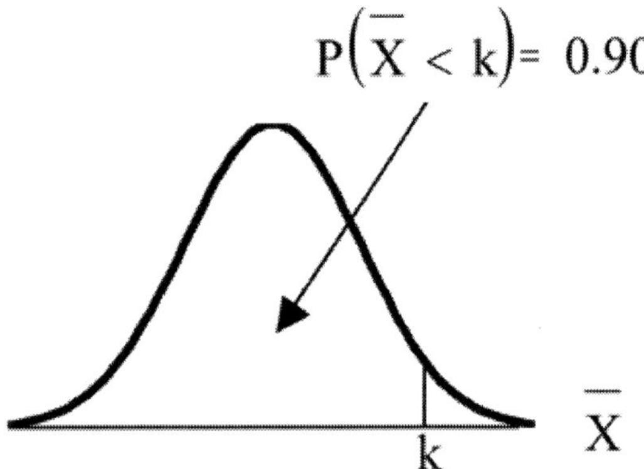

The 90th percentile for the average of 75 scores is about 3.2. This means that 90% of all the averages of 75 stress scores are at most 3.2 and 10% are at least 3.2.

`invNorm` $\left(.90, 3, \frac{1.15}{\sqrt{75}}\right) = 3.2$

For problems c and d, let ΣX = the sum of the 75 stress scores. Then, $\Sigma X \sim N\left[(75)\cdot(3), \sqrt{75}\cdot 1.15\right]$

Problem 3

Find $P(\Sigma X < 200)$. Draw the graph.

Solution

The mean of the sum of 75 stress scores is $75 \cdot 3 = 225$

The standard deviation of the sum of 75 stress scores is $\sqrt{75} \cdot 1.15 = 9.96$

$P(\Sigma X < 200) = 0$

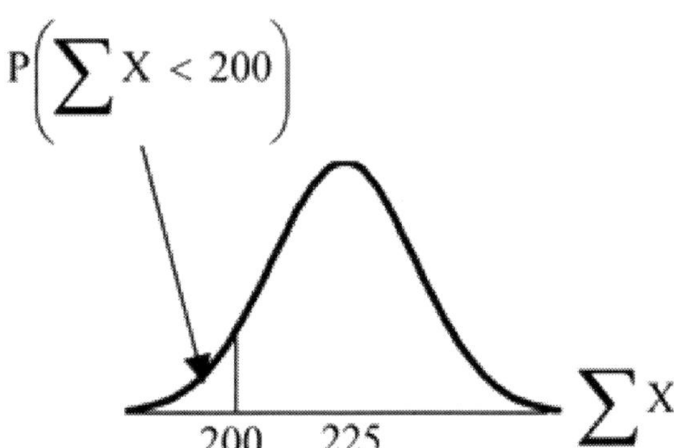

The probability that the total of 75 scores is less than 200 is about 0.

`normalcdf` $\left(75, 200, 75\cdot 3, \sqrt{75}\cdot 1.15\right) = 0.$

REMINDER: The smallest total of 75 stress scores is 75 since the smallest single score is 1.

Problem 4

Find the 90th percentile for the total of 75 stress scores. Draw a graph.

Solution
Let k = the 90th percentile.

Find k where $P(\Sigma X < k) = 0.90$.

$k = 237.8$

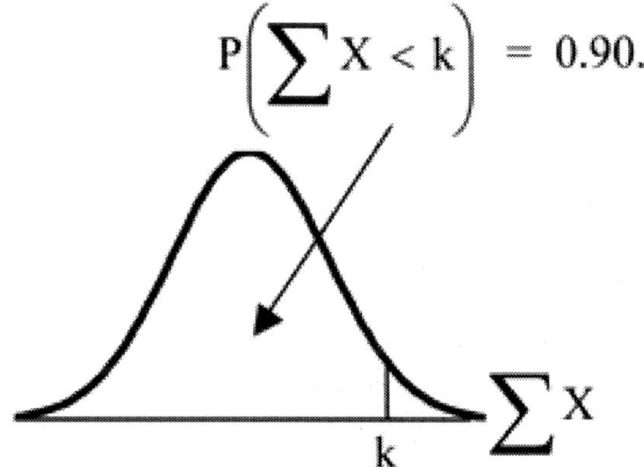

The 90th percentile for the sum of 75 scores is about 237.8. This means that 90% of all the sums of 75 scores are no more than 237.8 and 10% are no less than 237.8.

`invNorm` $\left(.90, 75 \cdot 3, \sqrt{75} \cdot 1.15\right) = 237.8$

Example 7.5
Suppose that a market research analyst for a cell phone company conducts a study of their customers who exceed the time allowance included on their basic cell phone contract. The analyst finds that for those customers who exceed the time included in their basic contract, the **excess time used** follows an **exponential distribution** with a mean of 22 minutes. Consider a random sample of 80 customers. Find

1. The probability that the **average excess time** used by the 80 customers in the sample is longer than 20 minutes. Draw a graph.
2. The 95th percentile for the **average excess time** for samples of 80 customers who exceed their basic contract time allowances. Draw a graph.

Let X = the excess time used by one individual cell phone customer who exceeds his contracted time allowance. Then $X \sim Exp\left(\frac{1}{22}\right)$ (Continuous Random Variables - Exponential). Because X is exponential, $\mu = 22$ and $\sigma = 22$. The sample size is $n = 80$.

Let $\overline{X}$ = the **average** excess time used by a sample of $n = 80$ customers who exceed their contracted time allowances. Then

$\overline{X} \sim N\left(22, \frac{22}{\sqrt{80}}\right)$ by the CLT for Sample Means or Averages

Problem 1
Find $P\left(\overline{X} > 20\right)$. Draw the graph.

Solution
$P\left(\overline{X} > 20\right) = 0.7919$

The probability that the average excess time used by a sample of 80 customers is longer than 20 minutes is 0.7919.

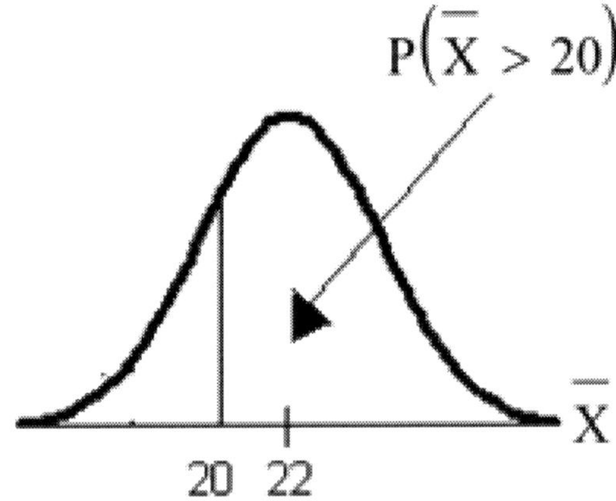

`normalcdf` $\left(20, 1E99, 22, \frac{22}{\sqrt{80}}\right)$

REMINDER: $1E99 = 10^{99}$ *and* $-1E99 = -10^{99}$. Press the `EE` key for E.

Problem 2

Find the 95th percentile for the **average excess time** for samples of 80 customers who exceed their basic contract time allowances. Draw a graph.

Solution

Let k = the 95th percentile for the average excess time.

Find k where $P\left(\overline{X} < k\right) = 0.95$

$k = 26.0$

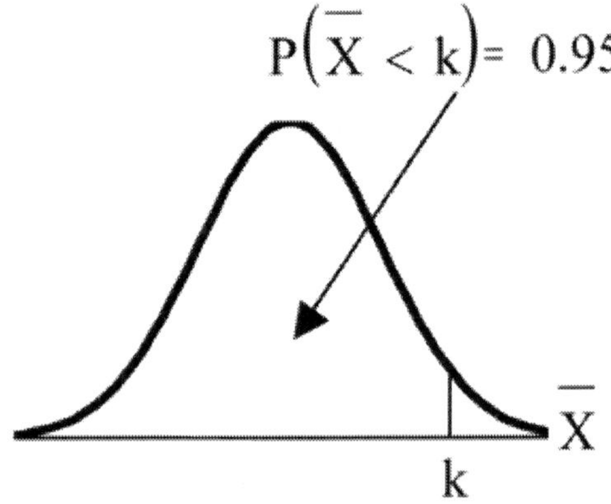

The 95th percentile for the average excess time for samples of 80 customers who exceed their basic contract time allowances is about 26 minutes. This means that 95% of the average excess times are at most 26 minutes and 10% are at least 26 minutes.

`invNorm` $\left(.95, 22, \frac{22}{\sqrt{80}}\right) = 26.0$

7.5 Summary of Formulas[5]

Formula 7.1: Central Limit Theorem for Sample Means (Averages)

$\overline{X} \sim N\left(\mu_X, \frac{\sigma_X}{\sqrt{n}}\right)$ **Mean for Averages** $(\overline{X})$**:** μ_X

Formula 7.2: Central Limit Theorem for Sample Means (Averages) Z-Score and Standard Error of the Mean

$z = \frac{\overline{x}-\mu_X}{\left(\frac{\sigma_X}{\sqrt{n}}\right)}$ **Standard Error of the Mean (Standard Deviation for Averages** $(\overline{X})$**):** $\frac{\sigma_X}{\sqrt{n}}$

Formula 7.3: Central Limit Theorem for Sums

$\Sigma X \sim N\left[(n) \cdot \mu_X, \sqrt{n} \cdot \sigma_X\right]$ **Mean for Sums** (ΣX)**:** $n \cdot \mu_X$

Formula 7.4: Central Limit Theorem for Sums Z-Score and Standard Deviation for Sums

$z = \frac{\Sigma x - n \cdot \mu_X}{\sqrt{n} \cdot \sigma_X}$ **Standard Deviation for Sums** (ΣX)**:** $\sqrt{n} \cdot \sigma_X$

[5]This content is available online at <http://cnx.org/content/m16956/1.6/>.

7.6 Practice: The Central Limit Theorem[6]

7.6.1 Student Learning Outcomes

- The student will explore the properties of data through the Central Limit Theorem.

7.6.2 Given

Yoonie is a personnel manager in a large corporation. Each month she must review 16 of the employees. From past experience, she has found that the reviews take her approximately 4 hours each to do with a population standard deviation of 1.2 hours. Let X be the random variable representing the time it takes her to complete one review. Assume X is normally distributed. Let $\overline{X}$ be the random variable representing the average time to complete the 16 reviews. Let ΣX be the total time it takes Yoonie to complete all of the month's reviews.

7.6.3 Distribution

Complete the distributions.

1. $X \sim$
2. $\overline{X} \sim$
3. $\Sigma X \sim$

7.6.4 Graphing Probability

For each problem below:

a. Sketch the graph. Label and scale the horizontal axis. Shade the region corresponding to the probability.
b. Calculate the value.

Exercise 7.6.1 *(Solution on p. 284.)*

Find the probability that **one** review will take Yoonie from 3.5 to 4.25 hours.

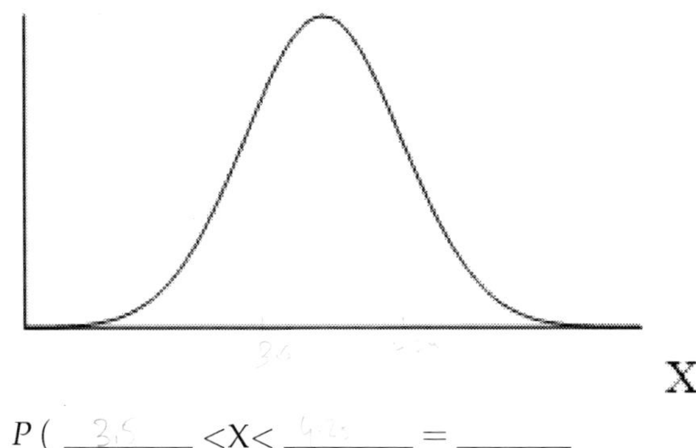

a.
b. $P($ ______ $<X<$ ______ $=$ ______

Exercise 7.6.2 *(Solution on p. 284.)*

Find the probability that the **average** of a month's reviews will take Yoonie from 3.5 to 4.25 hrs.

[6]This content is available online at <http://cnx.org/content/m16954/1.10/>.

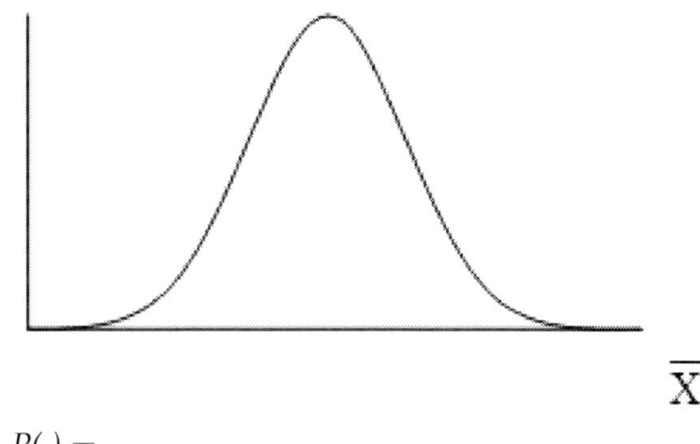

a.
b. $P(\ ) =$_______

Exercise 7.6.3 *(Solution on p. 284.)*

Find the 95th percentile for the **average** time to complete one month's reviews.

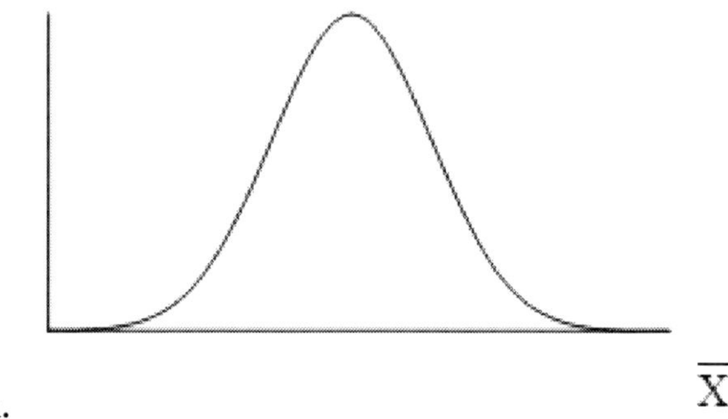

a.
b. The 95th Percentile=

Exercise 7.6.4 *(Solution on p. 284.)*

Find the probability that the **sum** of the month's reviews takes Yoonie from 60 to 65 hours.

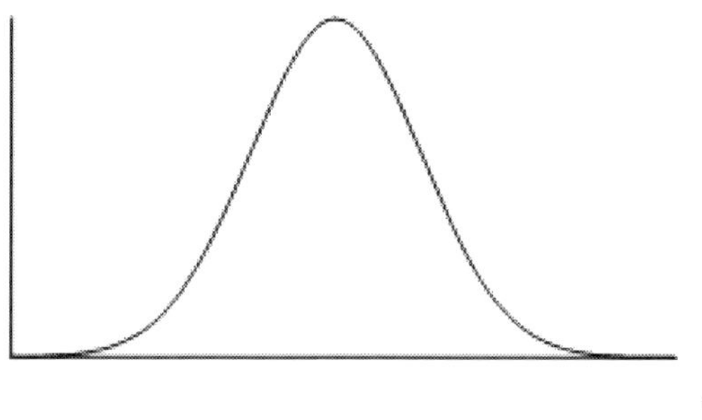

a.
b. The Probability=

Exercise 7.6.5 *(Solution on p. 284.)*

Find the 95th percentile for the **sum** of the month's reviews.

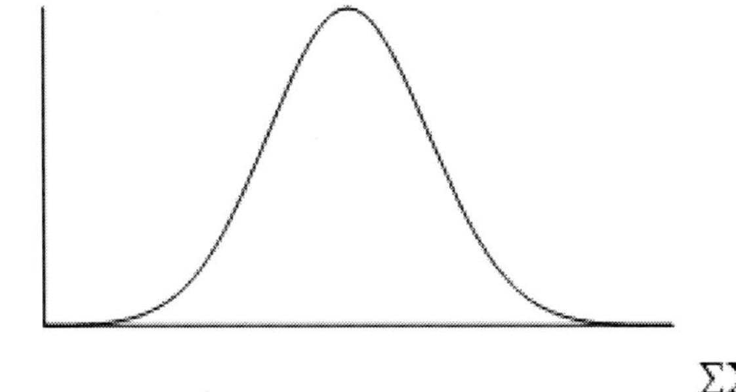

a.

b. The 95th percentile=

7.6.5 Discussion Question

Exercise 7.6.6

What causes the probabilities in Exercise 7.6.1 and Exercise 7.6.2 to differ?

7.7 Homework[7]

Exercise 7.7.1 *(Solution on p. 284.)*

$X \sim N(60, 9)$. Suppose that you form random samples of 25 from this distribution. Let $\overline{X}$ be the random variable of averages. Let ΣX be the random variable of sums. For **c - f**, sketch the graph, shade the region, label and scale the horizontal axis for $\overline{X}$, and find the probability.

- **a.** Sketch the distributions of X and $\overline{X}$ on the same graph.
- **b.** $\overline{X} \sim$
- **c.** $P(\overline{X} < 60) =$
- **d.** Find the 30th percentile.
- **e.** $P(56 < \overline{X} < 62) =$
- **f.** $P(18 < \overline{X} < 58) =$
- **g.** $\Sigma X \sim$
- **h.** Find the minimum value for the upper quartile.
- **i.** $P(1400 < \Sigma X < 1550) =$

Exercise 7.7.2

Determine which of the following are true and which are false. Then, in complete sentences, justify your answers.

- **a.** When the sample size is large, the mean of $\overline{X}$ is approximately equal to the mean of X.
- **b.** When the sample size is large, $\overline{X}$ is approximately normally distributed.
- **c.** When the sample size is large, the standard deviation of $\overline{X}$ is approximately the same as the standard deviation of X.

Exercise 7.7.3 *(Solution on p. 284.)*

The percent of fat calories that a person in America consumes each day is normally distributed with a mean of about 36 and a standard deviation of about 10. Suppose that 16 individuals are randomly chosen.

Let $\overline{X}$ =average percent of fat calories.

- **a.** $\overline{X}$~______ (______, ______)
- **b.** For the group of 16, find the probability that the average percent of fat calories consumed is more than 5. Graph the situation and shade in the area to be determined.
- **c.** Find the first quartile for the average percent of fat calories.

Exercise 7.7.4

Previously, De Anza statistics students estimated that the amount of change daytime statistics students carry is exponentially distributed with a mean of $0.88. Suppose that we randomly pick 25 daytime statistics students.

- **a.** In words, X =
- **b.** X~
- **c.** In words, $\overline{X}$ =
- **d.** $\overline{X}$~ ______ (______, ______)
- **e.** Find the probability that an individual had between $0.80 and $1.00. Graph the situation and shade in the area to be determined.
- **f.** Find the probability that the average of the 25 students was between $0.80 and $1.00. Graph the situation and shade in the area to be determined.

---

[7] This content is available online at <http://cnx.org/content/m16952/1.13/>.

g. Explain the why there is a difference in (e) and (f).

Exercise 7.7.5 *(Solution on p. 284.)*

Suppose that the distance of fly balls hit to the outfield (in baseball) is normally distributed with a mean of 250 feet and a standard deviation of 50 feet. We randomly sample 49 fly balls.

a. If $\overline{X}$ = average distance in feet for 49 fly balls, then $\overline{X}$~_______ (_______ , _______)
b. What is the probability that the 49 balls traveled an average of less than 240 feet? Sketch the graph. Scale the horizontal axis for $\overline{X}$. Shade the region corresponding to the probability. Find the probability.
c. Find the 80th percentile of the distribution of the average of 49 fly balls.

Exercise 7.7.6

Suppose that the weight of open boxes of cereal in a home with children is uniformly distributed from 2 to 6 pounds. We randomly survey 64 homes with children.

a. In words, $X =$
b. X~
c. $\mu_X =$
d. $\sigma_X =$
e. In words, $\Sigma X =$
f. ΣX~
g. Find the probability that the total weight of open boxes is less than 250 pounds.
h. Find the 35th percentile for the total weight of open boxes of cereal.

Exercise 7.7.7 *(Solution on p. 284.)*

Suppose that the duration of a particular type of criminal trial is known to have a mean of 21 days and a standard deviation of 7 days. We randomly sample 9 trials.

a. In words, $\Sigma X =$
b. ΣX~
c. Find the probability that the total length of the 9 trials is at least 225 days.
d. 90 percent of the total of 9 of these types of trials will last at least how long?

Exercise 7.7.8

According to the Internal Revenue Service, the average length of time for an individual to complete (record keep, learn, prepare, copy, assemble and send) IRS Form 1040 is 10.53 hours (without any attached schedules). The distribution is unknown. Let us assume that the standard deviation is 2 hours. Suppose we randomly sample 36 taxpayers.

a. In words, $X =$
b. In words, $\overline{X} =$
c. $\overline{X}$~
d. Would you be surprised if the 36 taxpayers finished their Form 1040s in an average of more than 12 hours? Explain why or why not in complete sentences.
e. Would you be surprised if one taxpayer finished his Form 1040 in more than 12 hours? In a complete sentence, explain why.

Exercise 7.7.9 *(Solution on p. 285.)*

Suppose that a category of world class runners are known to run a marathon (26 miles) in an average of 145 minutes with a standard deviation of 14 minutes. Consider 49 of the races.

Let $\overline{X}$ = the average of the 49 races.

a. $\overline{X}$~
b. Find the probability that the runner will average between 142 and 146 minutes in these 49 marathons.
c. Find the 80th percentile for the average of these 49 marathons.
d. Find the median of the average running times.

Exercise 7.7.10

The attention span of a two year-old is exponentially distributed with a mean of about 8 minutes. Suppose we randomly survey 60 two year-olds.

a. In words, $X =$
b. X~
c. In words, $\overline{X} =$
d. $\overline{X}$~
e. Before doing any calculations, which do you think will be higher? Explain why.

  i. the probability that an individual attention span is less than 10 minutes; or
  ii. the probability that the average attention span for the 60 children is less than 10 minutes? Why?

f. Calculate the probabilities in part (e).
g. Explain why the distribution for $\overline{X}$ is not exponential.

Exercise 7.7.11 ***(Solution on p. 285.)***

Suppose that the length of research papers is uniformly distributed from 10 to 25 pages. We survey a class in which 55 research papers were turned in to a professor. We are interested in the average length of the research papers.

a. In words, $X =$
b. X~
c. $\mu_X =$
d. $\sigma_X =$
e. In words, $\overline{X} =$
f. $\overline{X}$~
g. In words, $\Sigma X =$
h. ΣX~
i. Without doing any calculations, do you think that it's likely that the professor will need to read a total of more than 1050 pages? Why?
j. Calculate the probability that the professor will need to read a total of more than 1050 pages.
k. Why is it so unlikely that the average length of the papers will be less than 12 pages?

Exercise 7.7.12

The length of songs in a collector's CD collection is uniformly distributed from 2 to 3.5 minutes. Suppose we randomly pick 5 CDs from the collection. There is a total of 43 songs on the 5 CDs.

a. In words, $X =$
b. X~
c. In words, $\overline{X}$=
d. $\overline{X}$~
e. Find the first quartile for the average song length.
f. The IQR (interquartile range) for the average song length is from _______ to _______.

Exercise 7.7.13 *(Solution on p. 285.)*

Salaries for teachers in a particular elementary school district are normally distributed with a mean of $44,000 and a standard deviation of $6500. We randomly survey 10 teachers from that district.

a. In words, $X =$
b. In words, $\overline{X} =$
c. $\overline{X}\sim$
d. In words, $\Sigma X =$
e. $\Sigma X\sim$
f. Find the probability that the teachers earn a total of over $400,000.
g. Find the 90th percentile for an individual teacher's salary.
h. Find the 90th percentile for the average teachers' salary.
i. If we surveyed 70 teachers instead of 10, graphically, how would that change the distribution for $\overline{X}$?
j. If each of the 70 teachers received a $3000 raise, graphically, how would that change the distribution for $\overline{X}$?

Exercise 7.7.14

The distribution of income in some Third World countries is considered wedge shaped (many very poor people, very few middle income people, and few to many wealthy people). Suppose we pick a country with a wedge distribution. Let the average salary be $2000 per year with a standard deviation of $8000. We randomly survey 1000 residents of that country.

a. In words, $X =$
b. In words, $\overline{X} =$
c. $\overline{X}\sim$
d. How is it possible for the standard deviation to be greater than the average?
e. Why is it more likely that the average of the 1000 residents will be from $2000 to $2100 than from $2100 to $2200?

Exercise 7.7.15 *(Solution on p. 285.)*

The average length of a maternity stay in a U.S. hospital is said to be 2.4 days with a standard deviation of 0.9 days. We randomly survey 80 women who recently bore children in a U.S. hospital.

a. In words, $X =$
b. In words, $\overline{X} =$
c. $\overline{X}\sim$
d. In words, $\Sigma X =$
e. $\Sigma X\sim$
f. Is it likely that an individual stayed more than 5 days in the hospital? Why or why not?
g. Is it likely that the average stay for the 80 women was more than 5 days? Why or why not?
h. Which is more likely:

  i. an individual stayed more than 5 days; or
  ii. the average stay of 80 women was more than 5 days?

i. If we were to sum up the women's stays, is it likely that, collectively they spent more than a year in the hospital? Why or why not?

Exercise 7.7.16

In 1940 the average size of a U.S. farm was 174 acres. Let's say that the standard deviation was 55 acres. Suppose we randomly survey 38 farmers from 1940. (Source: U.S. Dept. of Agriculture)

a. In words, $X =$
b. In words, $\overline{X} =$
c. $\overline{X} \sim$
d. The IQR for $\overline{X}$ is from _______ acres to _______ acres.

Exercise 7.7.17 *(Solution on p. 285.)*

The stock closing prices of 35 U.S. semiconductor manufacturers are given below. (Source: **Wall Street Journal**)

8.625; 30.25; 27.625; 46.75; 32.875; 18.25; 5; 0.125; 2.9375; 6.875; 28.25; 24.25; 21; 1.5; 30.25; 71; 43.5; 49.25; 2.5625; 31; 16.5; 9.5; 18.5; 18; 9; 10.5; 16.625; 1.25; 18; 12.875; 7; 2.875; 2.875; 60.25; 29.25

a. In words, $X =$
b. i. $\overline{x} =$
  ii. $s_x =$
  iii. $n =$
c. Construct a histogram of the distribution of the averages. Start at $x = -0.0005$. Make bar widths of 10.
d. In words, describe the distribution of stock prices.
e. Randomly average 5 stock prices together. (Use a random number generator.) Continue averaging 5 pieces together until you have 10 averages. List those 10 averages.
f. Use the 10 averages from (e) to calculate:
  i. $\overline{x} =$
  ii. $s_{\overline{x}} =$
g. Construct a histogram of the distribution of the averages. Start at $x = -0.0005$. Make bar widths of 10.
h. Does this histogram look like the graph in (c)?
i. In 1 - 2 complete sentences, explain why the graphs either look the same or look different?
j. Based upon the theory of the Central Limit Theorem, $\overline{X} \sim$

Exercise 7.7.18

Use the Initial Public Offering data (Section 14.3.2: Stock Prices) (see "Table of Contents) to do this problem.

a. In words, $X =$
b. i. $\mu_X =$
  ii. $\sigma_X =$
  iii. $n =$
c. Construct a histogram of the distribution. Start at $x = -0.50$. Make bar widths of \$5.
d. In words, describe the distribution of stock prices.
e. Randomly average 5 stock prices together. (Use a random number generator.) Continue averaging 5 pieces together until you have 15 averages. List those 15 averages.
f. Use the 15 averages from (e) to calculate the following:
  i. $\overline{x} =$
  ii. $s_{\overline{x}} =$
g. Construct a histogram of the distribution of the averages. Start at $x = -0.50$. Make bar widths of \$5.
h. Does this histogram look like the graph in (c)? Explain any differences.
i. In 1 - 2 complete sentences, explain why the graphs either look the same or look different?
j. Based upon the theory of the Central Limit Theorem, $\overline{X} \sim$

7.7.1 Try these multiple choice questions.

The next two questions refer to the following information: The time to wait for a particular rural bus is distributed uniformly from 0 to 75 minutes. 100 riders are randomly sampled to learn how long they waited.

Exercise 7.7.19 *(Solution on p. 285.)*

The 90th percentile sample average wait time (in minutes) for a sample of 100 riders is:

A. 315.0
B. 40.3
C. 38.5
D. 65.2

Exercise 7.7.20 *(Solution on p. 285.)*

Would you be surprised, based upon numerical calculations, if the sample average wait time (in minutes) for 100 riders was less than 30 minutes?

A. Yes
B. No
C. There is not enough information.

Exercise 7.7.21 *(Solution on p. 285.)*

Which of the following is NOT TRUE about the distribution for averages?

A. The mean, median and mode are equal
B. The area under the curve is one
C. The curve never touches the x-axis
D. The curve is skewed to the right

The next two questions refer to the following information: The cost of unleaded gasoline in the Bay Area once followed an unknown distribution with a mean of $2.59 and a standard deviation of $0.10. Sixteen gas stations from the Bay Area are randomly chosen. We are interested in the average cost of gasoline for the 16 gas stations.

Exercise 7.7.22 *(Solution on p. 285.)*

The distribution to use for the average cost of gasoline for the 16 gas stations is

A. $\overline{X} \sim N(2.59, 0.10)$
B. $\overline{X} \sim N\left(2.59, \frac{0.10}{\sqrt{16}}\right)$
C. $\overline{X} \sim N\left(2.59, \frac{0.10}{16}\right)$
D. $\overline{X} \sim N\left(2.59, \frac{16}{0.10}\right)$

Exercise 7.7.23 *(Solution on p. 285.)*

What is the probability that the average price for 16 gas stations is over $2.69?

A. Almost zero
B. 0.1587
C. 0.0943
D. Unknown

7.8 Review[8]

The next three questions refer to the following information: Richard's Furniture Company delivers furniture from 10 A.M. to 2 P.M. continuously and uniformly. We are interested in how long (in hours) past the 10 A.M. start time that individuals wait for their delivery.

Exercise 7.8.1 *(Solution on p. 286.)*

$X \sim$

A. $U(0,4)$
B. $U(10,2)$
C. $Exp(2)$
D. $N(2,1)$

Exercise 7.8.2 *(Solution on p. 286.)*

The average wait time is:

A. 1 hour
B. 2 hour
C. 2.5 hour
D. 4 hour

Exercise 7.8.3 *(Solution on p. 286.)*

Suppose that it is now past noon on a delivery day. The probability that a person must wait at least $1\frac{1}{2}$ **more** hours is:

A. $\frac{1}{4}$
B. $\frac{1}{2}$
C. $\frac{3}{4}$
D. $\frac{3}{8}$

Exercise 7.8.4 *(Solution on p. 286.)*

Given: $X \sim Exp\left(\frac{1}{3}\right)$.

a. Find $P(X > 1)$
b. Calculate the minimum value for the upper quartile.
c. Find $P\left(X = \frac{1}{3}\right)$

Exercise 7.8.5 *(Solution on p. 286.)*

- 40% of full-time students took 4 years to graduate
- 30% of full-time students took 5 years to graduate
- 20% of full-time students took 6 years to graduate
- 10% of full-time students took 7 years to graduate

The expected time for full-time students to graduate is:

A. 4 years
B. 4.5 years
C. 5 years
D. 5.5 years

[8]This content is available online at <http://cnx.org/content/m16955/1.9/>.

Exercise 7.8.6 *(Solution on p. 286.)*

Which of the following distributions is described by the following example?

Many people can run a short distance of under 2 miles, but as the distance increases, fewer people can run that far.

A. Binomial
B. Uniform
C. Exponential
D. Normal

Exercise 7.8.7 *(Solution on p. 286.)*

The length of time to brush one's teeth is generally thought to be exponentially distributed with a mean of $\frac{3}{4}$ minutes. Find the probability that a randomly selected person brushes his/her teeth less than $\frac{3}{4}$ minutes.

A. 0.5
B. $\frac{3}{4}$
C. 0.43
D. 0.63

Exercise 7.8.8 *(Solution on p. 286.)*

Which distribution accurately describes the following situation?

The chance that a teenage boy regularly gives his mother a kiss goodnight (and he should!!) is about 20%. Fourteen teenage boys are randomly surveyed.

X =the number of teenage boys that regularly give their mother a kiss goodnight

A. $B\left(14, 0.20\right)$
B. $P\left(2.8\right)$
C. $N\left(2.8, 2.24\right)$
D. $Exp\left(\frac{1}{0.20}\right)$

7.9 Lab 1: Central Limit Theorem (Pocket Change)[9]

Class Time:

Names:

7.9.1 Student Learning Outcomes:

- The student will examine properties of the Central Limit Theorem.

NOTE: This lab works best when sampling from several classes and combining data.

7.9.2 Collect the Data

1. Count the change in your pocket. (Do not include bills.)
2. Randomly survey 30 classmates. Record the values of the change.

______	______	______	______	______
______	______	______	______	______
______	______	______	______	______
______	______	______	______	______
______	______	______	______	______
______	______	______	______	______

Table 7.1

3. Construct a histogram. Make 5 - 6 intervals. Sketch the graph using a ruler and pencil. Scale the axes.

[9]This content is available online at <http://cnx.org/content/m16950/1.9/>.

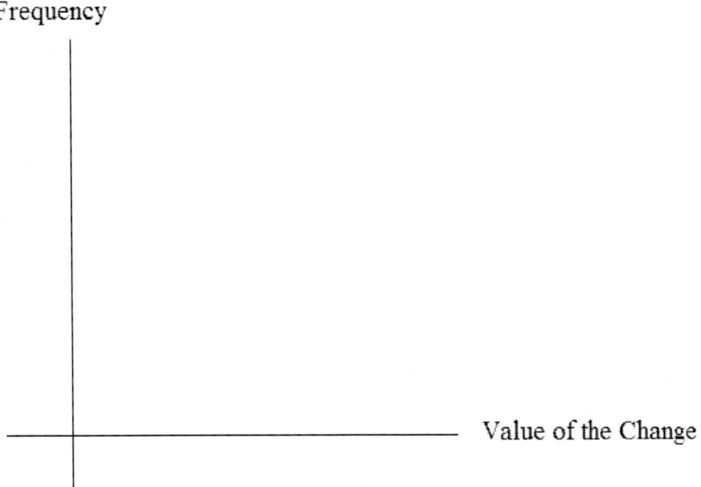

Figure 7.1

4. Calculate the following ($n = 1$; surveying one person at a time):
   a. $\overline{x} =$
   b. $s =$
5. Draw a smooth curve through the tops of the bars of the histogram. Use 1 – 2 complete sentences to describe the general shape of the curve.

7.9.3 Collecting Averages of Pairs

Repeat steps 1 - 5 (of the section above titled "Collect the Data") with one exception. Instead of recording the change of 30 classmates, record the average change of 30 pairs.

1. Randomly survey 30 **pairs** of classmates. Record the values of the average of their change.

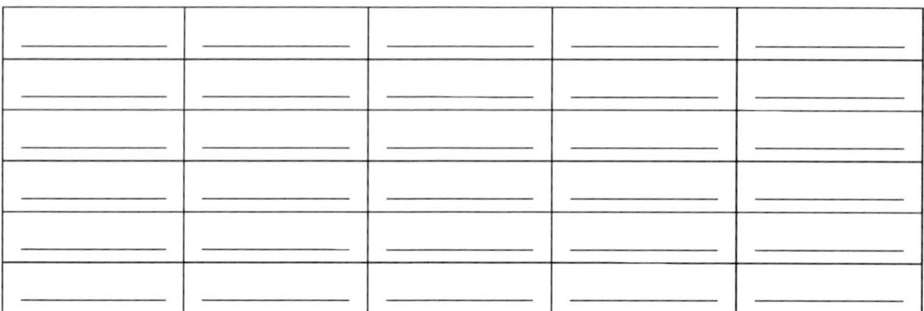

______	______	______	______	______
______	______	______	______	______
______	______	______	______	______
______	______	______	______	______
______	______	______	______	______
______	______	______	______	______

Table 7.2

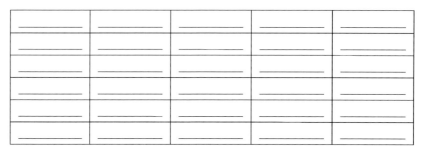

2. Construct a histogram. Scale the axes using the same scaling you did for the section titled "Collecting the Data". Sketch the graph using a ruler and a pencil.

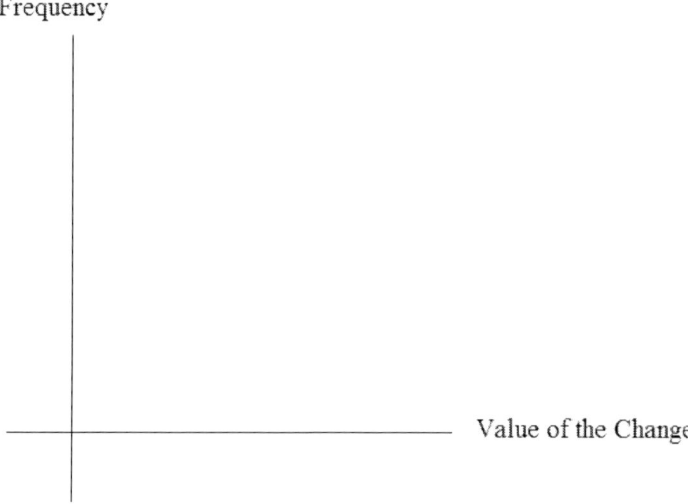

Figure 7.2

3. Calculate the following ($n = 2$; surveying two people at a time):
   a. $\overline{x} =$
   b. $s =$
4. Draw a smooth curve through tops of the bars of the histogram. Use 1 – 2 complete sentences to describe the general shape of the curve.

7.9.4 Collecting Averages of Groups of Five

Repeat steps 1 – 5 (of the section titled "Collect the Data") with one exception. Instead of recording the change of 30 classmates, record the average change of 30 groups of 5.

1. Randomly survey 30 **groups of 5** classmates. Record the values of the average of their change.

Table 7.3

2. Construct a histogram. Scale the axes using the same scaling you did for the section titled "Collect the Data". Sketch the graph using a ruler and a pencil.

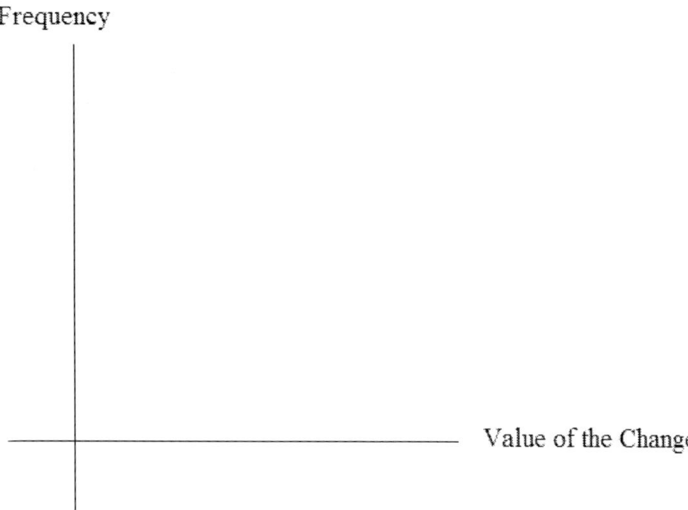

Figure 7.3

3. Calculate the following ($n = 5$; surveying five people at a time):
   - **a.** $\overline{x} =$
   - **b.** $s =$
4. Draw a smooth curve through tops of the bars of the histogram. Use 1 – 2 complete sentences to describe the general shape of the curve.

7.9.5 Discussion Questions

1. As n changed, why did the shape of the distribution of the data change? Use 1 – 2 complete sentences to explain what happened.
2. In the section titled "Collect the Data", what was the approximate distribution of the data? $X \sim$
3. In the section titled "Collecting Averages of Groups of Five", what was the approximate distribution of the averages? $\overline{X} \sim$
4. In 1 – 2 complete sentences, explain any differences in your answers to the previous two questions.

7.10 Lab 2: Central Limit Theorem (Cookie Recipes)[10]

Class Time:

Names:

7.10.1 Student Learning Outcomes:

- The student will examine properties of the Central Limit Theorem.

7.10.2 Given:

X = length of time (in days) that a cookie recipe lasted at the Olmstead Homestead. (Assume that each of the different recipes makes the same quantity of cookies.)

Recipe #	X		**Recipe #**	X		**Recipe #**	X		**Recipe #**	X
1	1		16	2		31	3		46	2
2	5		17	2		32	4		47	2
3	2		18	4		33	5		48	11
4	5		19	6		34	6		49	5
5	6		20	1		35	6		50	5
6	1		21	6		36	1		51	4
7	2		22	5		37	1		52	6
8	6		23	2		38	2		53	5
9	5		24	5		39	1		54	1
10	2		25	1		40	6		55	1
11	5		26	6		41	1		56	2
12	1		27	4		42	6		57	4
13	1		28	1		43	2		58	3
14	3		29	6		44	6		59	6
15	2		30	2		45	2		60	5

Table 7.4

Calculate the following:

a. $\mu_X =$
b. $\sigma_X =$

[10]This content is available online at <http://cnx.org/content/m16945/1.10/>.

7.10.3 Collect the Data

Use a random number generator to randomly select 4 samples of size $n = 5$ from the given population. Record your samples below. Then, for each sample, calculate the mean to the nearest tenth. Record them in the spaces provided. Record the sample means for the rest of the class.

1. Complete the table:

	Sample 1	**Sample 2**	**Sample 3**	**Sample 4**	**Sample means from other groups:**
Means:	$\overline{x} =$	$\overline{x} =$	$\overline{x} =$	$\overline{x} =$	

Table 7.5

2. Calculate the following:
   a. $\overline{x} =$
   b. $s_{\overline{x}} =$
3. Again, use a random number generator to randomly select 4 samples from the population. This time, make the samples of size $n = 10$. Record the samples below. As before, for each sample, calculate the mean to the nearest tenth. Record them in the spaces provided. Record the sample means for the rest of the class.

	Sample 1	**Sample 2**	**Sample 3**	**Sample 4**	**Sample means from other groups:**
Means:	$\overline{x} =$	$\overline{x} =$	$\overline{x} =$	$\overline{x} =$	

Table 7.6

4. Calculate the following:
   a. $\overline{x} =$
   b. $s_{\overline{x}} =$

5. For the original population, construct a histogram. Make intervals with bar width = 1 day. Sketch the graph using a ruler and pencil. Scale the axes.

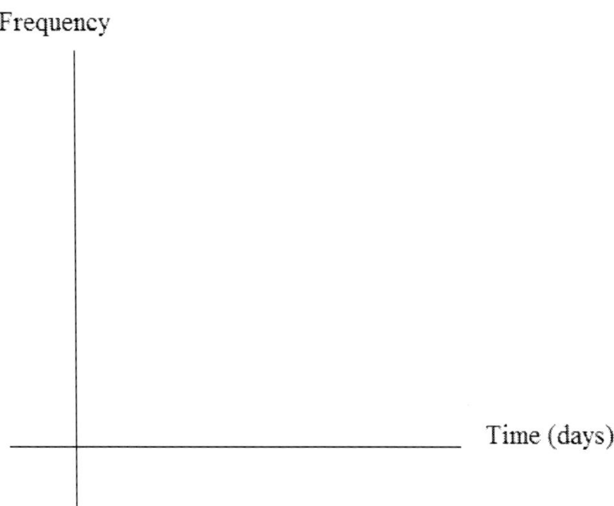

Figure 7.4

6. Draw a smooth curve through the tops of the bars of the histogram. Use 1 – 2 complete sentences to describe the general shape of the curve.

7.10.4 Repeat the Procedure for n=5

1. For the sample of $n = 5$ *days* averaged together, construct a histogram of the averages (your means together with the means of the other groups). Make intervals with *bar widths* $= \frac{1}{2}$*day*. Sketch the graph using a ruler and pencil. Scale the axes.

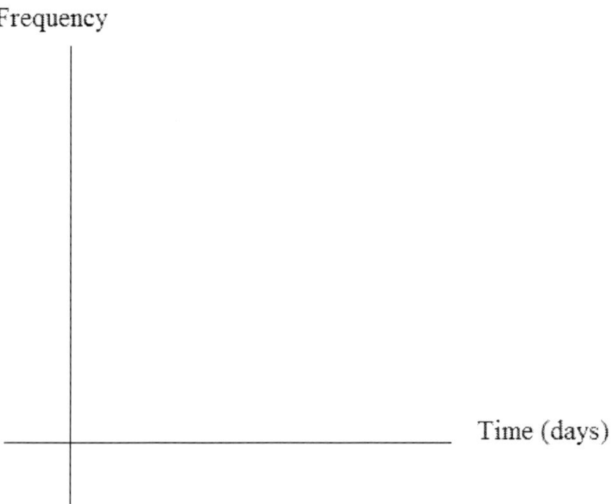

Figure 7.5

2. Draw a smooth curve through the tops of the bars of the histogram. Use 1 – 2 complete sentences to describe the general shape of the curve.

7.10.5 Repeat the Procedure for n=10

1. For the sample of $n = 10$ *days* averaged together, construct a histogram of the averages (your means together with the means of the other groups). Make intervals with *bar widths* $= \frac{1}{2}$*day*. Sketch the graph using a ruler and pencil. Scale the axes.

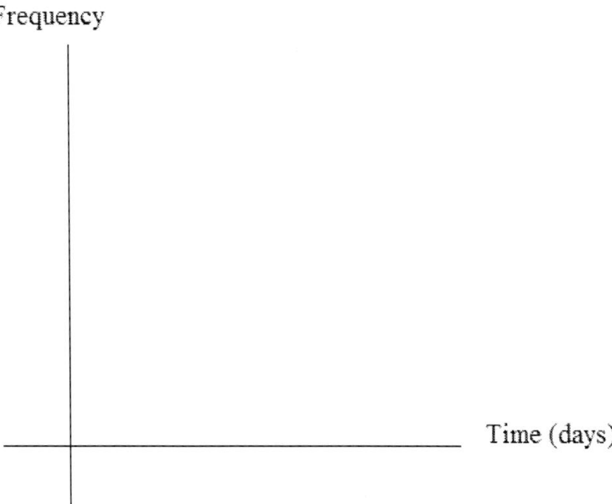

Figure 7.6

2. Draw a smooth curve through the tops of the bars of the histogram. Use 1 – 2 complete sentences to describe the general shape of the curve.

7.10.6 Discussion Questions

1. Compare the three histograms you have made, the one for the population and the two for the sample means. In three to five sentences, describe the similarities and differences.
2. State the theoretical (according to the CLT) distributions for the sample means.
   a. $n = 5$: $\overline{X} \sim$
   b. $n = 10$: $\overline{X} \sim$
3. Are the sample means for $n = 5$ and $n = 10$ "close" to the theoretical mean, μ_x? Explain why or why not.
4. Which of the two distributions of sample means has the smaller standard deviation? Why?
5. As n changed, why did the shape of the distribution of the data change? Use 1 – 2 complete sentences to explain what happened.

NOTE: *This lab was designed and contributed by Carol Olmstead.*

Solutions to Exercises in Chapter 7

Solution to Example 7.2, Problem 2 (p. 254)
$\mu_X = \mathbf{2}$, $\sigma_X = \mathbf{0.5}$, $n = \mathbf{50}$, and $X \sim N\left(2\,,\ \frac{0.5}{\sqrt{50}}\right)$

Solutions to Practice: The Central Limit Theorem

Solution to Exercise 7.6.1 (p. 263)

b. 3.5, 4.25, 0.2441

Solution to Exercise 7.6.2 (p. 263)

b. 0.7499

Solution to Exercise 7.6.3 (p. 264)

b. 4.49 hours

Solution to Exercise 7.6.4 (p. 264)

b. 0.3802

Solution to Exercise 7.6.5 (p. 264)

b: 71.90

Solutions to Homework

Solution to Exercise 7.7.1 (p. 266)

b. $Xbar{\sim}N\left(60, \frac{9}{\sqrt{25}}\right)$
c. 0.5000
d. 59.06
e. 0.8536
f. 0.1333
h. 1530.35
i. 0.8536

Solution to Exercise 7.7.3 (p. 266)

a. $N\left(36, \frac{10}{\sqrt{16}}\right)$
b. 1
c. 34.31

Solution to Exercise 7.7.5 (p. 267)

a. $N\left(250, \frac{50}{\sqrt{49}}\right)$
b. 0.0808
c. 256.01 feet

Solution to Exercise 7.7.7 (p. 267)

a. The total length of time for 9 criminal trials
b. $N\,(189, 21)$
c. 0.0432

d. 162.09

Solution to Exercise 7.7.9 (p. 267)

a. $N\left(145, \frac{14}{\sqrt{49}}\right)$
b. 0.6247
c. 146.68
d. 145 minutes

Solution to Exercise 7.7.11 (p. 268)

b. $U\left(10, 25\right)$
c. 17.5
d. $\sqrt{\frac{225}{12}} = 4.3301$
f. $N\left(17.5, 0.5839\right)$
h. $N\left(962.5, 32.11\right)$
j. 0.0032

Solution to Exercise 7.7.13 (p. 269)

c. $N\left(44,000, \frac{6500}{\sqrt{10}}\right)$
e. $N\left(440,000, \left(\sqrt{10}\right)\left(6500\right)\right)$
f. 0.9742
g. $52,330
h. $46,634

Solution to Exercise 7.7.15 (p. 269)

c. $N\left(2.4, \frac{0.9}{\sqrt{80}}\right)$
e. $N\left(192, 8.05\right)$
h. Individual

Solution to Exercise 7.7.17 (p. 270)

b. $20.71; $17.31; 35
d. Exponential distribution, $X \sim Exp\left(1/20.71\right)$
f. $20.71; $11.14
j. $N\left(20.71, \frac{17.31}{\sqrt{5}}\right)$

Solution to Exercise 7.7.19 (p. 271)
B
Solution to Exercise 7.7.20 (p. 271)
A
Solution to Exercise 7.7.21 (p. 271)
D
Solution to Exercise 7.7.22 (p. 271)
B
Solution to Exercise 7.7.23 (p. 271)
A

Solutions to Review

Solution to Exercise 7.8.1 (p. 272)
A
Solution to Exercise 7.8.2 (p. 272)
B
Solution to Exercise 7.8.3 (p. 272)
A
Solution to Exercise 7.8.4 (p. 272)

- **a.** 0.7165
- **b.** 4.16
- **c.** 0

Solution to Exercise 7.8.5 (p. 272)
C
Solution to Exercise 7.8.6 (p. 273)
C
Solution to Exercise 7.8.7 (p. 273)
D
Solution to Exercise 7.8.8 (p. 273)
A

Chapter 8

Confidence Intervals

8.1 Confidence Intervals[1]

8.1.1 Student Learning Objectives

By the end of this chapter, the student should be able to:

- Calculate and interpret confidence intervals for one population average and one population proportion.
- Interpret the student-t probability distribution as the sample size changes.
- Discriminate between problems applying the normal and the student-t distributions.

8.1.2 Introduction

Suppose you are trying to determine the average rent of a two-bedroom apartment in your town. You might look in the classified section of the newspaper, write down several rents listed, and average them together. You would have obtained a point estimate of the true mean. If you are trying to determine the percent of times you make a basket when shooting a basketball, you might count the number of shots you make and divide that by the number of shots you attempted. In this case, you would have obtained a point estimate for the true proportion.

We use sample data to make generalizations about an unknown population. This part of statistics is called **inferential statistics. The sample data help us to make an estimate of a population parameter**. We realize that the point estimate is most likely not the exact value of the population parameter, but close to it. After calculating point estimates, we construct confidence intervals in which we believe the parameter lies.

In this chapter, you will learn to construct and interpret confidence intervals. You will also learn a new distribution, the Student-t, and how it is used with these intervals.

If you worked in the marketing department of an entertainment company, you might be interested in the average number of compact discs (CD's) a consumer buys per month. If so, you could conduct a survey and calculate the sample average, $\overline{x}$, and the sample standard deviation, s. You would use $\overline{x}$ to estimate the population mean and s to estimate the population standard deviation. The sample mean, $\overline{x}$, is the **point estimate** for the population mean, μ. The sample standard deviation, s, is the point estimate for the population standard deviation, σ.

[1]This content is available online at <http://cnx.org/content/m16967/1.10/>.

A **confidence interval** is another type of estimate but, instead of being just one number, it is an interval of numbers. The interval of numbers is an estimated range of values calculated from a given set of sample data. The confidence interval is likely to include an unknown population parameter.

Suppose for the CD example we do not know the population mean μ but we do know that the population standard deviation is $\sigma = 1$ and our sample size is 100. Then by the Central Limit Theorem, the standard deviation for the sample mean is

$\frac{\sigma}{\sqrt{n}} = \frac{1}{\sqrt{100}} = 0.1.$

The **Empirical Rule**, which applies to bell-shaped distributions, says that in approximately 95% of the samples, the sample mean, $\overline{x}$, will be within two standard deviations of the population mean μ. For our CD example, two standard deviations is $(2)(0.1) = 0.2$. The sample mean $\overline{x}$ is within 0.2 units of μ.

Because $\overline{x}$ is within 0.2 units of μ, which is unknown, then μ is within 0.2 units of $\overline{x}$ in 95% of the samples. The population mean μ is contained in an interval whose lower number is calculated by taking the sample mean and subtracting two standard deviations $((2)(0.1))$ and whose upper number is calculated by taking the sample mean and adding two standard deviations. In other words, μ is between $\overline{x} - 0.2$ and $\overline{x} + 0.2$ in 95% of all the samples.

For the CD example, suppose that a sample produced a sample mean $\overline{x} = 2$. Then the unknown population mean μ is between

$\overline{x} - 0.2 = 2 - 0.2 = 1.8$ and $\overline{x} + 0.2 = 2 + 0.2 = 2.2$

We say that we are **95% confident** that the unknown population mean number of CDs is between 1.8 and 2.2. **The 95% confidence interval is (1.8, 2.2).**

The 95% confidence interval implies two possibilities. Either the interval (1.8, 2.2) contains the true mean μ or our sample produced an $\overline{x}$ that is not within 0.2 units of the true mean μ. The second possibility happens for only 5% of all the samples (100% - 95%).

Remember that a confidence interval is created for an unknown population parameter like the population mean, μ. A confidence interval has the form

(point estimate - margin of error, point estimate + margin of error)

The margin of error depends on the confidence level or percentage of confidence.

8.1.3 Optional Collaborative Classroom Activity

Have your instructor record the number of meals each student in your class eats out in a week. Assume that the standard deviation is known to be 3 meals. Construct an approximate 95% confidence interval for the true average number of meals students eat out each week.

1. Calculate the sample mean.
2. $\sigma = 3$ and $n =$ the number of students surveyed.
3. Construct the interval $\left(\overline{x} - 2 \cdot \frac{\sigma}{\sqrt{n}}, \overline{x} + 2 \cdot \frac{\sigma}{\sqrt{n}}\right)$

We say we are approximately 95% confident that the true average number of meals that students eat out in a week is between __________ and ___________.

8.2 Confidence Interval, Single Population Mean, Population Standard Deviation Known, Normal[2]

To construct a confidence interval for a single unknown population mean μ **where the population standard deviation is known,** we need $\overline{x}$ as an estimate for μ and a margin of error. Here, the margin of error is called the **error bound for a population mean** (abbreviated **EBM**). The margin of error depends on the **confidence level** (abbreviated **CL**). The confidence level is the probability that the confidence interval produced contains the true population parameter. Most often, it is the choice of the person constructing the confidence interval to choose a confidence level of 90% or higher because he wants to be reasonably certain of his conclusions.

There is another probability called **alpha** (α). α is the probability that the sample produced a point estimate that is not within the appropriate margin of error of the unknown population parameter.

Example 8.1
Suppose the sample mean is 7 and the error bound for the mean is 2.5.

Problem *(Solution on p. 327.)*
$\overline{x} =$ _______ and $EBM =$ _______.
The confidence interval is $(7 - 2.5, 7 + 2.5)$.

If the confidence level (CL) is 95%, then we say we are 95% confident that the true population mean is between 4.5 and 9.5.

A confidence interval for a population mean with a known standard deviation is based on the fact that the sample means follow an approximately normal distribution. Suppose we have constructed the 90% confidence interval (5, 15) where $\overline{x} = 10$ and $EBM = 5$. To get a 90% confidence interval, we must include the central 90% of the sample means. If we include the central 90%, we leave out a total of 10 % or 5% in each tail of the normal distribution. To capture the central 90% of the sample means, we must go out 1.645 standard deviations on either side of the calculated sample mean. The 1.645 is the z-score from a standard normal table that has area to the right equal to 0.05 (5% area in the right tail). The graph shows the general situation.

Confidence Level (CL) = 0.90

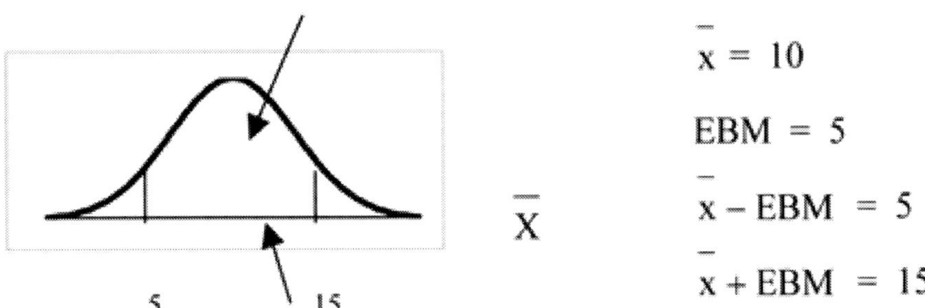

μ is believed to be in the interval (5, 15) with 90% confidence.

To summarize, resulting from the Central Limit Theorem,

$\overline{X}$ is normally distributed, that is, $\overline{X} \sim N\left(\mu_X, \frac{\sigma_X}{\sqrt{n}}\right)$

[2]This content is available online at <http://cnx.org/content/m16962/1.13/>.

Since the population standard deviation, σ, is known, we use a normal curve.

The confidence level, CL, is $CL = 1 - \alpha$. Each of the tails contains an area equal to $\frac{\alpha}{2}$.

The z-score that has area to the right of $\frac{\alpha}{2}$ is denoted by $z_{\frac{\alpha}{2}}$.

For example, if $\frac{\alpha}{2} = 0.025$, then area to the right $= 0.025$ and area to the left $= 1 - 0.025 = 0.975$ and $z_{\frac{\alpha}{2}} = z_{0.025} = 1.96$ using a calculator, computer or table. Using the TI83+ or 84 calculator function, `invNorm`, you can verify this result. `invNorm`$(.975, 0, 1) = 1.96$.

The error bound formula for a single population mean when the population standard deviation is known is

$EBM = z_{\frac{\alpha}{2}} \cdot \frac{\sigma}{\sqrt{n}}$

The confidence interval has the format $(\overline{x} - EBM, \overline{x} + EBM)$.

The graph gives a picture of the entire situation.

$CL + \frac{\alpha}{2} + \frac{\alpha}{2} = CL + \alpha = 1.$

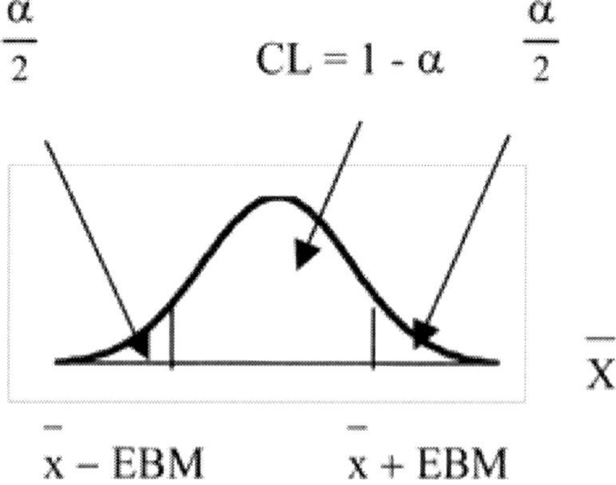

Example 8.2
Problem 1
Suppose scores on exams in statistics are normally distributed with an unknown population mean and a population standard deviation of 3 points. A sample of 36 scores is taken and gives a sample mean (sample average score) of 68. Find a 90% confidence interval for the true (population) mean of statistics exam scores.

- The first solution is step-by-step.
- The second solution uses the TI-83+ and TI-84 calculators.

Solution A
To find the confidence interval, you need the sample mean, $\overline{x}$, and the EBM.

a. $\overline{x} = 68$
b. $EBM = z_{\frac{\alpha}{2}} \cdot \left(\frac{\sigma}{\sqrt{n}}\right)$
c. $\sigma = 3$
d. $n = 36$

$CL = 0.90$ so $a = 1 - CL = 1 - 0.90 = 0.10$

Since $\frac{\alpha}{2} = 0.05$, then $z_{\frac{\alpha}{2}} = z_{.05} = 1.645$

from a calculator, computer or standard normal table. For the table, see the Table of Contents **15. Tables**.

Therefore, $EBM = 1.645 \cdot \left(\frac{3}{\sqrt{36}}\right) = 0.8225$

This gives $\overline{x} - EBM = 68 - 0.8225 = 67.18$

and $\overline{x} + EBM = 68 + 0.8225 = 68.82$

The 90% confidence interval is **(67.18, 68.82).**

Solution B
The TI-83+ and TI-84 caculators simplify this whole procedure. Press `STAT` and arrow over to `TESTS`. Arrow down to `7:ZInterval`. Press `ENTER`. Arrow to `Stats` and press `ENTER`. Arrow down and enter 3 for σ, 68 for $\overline{x}$, 36 for n, and .90 for `C-level`. Arrow down to `Calculate` and press `ENTER`. The confidence interval is (to 3 decimal places) (67.178, 68.822).

We can find the error bound from the confidence interval. From the upper value, subtract the sample mean **or** subtract the lower value from the upper value and divide by two. The result is the error bound for the mean (EBM).

$EBM = 68.822 - 68 = 0.822$ or $EBM = \frac{(68.822 - 67.178)}{2} = 0.822$

We can interpret the confidence interval in two ways:

1. We are 90% confident that the true population mean for statistics exam scores is between 67.178 and 68.822.
2. Ninety percent of all confidence intervals constructed in this way contain the true average statistics exam score. For example, if we constructed 100 of these confidence intervals, we would expect 90 of them to contain the true population mean exam score.

Now for the same problem, find a 95% confidence interval for the true (population) mean of scores. Draw the graph. The sample mean, standard deviation, and sample size are:

Problem 2 ***(Solution on p. 327.)***

a. $\overline{x} =$ 68
b. $\sigma =$ 3
c. $n =$ 36

The confidence level is $CL = 0.95$. Graph:

The confidence interval is (use technology)

Problem 3
$(\overline{x} - EBM, \overline{x} + EBM) =$ (_______, _______). The error bound $EBM =$ _______.

Solution
$(\overline{x} - EBM, \overline{x} + EBM) =$ *(67.02 , 68.98)*. The error bound $EBM = 0.98$.

We can say that we are 95 % confident that the true population mean for statistics exam scores is between 67.02 and 68.98 and that 95% of all confidence intervals constructed in this way contain the true average statistics exam score.

Example 8.3
Suppose we change the previous problem.

Problem 1
Leave everything the same except the sample size. For this problem, we can examine the impact of changing n to 100 or changing n to 25.

a. $\overline{x} = 68$
b. $\sigma = 3$
c. $z_{\frac{\alpha}{2}} = 1.645$

Solution A
If we **increase** the sample size n to 100, we **decrease** the error bound.

$$EBM = z_{\frac{\alpha}{2}} \cdot \left(\frac{\sigma}{\sqrt{n}}\right) = 1.645 \cdot \left(\frac{3}{\sqrt{100}}\right) = 0.4935$$

Solution B
If we **decrease** the sample size n to 25, we **increase** the error bound.

$$EBM = z_{\frac{\alpha}{2}} \cdot \left(\frac{\sigma}{\sqrt{n}}\right) = 1.645 \cdot \left(\frac{3}{\sqrt{25}}\right) = 0.987$$

Problem 2
Leave everything the same except for the confidence level. We increase the confidence level from 0.90 to 0.95.

a. $\overline{x} = 68$
b. $\sigma = 3$
c. $z_{\frac{\alpha}{2}}$ changes from 1.645 to 1.96.

Solution

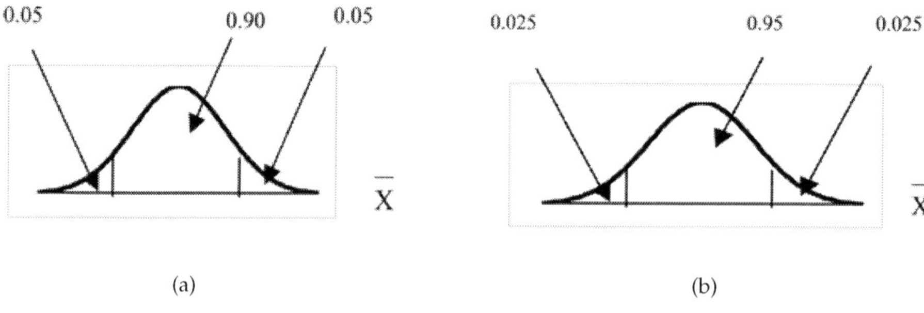

Figure 8.1

The 90% confidence interval is (67.18, 68.82). The 95% confidence interval is (67.02, 68.98). The 95% confidence interval is wider. **If you look at the graphs, because the area 0.95 is larger than the area 0.90, it makes sense that the 95% confidence interval is wider.**

8.2.1

Calculating the Sample Size n

If researchers desire a specific margin of error, then they can use the error bound formula to calculate the required sample size.

The error bound formula for a population mean when the population standard deviation is known is

- $EBM = z_{\frac{\alpha}{2}} \cdot \frac{\sigma}{\sqrt{n}}$
- Solving for n gives you an equation for the sample size.
- $n = \frac{z_{\frac{\alpha}{2}}{}^2 \cdot \sigma^2}{EBM^2}$

Example 8.4
The population standard deviation for the age of Foothill College students is 15 years. If we want to be 95% confident that the sample mean age is within 2 years of the true population mean age of Foothill College students , how many randomly selected Foothill College students must be surveyed?

From the problem, we know that

- $\sigma = 15$
- $EBM = 2$
- $z_{\frac{\alpha}{2}} = 1.96$ because the confidence level is 95%.

Using the equation for the sample size, we have

- $n = \frac{z_{\frac{\alpha}{2}}{}^2 \cdot \sigma^2}{EBM^2}$
- $n = \frac{1.96^2 \cdot 15^2}{2^2}$
- $n = 216.09$
- Round the answer to the next higher value to ensure that the sample size is as large as it should be. Therefore, 217 Foothill College students should be surveyed for us to be 95% confident that we are within 2 years of the true population age of Foothill College students.

NOTE: In reality, we usually do not know the population standard deviation so we estimate it with the sample standard deviation or use some other way of estimating it (for example, some statisticians use the results of some other earlier study as the estimate).

8.3 Confidence Interval, Single Population Mean, Standard Deviation Unknown, Student-T[3]

In practice, we rarely know the population **standard deviation**. In the past, when the sample size was large, this did not present a problem to statisticians. They used the sample standard deviation s as an estimate for σ and proceeded as before to calculate a **confidence interval** with close enough results. However, statisticians ran into problems when the sample size was small. A small sample size caused inaccuracies in the confidence interval. William S. Gossett of the Guinness brewery in Dublin, Ireland ran into this very problem. His experiments with hops and barley produced very few samples. Just replacing σ with s did not produce accurate results when he tried to calculate a confidence interval. He realized that he could not use a normal distribution for the calculation. This problem led him to "discover" what is called the **Student-t distribution**. The name comes from the fact that Gosset wrote under the pen name "Student."

Up until the mid 1990s, statisticians used the **normal distribution** approximation for large sample sizes and only used the Student-t distribution for sample sizes of at most 30. With the common use of graphing calculators and computers, the practice is to use the Student-t distribution whenever s is used as an estimate for σ.

If you draw a simple random sample of size n from a population that has approximately a normal distribution with mean μ and unknown population standard deviation σ and calculate the t-score $t = \frac{\overline{x}-\mu}{\left(\frac{s}{\sqrt{n}}\right)}$, then the t-scores follow a **Student-t distribution with** $n-1$ **degrees of freedom**. The t-score has the same interpretation as the **z-score**. It measures how far $\overline{x}$ is from its mean μ. For each sample size n, there is a different Student-t distribution.

The **degrees of freedom**, $n-1$, come from the sample standard deviation s. In Chapter 2, we used n deviations ($x - \overline{x}$ *values*) to calculate s. Because the sum of the deviations is 0, we can find the last deviation once we know the other $n-1$ deviations. The other $n-1$ deviations can change or vary freely. **We call the number** $n-1$ **the degrees of freedom (df).**

The following are some facts about the Student-t distribution:

1. The graph for the Student-t distribution is similar to the normal curve.
2. The Student-t distribution has more probability in its tails than the normal because the spread is somewhat greater than the normal.
3. The underlying population of observations is normal with unknown population mean μ and unknown population standard deviation σ. In the real world, however, as long as the underlying population is large and bell-shaped, and the data are a simple random sample, practitioners often consider the assumptions met.

A Student-t table (See the Table of Contents **15. Tables**) gives t-scores given the degrees of freedom and the right-tailed probability. The table is very limited. **Calculators and computers can easily calculate any Student-t probabilities.**

The notation for the Student-t distribution is (using T as the random variable)

$T \sim t_{df}$ where $df = n - 1$.

If the population standard deviation is **not known**, then the **error bound for a population mean** formula is:

$EBM = t_{\frac{\alpha}{2}} \cdot \left(\frac{s}{\sqrt{n}}\right)$ $\quad$ $t_{\frac{\alpha}{2}}$ is the t-score with area to the right equal to $\frac{\alpha}{2}$.

[3]This content is available online at <http://cnx.org/content/m16959/1.11/>.

s = the sample standard deviation

The mechanics for calculating the error bound and the confidence interval are the same as when σ is known.

Example 8.5
Suppose you do a study of acupuncture to determine how effective it is in relieving pain. You measure sensory rates for 15 subjects with the results given below. Use the sample data to construct a 95% confidence interval for the mean sensory rate for the population (assumed normal) from which you took the data.

8.6; 9.4; 7.9; 6.8; 8.3; 7.3; 9.2; 9.6; 8.7; 11.4; 10.3; 5.4; 8.1; 5.5; 6.9

Note:

- The first solution is step-by-step.
- The second solution uses the TI-83+ and TI-84 calculators.

Solution A
To find the confidence interval, you need the sample mean, $\overline{x}$, and the EBM.

$\overline{x} = 8.2267 \qquad s = 1.6722 \qquad n = 15$

$CL = 0.95 \text{ so } \alpha = 1 - CL = 1 - 0.95 = 0.05$

$EBM = t_{\frac{\alpha}{2}} \cdot \left(\frac{s}{\sqrt{n}}\right)$

$\frac{\alpha}{2} = 0.025 \qquad t_{\frac{\alpha}{2}} = t_{.025} = 2.14$

(Student-t table with $df = 15 - 1 = 14$)

Therefore, $EBM = 2.14 \cdot \left(\frac{1.6722}{\sqrt{15}}\right) = 0.924$

This gives $\overline{x} - EBM = 8.2267 - 0.9240 = 7.3$

and $\overline{x} + EBM = 8.2267 + 0.9240 = 9.15$

The 95% confidence interval is **(7.30, 9.15)**.

You are 95% confident or sure that the true population average sensory rate is between 7.30 and 9.15.

Solution B
TI-83+ or TI-84: Use the function `8:TInterval` in `STAT TESTS`. Once you are in `TESTS`, press `8:TInterval` and arrow to `Data`. Press `ENTER`. Arrow down and enter the list name where you put the data for `List`, enter 1 for `Freq`, and enter .95 for `C-level`. Arrow down to `Calculate` and press `ENTER`. The confidence interval is (7.3006, 9.1527)

8.4 Confidence Interval for a Population Proportion[4]

During an election year, we see articles in the newspaper that state **confidence intervals** in terms of proportions or percentages. For example, a poll for a particular candidate running for president might show

[4] This content is available online at <http://cnx.org/content/m16963/1.10/>.

that the candidate has 40% of the vote within 3 percentage points. Often, election polls are calculated with 95% confidence. So, the pollsters would be 95% confident that the true proportion of voters who favored the candidate would be between 0.37 and 0.43 $(0.40 - 0.03, 0.40 + 0.03)$.

Investors in the stock market are interested in the true proportion of stocks that go up and down each week. Businesses that sell personal computers are interested in the proportion of households in the United States that own personal computers. Confidence intervals can be calculated for the true proportion of stocks that go up or down each week and for the true proportion of households in the United States that own personal computers.

The procedure to find the confidence interval, the sample size, the **error bound,** and the **confidence level** for a proportion is similar to that for the population mean. The formulas are different.

How do you know you are dealing with a proportion problem? First, the underlying **distribution is binomial.** (There is no mention of a mean or average.) If X is a binomial random variable, then $X \sim B(n, p)$ where n = the number of trials and p = the probability of a success. To form a proportion, take X, the random variable for the number of successes and divide it by n, the number of trials (or the sample size). The random variable P' (read "P prime") is that proportion,

$$P' = \frac{X}{n}$$

(Sometimes the random variable is denoted as $\hat{P}$, read "P hat".)

When n is large, we can use the **normal distribution** to approximate the binomial.

$$X \sim N\left(n \cdot p, \sqrt{n \cdot p \cdot q}\right)$$

If we divide the random variable by n, the mean by n, and the standard deviation by n, we get a normal distribution of proportions with P', called the estimated proportion, as the random variable. (Recall that a proportion = the number of successes divided by n.)

$$\frac{X}{n} = P' \sim N\left(\frac{n \cdot p}{n}, \frac{\sqrt{n \cdot p \cdot q}}{n}\right)$$

By algebra, $\frac{\sqrt{n \cdot p \cdot q}}{n} = \sqrt{\frac{p \cdot q}{n}}$

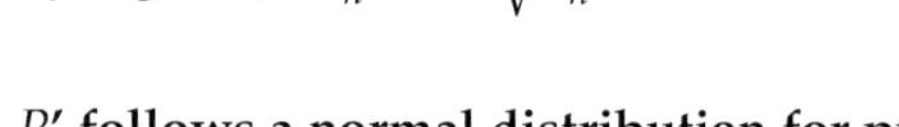

P' **follows a normal distribution for proportions**: $P' \sim N\left(p, \sqrt{\frac{p \cdot q}{n}}\right)$

The confidence interval has the form $(p' - EBP, p' + EBP)$.

$$p' = \frac{x}{n}$$

p' = the **estimated proportion** of successes (p' is a **point estimate** for p, the true proportion)

x = the **number** of successes.

n = the size of the sample

The error bound for a proportion is

$$EBP = z_{\frac{\alpha}{2}} \cdot \sqrt{\frac{p' \cdot q'}{n}} \qquad q' = 1 - p'$$

This formula is actually very similar to the error bound formula for a mean. The difference is the standard deviation. For a mean where the population standard deviation is known, the standard deviation is $\frac{\sigma}{\sqrt{n}}$.

For a proportion, the standard deviation is $\sqrt{\frac{p \cdot q}{n}}$.

However, in the error bound formula, the standard deviation is $\sqrt{\frac{p' \cdot q'}{n}}$.

In the error bound formula, p' **and** q' **are estimates of** p **and** q. The estimated proportions p' and q' are used because p and q are not known. p' and q' are calculated from the data. p' is the estimated proportion of successes. q' is the estimated proportion of failures.

NOTE: For the normal distribution of proportions, the z-score formula is as follows.

If $P' \sim N\left(p, \sqrt{\frac{p \cdot q}{n}}\right)$ then the z-score formula is $z = \frac{p'-p}{\sqrt{\frac{p \cdot q}{n}}}$

Example 8.6
Suppose that a sample of 500 households in Phoenix was taken last May to determine whether the oldest child had given his/her mother a Mother's Day card. Of the 500 households, 421 responded yes. Compute a 95% confidence interval for the true proportion of all Phoenix households whose oldest child gave his/her mother a Mother's Day card.

Note:

- The first solution is step-by-step.
- The second solution uses the TI-83+ and TI-84 calculators.

Solution A
Let X = the number of oldest children who gave their mothers Mother's Day card last May. X is binomial. $X \sim B\left(500, \frac{421}{500}\right)$.

To calculate the confidence interval, you must find p', q', and EBP.

$n = 500$ $\quad$ x = the number of successes $= 421$

$p' = \frac{x}{n} = \frac{421}{500} = 0.842$

$q' = 1 - p' = 1 - 0.842 = 0.158$

Since $CL = 0.95$, then $\alpha = 1 - CL = 1 - 0.95 = 0.05$ $\quad$ $\frac{\alpha}{2} = 0.025$.

Then $z_{\frac{\alpha}{2}} = z_{.025} = 1.96$ using a calculator, computer, or standard normal table.

Remember that the area to the right = 0.025 and therefore, area to the left is 0.975.

The z-score that corresponds to 0.975 is 1.96.

$$EBP = z_{\frac{\alpha}{2}} \cdot \sqrt{\frac{p' \cdot q'}{n}} = 1.96 \cdot \sqrt{\left[\frac{(.842) \cdot (.158)}{500}\right]} = 0.032$$

$p' - EBP = 0.842 - 0.032 = 0.81$

$p' + EBP = 0.842 + 0.032 = 0.874$

The confidence interval for the true binomial population proportion is $(p' - EBP, p' + EBP) = (0.810, 0.874)$.

We are 95% confident that between 81% and 87.4% of the oldest children in households in Phoenix gave their mothers a Mother's Day card last May.

We can also say that 95% of the confidence intervals constructed in this way contain the true proportion of oldest children in Phoenix who gave their mothers a Mother's Day card last May.

Solution B
TI-83+ and TI-84: Press `STAT` and arrow over to `TESTS`. Arrow down to `A:PropZint`. Press `ENTER`. Enter 421 for x, 500 for n, and .95 for `C-Level`. Arrow down to `Calculate` and press `ENTER`. The confidence interval is (0.81003, 0.87397).

Example 8.7
For a class project, a political science student at a large university wants to determine the percent of students that are registered voters. He surveys 500 students and finds that 300 are registered voters. Compute a 90% confidence interval for the true percent of students that are registered voters and interpret the confidence interval.

Solution
$x = 300$ and $n = 500$. Using a TI-83+ or 84 calculator, the 90% confidence interval for the true percent of students that are registered voters is (0.564, 0.636).

Interpretation:

- We are 90% confident that the true percent of students that are registered voters is between 56.4% and 63.6%.
- Ninety percent (90 %) of all confidence intervals constructed in this way contain the true percent of students that are registered voters.

8.4.1

Calculating the Sample Size n

If researchers desire a specific margin of error, then they can use the error bound formula to calculate the required sample size.

The error bound formula for a population proportion is

- $EBM = z_{\frac{\alpha}{2}} \cdot \sqrt{\frac{p'q'}{n}}$
- Solving for n gives you an equation for the sample size.
- $n = \frac{z_{\frac{\alpha}{2}}^{2} \cdot p'q'}{EBM^2}$

Example 8.8
Suppose a mobile phone company wants to determine the current percentage of customers aged 50+ that use text messaging on their cell phone. How many customers aged 50+ should the company survey in order to be 90% confident that the estimated (sample) proportion is within 3 percentage points of the true population proportion of customers aged 50+ that use text messaging on their cell phone.

From the problem, we know that

- $EBP = 0.03$ $(3\% = 0.03)$

- $z_{\frac{\alpha}{2}} = 1.645$ because the confidence level is 90%.

However, in order to find n , we need to know the estimated (sample) proportion p'. Remember that $q' = 1 - p'$. But, we do not know p'. Since we multiply p' and q' together, we make them both equal to 0.5 because $p'q' = (.5)(.5) = 25$ results in the largest possible product. (Try other products: $(.6)(.4) = 24$; $(.3)(.7) = 21$; $(.2)(.8) = 16$; and so on). The largest possible product gives us the largest n. This gives us a large enough sample so that we can be 90% confident that we are within 3 percentage points of the true proportion of customers aged 50+ that use text messaging on their cell phone. To calculate the sample size n, use the formula and make the substitutions.

- $n = \frac{{z_{\frac{\alpha}{2}}}^2 \cdot p'q'}{EBM^2}$
- $n = \frac{(1.645^2) \cdot (.5)(.5)}{.03^2}$
- $n = 751.7$
- Round the answer to the next higher value. The sample size should be 758 cell phone customers aged 50+ in order to be 90% confident that the estimated (sample) proportion is within 3 percentage points of the true population proportion of customers aged 50+ that use text messaging on their cell phone.

8.5 Summary of Formulas[5]

Formula 8.1: General form of a confidence interval
$(lower\ value, upper\ value) = (point\ estimate - error\ bound, point\ estimate + error\ bound)$

Formula 8.2: To find the error bound when you know the confidence interval
$error\ bound = upper\ value - point\ estimate$ OR $error\ bound = \frac{upper\ value - lower\ value}{2}$

Formula 8.3: Single Population Mean, Known Standard Deviation, Normal Distribution
Use the Normal Distribution for Means (Section 7.2) $EBM = z_{\frac{\alpha}{2}} \cdot \frac{\sigma}{\sqrt{n}}$
The confidence interval has the format $(\overline{x} - EBM, \overline{x} + EBM)$.

Formula 8.4: Single Population Mean, Unknown Standard Deviation, Student-t Distribution
Use the Student-t Distribution with degrees of freedom $df = n - 1$. $EBM = t_{\frac{\alpha}{2}} \cdot \frac{s}{\sqrt{n}}$

Formula 8.5: Single Population Proportion, Normal Distribution
Use the Normal Distribution for a single population proportion $p' = \frac{x}{n}$
$EBP = z_{\frac{\alpha}{2}} \cdot \sqrt{\frac{p' \cdot q'}{n}}$ $p' + q' = 1$
The confidence interval has the format $(p' - EBP, p' + EBP)$.

Formula 8.6: Point Estimates
$\overline{x}$ is a point estimate for μ
p' is a point estimate for ρ

s is a point estimate for σ

[5]This content is available online at <http://cnx.org/content/m16973/1.7/>.

8.6 Practice 1: Confidence Intervals for Averages, Known Population Standard Deviation[6]

8.6.1 Student Learning Outcomes

- The student will explore the properties of Confidence Intervals for Averages when the population standard deviation is known.

8.6.2 Given

The average age for all Foothill College students for Fall 2005 was 32.7. The population standard deviation has been pretty consistent at 15. Twenty-five Winter 2006 students were randomly selected. The average age for the sample was 30.4. We are interested in the true average age for Winter 2006 Foothill College students. (http://research.fhda.edu/factbook/FHdemofs/Fact_sheet_fh_2005f.pdf[7])

Let X = the age of a Winter 2006 Foothill College student

8.6.3 Calculating the Confidence Interval

Exercise 8.6.1 *(Solution on p. 327.)*
$\overline{x} =$

Exercise 8.6.2 *(Solution on p. 327.)*
$n=$

Exercise 8.6.3 *(Solution on p. 327.)*
15=(insert symbol here)

Exercise 8.6.4 *(Solution on p. 327.)*
Define the Random Variable, $\overline{X}$, in words.

$\overline{X} =$

Exercise 8.6.5 *(Solution on p. 327.)*
What is $\overline{x}$ estimating?

Exercise 8.6.6 *(Solution on p. 327.)*
Is σ_x known?

Exercise 8.6.7 *(Solution on p. 327.)*
As a result of your answer to (4), state the exact distribution to use when calculating the Confidence Interval.

8.6.4 Explaining the Confidence Interval

Construct a 95% Confidence Interval for the true average age of Winter 2006 Foothill College students.

Exercise 8.6.8 *(Solution on p. 327.)*
How much area is in both tails (combined)? $\alpha =$ ________

Exercise 8.6.9 *(Solution on p. 327.)*
How much area is in each tail? $\frac{\alpha}{2} =$ ________

Exercise 8.6.10 *(Solution on p. 327.)*
Identify the following specifications:

[6]This content is available online at <http://cnx.org/content/m16970/1.9/>.
[7]http://research.fhda.edu/factbook/FHdemofs/Fact_sheet_fh_2005f.pdf

a. lower limit =
b. upper limit =
c. error bound =

Exercise 8.6.11 *(Solution on p. 327.)*

The 95% Confidence Interval is:__________________

Exercise 8.6.12

Fill in the blanks on the graph with the areas, upper and lower limits of the Confidence Interval, and the sample mean.

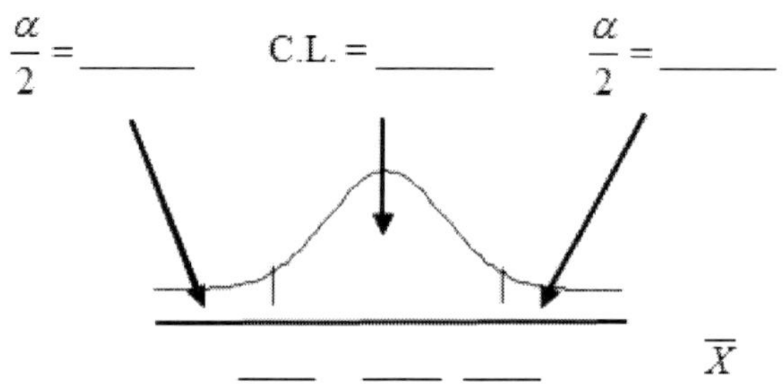

Figure 8.2

Exercise 8.6.13

In one complete sentence, explain what the interval means.

8.6.5 Discussion Questions

Exercise 8.6.14

Using the same mean, standard deviation and level of confidence, suppose that n were 69 instead of 25. Would the error bound become larger or smaller? How do you know?

Exercise 8.6.15

Using the same mean, standard deviation and sample size, how would the error bound change if the confidence level were reduced to 90%? Why?

8.7 Practice 2: Confidence Intervals for Averages, Unknown Population Standard Deviation[8]

8.7.1 Student Learning Outcomes

- The student will explore the properties of confidence intervals for averages, as well as the properties of an unknown population standard deviation.

8.7.2 Given

The following real data are the result of a random survey of 39 national flags (with replacement between picks) from various countries. We are interested in finding a confidence interval for the true average number of colors on a national flag. Let $X =$ the number of colors on a national flag.

X	Freq.
1	1
2	7
3	18
4	7
5	6

Table 8.1

8.7.3 Calculating the Confidence Interval

Exercise 8.7.1 *(Solution on p. 327.)*
Calculate the following:

a. $\overline{x} =$
b. $s_x =$
c. $n =$

Exercise 8.7.2 *(Solution on p. 327.)*
Define the Random Variable, $\overline{X}$, in words. $\overline{X} =$ ______________________________

Exercise 8.7.3 *(Solution on p. 327.)*
What is $\overline{x}$ estimating?

Exercise 8.7.4 *(Solution on p. 327.)*
Is σ_x known?

Exercise 8.7.5 *(Solution on p. 328.)*
As a result of your answer to (4), state the exact distribution to use when calculating the Confidence Interval.

[8]This content is available online at <http://cnx.org/content/m16971/1.10/>.

8.7.4 Confidence Interval for the True Average Number

Construct a 95% Confidence Interval for the true average number of colors on national flags.

Exercise 8.7.6 *(Solution on p. 328.)*
How much area is in both tails (combined)? $\alpha =$

Exercise 8.7.7 *(Solution on p. 328.)*
How much area is in each tail? $\frac{\alpha}{2} =$

Exercise 8.7.8 *(Solution on p. 328.)*
Calculate the following:

- **a.** lower limit =
- **b.** upper limit =
- **c.** error bound =

Exercise 8.7.9 *(Solution on p. 328.)*
The 95% Confidence Interval is:

Exercise 8.7.10
Fill in the blanks on the graph with the areas, upper and lower limits of the Confidence Interval, and the sample mean.

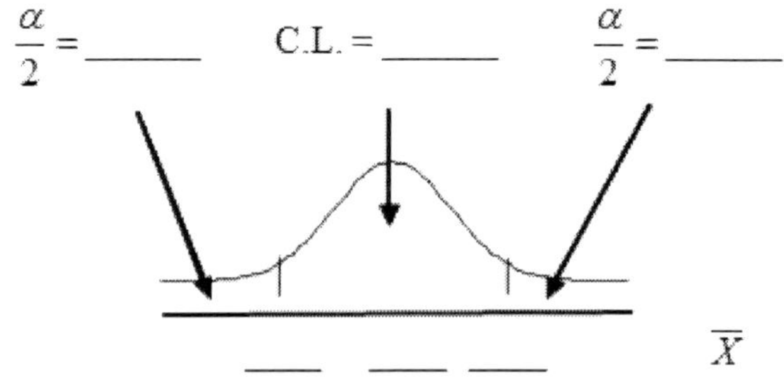

Figure 8.3

Exercise 8.7.11
In one complete sentence, explain what the interval means.

8.7.5 Discussion Questions

Exercise 8.7.12
Using the same $\overline{x}$, s_x, and level of confidence, suppose that n were 69 instead of 39. Would the error bound become larger or smaller? How do you know?

Exercise 8.7.13
Using the same $\overline{x}$, s_x, and $n = 39$, how would the error bound change if the confidence level were reduced to 90%? Why?

8.8 Practice 3: Confidence Intervals for Proportions[9]

8.8.1 Student Learning Outcomes

- The student will explore the properties of the confidence intervals for proportions.

8.8.2 Given

The Ice Chalet offers dozens of different beginning ice-skating classes. All of the class names are put into a bucket. The 5 P.M., Monday night, ages 8 - 12, beginning ice-skating class was picked. In that class were 64 girls and 16 boys. Suppose that we are interested in the true proportion of girls, ages 8 - 12, in all beginning ice-skating classes at the Ice Chalet.

8.8.3 Estimated Distribution

Exercise 8.8.1
What is being counted?

Exercise 8.8.2 *(Solution on p. 328.)*
In words, define the Random Variable X. $X =$

Exercise 8.8.3 *(Solution on p. 328.)*
Calculate the following:

a. $x =$
b. $n =$
c. $p' =$

Exercise 8.8.4 *(Solution on p. 328.)*
State the estimated distribution of X. $X \sim$

Exercise 8.8.5 *(Solution on p. 328.)*
Define a new Random Variable P'. What is p' estimating?

Exercise 8.8.6 *(Solution on p. 328.)*
In words, define the Random Variable P' . $P' =$

Exercise 8.8.7
State the estimated distribution of P'. $P' \sim$

8.8.4 Explaining the Confidence Interval

Construct a 92% Confidence Interval for the true proportion of girls in the age 8 - 12 beginning ice-skating classes at the Ice Chalet.

Exercise 8.8.8 *(Solution on p. 328.)*
How much area is in both tails (combined)? $\alpha =$

Exercise 8.8.9 *(Solution on p. 328.)*
How much area is in each tail? $\frac{\alpha}{2} =$

Exercise 8.8.10 *(Solution on p. 328.)*
Calculate the following:

a. lower limit =

[9] This content is available online at <http://cnx.org/content/m16968/1.10/>.

b. upper limit =
c. error bound =

Exercise 8.8.11 *(Solution on p. 328.)*

The 92% Confidence Interval is:

Exercise 8.8.12

Fill in the blanks on the graph with the areas, upper and lower limits of the Confidence Interval, and the sample proportion.

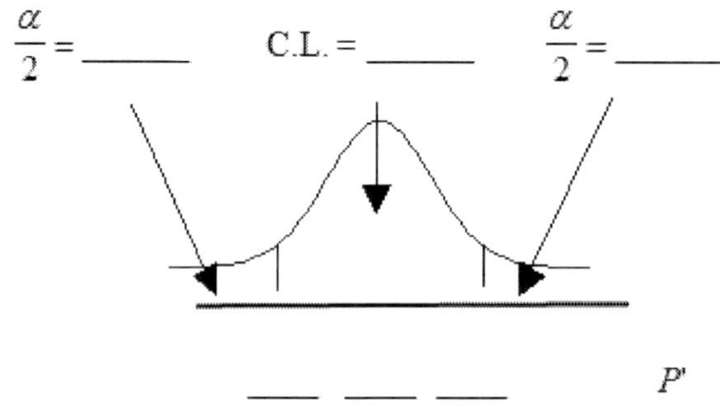

Figure 8.4

Exercise 8.8.13

In one complete sentence, explain what the interval means.

8.8.5 Discussion Questions

Exercise 8.8.14

Using the same p' and level of confidence, suppose that n were increased to 100. Would the error bound become larger or smaller? How do you know?

Exercise 8.8.15

Using the same p' and $n = 80$, how would the error bound change if the confidence level were increased to 98%? Why?

Exercise 8.8.16

If you decreased the allowable error bound, why would the minimum sample size increase (keeping the same level of confidence)?

8.9 Homework[10]

NOTE: If you are using a student-t distribution for a homework problem below, you may assume that the underlying population is normally distributed. (In general, you must first prove that assumption, though.)

Exercise 8.9.1 *(Solution on p. 328.)*

Among various ethnic groups, the standard deviation of heights is known to be approximately 3 inches. We wish to construct a 95% confidence interval for the average height of male Swedes. 48 male Swedes are surveyed. The sample mean is 71 inches. The sample standard deviation is 2.8 inches.

a. **i.** $\overline{x} =$________
  ii. $\sigma =$ ________
  iii. $s_x =$________
  iv. $n =$________
  v. $n - 1 =$________

b. Define the Random Variables X and $\overline{X}$, in words.

c. Which distribution should you use for this problem? Explain your choice.

d. Construct a 95% confidence interval for the population average height of male Swedes.
  i. State the confidence interval.
  ii. Sketch the graph.
  iii. Calculate the error bound.

e. What will happen to the level of confidence obtained if 1000 male Swedes are surveyed instead of 48? Why?

Exercise 8.9.2

In six packages of "The Flintstones® Real Fruit Snacks" there were 5 Bam-Bam snack pieces. The total number of snack pieces in the six bags was 68. We wish to calculate a 96% confidence interval for the population proportion of Bam-Bam snack pieces.

a. Define the Random Variables X and P', in words.

b. Which distribution should you use for this problem? Explain your choice

c. Calculate p'.

d. Construct a 96% confidence interval for the population proportion of Bam-Bam snack pieces per bag.
  i. State the confidence interval.
  ii. Sketch the graph.
  iii. Calculate the error bound.

e. Do you think that six packages of fruit snacks yield enough data to give accurate results? Why or why not?

Exercise 8.9.3 *(Solution on p. 329.)*

A random survey of enrollment at 35 community colleges across the United States yielded the following figures (source: **Microsoft Bookshelf**): 6414; 1550; 2109; 9350; 21828; 4300; 5944; 5722; 2825; 2044; 5481; 5200; 5853; 2750; 10012; 6357; 27000; 9414; 7681; 3200; 17500; 9200; 7380; 18314; 6557; 13713; 17768; 7493; 2771; 2861; 1263; 7285; 28165; 5080; 11622. Assume the underlying population is normal.

a. **i.** $\overline{x} =$

[10] This content is available online at <http://cnx.org/content/m16966/1.11/>.

ii. $s_x =$ ________
iii. $n =$ ________
iv. $n - 1 =$ ________

b. Define the Random Variables X and $\overline{X}$, in words.
c. Which distribution should you use for this problem? Explain your choice.
d. Construct a 95% confidence interval for the population average enrollment at community colleges in the United States.

i. State the confidence interval.
ii. Sketch the graph.
iii. Calculate the error bound.

e. What will happen to the error bound and confidence interval if 500 community colleges were surveyed? Why?

Exercise 8.9.4
From a stack of **IEEE Spectrum** magazines, announcements for 84 upcoming engineering conferences were randomly picked. The average length of the conferences was 3.94 days, with a standard deviation of 1.28 days. Assume the underlying population is normal.

a. Define the Random Variables X and $\overline{X}$, in words.
b. Which distribution should you use for this problem? Explain your choice.
c. Construct a 95% confidence interval for the population average length of engineering conferences.

i. State the confidence interval.
ii. Sketch the graph.
iii. Calculate the error bound.

Exercise 8.9.5 *(Solution on p. 329.)*
Suppose that a committee is studying whether or not there is waste of time in our judicial system. It is interested in the average amount of time individuals waste at the courthouse waiting to be called for service. The committee randomly surveyed 81 people. The sample average was 8 hours with a sample standard deviation of 4 hours.

a. i. $\overline{x} =$ ________
ii. $s_x =$ ________
iii. $n =$ ________
iv. $n - 1 =$ ________

b. Define the Random Variables X and $\overline{X}$, in words.
c. Which distribution should you use for this problem? Explain your choice.
d. Construct a 95% confidence interval for the population average time wasted.

a. State the confidence interval.
b. Sketch the graph.
c. Calculate the error bound.

e. Explain in a complete sentence what the confidence interval means.

Exercise 8.9.6
Suppose that an accounting firm does a study to determine the time needed to complete one person's tax forms. It randomly surveys 100 people. The sample average is 23.6 hours. There is a known standard deviation of 7.0 hours. The population distribution is assumed to be normal.

a. i. $\overline{x} =$ ________
ii. $\sigma =$ ________

iii. s_x =________
iv. n = ________
v. $n - 1$ =________

b. Define the Random Variables X and $\overline{X}$, in words.
c. Which distribution should you use for this problem? Explain your choice.
d. Construct a 90% confidence interval for the population average time to complete the tax forms.

i. State the confidence interval.
ii. Sketch the graph.
iii. Calculate the error bound.

e. If the firm wished to increase its level of confidence and keep the error bound the same by taking another survey, what changes should it make?
f. If the firm did another survey, kept the error bound the same, and only surveyed 49 people, what would happen to the level of confidence? Why?
g. Suppose that the firm decided that it needed to be at least 96% confident of the population average length of time to within 1 hour. How would the number of people the firm surveys change? Why?

Exercise 8.9.7 *(Solution on p. 329.)*

A sample of 16 small bags of the same brand of candies was selected. Assume that the population distribution of bag weights is normal. The weight of each bag was then recorded. The mean weight was 2 ounces with a standard deviation of 0.12 ounces. The population standard deviation is known to be 0.1 ounce.

a. i. $\overline{x}$ = ________
ii. σ = ________
iii. s_x =________
iv. n = ________
v. $n - 1$ = ________

b. Define the Random Variable X, in words.
c. Define the Random Variable $\overline{X}$, in words.
d. Which distribution should you use for this problem? Explain your choice.
e. Construct a 90% confidence interval for the population average weight of the candies.

i. State the confidence interval.
ii. Sketch the graph.
iii. Calculate the error bound.

f. Construct a 98% confidence interval for the population average weight of the candies.

i. State the confidence interval.
ii. Sketch the graph.
iii. Calculate the error bound.

g. In complete sentences, explain why the confidence interval in (f) is larger than the confidence interval in (e).
h. In complete sentences, give an interpretation of what the interval in (f) means.

Exercise 8.9.8

A pharmaceutical company makes tranquilizers. It is assumed that the distribution for the length of time they last is approximately normal. Researchers in a hospital used the drug on a random sample of 9 patients. The effective period of the tranquilizer for each patient (in hours) was as follows: 2.7; 2.8; 3.0; 2.3; 2.3; 2.2; 2.8; 2.1; and 2.4 .

a. i. $\overline{x} =$ ________
  ii. $s_x =$ ________
  iii. $n =$ ________
  iv. $n - 1 =$ ________

b. Define the Random Variable X, in words.
c. Define the Random Variable $\overline{X}$, in words.
d. Which distribution should you use for this problem? Explain your choice.
e. Construct a 95% confidence interval for the population average length of time.
  i. State the confidence interval.
  ii. Sketch the graph.
  iii. Calculate the error bound.

f. What does it mean to be "95% confident" in this problem?

Exercise 8.9.9 *(Solution on p. 329.)*

Suppose that 14 children were surveyed to determine how long they had to use training wheels. It was revealed that they used them an average of 6 months with a sample standard deviation of 3 months. Assume that the underlying population distribution is normal.

a. i. $\overline{x} =$ ________
  ii. $s_x =$ ________
  iii. $n =$ ________
  iv. $n - 1 =$ ________

b. Define the Random Variable X, in words.
c. Define the Random Variable $\overline{X}$, in words.
d. Which distribution should you use for this problem? Explain your choice.
e. Construct a 99% confidence interval for the population average length of time using training wheels.
  i. State the confidence interval.
  ii. Sketch the graph.
  iii. Calculate the error bound.

f. Why would the error bound change if the confidence level was lowered to 90%?

Exercise 8.9.10

Insurance companies are interested in knowing the population percent of drivers who always buckle up before riding in a car.

a. When designing a study to determine this population proportion, what is the minimum number you would need to survey to be 95% confident that the population proportion is estimated to within 0.03?
b. If it was later determined that it was important to be more than 95% confident and a new survey was commissioned, how would that affect the minimum number you would need to survey? Why?

Exercise 8.9.11 *(Solution on p. 330.)*

Suppose that the insurance companies did do a survey. They randomly surveyed 400 drivers and found that 320 claimed to always buckle up. We are interested in the population proportion of drivers who claim to always buckle up.

a. i. $x =$ ________
  ii. $n =$ ________
  iii. $p' =$ ________

b. Define the Random Variables X and P', in words.
c. Which distribution should you use for this problem? Explain your choice.
d. Construct a 95% confidence interval for the population proportion that claim to always buckle up.

- **i.** State the confidence interval.
- **ii.** Sketch the graph.
- **iii.** Calculate the error bound.

e. If this survey were done by telephone, list 3 difficulties the companies might have in obtaining random results.

Exercise 8.9.12

Unoccupied seats on flights cause airlines to lose revenue. Suppose a large airline wants to estimate its average number of unoccupied seats per flight over the past year. To accomplish this, the records of 225 flights are randomly selected and the number of unoccupied seats is noted for each of the sampled flights. The sample mean is 11.6 seats and the sample standard deviation is 4.1 seats.

a. **i.** $\overline{x} =$ ________
- **ii.** $s_x =$ ________
- **iii.** $n =$ ________
- **iv.** $n - 1 =$ ________

b. Define the Random Variables X and $\overline{X}$, in words.
c. Which distribution should you use for this problem? Explain your choice.
d. Construct a 92% confidence interval for the population average number of unoccupied seats per flight.

- **i.** State the confidence interval.
- **ii.** Sketch the graph.
- **iii.** Calculate the error bound.

Exercise 8.9.13 *(Solution on p. 330.)*

According to a recent survey of 1200 people, 61% feel that the president is doing an acceptable job. We are interested in the population proportion of people who feel the president is doing an acceptable job.

a. Define the Random Variables X and P', in words.
b. Which distribution should you use for this problem? Explain your choice.
c. Construct a 90% confidence interval for the population proportion of people who feel the president is doing an acceptable job.

- **i.** State the confidence interval.
- **ii.** Sketch the graph.
- **iii.** Calculate the error bound.

Exercise 8.9.14

A survey of the average amount of cents off that coupons give was done by randomly surveying one coupon per page from the coupon sections of a recent San Jose Mercury News. The following data were collected: 20¢; 75¢; 50¢; 65¢; 30¢; 55¢; 40¢; 40¢; 30¢; 55¢; $1.50; 40¢; 65¢; 40¢. Assume the underlying distribution is approximately normal.

a. **i.** $\overline{x} =$ ________
- **ii.** $s_x =$ ________
- **iii.** $n =$ ________

iv. $n - 1 =$ ________

b. Define the Random Variables X and $\overline{X}$, in words.
c. Which distribution should you use for this problem? Explain your choice.
d. Construct a 95% confidence interval for the population average worth of coupons.

i. State the confidence interval.
ii. Sketch the graph.
iii. Calculate the error bound.

e. If many random samples were taken of size 14, what percent of the confident intervals constructed should contain the population average worth of coupons? Explain why.

Exercise 8.9.15 *(Solution on p. 330.)*

An article regarding interracial dating and marriage recently appeared in the **Washington Post**. Of the 1709 randomly selected adults, 315 identified themselves as Latinos, 323 identified themselves as blacks, 254 identified themselves as Asians, and 779 identified themselves as whites. In this survey, 86% of blacks said that their families would welcome a white person into their families. Among Asians, 77% would welcome a white person into their families, 71% would welcome a Latino, and 66% would welcome a black person.

a. We are interested in finding the 95% confidence interval for the percent of black families that would welcome a white person into their families. Define the Random Variables X and P', in words.
b. Which distribution should you use for this problem? Explain your choice.
c. Construct a 95% confidence interval

i. State the confidence interval.
ii. Sketch the graph.
iii. Calculate the error bound.

Exercise 8.9.16

Refer to the problem above.

a. Construct the 95% confidence intervals for the three Asian responses.
b. Even though the three point estimates are different, do any of the confidence intervals overlap? Which?
c. For any intervals that do overlap, in words, what does this imply about the significance of the differences in the true proportions?
d. For any intervals that do not overlap, in words, what does this imply about the significance of the differences in the true proportions?

Exercise 8.9.17 *(Solution on p. 330.)*

A camp director is interested in the average number of letters each child sends during his/her camp session. The population standard deviation is known to be 2.5. A survey of 20 campers is taken. The average from the sample is 7.9 with a sample standard deviation of 2.8.

a. i. $\overline{x} =$ ________
ii. $\sigma =$ ________
iii. $s_x =$ ________
iv. $n =$ ________
v. $n - 1 =$ ________

b. Define the Random Variables X and $\overline{X}$, in words.
c. Which distribution should you use for this problem? Explain your choice.
d. Construct a 90% confidence interval for the population average number of letters campers send home.

i. State the confidence interval.
ii. Sketch the graph.
iii. Calculate the error bound.

e. What will happen to the error bound and confidence interval if 500 campers are surveyed? Why?

Exercise 8.9.18

Stanford University conducted a study of whether running is healthy for men and women over age 50. During the first eight years of the study, 1.5% of the 451 members of the 50-Plus Fitness Association died. We are interested in the proportion of people over 50 who ran and died in the same eight–year period.

a. Define the Random Variables X and P', in words.
b. Which distribution should you use for this problem? Explain your choice.
c. Construct a 97% confidence interval for the population proportion of people over 50 who ran and died in the same eight–year period.

i. State the confidence interval.
ii. Sketch the graph.
iii. Calculate the error bound.

d. Explain what a "97% confidence interval" means for this study.

Exercise 8.9.19 ***(Solution on p. 330.)***

In a recent sample of 84 used cars sales costs, the sample mean was $6425 with a standard deviation of $3156. Assume the underlying distribution is approximately normal.

a. Which distribution should you use for this problem? Explain your choice.
b. Define the Random Variable $\overline{X}$, in words.
c. Construct a 95% confidence interval for the population average cost of a used car.

i. State the confidence interval.
ii. Sketch the graph.
iii. Calculate the error bound.

d. Explain what a "95% confidence interval" means for this study.

Exercise 8.9.20

A telephone poll of 1000 adult Americans was reported in an issue of **Time Magazine**. One of the questions asked was "What is the main problem facing the country?" 20% answered "crime". We are interested in the population proportion of adult Americans who feel that crime is the main problem.

a. Define the Random Variables X and P', in words.
b. Which distribution should you use for this problem? Explain your choice.
c. Construct a 95% confidence interval for the population proportion of adult Americans who feel that crime is the main problem.

i. State the confidence interval.
ii. Sketch the graph.
iii. Calculate the error bound.

d. Suppose we want to lower the sampling error. What is one way to accomplish that?
e. The sampling error given by Yankelovich Partners, Inc. (which conducted the poll) is $\pm$ 3%. In 1-3 complete sentences, explain what the $\pm$ 3% represents.

Exercise 8.9.21 *(Solution on p. 330.)*

Refer to the above problem. Another question in the poll was "[How much are] you worried about the quality of education in our schools?" 63% responded "a lot". We are interested in the population proportion of adult Americans who are worried a lot about the quality of education in our schools.

1. Define the Random Variables X and P', in words.
2. Which distribution should you use for this problem? Explain your choice.
3. Construct a 95% confidence interval for the population proportion of adult Americans worried a lot about the quality of education in our schools.
   - **i.** State the confidence interval.
   - **ii.** Sketch the graph.
   - **iii.** Calculate the error bound.
4. The sampling error given by Yankelovich Partners, Inc. (which conducted the poll) is $\pm$ 3%. In 1-3 complete sentences, explain what the $\pm$ 3% represents.

Exercise 8.9.22

Six different national brands of chocolate chip cookies were randomly selected at the supermarket. The grams of fat per serving are as follows: 8; 8; 10; 7; 9; 9. Assume the underlying distribution is approximately normal.

- **a.** Calculate a 90% confidence interval for the population average grams of fat per serving of chocolate chip cookies sold in supermarkets.
  - **i.** State the confidence interval.
  - **ii.** Sketch the graph.
  - **iii.** Calculate the error bound.
- **b.** If you wanted a smaller error bound while keeping the same level of confidence, what should have been changed in the study before it was done?
- **c.** Go to the store and record the grams of fat per serving of six brands of chocolate chip cookies.
- **d.** Calculate the average.
- **e.** Is the average within the interval you calculated in part (a)? Did you expect it to be? Why or why not?

Exercise 8.9.23

A confidence interval for a proportion is given to be (– 0.22, 0.34). Why doesn't the lower limit of the confidence interval make practical sense? How should it be changed? Why?

8.9.1 Try these multiple choice questions.

The next three problems refer to the following: According a Field Poll conducted February 8 – 17, 2005, 79% of California adults (actual results are 400 out of 506 surveyed) feel that "education and our schools" is one of the top issues facing California. We wish to construct a 90% confidence interval for the true proportion of California adults who feel that education and the schools is one of the top issues facing California.

Exercise 8.9.24 *(Solution on p. 330.)*

A point estimate for the true population proportion is:

- **A.** 0.90
- **B.** 1.27

C. 0.79
D. 400

Exercise 8.9.25 *(Solution on p. 330.)*
A 90% confidence interval for the population proportion is:

A. (0.761, 0.820)
B. (0.125, 0.188)
C. (0.755, 0.826)
D. (0.130, 0.183)

Exercise 8.9.26 *(Solution on p. 330.)*
The error bound is approximately

A. 1.581
B. 0.791
C. 0.059
D. 0.030

The next two problems refer to the following:

A quality control specialist for a restaurant chain takes a random sample of size 12 to check the amount of soda served in the 16 oz. serving size. The sample average is 13.30 with a sample standard deviation is 1.55. Assume the underlying population is normally distributed.

Exercise 8.9.27 *(Solution on p. 330.)*
Find the 95% Confidence Interval for the true population mean for the amount of soda served.

A. (12.42, 14.18)
B. (12.32, 14.29)
C. (12.50, 14.10)
D. Impossible to determine

Exercise 8.9.28 *(Solution on p. 330.)*
What is the error bound?

A. 0.87
B. 1.98
C. 0.99
D. 1.74

Exercise 8.9.29 *(Solution on p. 331.)*
What is meant by the term "90% confident" when constructing a confidence interval for a mean?

A. If we took repeated samples, approximately 90% of the samples would produce the same confidence interval.
B. If we took repeated samples, approximately 90% of the confidence intervals calculated from those samples would contain the sample mean.
C. If we took repeated samples, approximately 90% of the confidence intervals calculated from those samples would contain the true value of the population mean.
D. If we took repeated samples, the sample mean would equal the population mean in approximately 90% of the samples.

The next two problems refer to the following:

Five hundred and eleven (511) homes in a certain southern California community are randomly surveyed to determine if they meet minimal earthquake preparedness recommendations. One hundred seventy-three (173) of the homes surveyed met the minimum recommendations for earthquake preparedness and 338 did not.

Exercise 8.9.30 *(Solution on p. 331.)*

Find the Confidence Interval at the 90% Confidence Level for the true population proportion of southern California community homes meeting at least the minimum recommendations for earthquake preparedness.

A. (0.2975, 0.3796)
B. (0.6270, 6959)
C. (0.3041, 0.3730)
D. (0.6204, 0.7025)

Exercise 8.9.31 *(Solution on p. 331.)*

The point estimate for the population proportion of homes that do not meet the minimum recommendations for earthquake preparedness is:

A. 0.6614
B. 0.3386
C. 173
D. 338

8.10 Review[11]

The next three problems refer to the following situation: Suppose that a sample of 15 randomly chosen people were put on a special weight loss diet. The amount of weight lost, in pounds, follows an unknown distribution with mean equal to 12 pounds and standard deviation equal to 3 pounds.

Exercise 8.10.1 *(Solution on p. 331.)*

To find the probability that the average of the 15 people lose no more than 14 pounds, the random variable should be:

A. The number of people who lost weight on the special weight loss diet
B. The number of people who were on the diet
C. The average amount of weight lost by 15 people on the special weight loss diet
D. The total amount of weight lost by 15 people on the special weight loss diet

Exercise 8.10.2 *(Solution on p. 331.)*

Find the probability asked for in the previous problem.

Exercise 8.10.3 *(Solution on p. 331.)*

Find the 90th percentile for the average amount of weight lost by 15 people.

The next three questions refer to the following situation: The time of occurrence of the first accident during rush-hour traffic at a major intersection is uniformly distributed between the three hour interval 4 p.m. to 7 p.m. Let X = the amount of time (hours) it takes for the first accident to occur.

- So, if an accident occurs at 4 p.m., the amount of time, in hours, it took for the accident to occur is _______.
- $\mu =$ _______
- $\sigma^2 =$ _______

Exercise 8.10.4 *(Solution on p. 331.)*

What is the probability that the time of occurrence is within the first half-hour or the last hour of the period from 4 to 7 p.m.?

A. Cannot be determined from the information given
B. $\frac{1}{6}$
C. $\frac{1}{2}$
D. $\frac{1}{3}$

Exercise 8.10.5 *(Solution on p. 331.)*

The 20th percentile occurs after how many hours?

A. 0.20
B. 0.60
C. 0.50
D. 1

Exercise 8.10.6 *(Solution on p. 331.)*

Assume Ramon has kept track of the times for the first accidents to occur for 40 different days. Let C = the total cumulative time. Then C follows which distribution?

A. $U\,(0,3)$
B. $Exp\left(\frac{1}{3}\right)$

[11]This content is available online at <http://cnx.org/content/m16972/1.8/>.

C. $N(60, 30)$
D. $N(1.5, 0.01875)$

Exercise 8.10.7 *(Solution on p. 331.)*

Using the information in question #6, find the probability that the total time for all first accidents to occur is more than 43 hours.

The next two questions refer to the following situation: The length of time a parent must wait for his children to clean their rooms is uniformly distributed in the time interval from 1 to 15 days.

Exercise 8.10.8 *(Solution on p. 331.)*

How long must a parent expect to wait for his children to clean their rooms?

A. 8 days
B. 3 days
C. 14 days
D. 6 days

Exercise 8.10.9 *(Solution on p. 331.)*

What is the probability that a parent will wait more than 6 days given that the parent has already waited more than 3 days?

A. 0.5174
B. 0.0174
C. 0.7500
D. 0.2143

The next five problems refer to the following study: Twenty percent of the students at a local community college live in within five miles of the campus. Thirty percent of the students at the same community college receive some kind of financial aid. Of those who live within five miles of the campus, 75% receive some kind of financial aid.

Exercise 8.10.10 *(Solution on p. 331.)*

Find the probability that a randomly chosen student at the local community college does not live within five miles of the campus.

A. 80%
B. 20%
C. 30%
D. Cannot be determined

Exercise 8.10.11 *(Solution on p. 331.)*

Find the probability that a randomly chosen student at the local community college lives within five miles of the campus or receives some kind of financial aid.

A. 50%
B. 35%
C. 27.5%
D. 75%

Exercise 8.10.12 *(Solution on p. 331.)*

Based upon the above information, are living in student housing within five miles of the campus and receiving some kind of financial aid mutually exclusive?

A. Yes

B. No
C. Cannot be determined

Exercise 8.10.13 *(Solution on p. 331.)*
The interest rate charged on the financial aid is _______ data.

A. quantitative discrete
B. quantitative continuous
C. qualitative discrete
D. qualitative

Exercise 8.10.14 *(Solution on p. 331.)*
What follows is information about the students who receive financial aid at the local community college.

- 1st quartile = $250
- 2nd quartile = $700
- 3rd quartile = $1200

(These amounts are for the school year.) If a sample of 200 students is taken, how many are expected to receive $250 or more?

A. 50
B. 250
C. 150
D. Cannot be determined

The next two problems refer to the following information: $P(A) = 0.2$, $P(B) = 0.3$, A and B are independent events.

Exercise 8.10.15 *(Solution on p. 331.)*
$P(A\ AND\ B) =$

A. 0.5
B. 0.6
C. 0
D. 0.06

Exercise 8.10.16 *(Solution on p. 331.)*
$P(A\ OR\ B) =$

A. 0.56
B. 0.5
C. 0.44
D. 1

Exercise 8.10.17 *(Solution on p. 331.)*
If H and D are mutually exclusive events, $P(H) = 0.25$, $P(D) = 0.15$, then $P(H|D)$

A. 1
B. 0
C. 0.40
D. 0.0375

8.11 Lab 1: Confidence Interval (Home Costs)[12]

Class Time:

Names:

8.11.1 Student Learning Outcomes:

- The student will calculate the 90% confidence interval for the average cost of a home in the area in which this school is located.
- The student will interpret confidence intervals.
- The student will examine the effects that changing conditions has on the confidence interval.

8.11.2 Collect the Data

Check the Real Estate section in your local newspaper. (Note: many papers only list them one day per week. Also, we will assume that homes come up for sale randomly.) Record the sales prices for 35 randomly selected homes recently listed in the county.

1. Complete the table:

______	______	______	______	______
______	______	______	______	______
______	______	______	______	______
______	______	______	______	______
______	______	______	______	______
______	______	______	______	______
______	______	______	______	______

Table 8.2

8.11.3 Describe the Data

1. Compute the following:
   a. $\overline{x} =$
   b. $s_x =$
   c. $n =$
2. Define the Random Variable $\overline{X}$, in words. $\overline{X} =$
3. State the estimated distribution to use. Use both words and symbols.

[12]This content is available online at <http://cnx.org/content/m16960/1.9/>.

8.11.4 Find the Confidence Interval

1. Calculate the confidence interval and the error bound.
    - **a.** Confidence Interval:
    - **b.** Error Bound:
2. How much area is in both tails (combined)? $\alpha =$
3. How much area is in each tail? $\frac{\alpha}{2} =$
4. Fill in the blanks on the graph with the area in each section. Then, fill in the number line with the upper and lower limits of the confidence interval and the sample mean.

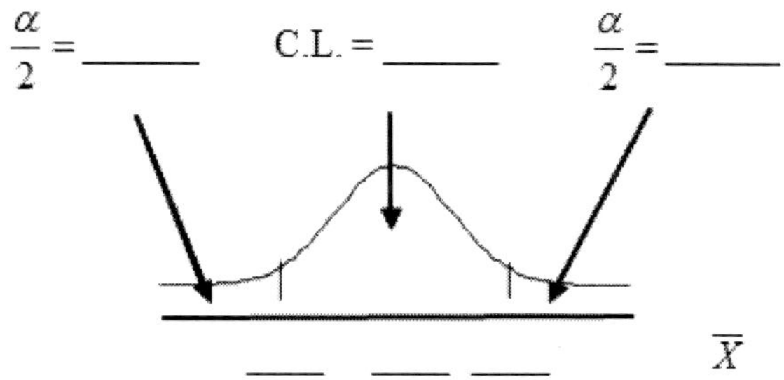

Figure 8.5

5. Some students think that a 90% confidence interval contains 90% of the data. Use the list of data on the first page and count how many of the data values lie within the confidence interval. What percent is this? Is this percent close to 90%? Explain why this percent should or should not be close to 90%.

8.11.5 Describe the Confidence Interval

1. In two to three complete sentences, explain what a Confidence Interval means (in general), as if you were talking to someone who has not taken statistics.
2. In one to two complete sentences, explain what this Confidence Interval means for this particular study.

8.11.6 Use the Data to Construct Confidence Intervals

1. Using the above information, construct a confidence interval for each confidence level given.

Confidence level	EBM / Error Bound	Confidence Interval
50%		
80%		
95%		
99%		

Table 8.3

2. What happens to the EBM as the confidence level increases? Does the width of the confidence interval increase or decrease? Explain why this happens.

8.12 Lab 2: Confidence Interval (Place of Birth)[13]

Class Time:

Names:

8.12.1 Student Learning Outcomes:

- The student will calculate the 90% confidence interval for proportion of students in this school that were born in this state.
- The student will interpret confidence intervals.
- The student will examine the effects that changing conditions have on the confidence interval.

8.12.2 Collect the Data

1. Survey the students in your class, asking them if they were born in this state. Let X = the number that were born in this state.
   a. n =____________
   b. x =____________
2. Define the Random Variable P' in words.
3. State the estimated distribution to use.

8.12.3 Find the Confidence Interval and Error Bound

1. Calculate the confidence interval and the error bound.
   a. Confidence Interval:
   b. Error Bound:
2. How much area is in both tails (combined)? α=
3. How much area is in each tail? $\frac{\alpha}{2}$ =
4. Fill in the blanks on the graph with the area in each section. Then, fill in the number line with the upper and lower limits of the confidence interval and the sample proportion.

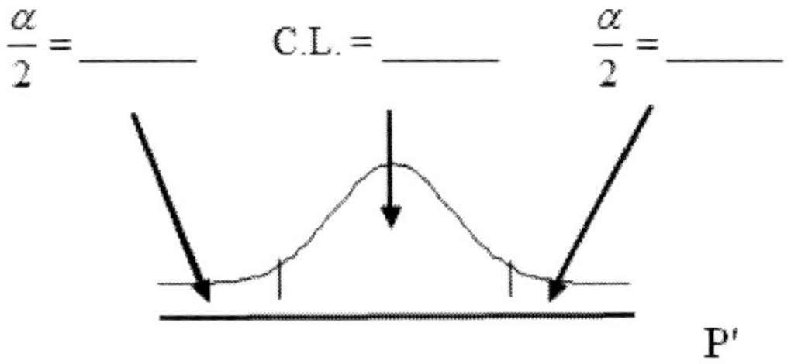

Figure 8.6

[13]This content is available online at <http://cnx.org/content/m16961/1.10/>.

8.12.4 Describe the Confidence Interval

1. In two to three complete sentences, explain what a Confidence Interval means (in general), as if you were talking to someone who has not taken statistics.
2. In one to two complete sentences, explain what this Confidence Interval means for this particular study.
3. Using the above information, construct a confidence interval for each given confidence level given.

Confidence level	EBP / Error Bound	Confidence Interval
50%		
80%		
95%		
99%		

Table 8.4

4. What happens to the EBP as the confidence level increases? Does the width of the confidence interval increase or decrease? Explain why this happens.

8.13 Lab 3: Confidence Interval (Womens' Heights)[14]

Class Time:

Names:

8.13.1 Student Learning Outcomes:

- The student will calculate a 90% confidence interval using the given data.
- The student will examine the relationship between the confidence level and the percent of constructed intervals that contain the population average.

8.13.2 Given:

1. **Heights of 100 Women (in Inches)**

59.4	71.6	69.3	65.0	62.9
66.5	61.7	55.2	67.5	67.2
63.8	62.9	63.0	63.9	68.7
65.5	61.9	69.6	58.7	63.4
61.8	60.6	69.8	60.0	64.9
66.1	66.8	60.6	65.6	63.8
61.3	59.2	64.1	59.3	64.9
62.4	63.5	60.9	63.3	66.3
61.5	64.3	62.9	60.6	63.8
58.8	64.9	65.7	62.5	70.9
62.9	63.1	62.2	58.7	64.7
66.0	60.5	64.7	65.4	60.2
65.0	64.1	61.1	65.3	64.6
59.2	61.4	62.0	63.5	61.4
65.5	62.3	65.5	64.7	58.8
66.1	64.9	66.9	57.9	69.8
58.5	63.4	69.2	65.9	62.2
60.0	58.1	62.5	62.4	59.1
66.4	61.2	60.4	58.7	66.7
67.5	63.2	56.6	67.7	62.5

Table 8.5

Listed above are the heights of 100 women. Use a random number generator to randomly select 10 data values.

[14]This content is available online at <http://cnx.org/content/m16964/1.9/>.

2. Calculate the sample mean and sample standard deviation. Assume that the population standard deviation is known to be 3.3 inches. With these values, construct a 90% confidence interval for your sample of 10 values. Write the confidence interval you obtained in the first space of the table below.
3. Now write your confidence interval on the board. As others in the class write their confidence intervals on the board, copy them into the table below:

90% Confidence Intervals

______	______	______	______	______
______	______	______	______	______
______	______	______	______	______
______	______	______	______	______
______	______	______	______	______
______	______	______	______	______
______	______	______	______	______
______	______	______	______	______

Table 8.6

8.13.3 Discussion Questions

1. The actual population mean for the 100 heights given above is $\mu = 63.4$. Using the class listing of confidence intervals, count how many of them contain the population mean μ; i.e., for how many intervals does the value of μ lie between the endpoints of the confidence interval?
2. Divide this number by the total number of confidence intervals generated by the class to determine the percent of confidence intervals that contain the mean μ. Write this percent below.
3. Is the percent of confidence intervals that contain the population mean μ close to 90%?
4. Suppose we had generated 100 confidence intervals. What do you think would happen to the percent of confidence intervals that contained the population mean?
5. When we construct a 90% confidence interval, we say that we are **90% confident that the true population mean lies within the confidence interval.** Using complete sentences, explain what we mean by this phrase.
6. Some students think that a 90% confidence interval contains 90% of the data. Use the list of data given (the heights of women) and count how many of the data values lie within the confidence interval that you generated on that page. How many of the 100 data values lie within your confidence interval? What percent is this? Is this percent close to 90%?
7. Explain why it does not make sense to count data values that lie in a confidence interval. Think about the random variable that is being used in the problem.
8. Suppose you obtained the heights of 10 women and calculated a confidence interval from this information. Without knowing the population mean μ, would you have any way of knowing **for certain** if your interval actually contained the value of μ? Explain.

NOTE: *This lab was designed and contributed by Diane Mathios.*

Solutions to Exercises in Chapter 8

Solution to Example 8.1 (p. 289)
$\overline{x} = 7$ and $EBM = 2.5$.
Solution to Example 8.2, Problem 2 (p. 291)

a. $\overline{x} = 68$
b. $\sigma = 3$
c. $n = 36$

Solutions to Practice 1: Confidence Intervals for Averages, Known Population Standard Deviation

Solution to Exercise 8.6.1 (p. 301)
30.4
Solution to Exercise 8.6.2 (p. 301)
25
Solution to Exercise 8.6.3 (p. 301)
σ
Solution to Exercise 8.6.4 (p. 301)
the age of Winter 2006 Foothill students
Solution to Exercise 8.6.5 (p. 301)
μ
Solution to Exercise 8.6.6 (p. 301)
yes
Solution to Exercise 8.6.7 (p. 301)
Normal
Solution to Exercise 8.6.8 (p. 301)
0.05
Solution to Exercise 8.6.9 (p. 301)
0.025
Solution to Exercise 8.6.10 (p. 301)

a. 24.52
b. 36.28
c. 5.88

Solution to Exercise 8.6.11 (p. 302)
$(24.52, 36.28)$

Solutions to Practice 2: Confidence Intervals for Averages, Unknown Population Standard Deviation

Solution to Exercise 8.7.1 (p. 303)

a. 3.26
b. 1.02
c. 39

Solution to Exercise 8.7.2 (p. 303)
the average number of colors of 39 flags
Solution to Exercise 8.7.3 (p. 303)
μ

Solution to Exercise 8.7.4 (p. 303)
No
Solution to Exercise 8.7.5 (p. 303)
t_{38}
Solution to Exercise 8.7.6 (p. 304)
0.05
Solution to Exercise 8.7.7 (p. 304)
0.025
Solution to Exercise 8.7.8 (p. 304)

- **a.** 2.93
- **b.** 3.59
- **c.** 0.33

Solution to Exercise 8.7.9 (p. 304)
2.93; 3.59

Solutions to Practice 3: Confidence Intervals for Proportions

Solution to Exercise 8.8.2 (p. 305)
The number of girls, age 8-12, in the beginning ice skating class
Solution to Exercise 8.8.3 (p. 305)

- **a.** 64
- **b.** 80
- **c.** 0.8

Solution to Exercise 8.8.4 (p. 305)
$B(80, 0.80)$
Solution to Exercise 8.8.5 (p. 305)
p'
Solution to Exercise 8.8.6 (p. 305)
The proportion of girls, age 8-12, in the beginning ice skating class.
Solution to Exercise 8.8.8 (p. 305)
0.80
Solution to Exercise 8.8.9 (p. 305)
0.04
Solution to Exercise 8.8.10 (p. 305)

- **a.** 0.72
- **b.** 0.88
- **c.** 0.08

Solution to Exercise 8.8.11 (p. 306)
0.72; 0.88

Solutions to Homework

Solution to Exercise 8.9.1 (p. 307)

- **a.**
  - **i.** 71
  - **ii.** 3
  - **iii.** 2.8
  - **iv.** 48

v. 47

c. $N\left(71, \frac{3}{\sqrt{48}}\right)$

d. i. CI: (70.15,71.85)

iii. EB = 0.85

Solution to Exercise 8.9.3 (p. 307)

a. i. 8629

ii. 6944

iii. 35

iv. 34

c. t_{34}

d. i. CI: (6243, 11,014)

iii. EB = 2385

e. It will become smaller

Solution to Exercise 8.9.5 (p. 308)

a. i. 8

ii. 4

iii. 81

iv. 80

c. t_{80}

d. i. CI: (7.12, 8.88)

iii. EB = 0.88

Solution to Exercise 8.9.7 (p. 309)

a. i. 2

ii. 0.1

iii . 0.12

iv. 16

v. 15

b. the weight of 1 small bag of candies

c. the average weight of 16 small bags of candies

d. $N\left(2, \frac{0.1}{\sqrt{16}}\right)$

e. i. CI: (1.96, 2.04)

iii. EB = 0.04

f. i. CI: (1.94, 2.06)

iii. EB = 0.06

Solution to Exercise 8.9.9 (p. 310)

a. i. 6

ii. 3

iii. 14

iv. 13

b. the time for a child to remove his training wheels

c. the average time for 14 children to remove their training wheels.

d. t_{13}

e. i. CI: (3.58, 8.42)

iii. EB = 2.42

Solution to Exercise 8.9.11 (p. 310)

- **a. i.** 320
  - **ii .** 400
  - **iii.** 0.80
- **c.** $N\left(0.80, \sqrt{\frac{(0.80)(0.20)}{400}}\right)$
- **d. i.** CI: (0.76, 0.84)
  - **iii.** EB = 0.02

Solution to Exercise 8.9.13 (p. 311)

- **b.** $N\left(0.61, \sqrt{\frac{(0.61)(0.39)}{1200}}\right)$
- **c. i.** CI: (0.59, 0.63)
  - **iii.** EB = 0.02

Solution to Exercise 8.9.15 (p. 312)

- **b.** $N\left(0.86, \sqrt{\frac{(0.86)(0.14)}{323}}\right)$
- **c. i.** CI: (0.8229, 0.8984)
  - **iii.** EB = 0.038

Solution to Exercise 8.9.17 (p. 312)

- **a. i.** 7.9
  - **ii.** 2.5
  - **iii.** 2.8
  - **iv.** 20
  - **v.** 19
- **c.** $N\left(7.9, \frac{2.5}{\sqrt{20}}\right)$
- **d. i.** CI: (6.98, 8.82)
  - **iii.** EB: 0.92

Solution to Exercise 8.9.19 (p. 313)

- **a.** t_{83}
- **b.** average cost of 84 used cars
- **c. i.** CI: (5740.10, 7109.90)
  - **iii.** EB = 684.90

Solution to Exercise 8.9.21 (p. 314)

- **b.** $N\left(0.63, \sqrt{\frac{(0.63)(0.37)}{1000}}\right)$
- **c. i.** CI: (0.60, 0.66)
  - **iii.** EB = 0.03

Solution to Exercise 8.9.24 (p. 314)
C

Solution to Exercise 8.9.25 (p. 315)
A

Solution to Exercise 8.9.26 (p. 315)
D

Solution to Exercise 8.9.27 (p. 315)
B

Solution to Exercise 8.9.28 (p. 315)
C
Solution to Exercise 8.9.29 (p. 315)
C
Solution to Exercise 8.9.30 (p. 316)
C
Solution to Exercise 8.9.31 (p. 316)
A

Solutions to Review

Solution to Exercise 8.10.1 (p. 317)
C
Solution to Exercise 8.10.2 (p. 317)
0.9951
Solution to Exercise 8.10.3 (p. 317)
12.99
Solution to Exercise 8.10.4 (p. 317)
C
Solution to Exercise 8.10.5 (p. 317)
B
Solution to Exercise 8.10.6 (p. 317)
C
Solution to Exercise 8.10.7 (p. 318)
0.9990
Solution to Exercise 8.10.8 (p. 318)
A
Solution to Exercise 8.10.9 (p. 318)
C
Solution to Exercise 8.10.10 (p. 318)
A
Solution to Exercise 8.10.11 (p. 318)
B
Solution to Exercise 8.10.12 (p. 318)
B
Solution to Exercise 8.10.13 (p. 319)
B
Solution to Exercise 8.10.14 (p. 319)

C. 150

Solution to Exercise 8.10.15 (p. 319)
D
Solution to Exercise 8.10.16 (p. 319)
C
Solution to Exercise 8.10.17 (p. 319)
B

Chapter 9

Hypothesis Testing: Single Mean and Single Proportion

9.1 Hypothesis Testing: Single Mean and Single Proportion[1]

9.1.1 Student Learning Objectives

By the end of this chapter, the student should be able to:

- Differentiate between Type I and Type II Errors
- Describe hypothesis testing in general and in practice
- Conduct and interpret hypothesis tests for a single population mean, population standard deviation known.
- Conduct and interpret hypothesis tests for a single population mean, population standard deviation unknown.
- Conduct and interpret hypothesis tests for a single population proportion.

9.1.2 Introduction

One job of a statistician is to make statistical inferences about populations based on samples taken from the population. **Confidence intervals** are one way to estimate a population parameter. Another way to make a statistical inference is to make a decision about a parameter. For instance, a car dealer advertises that its new small truck gets 35 miles per gallon, on the average. A tutoring service claims that its method of tutoring helps 90% of its students get an A or a B. A company says that women managers in their company earn an average of $60,000 per year.

A statistician will make a decision about these claims. This process is called **"hypothesis testing."** A hypothesis test involves collecting data from a sample and evaluating the data. Then, the statistician makes a decision as to whether or not the data supports the claim that is made about the population.

In this chapter, you will conduct hypothesis tests on single means and single proportions. You will also learn about the errors associated with these tests.

Hypothesis testing consists of two contradictory hypotheses or statements, a decision based on the data, and a conclusion. To perform a hypothesis test, a statistician will:

[1]This content is available online at <http://cnx.org/content/m16997/1.8/>.

1. Set up two contradictory hypotheses.
2. Collect sample data (in homework problems, the data or summary statistics will be given to you).
3. Determine the correct distribution to perform the hypothesis test.
4. Analyze sample data by performing the calculations that ultimately will support one of the hypotheses.
5. Make a decision and write a meaningful conclusion.

NOTE: To do the hypothesis test homework problems for this chapter and later chapters, make copies of the appropriate special solution sheets. See the Table of Contents topic "Solution Sheets".

9.2 Null and Alternate Hypotheses[2]

The actual test begins by considering two **hypotheses**. They are called the **null hypothesis** and the **alternate hypothesis**. These hypotheses contain opposing viewpoints.

H_o: **The null hypothesis:** It is a statement about the population that will be assumed to be true unless it can be shown to be incorrect beyond a reasonable doubt.

H_a: **The alternate hypothesis:** It is a claim about the population that is contradictory to H_o and what we conclude when we reject H_o.

Example 9.1
H_o: No more than 30% of the registered voters in Santa Clara County voted in the primary election.

H_a: More than 30% of the registered voters in Santa Clara County voted in the primary election.

Example 9.2
We want to test whether the average grade point average in American colleges is 2.0 (out of 4.0) or not.

H_a: $\mu = 2.0$ H_o: $\mu \neq 2.0$

Example 9.3
We want to test if college students take less than five years to graduate from college, on the average.

H_o: $\mu \geq 5$ H_a: $\mu < 5$

Example 9.4
In an issue of **U. S. News and World Report**, an article on school standards stated that about half of all students in France, Germany, and Israel take advanced placement exams and a third pass. The same article stated that 6.6% of U. S. students take advanced placement exams and 4.4 % pass. Test if the percentage of U. S. students who take advanced placement exams is more than 6.6%.

H_o: $p= 0.066$ H_a: $p > 0.066$

Since the null and alternate hypotheses are contradictory, you must examine evidence to decide which hypothesis the evidence supports. The evidence is in the form of sample data. The sample might support either the null hypothesis or the alternate hypothesis but not both.

After you have determined which hypothesis the sample supports, you make a **decision.** There are two options for a decision. They are "reject H_o" if the sample information favors the alternate hypothesis or

[2]This content is available online at <http://cnx.org/content/m16998/1.7/>.

"do not reject H_0" if the sample information favors the null hypothesis, meaning that there is not enough information to reject the null.

Mathematical Symbols Used in H_0 and H_a:

H_0	H_a
equal (=)	not equal (≠) **or** greater than (>) **or** less than (<)
greater than or equal to (≥)	less than (<)
less than or equal to (≤)	more than (>)

Table 9.1

NOTE: H_0 always has a symbol with an equal in it. H_a never has a symbol with an equal in it. The choice of symbol depends on the wording of the hypothesis test.

9.2.1 Optional Collaborative Classroom Activity

Bring to class a newspaper, some news magazines, and some Internet articles . In groups, find articles from which your group can write a null and alternate hypotheses. Discuss your hypotheses with the rest of the class.

9.3 Outcomes and the Type I and Type II Errors[3]

When you perform a hypothesis test, there are four outcomes depending on the actual truth (or falseness) of the null hypothesis H_0 and the decision to reject or not. The outcomes are summarized in the following table:

ACTION	**H_0 IS ACTUALLY**	**...**
	True	False
Do not reject H_0	Correct Outcome	Type II error
Reject H_0	Type I Error	Correct Outcome

Table 9.2

The four outcomes in the table are:

- The decision is to **not reject** H_0 when, in fact, H_0 **is true (correct decision).**
- The decision is to **reject** H_0 when, in fact, H_0 **is true** (incorrect decision known as a **Type I error**).
- The decision is to **not reject** H_0 when, in fact, H_0 **is false** (incorrect decision known as a **Type II error**).
- The decision is to **reject** H_0 when, in fact, H_0 **is false** (**correct decision** whose probability is called the **Power of the Test**).

Each of the errors occurs with a particular probability. The Greek letters α and β represent the probabilities.

α = probability of a Type I error = **P(Type I error)** = probability of rejecting the null hypothesis when the null hypothesis is true.

[3]This content is available online at <http://cnx.org/content/m17006/1.6/>.

β = probability of a Type II error = **P(Type II error)** = probability of not rejecting the null hypothesis when the null hypothesis is false.

α and β should be as small as possible because they are probabilities of errors. They are rarely 0.

The Power of the Test is $1 - \beta$. Ideally, we want a high power that is as close to 1 as possible.

The following are examples of Type I and Type II errors.

Example 9.5
Suppose the null hypothesis, H_o, is: Frank's rock climbing equipment is safe.

Type I error: Frank concludes that his rock climbing equipment may not be safe when, in fact, it really is safe. **Type II error**: Frank concludes that his rock climbing equipment is safe when, in fact, it is not safe.

α = **probability** that Frank thinks his rock climbing equipment may not be safe when, in fact, it really is. β = **probability** that Frank thinks his rock climbing equipment is safe when, in fact, it is not.

Notice that, in this case, the error with the greater consequence is the Type II error. (If Frank thinks his rock climbing equipment is safe, he will go ahead and use it.)

Example 9.6
Suppose the null hypothesis, H_o, is: The victim of an automobile accident is alive when he arrives at the emergency room of a hospital.

Type I error: The emergency crew concludes that the victim is dead when, in fact, the victim is alive. **Type II error**: The emergency crew concludes that the victim is alive when, in fact, the victim is dead.

α = **probability** that the emergency crew thinks the victim is dead when, in fact, he is really alive = *P(Type I error)*. β = **probability** that the emergency crew thinks the victim is alive when, in fact, he is dead = *P(Type II error)*.

The error with the greater consequence is the Type I error. (If the emergency crew thinks the victim is dead, they will not treat him.)

9.4 Distribution Needed for Hypothesis Testing[4]

Earlier in the course, we discussed sampling distributions. **Particular distributions are associated with hypothesis testing.** Perform tests of a population mean using a **normal distribution** or a **student-t distribution.** (Remember, use a student-t distribution when the population **standard deviation** is unknown and the population from which the sample is taken is normal.) In this chapter we perform tests of a population proportion using a normal distribution (usually n is large or the sample size is large).

If you are testing a **single population mean**, the distribution for the test is for **averages**:

$$\overline{X} \sim N\left(\mu_X, \frac{\sigma_X}{\sqrt{n}}\right) \qquad \text{or} \qquad t_{df}$$

The population parameter is μ. The estimated value (point estimate) for μ is $\overline{x}$, the sample mean.

[4]This content is available online at <http://cnx.org/content/m17017/1.6/>.

If you are testing a **single population proportion**, the distribution for the test is for proportions or percentages:

$$P' \sim N\left(p, \sqrt{\frac{p \cdot q}{n}}\right)$$

The population parameter is p. The estimated value (point estimate) for p is p'. $p' = \frac{x}{n}$ where x is the number of successes and n is the sample size.

9.5 Assumption[5]

When you perform a **hypothesis test of a single population mean** μ using a **Student-t distribution** (often called a t-test), there are fundamental assumptions that need to be met in order for the test to work properly. Your data should be a **simple random sample** that comes from a population that is approximately **normally distributed**. You use the sample **standard deviation** to approximate the population standard deviation. (Note that if the sample size is larger than 30, a t-test will work even if the population is not approximately normally distributed).

When you perform a **hypothesis test of a single population mean** μ using a normal distribution (often called a z-test), you take a simple random sample from the population. The population you are testing is normally distributed or your sample size is larger than 30 or both. You know the value of the population standard deviation.

When you perform a **hypothesis test of a single population proportion** p, you take a simple random sample from the population. You must meet the conditions for a **binomial distribution** which are there are a certain number n of independent trials, the outcomes of any trial are success or failure, and each trial has the same probability of a success p. The shape of the binomial distribution needs to be similar to the shape of the normal distribution. To ensure this, the quantities np and nq must both be greater than five ($np > 5$ and $nq > 5$). Then the binomial distribution of sample (estimated) proportion can be approximated by the normal distribution with $\mu = p$ and $\sigma = \sqrt{\frac{p \cdot q}{n}}$. Remember that $q = 1 - p$.

9.6 Rare Events[6]

Suppose you make an assumption about a property of the population (this assumption is the **null hypothesis**). Then you gather sample data randomly. If the sample has properties that would be very **unlikely** to occur if the assumption is true, then you would conclude that your assumption about the population is probably incorrect. (Remember that your assumption is just an **assumption** - it is not a fact and it may or may not be true. But your sample data is real and it is showing you a fact that seems to contradict your assumption.)

For example, Didi and Ali are at a birthday party of a very wealthy friend. They hurry to be first in line to grab a prize from a tall basket that they cannot see inside because they will be blindfolded. There are 200 plastic bubbles in the basket and Didi and Ali have been told that there is only one with a \$100 bill. Didi is the first person to reach into the basket and pull out a bubble. Her bubble contains a \$100 bill. The probability of this happening is $\frac{1}{200} = 0.005$. Because this is so unlikely, Ali is hoping that what the two of them were told is wrong and there are more \$100 bills in the basket. A "rare event" has occurred (Didi getting the \$100 bill) so Ali doubts the assumption about only one \$100 bill being in the basket.

---

[5]This content is available online at <http://cnx.org/content/m17002/1.7/>.
[6]This content is available online at <http://cnx.org/content/m16994/1.5/>.

9.7 Using the Sample to Support One of the Hypotheses[7]

Use the sample (data) to calculate the actual probability of getting the test result, called the **p-value**. The p-value is the **probability that an outcome of the data (for example, the sample mean) will happen purely by chance when the null hypothesis is true**.

A large p-value calculated from the data indicates that the sample result is likely happening purely by chance. The data supports the **null hypothesis** so we do not reject it. The smaller the p-value, the more unlikely the outcome, and the stronger the evidence is against the null hypothesis. We would reject the null hypothesis if the evidence is strongly against the null hypothesis.

The p-value is sometimes called the **computed** α because it is calculated from the data. You can think of it as the probability of (incorrectly) rejecting the null hypothesis when the null hypothesis is actually true.

Draw a graph that shows the p-value. The hypothesis test is easier to perform if you use a graph because you see the problem more clearly.

Example 9.7: (to illustrate the p-value)
Suppose a baker claims that his bread height is more than 15 cm, on the average. Several of his customers do not believe him. To persuade his customers that he is right, the baker decides to do a hypothesis test. He bakes 10 loaves of bread. The average height of the sample loaves is 17 cm. The baker knows from baking hundreds of loaves of bread that the **standard deviation** for the height is 0.5 cm.

The null hypothesis could be H_o: $\mu \leq 15$ The alternate hypothesis is H_a: $\mu > 15$

The words **"is more than"** translates as a "$>$ " so "$\mu > 15$" goes into the alternate hypothesis. The null hypothesis must contradict the alternate hypothesis.

Since σ **is known** ($\sigma = 0.5$ cm.), the distribution for the test is normal with mean $\mu = 15$ and standard deviation $\frac{\sigma}{\sqrt{n}} = \frac{0.5}{\sqrt{10}} = 0.16$.

Suppose the null hypothesis is true (the average height of the loaves is no more than 15 cm). Then is the average height (17 cm) calculated from the sample unexpectedly large? The hypothesis test works by asking the question how **unlikely** the sample average would be if the null hypothesis were true. The graph shows how far out the sample average is on the normal curve. How far out the sample average is on the normal curve is measured by the p-value. The p-value is the probability that, if we were to take other samples, any other sample average would fall at least as far out as 17 cm.

The p-value, then, is the probability that a sample average is the same or greater than 17 cm. when the population mean is, in fact, 15 cm. We can calculate this probability using the normal distribution for averages from Chapter 7.

[7]This content is available online at <http://cnx.org/content/m16995/1.8/>.

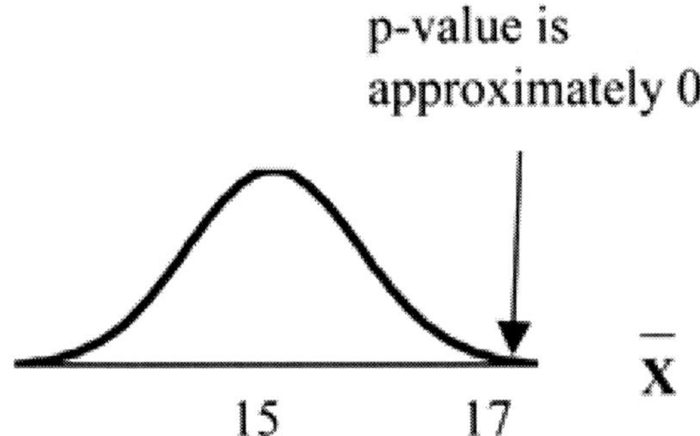

p-value $= P\left(\overline{X} > 17\right)$ which is approximately 0.

A p-value of approximately 0 tells us that it is highly unlikely that a loaf of bread rises no more than 15 cm, on the average. That is, almost 0% of all loaves of bread would be at least as high as 17 cm. **purely by CHANCE**. Because the outcome of 17 cm. is so **unlikely (meaning it is happening NOT by chance alone)**, we conclude that the evidence is strongly against the null hypothesis (the average height is at most 15 cm.). There is sufficient evidence that the true average height for the population of the baker's loaves of bread is greater than 15 cm.

9.8 Decision and Conclusion[8]

A systematic way to make a decision of whether to reject or not reject the **null hypothesis** is to compare the **p-value** and a **preset or preconceived** α **(also called a "significance level")**. A preset α is the probability of a **Type I error** (rejecting the null hypothesis when the null hypothesis is true). It may or may not be given to you at the beginning of the problem.

When you make a **decision** to reject or not reject H_o, do as follows:

- If $\alpha >$ *p-value*, reject H_o. The results of the sample data are significant. There is sufficient evidence to conclude that H_o is an incorrect belief and that the **alternative hypothesis**, H_a, may be correct.
- If $\alpha \leq$ *p-value*, do not reject H_o. The results of the sample data are not significant. There is not sufficient evidence to conclude that the alternative hypothesis, H_a, may be correct.
- When you "do not reject H_o", it does not mean that you should believe that H_o is true. It simply means that the sample data has **failed** to provide sufficient evidence to cast serious doubt about the truthfulness of H_o.

Conclusion: After you make your decision, write a thoughtful **conclusion** about the hypotheses in terms of the given problem.

9.9 Additional Information[9]

- In a **hypothesis test** problem, you may see words such as "the level of significance is 1%." The "1%" is the preconceived or preset α.
- The statistician setting up the hypothesis test selects the value of α to use **before** collecting the sample data.
- **If no level of significance is given, we generally can use** $\alpha = 0.05$.

[8]This content is available online at <http://cnx.org/content/m16992/1.7/>.
[9]This content is available online at <http://cnx.org/content/m16999/1.6/>.

- When you calculate the **p-value** and draw the picture, the p-value is in the left tail, the right tail, or split evenly between the two tails. For this reason, we call the hypothesis test left, right, or two tailed.
- The **alternate hypothesis**, H_a, tells you if the test is left, right, or two-tailed. It is the **key** to conducting the appropriate test.
- H_a **never** has a symbol that contains an equal sign.

The following examples illustrate a left, right, and two-tailed test.

Example 9.8
H_o: $\mu = 5$ H_a: $\mu < 5$

Test of a single population mean. H_a tells you the test is left-tailed. The picture of the p-value is as follows:

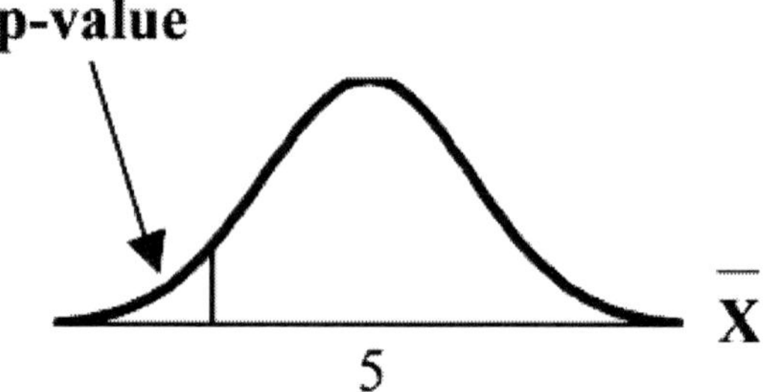

Example 9.9
H_o: $p \leq 0.2$ H_a: $p > 0.2$

This is a test of a single population proportion. H_a tells you the test is **right-tailed**. The picture of the p-value is as follows:

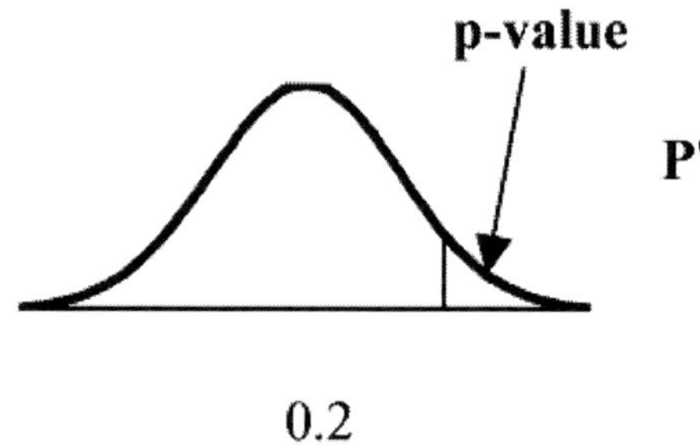

Example 9.10
H_o: $\mu = 50$ H_a: $\mu \neq 50$

This is a test of a single population mean. H_a tells you the test is **two-tailed**. The picture of the p-value is as follows.

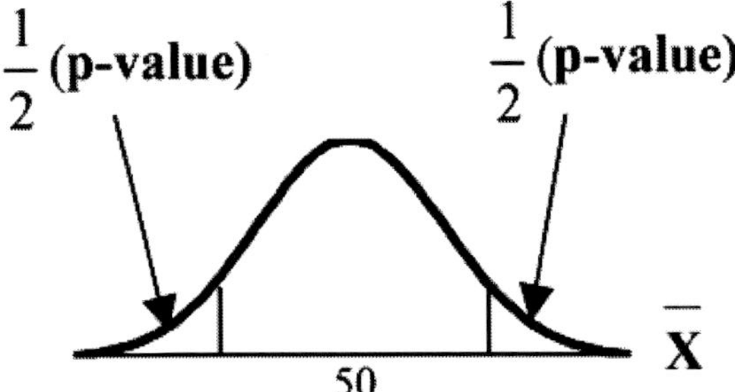

$\frac{1}{2}$ (p-value)
$\frac{1}{2}$ (p-value)
50
$\overline{X}$

9.10 Summary of the Hypothesis Test[10]

The **hypothesis test** itself has an established process. This can be summarized as follows:

1. Determine H_o and H_a. Remember, they are contradictory.
2. Determine the random variable.
3. Determine the distribution for the test.
4. Draw a graph, calculate the test statistic, and use the test statistic to calculate the **p-value**. (A z-score and a t-score are examples of test statistics.)
5. Compare the preconceived α with the p-value, make a decision (reject or cannot reject H_o), and write a clear conclusion using English sentences.

Notice that in performing the hypothesis test, you use α and not β. β is needed to help determine the sample size of the data that is used in calculating the p-value. Remember that the quantity $1 - \beta$ is called the **Power of the Test**. A high power is desirable. If the power is too low, statisticians typically increase the sample size while keeping α the same. If the power is low, the null hypothesis might not be rejected when it should be.

9.11 Examples[11]

Example 9.11

Jeffrey, as an eight-year old, **established an average time of 16.43 seconds** for swimming the 25-yard freestyle, with a **standard deviation of 0.8 seconds**. His dad, Frank, thought that Jeffrey could swim the 25-yard freestyle faster by using goggles. Frank bought Jeffrey a new pair of expensive goggles and timed Jeffrey for **15 25-yard freestyle swims**. For the 15 swims, **Jeffrey's average time was 16 seconds. Frank thought that the goggles helped Jeffrey to swim faster than the 16.43 seconds.** Conduct a hypothesis test using a preset $\alpha = 0.05$. Assume that the swim times for the 25-yard freestyle are normal.

Solution

Set up the Hypothesis Test:

Since the problem is about a mean (average), this is a **test of a single population mean**.

H_o: $\mu = 16.43$ H_a: $\mu < 16.43$

For Jeffrey to swim faster, his time will be less than 16.43 seconds. The "<" tells you this is left-tailed.

Determine the distribution needed:

Random variable: $\overline{X}$ = the average time to swim the 25-yard freestyle.

Distribution for the test: $\overline{X}$ is normal (population **standard deviation** is known: $\sigma = 0.8$)

$\overline{X} \sim N\left(\mu, \frac{\sigma_X}{\sqrt{n}}\right)$ Therefore, $\overline{X} \sim N\left(16.43, \frac{0.8}{\sqrt{15}}\right)$

$\mu = 16.43$ comes from H_0 and not the data. $\sigma = 0.8$, and $n = 15$.

Calculate the p-value using the normal distribution for a mean:

[10]This content is available online at <http://cnx.org/content/m16993/1.3/>.

[11]This content is available online at <http://cnx.org/content/m17005/1.12/>.

p-value $= P\left(\overline{X} < 16\right) = 0.0187$ where the sample mean in the problem is given s 16.

p-value $= 0.0187$ (This is called the **actual level of significance**.) The p-value is the area to the left of the sample mean is given as 16.

Graph:

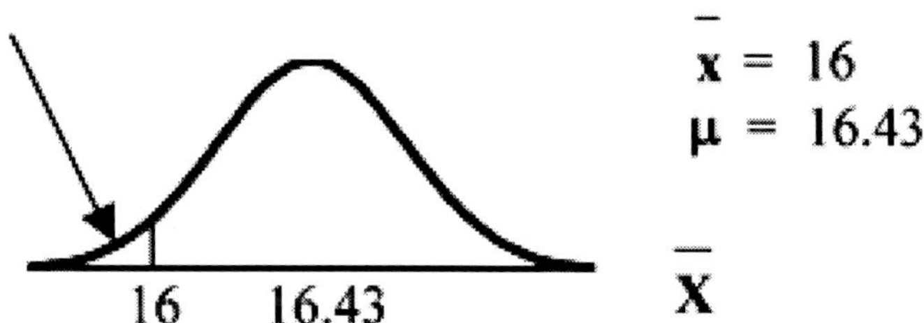

Figure 9.1

$\mu = 16.43$ comes from H_0. Our assumption is $\mu = 16.43$.

Interpretation of the p-value: If H_o **is true,** there is a 0.0187 probability (1.87%) that Jeffrey's mean (or average) time to swim the 25-yard freestyle is 16 seconds or less. Because a 1.87% chance is small, the mean time of 16 seconds or less is not happening randomly. It is a rare event.

Compare α and the p-value:

$\alpha = 0.05$ *p-value* $= 0.0187$ $\alpha >$ *p-value*

Make a decision: Since $\alpha >$ *p-value*, reject H_o.

This means that you reject $\mu = 16.43$. In other words, you do not think Jeffrey swims the 25-yard freestyle in 16.43 seconds but faster with the new goggles.

Conclusion: At the 5% significance level, we conclude that Jeffrey swims faster using the new goggles. The sample data show there is sufficient evidence that Jeffrey's mean time to swim the 25-yard freestyle is less than 16.43 seconds.

The p-value can easily be calculated using the TI-83+ and the TI-84 calculators:

Press `STAT` and arrow over to `TESTS`. Press `1:Z-Test`. Arrow over to `Stats` and press `ENTER`. Arrow down and enter 16.43 for μ_0 (null hypothesis), .8 for σ, 16 for the sample mean, and 15 for n. Arrow down to μ: (alternate hypothesis) and arrow over to $<\mu_0$. Press `ENTER`. Arrow down to `Calculate` and press `ENTER`. The calculator not only calculates the p-value ($p = 0.0187$) but it also calculates the test statistic (z-score) for the sample mean. $\mu < 16.43$ is the alternate hypothesis. Do this set of instructions again except arrow to `Draw` (instead of `Calculate`). Press `ENTER`. A shaded graph appears with $z = -2.08$ (test statistic) and $p = 0.0187$ (p-value). Make sure when you use `Draw` that no other equations are highlighted in $Y =$ and the plots are turned off.

When the calculator does a Z-Test, the Z-Test function finds the p-value by doing a normal probability calculation using the **Central Limit Theorem**:

$P\left(\overline{X} < 16 = \texttt{2nd DISTR normcdf} \left(-10\text{^}99, 16, 16.43, 0.8/\sqrt{15}\right)\right.$.

The Type I and Type II errors for this problem are as follows:

The Type I error is to conclude that Jeffrey swims the 25-yard freestyle, on average, in less than 16.43 seconds when, in fact, he actually swims the 25-yard freestyle, on average, in 16.43 seconds. (Reject the null hypothesis when the null hypothesis is true.)

The Type II error is to conclude that Jeffrey swims the 25-yard freestyle, on average, in 16.43 seconds when, in fact, he actually swims the 25-yard freestyle, on average, in less than 16.43 seconds. (Do not reject the null hypothesis when the null hypothesis is false.)

Historical Note: The traditional way to compare the two probabilities, α and the p-value, is to compare their test statistics (z-scores). The calculated test statistic for the p-value is -2.08. (From the Central Limit Theorem, the test statistic formula is $z = \frac{\overline{x}-\mu_X}{\left(\frac{\sigma_X}{\sqrt{n}}\right)}$. For this problem, $\overline{x} = 16$, $\mu_X = 16.43$ from the null hypothesis, $\sigma_X = 0.8$, and $n = 15$.) You can find the test statistic for $\alpha = 0.05$ in the normal table (see **15.Tables** in the Table of Contents). The z-score for an area to the left equal to 0.05 is midway between -1.65 and -1.64 (0.05 is midway between 0.0505 and 0.0495). The z-score is -1.645. Since $-1.645 > -2.08$ (which demonstrates that $\alpha > \textit{p-value}$), reject H_0. Traditionally, the decision to reject or not reject was done in this way. Today, comparing the two probabilities α and the p-value is very common and advantageous. For this problem, the p-value, 0.0187 is considerably smaller than α, 0.05. You can be confident about your decision to reject. It is difficult to know that the p-value is traditionally smaller than α by just examining the test statistics. The graph shows α, the p-value, and the two test statistics (z scores).

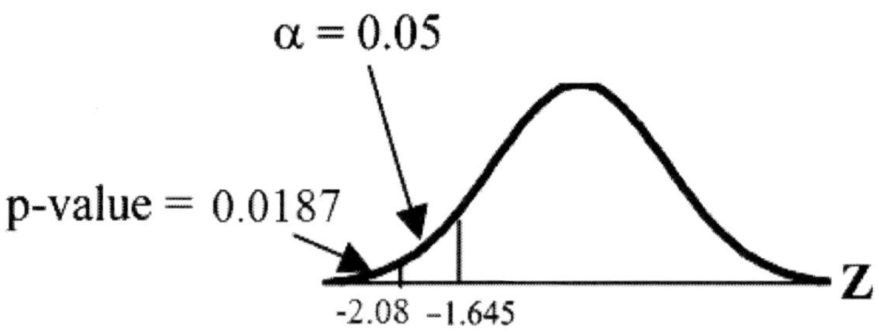

Figure 9.2

Example 9.12

A college football coach thought that his players could bench press an **average of 275 pounds**. It is known that the **standard deviation is 55 pounds**. Three of his players thought that the average was **more than** that amount. They asked **30** of their teammates for their estimated maximum lift on the bench press exercise. The data ranged from 205 pounds to 385 pounds. The actual different weights were (frequencies are in parentheses) 205(3); 215(3); 225(1); 241(2); 252(2); 265(2); 275(2); 313(2); 316(5); 338(2); 341(1); 345(2); 368(2); 385(1). (Source: data from Reuben Davis, Kraig Evans, and Scott Gunderson.)

Conduct a hypothesis test using a 2.5% level of significance to determine if the bench press average is **more than 275 pounds**.

Solution
Set up the Hypothesis Test:

Since the problem is about a mean (average), this is a **test of a single population mean**.

H_o: $\mu = 275$ H_a: $\mu > 275$ This is a right-tailed test.

Calculating the distribution needed:

Random variable: $\overline{X}$ = the average weight lifted by the football players.

Distribution for the test: It is normal because σ is known.

$\overline{X} \sim N\left(275, \frac{55}{\sqrt{30}}\right)$

$\overline{x} = 286.2$ pounds (from the data).

$\sigma = 55$ pounds **(Always use σ if you know it.)** We assume $\mu = 275$ pounds unless our data shows us otherwise.

Calculate the p-value using the normal distribution for a mean:

$p\text{-}value = P(\ \overline{X} > 286.2 = 0.1323$ where the sample mean is calculated as 286.2 pounds from the data.

Interpretation of the p-value: If H_o is true, then there is a 0.1323 probability (13.23%) that the football players can lift a mean (or average) weight of 286.2 pounds or more. Because a 13.23% chance is large enough, a mean weight lift of 286.2 pounds or more is happening randomly and is not a rare event.

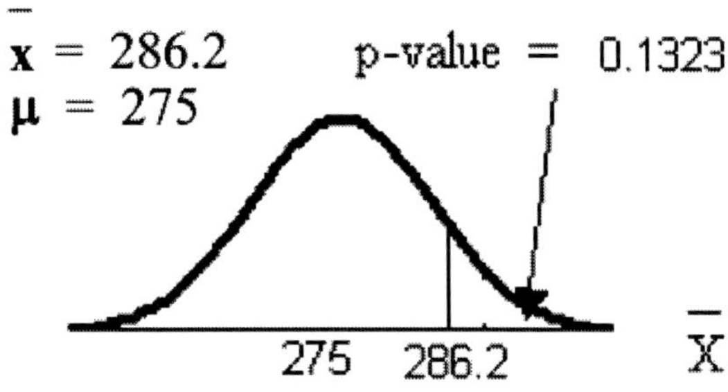

Figure 9.3

Compare α and the p-value:

$\alpha = 0.025$ $p\text{-}value = 0.1323$

Make a decision: Since α<*p-value*, do not reject H_o.

Conclusion: At the 2.5% level of significance, from the sample data, there is not sufficient evidence to conclude that the true mean weight lifted is more than 275 pounds.

The p-value can easily be calculated using the TI-83+ and the TI-84 calculators:

Put the data and frequencies into lists. Press `STAT` and arrow over to `TESTS`. Press `1:Z-Test`. Arrow over to `Data` and press `ENTER`. Arrow down and enter 275 for μ_0, 55 for σ, the name of the list where you put the data, and the name of the list where you put the frequencies. Arrow down to μ : and arrow over to $> \mu_0$. Press `ENTER`. Arrow down to `Calculate` and press `ENTER`. The calculator not only calculates the p-value ($p = 0.1331$, a little different from the above calculation - in it we used the sample mean rounded to one decimal place instead of the data) but it also calculates the test statistic (z-score) for the sample mean, the sample mean, and the sample standard deviation. $\mu > 275$ is the alternate hypothesis. Do this set of instructions again except arrow to `Draw` (instead of `Calculate`). Press `ENTER`. A shaded graph appears with $z = 1.112$ (test statistic) and $p = 0.1331$ (p-value). Make sure when you use `Draw` that no other equations are highlighted in $Y =$ and the plots are turned off.

Example 9.13
Statistics students believe that the average score on the first statistics test is 65. A statistics instructor thinks the average score is higher than 65. He samples ten statistics students and obtains the scores 65; 65; 70; 67; 66; 63; 63; 68; 72; 71. He performs a hypothesis test using a 5% level of significance. The data are from a normal distribution.

Solution
Set up the Hypothesis Test:

A 5% level of significance means that $\alpha = 0.05$. This is a test of a **single population mean**.

H_o: $\mu = 65$ $\qquad$ H_a: $\mu > 65$

Since the instructor thinks the average score is higher, use a "$>$ ". The "$>$ " means the test is right-tailed.

Determine the distribution needed:

Random variable: $\overline{X}$ = average score on the first statistics test.

Distribution for the test: If you read the problem carefully, you will notice that there is **no population standard deviation given**. You are only given $n = 10$ sample data values. Notice also that the data come from a normal distribution. This means that the distribution for the test is a student-t.

Use t_{df}. Therefore, the distribution for the test is t_9 where $n = 10$ and $df = 10 - 1 = 9$.

Calculate the p-value using the Student-t distribution:

p-value $= P(\ \overline{X} > 67 = 0.0396$ where the sample mean and sample standard deviation are calculated as 67 and 3.1972 from the data.

Interpretation of the p-value: If the null hypothesis is true, then there is a 0.0396 probability (3.96%) that the sample mean is 67 pounds or more.

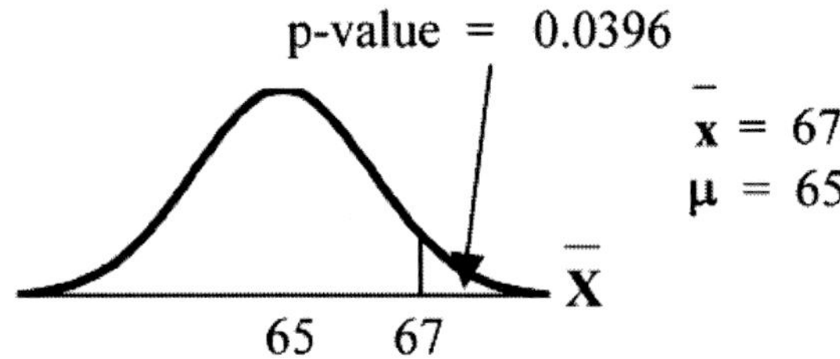

Figure 9.4

Compare α and the p-value:

Since $\alpha = .05$ and *p-value* $= 0.0396$. Therefore, $\alpha >$ *p-value*.

Make a decision: Since $\alpha >$ *p-value*, reject H_o.

This means you reject $\mu = 65$. In other words, you believe the average test score is more than 65.

Conclusion: At a 5% level of significance, the sample data show sufficient evidence that the mean (average) test score is more than 65, just as the math instructor thinks.

The p-value can easily be calculated using the TI-83+ and the TI-84 calculators:

Put the data into a list. Press `STAT` and arrow over to `TESTS`. Press `2:T-Test`. Arrow over to `Data` and press `ENTER`. Arrow down and enter 65 for μ_0, the name of the list where you put the data, and 1 for `Freq:`. Arrow down to μ : and arrow over to $> \mu_0$. Press `ENTER`. Arrow down to `Calculate` and press `ENTER`. The calculator not only calculates the p-value ($p = 0.0396$) but it also calculates the test statistic (t-score) for the sample mean, the sample mean, and the sample standard deviation. $\mu > 65$ is the alternate hypothesis. Do this set of instructions again except arrow to `Draw` (instead of `Calculate`). Press `ENTER`. A shaded graph appears with $t = 1.9781$ (test statistic) and $p = 0.0396$ (p-value). Make sure when you use `Draw` that no other equations are highlighted in $Y =$ and the plots are turned off.

Example 9.14
Joon believes that 50% of first-time brides in the United States are younger than their grooms. She performs a hypothesis test to determine if the percentage is **the same or different from 50%**. Joon samples **100 first-time brides** and **53** reply that they are younger than their grooms. For the hypothesis test, she uses a 1% level of significance.

Solution
Set up the Hypothesis Test:

The 1% level of significance means that $\alpha = 0.01$. This is a **test of a single population proportion**.

H_o: $p = 0.50$ $\qquad H_a$: $p \neq 0.50$

The words **"is the same or different from"** tell you this is a two-tailed test.

Calculate the distribution needed:

Random variable: P' = the percent of of first-time brides who are younger than their grooms. Distribution

Distribution for the test: The problem contains no mention of an average. The information is given in terms of percentages. Use the distribution for P', the estimated proportion.

$P' \sim N\left(p, \sqrt{\frac{p \cdot q}{n}}\right)$ Therefore, $P' \sim N\left(0.5, \sqrt{\frac{0.5 \cdot 0.5}{100}}\right)$ where $p = 0.50$, $q = 1 - p = 0.50$, and $n = 100$.

Calculate the p-value using the normal distribution for proportions:

$p\text{-}value = P\left(P' < 0.47 \text{ or } P' > 0.53 = 0.5485\right.$

where $x = 53$, $p' = \frac{x}{n} = \frac{53}{100} = 0.53$.

Interpretation of the p-value: If the null hypothesis is true, there is 0.5485 probability (54.85%) that the sample (estimated) proportion p' is 0.53 or more OR 0.47 or less (see the graph below).

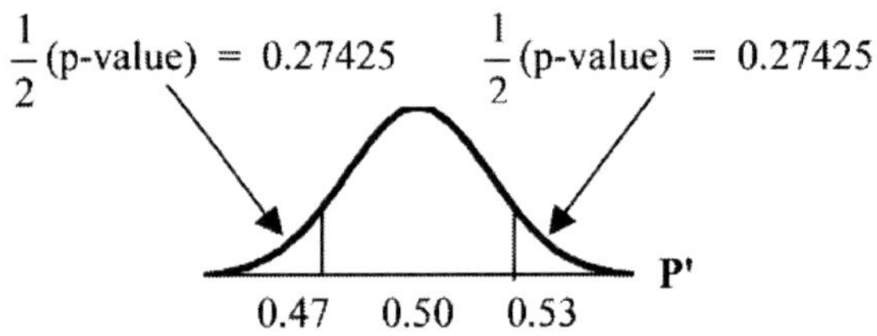

Figure 9.5

$\mu = p = 0.50$ comes from H_o, the null hypothesis.

p'= 0.53. Since the curve is symmetrical and the test is two-tailed, the p' for the left tail is equal to $0.50 - 0.03 = 0.47$ where $\mu = p = 0.50$. (0.03 is the difference between 0.53 and 0.50.)

Compare α and the p-value:

Since $\alpha = 0.01$ and $p\text{-}value = 0.5485$. Therefore, $\alpha < p\text{-}value$.

Make a decision: Since $\alpha < p\text{-}value$, you cannot reject H_o.

Conclusion: At the 1% level of significance, the sample data do not show sufficient evidence that the percentage of first-time brides who are younger than their grooms is different from 50%.

The p-value can easily be calculated using the TI-83+ and the TI-84 calculators:

Press `STAT` and arrow over to `TESTS`. Press `5:1-PropZTest`. Enter .5 for p_0 and 100 for n. Arrow down to `Prop` and arrow to `not equals` p_0. Press `ENTER`. Arrow down to `Calculate` and press `ENTER`. The calculator calculates the p-value ($p = 0.5485$) and the test statistic (z-score). `Prop not equals` .5 is the alternate hypothesis. Do this set of instructions again except arrow to `Draw` (instead of `Calculate`). Press `ENTER`. A shaded graph appears with $z = 0.6$ (test statistic) and $p = 0.5485$

(p-value). Make sure when you use `Draw` that no other equations are highlighted in $Y =$ and the plots are turned off.

The Type I and Type II errors are as follows:

The Type I error is to conclude that the proportion of first-time brides that are younger than their grooms is different from 50% when, in fact, the proportion is actually 50%. (Reject the null hypothesis when the null hypothesis is true).

The Type II error is to conclude that the proportion of first-time brides that are younger than their grooms is equal to 50% when, in fact, the proportion is different from 50%. (Do not reject the null hypothesis when the null hypothesis is false.)

Example 9.15
Problem 1
Suppose a consumer group suspects that the proportion of households that have three cell phones is not known to be 30%. A cell phone company has reason to believe that the proportion is 30%. Before they start a big advertising campaign, they conduct a hypothesis test. Their marketing people survey 150 households with the result that 43 of the households have three cell phones.

Solution
Set up the Hypothesis Test:

H_o: $p = 0.30$ H_a: $p \neq 0.30$

Determine the distribution needed:

The **random variable** is P' = proportion of households that have three cell phones.

The **distribution** for the hypothesis test is $P' \sim N\left(0.30, \sqrt{\frac{0.30 \cdot 0.70}{150}}\right)$

Problem 2 *(Solution on p. 380.)*
The value that helps determine the p-value is p'. Calculate p'.

Problem 3 *(Solution on p. 380.)*
What is a **success** for this problem?

Problem 4 *(Solution on p. 380.)*
What is the level of significance?

Draw the graph for this problem. Draw the horizontal axis. Label and shade appropriately.

Problem 5 *(Solution on p. 380.)*
Calculate the p-value.

Problem 6 *(Solution on p. 380.)*
Make a decision. ____________(Reject/Do not reject) H_0 because____________.

The next example is a poem written by a statistics student named Nicole Hart. The solution to the problem follows the poem. Notice that the hypothesis test is for a single population proportion. This means that the null and alternate hypotheses use the parameter p. The distribution for the test is normal. The estimated proportion p' is the proportion of fleas killed to the total fleas found on Fido. This is sample information. The problem gives a preconceived $\alpha = 0.01$, for comparison, and a 95% confidence interval computation. The poem is clever and humorous, so please enjoy it!

NOTE: Notice the solution sheet that has the solution. Look in the Table of Contents for the topic "Solution Sheets." Use copies of the appropriate solution sheet for homework problems.

Example 9.16

```
    My dog has so many fleas,
They do not come off with ease.
As for shampoo, I have tried many types
Even one called Bubble Hype,
Which only killed 25% of the fleas,
Unfortunately I was not pleased.

I've used all kinds of soap,
Until I had give up hope
Until one day I saw
An ad that put me in awe.

A shampoo used for dogs
Called GOOD ENOUGH to Clean a Hog
Guaranteed to kill more fleas.

I gave Fido a bath
And after doing the math
His number of fleas
Started dropping by 3's!

Before his shampoo
I counted 42.
At the end of his bath,
I redid the math
And the new shampoo had killed 17 fleas.
So now I was pleased.

Now it is time for you to have some fun
With the level of significance being .01,
You must help me figure out
Use the new shampoo or go without?
```

Solution

Set up the Hypothesis Test:

H_o: $p = 0.25$ H_a: $p > 0.25$

Determine the distribution needed:

In words, CLEARLY state what your random variable $\overline{X}$ or P' represents.

P' = The proportion of fleas that are killed by the new shampoo

State the distribution to use for the test.

Normal: $N\left(0.25, \sqrt{\frac{(0.25)(1-0.25)}{42}}\right)$

Test Statistic: $z = 2.3163$

Calculate the p-value using the normal distribution for proportions:

p-value =0.0103

In 1 – 2 complete sentences, explain what the p-value means for this problem.

If the null hypothesis is true (the proportion is 0.25), then there is a 0.0103 probability that the sample (estimated) proportion is 0.4048 $\left(\frac{17}{42}\right)$ or more.

Use the previous information to sketch a picture of this situation. CLEARLY, label and scale the horizontal axis and shade the region(s) corresponding to the p-value.

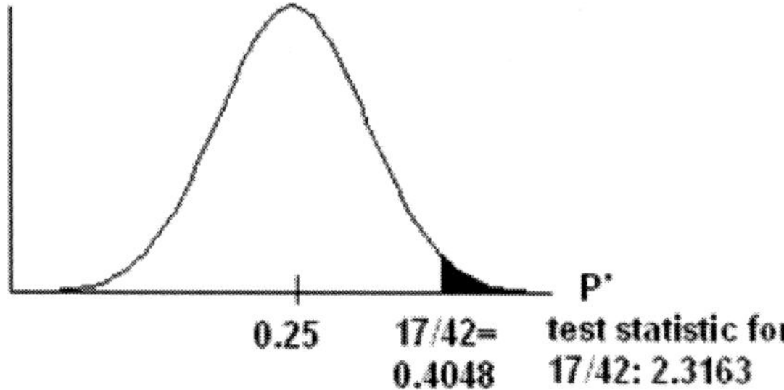

Figure 9.6

Compare α and the p-value:

Indicate the correct decision ("reject" or "do not reject" the null hypothesis), the reason for it, and write an appropriate conclusion, using COMPLETE SENTENCES.

alpha	decision	reason for decision
0.01	Do not reject H_o	α<*p-value*

Table 9.3

Conclusion: At the 1% level of significance, the sample data do not show sufficient evidence that the percentage of fleas that are killed by the new shampoo is more than 25%.

Construct a 95% Confidence Interval for the true mean or proportion. Include a sketch of the graph of the situation. Label the point estimate and the lower and upper bounds of the Confidence Interval.

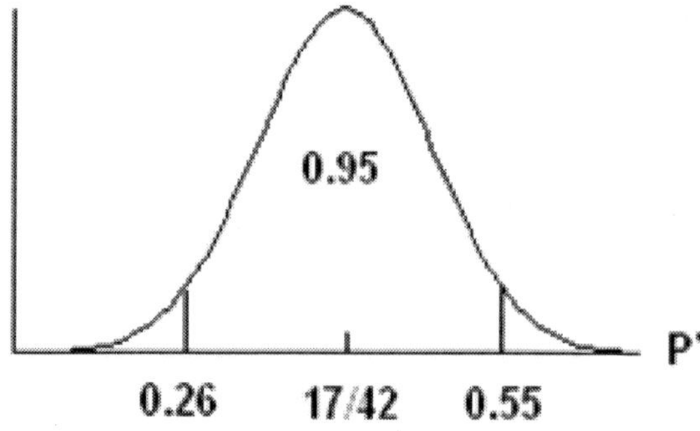

Figure 9.7

Confidence Interval: $(0.26, 0.55)$ We are 95% confident that the true population proportion p of fleas that are killed by the new shampoo is between 26% and 55%.

NOTE: This test result is not very definitive since the p-value is very close to alpha. In reality, one would probably do more tests by giving the dog another bath after the fleas have had a chance to return.

9.12 Summary of Formulas[12]

H_o and H_a are contradictory.

If H_o has:	equal ($=$)	greater than or equal to ($\geq$)	less than or equal to ($\leq$)
then H_a has:	not equal ($\neq$) **or** greater than ($>$) **or** less than ($<$)	less than ($<$)	greater than ($>$)

Table 9.4

If $\alpha \leq$ p-value, then do not reject H_o.

If $\alpha >$ p-value, then reject H_o.

α is preconceived. Its value is set before the hypothesis test starts. The p-value is calculated from the data.

α = probability of a Type I error = P(Type I error) = probability of rejecting the null hypothesis when the null hypothesis is true.

β = probability of a Type II error = P(Type II error) = probability of not rejecting the null hypothesis when the null hypothesis is false.

If there is no given preconceived α, then use $\alpha = 0.05$.

Types of Hypothesis Tests

- Single population mean, **known** population variance (or standard deviation): **Normal test**.
- Single population mean, **unknown** population variance (or standard deviation): **Student-t test**.
- Single population proportion: **Normal test**.

[12] This content is available online at <http://cnx.org/content/m16996/1.7/>.

9.13 Practice 1: Single Mean, Known Population Standard Deviation[13]

9.13.1 Student Learning Outcomes

- The student will explore hypothesis testing with single mean and known population standard deviation data.

9.13.2 Given

Suppose that a recent article stated that the average time spent in jail by a first–time convicted burglar is 2.5 years. A study was then done to see if the average time has increased in the new century. A random sample of 26 first–time convicted burglars in a recent year was picked. The average length of time in jail from the survey was 3 years with a standard deviation of 1.8 years. Suppose that it is somehow known that the population standard deviation is 1.5. Conduct a hypothesis test to determine if the average length of jail time has increased.

9.13.3 Hypothesis Testing: Single Average

Exercise 9.13.1 *(Solution on p. 380.)*
Is this a test of averages or proportions?

Exercise 9.13.2 *(Solution on p. 380.)*
State the null and alternative hypotheses.

a. H_0:
b. H_a:

Exercise 9.13.3 *(Solution on p. 380.)*
Is this a right-tailed, left-tailed, or two-tailed test? How do you know?

Exercise 9.13.4 *(Solution on p. 380.)*
What symbol represents the Random Variable for this test?

Exercise 9.13.5 *(Solution on p. 380.)*
In words, define the Random Variable for this test.

Exercise 9.13.6 *(Solution on p. 380.)*
Is the population standard deviation known and, if so, what is it?

Exercise 9.13.7 *(Solution on p. 380.)*
Calculate the following:

a. $\overline{x} =$
b. $\sigma =$
c. $s_x =$
d. $n =$

Exercise 9.13.8 *(Solution on p. 380.)*
Since both σ and s_x are given, which should be used? In 1 -2 complete sentences, explain why.

Exercise 9.13.9 *(Solution on p. 380.)*
State the distribution to use for the hypothesis test.

Exercise 9.13.10
Sketch a graph of the situation. Label the horizontal axis. Mark the hypothesized mean and the sample mean $\overline{x}$. Shade the area corresponding to the p-value.

[13]This content is available online at <http://cnx.org/content/m17004/1.8/>.

Exercise 9.13.11 *(Solution on p. 380.)*

Find the p-value.

Exercise 9.13.12 *(Solution on p. 380.)*

At a pre-conceived $\alpha = 0.05$, what is your:

- **a.** Decision:
- **b.** Reason for the decision:
- **c.** Conclusion (write out in a complete sentence):

9.13.4 Discussion Questions

Exercise 9.13.13

Does it appear that the average jail time spent for first time convicted burglars has increased? Why or why not?

9.14 Practice 2: Single Mean, Unknown Population Standard Deviation[14]

9.14.1 Student Learning Outcomes

- The student will explore the properties of hypothesis testing with a single mean and unknown population standard deviation.

9.14.2 Given

A random survey of 75 death row inmates revealed that the average length of time on death row is 17.4 years with a standard deviation of 6.3 years. Conduct a hypothesis test to determine if the population average time on death row could likely be 15 years.

9.14.3 Hypothesis Testing: Single Average

Exercise 9.14.1 *(Solution on p. 381.)*
Is this a test of averages or proportions?

Exercise 9.14.2 *(Solution on p. 381.)*
State the null and alternative hypotheses.

a. H_o :
b. H_a :

Exercise 9.14.3 *(Solution on p. 381.)*
Is this a right-tailed, left-tailed, or two-tailed test? How do you know?

Exercise 9.14.4 *(Solution on p. 381.)*
What symbol represents the Random Variable for this test?

Exercise 9.14.5 *(Solution on p. 381.)*
In words, define the Random Variable for this test.

Exercise 9.14.6 *(Solution on p. 381.)*
Is the population standard deviation known and, if so, what is it?

Exercise 9.14.7 *(Solution on p. 381.)*
Calculate the following:

a. $\overline{x} =$
b. $6.3 =$
c. $n =$

Exercise 9.14.8 *(Solution on p. 381.)*
Which test should be used? In 1 -2 complete sentences, explain why.

Exercise 9.14.9 *(Solution on p. 381.)*
State the distribution to use for the hypothesis test.

Exercise 9.14.10
Sketch a graph of the situation. Label the horizontal axis. Mark the hypothesized mean and the sample mean, $\overline{x}$. Shade the area corresponding to the p-value.

[14] This content is available online at <http://cnx.org/content/m17016/1.8/>.

Figure 9.8

Exercise 9.14.11 *(Solution on p. 381.)*

Find the p-value.

Exercise 9.14.12 *(Solution on p. 381.)*

At a pre-conceived $\alpha = 0.05$, what is your:

a. Decision:
b. Reason for the decision:
c. Conclusion (write out in a complete sentence):

9.14.4 Discussion Question

Does it appear that the average time on death row could be 15 years? Why or why not?

9.15 Practice 3: Single Proportion[15]

9.15.1 Student Learning Outcomes

- The student will explore the properties of hypothesis testing with a single proportion.

9.15.2 Given

The National Institute of Mental Health published an article stating that in any one-year period, approximately 9.5 percent of American adults suffer from depression or a depressive illness. (http://www.nimh.nih.gov/publicat/depression.cfm) Suppose that in a survey of 100 people in a certain town, seven of them suffered from depression or a depressive illness. Conduct a hypothesis test to determine if the true proportion of people in that town suffering from depression or a depressive illness is lower than the percent in the general adult American population.

9.15.3 Hypothesis Testing: Single Proportion

Exercise 9.15.1 *(Solution on p. 381.)*
Is this a test of averages or proportions?

Exercise 9.15.2 *(Solution on p. 381.)*
State the null and alternative hypotheses.

a. H_o :
b. H_a :

Exercise 9.15.3 *(Solution on p. 381.)*
Is this a right-tailed, left-tailed, or two-tailed test? How do you know?

Exercise 9.15.4 *(Solution on p. 381.)*
What symbol represents the Random Variable for this test?

Exercise 9.15.5 *(Solution on p. 381.)*
In words, define the Random Variable for this test.

Exercise 9.15.6 *(Solution on p. 381.)*
Calculate the following:

a: $x =$
b: $n =$
c: $p\text{-}hat =$

Exercise 9.15.7 *(Solution on p. 382.)*
Calculate σ_x. Make sure to show how you set up the formula.

Exercise 9.15.8 *(Solution on p. 382.)*
State the distribution to use for the hypothesis test.

Exercise 9.15.9
Sketch a graph of the situation. Label the horizontal axis. Mark the hypothesized mean and the sample proportion, p-hat. Shade the area corresponding to the p-value.

[15] This content is available online at <http://cnx.org/content/m17003/1.8/>.

Exercise 9.15.10 *(Solution on p. 382.)*

Find the p-value

Exercise 9.15.11 *(Solution on p. 382.)*

At a pre-conceived $\alpha = 0.05$, what is your:

- **a.** Decision:
- **b.** Reason for the decision:
- **c.** Conclusion (write out in a complete sentence):

9.15.4 Discusion Question

Exercise 9.15.12

Does it appear that the proportion of people in that town with depression or a depressive illness is lower than general adult American population? Why or why not?

9.16 Homework[16]

Exercise 9.16.1 *(Solution on p. 382.)*

Some of the statements below refer to the null hypothesis, some to the alternate hypothesis.

State the null hypothesis, H_o, and the alternative hypothesis, H_a, in terms of the appropriate parameter (μ or p).

a. Americans work an average of 34 years before retiring.
b. At most 60% of Americans vote in presidential elections.
c. The average starting salary for San Jose State University graduates is at least $100,000 per year.
d. 29% of high school seniors get drunk each month.
e. Fewer than 5% of adults ride the bus to work in Los Angeles.
f. The average number of cars a person owns in her lifetime is not more than 10.
g. About half of Americans prefer to live away from cities, given the choice.
h. Europeans have an average paid vacation each year of six weeks.
i. The chance of developing breast cancer is under 11% for women.
j. Private universities cost, on average, more than $20,000 per year for tuition.

Exercise 9.16.2 *(Solution on p. 382.)*

For (a) - (j) above, state the Type I and Type II errors in complete sentences.

Exercise 9.16.3

For (a) - (j) above, in complete sentences:

a. State a consequence of committing a Type I error.
b. State a consequence of committing a Type II error.

DIRECTIONS: For each of the word problems, use a solution sheet to do the hypothesis test. The solution sheet is found in the Appendix. Please feel free to make copies of it. For the online version of the book, it is suggested that you copy the .doc or the .pdf files.

NOTE: If you are using a student-t distribution for a homework problem below, you may assume that the underlying population is normally distributed. (In general, you must first prove that assumption, though.)

Exercise 9.16.4

A particular brand of tires claims that its deluxe tire averages at least 50,000 miles before it needs to be replaced. From past studies of this tire, the standard deviation is known to be 8000. A survey of owners of that tire design is conducted. From the 28 tires surveyed, the average lifespan was 46,500 miles with a standard deviation of 9800 miles. Do the data support the claim at the 5% level?

Exercise 9.16.5 *(Solution on p. 382.)*

From generation to generation, the average age when smokers first start to smoke varies. However, the standard deviation of that age remains constant of around 2.1 years. A survey of 40 smokers of this generation was done to see if the average starting age is at least 19. The sample average was 18.1 with a sample standard deviation of 1.3. Do the data support the claim at the 5% level?

[16]This content is available online at <http://cnx.org/content/m17001/1.10/>.

Exercise 9.16.6

The cost of a daily newspaper varies from city to city. However, the variation among prices remains steady with a standard deviation of 6¢. A study was done to test the claim that the average cost of a daily newspaper is 35¢. Twelve costs yield an average cost of 30¢ with a standard deviation of 4¢. Do the data support the claim at the 1% level?

Exercise 9.16.7 *(Solution on p. 382.)*

An article in the **San Jose Mercury News** stated that students in the California state university system take an average of 4.5 years to finish their undergraduate degrees. Suppose you believe that the average time is longer. You conduct a survey of 49 students and obtain a sample mean of 5.1 with a sample standard deviation of 1.2. Do the data support your claim at the 1% level?

Exercise 9.16.8

The average number of sick days an employee takes per year is believed to be about 10. Members of a personnel department do not believe this figure. They randomly survey 8 employees. The number of sick days they took for the past year are as follows: 12; 4; 15; 3; 11; 8; 6; 8. Let x = the number of sick days they took for the past year. Should the personnel team believe that the average number is about 10?

Exercise 9.16.9 *(Solution on p. 382.)*

In 1955, **Life Magazine** reported that the 25 year-old mother of three worked [on average] an 80 hour week. Recently, many groups have been studying whether or not the women's movement has, in fact, resulted in an increase in the average work week for women (combining employment and at-home work). Suppose a study was done to determine if the average work week has increased. 81 women were surveyed with the following results. The sample average was 83; the sample standard deviation was 10. Does it appear that the average work week has increased for women at the 5% level?

Exercise 9.16.10

Your statistics instructor claims that 60 percent of the students who take her Elementary Statistics class go through life feeling more enriched. For some reason that she can't quite figure out, most people don't believe her. You decide to check this out on your own. You randomly survey 64 of her past Elementary Statistics students and find that 34 feel more enriched as a result of her class. Now, what do you think?

Exercise 9.16.11 *(Solution on p. 382.)*

A Nissan Motor Corporation advertisement read, "The average man's I.Q. is 107. The average brown trout's I.Q. is 4. So why can't man catch brown trout?" Suppose you believe that the average brown trout's I.Q. is greater than 4. You catch 12 brown trout. A fish psychologist determines the I.Q.s as follows: 5; 4; 7; 3; 6; 4; 5; 3; 6; 3; 8; 5. Conduct a hypothesis test of your belief.

Exercise 9.16.12

Refer to the previous problem. Conduct a hypothesis test to see if your decision and conclusion would change if your belief were that the average brown trout's I.Q. is **not** 4.

Exercise 9.16.13 *(Solution on p. 383.)*

According to an article in **Newsweek**, the natural ratio of girls to boys is 100:105. In China, the birth ratio is 100: 114 (46.7% girls). Suppose you don't believe the reported figures of the percent of girls born in China. You conduct a study. In this study, you count the number of girls and boys born in 150 randomly chosen recent births. There are 60 girls and 90 boys born of the 150. Based on your study, do you believe that the percent of girls born in China is 46.7?

Exercise 9.16.14

A poll done for **Newsweek** found that 13% of Americans have seen or sensed the presence of an angel. A contingent doubts that the percent is really that high. It conducts its own survey. Out of 76 Americans surveyed, only 2 had seen or sensed the presence of an angel. As a result of the contingent's survey, would you agree with the **Newsweek** poll? In complete sentences, also give three reasons why the two polls might give different results.

Exercise 9.16.15 *(Solution on p. 383.)*

The average work week for engineers in a start-up company is believed to be about 60 hours. A newly hired engineer hopes that it's shorter. She asks 10 engineering friends in start-ups for the lengths of their average work weeks. Based on the results that follow, should she count on the average work week to be shorter than 60 hours?

Data (length of average work week): 70; 45; 55; 60; 65; 55; 55; 60; 50; 55.

Exercise 9.16.16

Use the "Lap time" data for Lap 4 (see Table of Contents) to test the claim that Terri finishes Lap 4 on average in less than 129 seconds. Use all twenty races given.

Exercise 9.16.17

Use the "Initial Public Offering" data (see Table of Contents) to test the claim that the average offer price was $18 per share. Do not use all the data. Use your random number generator to randomly survey 15 prices.

NOTE: The following questions were written by past students. They are excellent problems!

Exercise 9.16.18

18. "Asian Family Reunion" by Chau Nguyen

```
    Every two years it comes around
We all get together from different towns.
In my honest opinion
It's not a typical family reunion
Not forty, or fifty, or sixty,
But how about seventy companions!
The kids would play, scream, and shout
One minute they're happy, another they'll pout.
The teenagers would look, stare, and compare
From how they look to what they wear.
The men would chat about their business
That they make more, but never less.
Money is always their subject
And there's always talk of more new projects.
The women get tired from all of the chats
They head to the kitchen to set out the mats.
Some would sit and some would stand
Eating and talking with plates in their hands.
Then come the games and the songs
And suddenly, everyone gets along!
With all that laughter, it's sad to say
That it always ends in the same old way.
They hug and kiss and say "good-bye"
And then they all begin to cry!
I say that 60 percent shed their tears
But my mom counted 35 people this year.
She said that boys and men will always have their pride,
So we won't ever see them cry.
I myself don't think she's correct,
```

```
So could you please try this problem to see if you object?
```

Exercise 9.16.19 *(Solution on p. 383.)*

"The Problem with Angels" by Cyndy Dowling

```
    Although this problem is wholly mine,
The catalyst came from the magazine, Time.
On the magazine cover I did find
The realm of angels tickling my mind.

    Inside, 69% I found to be
In angels, Americans do believe.

    Then, it was time to rise to the task,
Ninety-five high school and college students I did ask.
Viewing all as one group,
Random sampling to get the scoop.

    So, I asked each to be true,
"Do you believe in angels?"  Tell me, do!

    Hypothesizing at the start,
Totally believing in my heart
That the proportion who said yes
Would be equal on this test.

    Lo and behold, seventy-three did arrive,
Out of the sample of ninety-five.
Now your job has just begun,
Solve this problem and have some fun.
```

Exercise 9.16.20

"Blowing Bubbles" by Sondra Prull

```
    Studying stats just made me tense,
I had to find some sane defense.
Some light and lifting simple play
To float my math anxiety away.

    Blowing bubbles lifts me high
Takes my troubles to the sky.
POIK! They're gone, with all my stress
Bubble therapy is the best.

    The label said each time I blew
The average number of bubbles would be at least 22.
I blew and blew and this I found
From 64 blows, they all are round!
```

```
    But the number of bubbles in 64 blows
Varied widely, this I know.
20 per blow became the mean
They deviated by 6, and not 16.


    From counting bubbles, I sure did relax
But now I give to you your task.
Was 22 a reasonable guess?
Find the answer and pass this test!
```

Exercise 9.16.21 *(Solution on p. 383.)*

21. "Dalmatian Darnation" by Kathy Sparling

```
    A greedy dog breeder named Spreckles
Bred puppies with numerous freckles
The Dalmatians he sought
Possessed spot upon spot
The more spots, he thought, the more shekels.


    His competitors did not agree
That freckles would increase the fee.
They said, ``Spots are quite nice
But they don't affect price;
One should breed for improved pedigree.''


    The breeders decided to prove
This strategy was a wrong move.
Breeding only for spots
Would wreak havoc, they thought.
His theory they want to disprove.


    They proposed a contest to Spreckles
Comparing dog prices to freckles.
In records they looked up
One hundred one pups:
Dalmatians that fetched the most shekels.


    They asked Mr. Spreckles to name
An average spot count he'd claim
To bring in big bucks.
Said Spreckles, ``Well, shucks,
It's for one hundred one that I aim.''


    Said an amateur statistician
Who wanted to help with this mission.
``Twenty-one for the sample
Standard deviation's ample:
```

```
    They examined one hundred and one
Dalmatians that fetched a good sum.
They counted each spot,
Mark, freckle and dot
And tallied up every one.

    Instead of one hundred one spots
They averaged ninety six dots
Can they muzzle Spreckles'
Obsession with freckles
Based on all the dog data they've got?
```

Exercise 9.16.22
"Macaroni and Cheese, please!!" by Nedda Misherghi and Rachelle Hall

As a poor starving student I don't have much money to spend for even the bare necessities. So my favorite and main staple food is macaroni and cheese. It's high in taste and low in cost and nutritional value.

One day, as I sat down to determine the meaning of life, I got a serious craving for this, oh, so important, food of my life. So I went down the street to Greatway to get a box of macaroni and cheese, but it was SO expensive! $2.02 !!! Can you believe it? It made me stop and think. The world is changing fast. I had thought that the average cost of a box (the normal size, not some super-gigantic-family-value-pack) was at most $1, but now I wasn't so sure. However, I was determined to find out. I went to 53 of the closest grocery stores and surveyed the prices of macaroni and cheese. Here are the data I wrote in my notebook:

Price per box of Mac and Cheese:

- 5 stores @ $2.02
- 15 stores @ $0.25
- 3 stores @ $1.29
- 6 stores @ $0.35
- 4 stores @ $2.27
- 7 stores @ $1.50
- 5 stores @ $1.89
- 8 stores @ 0.75.

I could see that the costs varied but I had to sit down to figure out whether or not I was right. If it does turn out that this mouth-watering dish is at most $1, then I'll throw a big cheesy party in our next statistics lab, with enough macaroni and cheese for just me. (After all, as a poor starving student I can't be expected to feed our class of animals!)

Exercise 9.16.23 ***(Solution on p. 383.)***
"William Shakespeare: The Tragedy of Hamlet, Prince of Denmark" by Jacqueline Ghodsi

THE CHARACTERS (in order of appearance):

- HAMLET, Prince of Denmark and student of Statistics
- POLONIUS, Hamlet's tutor
- HOROTIO, friend to Hamlet and fellow student

Scene: The great library of the castle, in which Hamlet does his lessons

Act I

(The day is fair, but the face of Hamlet is clouded. He paces the large room. His tutor, Polonius, is reprimanding Hamlet regarding the latter's recent experience. Horatio is seated at the large table at right stage.)

POLONIUS: My Lord, how cans't thou admit that thou hast seen a ghost! It is but a figment of your imagination!

HAMLET: I beg to differ; I know of a certainty that five-and-seventy in one hundred of us, condemned to the whips and scorns of time as we are, have gazed upon a spirit of health, or goblin damn'd, be their intents wicked or charitable.

POLONIUS If thou doest insist upon thy wretched vision then let me invest your time; be true to thy work and speak to me through the reason of the null and alternate hypotheses. (He turns to Horatio.) Did not Hamlet himself say, "What piece of work is man, how noble in reason, how infinite in faculties? Then let not this foolishness persist. Go, Horatio, make a survey of three-and-sixty and discover what the true proportion be. For my part, I will never succumb to this fantasy, but deem man to be devoid of all reason should thy proposal of at least five-and-seventy in one hundred hold true.

HORATIO (to Hamlet): What should we do, my Lord?

HAMLET: Go to thy purpose, Horatio.

HORATIO: To what end, my Lord?

HAMLET: That you must teach me. But let me conjure you by the rights of our fellowship, by the consonance of our youth, but the obligation of our ever-preserved love, be even and direct with me, whether I am right or no.

(Horatio exits, followed by Polonius, leaving Hamlet to ponder alone.)

Act II

(The next day, Hamlet awaits anxiously the presence of his friend, Horatio. Polonius enters and places some books upon the table just a moment before Horatio enters.)

POLONIUS: So, Horatio, what is it thou didst reveal through thy deliberations?

HORATIO: In a random survey, for which purpose thou thyself sent me forth, I did discover that one-and-forty believe fervently that the spirits of the dead walk with us. Before my God, I might not this believe, without the sensible and true avouch of mine own eyes.

POLONIUS: Give thine own thoughts no tongue, Horatio. (Polonius turns to Hamlet.) But look to't I charge you, my Lord. Come Horatio, let us go together, for this is not our test. (Horatio and Polonius leave together.)

HAMLET: To reject, or not reject, that is the question: whether 'tis nobler in the mind to suffer the slings and arrows of outrageous statistics, or to take arms against a sea of data, and, by opposing, end them. (Hamlet resignedly attends to his task.)

(Curtain falls)

Exercise 9.16.24
"Untitled" by Stephen Chen

I've often wondered how software is released and sold to the public. Ironically, I work for a company that sells products with known problems. Unfortunately, most of the problems are difficult to create, which makes them difficult to fix. I usually use the test program X, which tests the product, to try to create a specific problem. When the test program is run to make an error occur, the likelihood of generating an error is 1%.

So, armed with this knowledge, I wrote a new test program Y that will generate the same error that test program X creates, but more often. To find out if my test program is better than the original, so that I can convince the management that I'm right, I ran my test program to find out how often I can generate the same error. When I ran my test program 50 times, I generated the error twice. While this may not seem much better, I think that I can convince the management to use my test program instead of the original test program. Am I right?

Exercise 9.16.25 *(Solution on p. 383.)*

Japanese Girls' Names

by Kumi Furuichi

It used to be very typical for Japanese girls' names to end with "ko." (The trend might have started around my grandmothers' generation and its peak might have been around my mother's generation.) "Ko" means "child" in Chinese character. Parents would name their daughters with "ko" attaching to other Chinese characters which have meanings that they want their daughters to become, such as Sachiko – a happy child, Yoshiko – a good child, Yasuko – a healthy child, and so on.

However, I noticed recently that only two out of nine of my Japanese girlfriends at this school have names which end with "ko." More and more, parents seem to have become creative, modernized, and, sometimes, westernized in naming their children.

I have a feeling that, while 70 percent or more of my mother's generation would have names with "ko" at the end, the proportion has dropped among my peers. I wrote down all my Japanese friends', ex-classmates', co-workers, and acquaintances' names that I could remember. Below are the names. (Some are repeats.) Test to see if the proportion has dropped for this generation.

Ai, Akemi, Akiko, Ayumi, Chiaki, Chie, Eiko, Eri, Eriko, Fumiko, Harumi, Hitomi, Hiroko, Hiroko, Hidemi, Hisako, Hinako, Izumi, Izumi, Junko, Junko, Kana, Kanako, Kanayo, Kayo, Kayoko, Kazumi, Keiko, Keiko, Kei, Kumi, Kumiko, Kyoko, Kyoko, Madoka, Maho, Mai, Maiko, Maki, Miki, Miki, Mikiko, Mina, Minako, Miyako, Momoko, Nana, Naoko, Naoko, Naoko, Noriko, Rieko, Rika, Rika, Rumiko, Rei, Reiko, Reiko, Sachiko, Sachiko, Sachiyo, Saki, Sayaka, Sayoko, Sayuri, Seiko, Shiho, Shizuka, Sumiko, Takako, Takako, Tomoe, Tomoe, Tomoko, Touko, Yasuko, Yasuko, Yasuyo, Yoko, Yoko, Yoko, Yoshiko, Yoshiko, Yoshiko, Yuka, Yuki, Yuki, Yukiko, Yuko, Yuko.

Exercise 9.16.26

Phillip's Wish by Suzanne Osorio

```
    My nephew likes to play
Chasing the girls makes his day.
He asked his mother
If it is okay
To get his ear pierced.
She said, ''No way!''
To poke a hole through your ear,
Is not what I want for you, dear.
He argued his point quite well,
```

```
Says even my macho pal,  Mel,
Has gotten this done.
It's all just for fun.
C'mon please, mom, please, what the hell.
Again Phillip complained to his mother,
Saying half his friends (including their brothers)
Are piercing their ears
And they have no fears
He wants to be like the others.
She said, ''I think it's much less.
We must do a hypothesis test.
And if you are right,
I won't put up a fight.
But, if not, then my case will rest.''
We proceeded to call fifty guys
To see whose prediction would fly.
Nineteen of the fifty
Said piercing was nifty
And earrings they'd occasionally buy.
Then there's the other thirty-one,
Who said they'd never have this done.
So now this poem's finished.
Will his hopes be diminished,
Or will my nephew have his fun?
```

Exercise 9.16.27 *(Solution on p. 383.)*

The Craven by Mark Salangsang

```
    Once upon a morning dreary
In stats class I was weak and weary.
Pondering over last night's homework
Whose answers were now on the board
This I did and nothing more.

    While I nodded nearly napping
Suddenly, there came a tapping.
As someone gently rapping,
Rapping my head as I snore.
Quoth the teacher, ''Sleep no more.''

    ''In every class you fall asleep,''
The teacher said, his voice was deep.
''So a tally I've begun to keep
Of every class you nap and snore.
The percentage being forty-four.''

    ''My dear teacher I must confess,
While sleeping is what I do best.
The percentage, I think, must be less,
A percentage less than forty-four.''
This I said and nothing more.
```

```
    ''We'll see,'' he said and walked away,
And fifty classes from that day
He counted till the month of May
The classes in which I napped and snored.
The number he found was twenty-four.


    At a significance level of 0.05,
Please tell me am I still alive?
Or did my grade just take a dive
Plunging down beneath the floor?
Upon thee I hereby implore.
```

Exercise 9.16.28

Toastmasters International cites a February 2001 report by Gallop Poll that 40% of Americans fear public speaking. A student believes that less than 40% of students at her school fear public speaking. She randomly surveys 361 schoolmates and finds that 135 report they fear public speaking. Conduct a hypothesis test to determine if the percent at her school is less than 40%. (*Source: http://toastmasters.org/artisan/detail.asp?CategoryID=1&SubCategoryID=10&ArticleID=429&Page=1*[17])

Exercise 9.16.29 ***(Solution on p. 383.)***

In 2004, 68% of online courses taught at community colleges nationwide were taught by full-time faculty. To test if 68% also represents California's percent for full-time faculty teaching the online classes, Long Beach City College (LBCC), CA, was randomly selected for comparison. In 2004, 34 of the 44 online courses LBCC offered were taught by full-time faculty. Conduct a hypothesis test to determine if 68% represents CA. NOTE: For a true test, use more CA community colleges. (Sources: **Growing by Degrees** by Allen and Seaman; Amit Schitai, Director of Instructional Technology and Distance Learning, LBCC).

NOTE: For a true test, use more CA community colleges.

Exercise 9.16.30

According to an article in **The New York Times** (5/12/2004), 19.3% of New York City adults smoked in 2003. Suppose that a survey is conducted to determine this year's rate. Twelve out of 70 randomly chosen N.Y. City residents reply that they smoke. Conduct a hypothesis test to determine is the rate is still 19.3%.

Exercise 9.16.31 ***(Solution on p. 384.)***

The average age of De Anza College students in Winter 2006 term was 26.6 years old. An instructor thinks the average age for online students is older than 26.6. She randomly surveys 56 online students and finds that the sample average is 29.4 with a standard deviation of 2.1. Conduct a hypothesis test. (*Source: http://research.fhda.edu/factbook/DAdemofs/Fact_sheet_da_2006w.pdf*[18])

Exercise 9.16.32

In 2004, registered nurses earned an average annual salary of $52,330. A survey was conducted of 41 California nursed to determine if the annual salary is higher than $52,330 for California nurses. The sample average was $61,121 with a sample standard deviation of $7,489. Conduct a hypothesis test. (*Source: http://stats.bls.gov/oco/ocos083.htm#earnings*[19])

[17]http://toastmasters.org/artisan/detail.asp?CategoryID=1&SubCategoryID=10&ArticleID=429&Page=1
[18]http://research.fhda.edu/factbook/DAdemofs/Fact_sheet_da_2006w.pdf
[19]http://stats.bls.gov/oco/ocos083.htm#earnings

Exercise 9.16.33 *(Solution on p. 384.)*

La Leche League International reports that the average age of weaning a child from breastfeeding is age 4 to 5 worldwide. In America, most nursing mothers wean their children much earlier. Suppose a random survey is conducted of 21 U.S. mothers who recently weaned their children. The average weaning age was 9 months (3/4 year) with a standard deviation of 4 months. Conduct a hypothesis test to determine is the average weaning age in the U.S. is less than 4 years old. (*Source: http://www.lalecheleague.org/Law/BAFeb01.html*[20])

9.16.1 Try these multiple choice questions.

Exercise 9.16.34 *(Solution on p. 384.)*

When a new drug is created, the pharmaceutical company must subject it to testing before receiving the necessary permission from the Food and Drug Administration (FDA) to market the drug. Suppose the null hypothesis is "the drug is unsafe." What is the Type II Error?

A. To claim the drug is safe when in, fact, it is unsafe
B. To claim the drug is unsafe when, in fact, it is safe.
C. To claim the drug is safe when, in fact, it is safe.
D. To claim the drug is unsafe when, in fact, it is unsafe

The next two questions refer to the following information: Over the past few decades, public health officials have examined the link between weight concerns and teen girls smoking. Researchers surveyed a group of 273 randomly selected teen girls living in Massachusetts (between 12 and 15 years old). After four years the girls were surveyed again. Sixty-three (63) said they smoked to stay thin. Is there good evidence that more than thirty percent of the teen girls smoke to stay thin?

Exercise 9.16.35 *(Solution on p. 384.)*

The alternate hypothesis is

A. $p < 0.30$
B. $p \leq 0.30$
C. $p \geq 0.30$
D. $p > 0.30$

Exercise 9.16.36 *(Solution on p. 384.)*

After conducting the test, your decision and conclusion are

A. Reject H_0: More than 30% of teen girls smoke to stay thin.
B. Do not reject H_0: Less than 30% of teen girls smoke to stay thin.
C. Do not reject H_0: At most 30% of teen girls smoke to stay thin.
D. Reject H_0: Less than 30% of teen girls smoke to stay thin.

The next three questions refer to the following information: A statistics instructor believes that fewer than 20% of Evergreen Valley College (EVC) students attended the opening night midnight showing of the latest Harry Potter movie. She surveys 84 of her students and finds that 11 of attended the midnight showing.

Exercise 9.16.37 *(Solution on p. 384.)*

An appropriate alternative hypothesis is

[20] http://www.lalecheleague.org/Law/BAFeb01.html

A. $p = 0.20$
B. $p > 0.20$
C. $p < 0.20$
D. $p \leq 0.20$

Exercise 9.16.38 *(Solution on p. 384.)*
At a 1% level of significance, an appropriate conclusion is:

A. The percent of EVC students who attended the midnight showing of Harry Potter is at least 20%.
B. The percent of EVC students who attended the midnight showing of Harry Potter is more than 20%.
C. The percent of EVC students who attended the midnight showing of Harry Potter is less than 20%.
D. There is not enough information to make a decision.

Exercise 9.16.39 *(Solution on p. 384.)*
The Type I error is believing that the percent of EVC students who attended is:

A. at least 20%, when in fact, it is less than 20%.
B. 20%, when in fact, it is 20%.
C. less than 20%, when in fact, it is at least 20%.
D. less than 20%, when in fact, it is less than 20%.

The next two questions refer to the following information:

It is believed that Lake Tahoe Community College (LTCC) Intermediate Algebra students get less than 7 hours of sleep per night, on average. A survey of 22 LTCC Intermediate Algebra students generated an average of 7.24 hours with a standard deviation of 1.93 hours. At a level of significance of 5%, do LTCC Intermediate Algebra students get less than 7 hours of sleep per night, on average?

Exercise 9.16.40 *(Solution on p. 384.)*
The distribution to be used for this test is $\overline{X} \sim$

A. $N\left(7.24, \frac{1.93}{\sqrt{22}}\right)$
B. $N(7.24, 1.93)$
C. t_{22}
D. t_{21}

Exercise 9.16.41 *(Solution on p. 384.)*
The Type II error is "I believe that the average number of hours of sleep LTCC students get per night

A. is less than 7 hours when, in fact, it is at least 7 hours."
B. is less than 7 hours when, in fact, it is less than 7 hours."
C. is at least 7 hours when, in fact, it is at least 7 hours."
D. is at least 7 hours when, in fact, it is less than 7 hours."

The next three questions refer to the following information: An organization in 1995 reported that teenagers spent an average of 4.5 hours per week on the telephone. The organization thinks that, in 2007, the average is higher. Fifteen (15) randomly chosen teenagers were asked how many hours per week they spend on the telephone. The sample mean was 4.75 hours with a sample standard deviation of 2.0.

Exercise 9.16.42 *(Solution on p. 384.)*
The null and alternate hypotheses are:

A. $H_o : \overline{x} = 4.5$, $H_a : \overline{x} > 4.5$
B. $H_o : \mu \geq 4.5$ $H_a : \mu < 4.5$
C. $H_o : \mu = 4.75$ $H_{a:} \mu > 4.75$
D. $H_o : \mu = 4.5$ $H_a : \mu > 4.5$

Exercise 9.16.43 *(Solution on p. 384.)*

At a significance level of $a = 0.05$, the correct conclusion is:

A. The average in 2007 is higher than it was in 1995.
B. The average in 1995 is higher than in 2007.
C. The average is still about the same as it was in 1995.
D. The test is inconclusive.

Exercise 9.16.44 *(Solution on p. 384.)*

The Type I error is:

A. To conclude the average hours per week in 2007 is higher than in 1995, when in fact, it is higher.
B. To conclude the average hours per week in 2007 is higher than in 1995, when in fact, it is the same.
C. To conclude the average hours per week in 2007 is the same as in 1995, when in fact, it is higher.
D. To conclude the average hours per week in 2007 is no higher than in 1995, when in fact, it is not higher.

9.17 Review[21]

Exercise 9.17.1 *(Solution on p. 384.)*

1. Rebecca and Matt are 14 year old twins. Matt's height is 2 standard deviations below the mean for 14 year old boys' height. Rebecca's height is 0.10 standard deviations above the mean for 14 year old girls' height. Interpret this.

A. Matt is 2.1 inches shorter than Rebecca
B. Rebecca is very tall compared to other 14 year old girls.
C. Rebecca is taller than Matt.
D. Matt is shorter than the average 14 year old boy.

2. Construct a histogram of the IPO data (see Table of Contents, 14. Appendix, Data Sets). Use 5 intervals.

The next six questions refer to the following information: Ninety homeowners were asked the number of estimates they obtained before having their homes fumigated. X = the number of estimates.

X	**Rel. Freq.**	**Cumulative Rel. Freq.**
1	0.3	
2	0.2	
4	0.4	
5	0.1	

Table 9.5

3. Calculate the frequencies.

4. Complete the cumulative relative frequency column. What percent of the estimates fell at or below 4?

Exercise 9.17.2 *(Solution on p. 384.)*

5. Calculate the sample mean (a) and sample standard deviation (b).

Exercise 9.17.3 *(Solution on p. 384.)*

6. Calculate the median, M, the first quartile, Q1, the third quartile, Q3.

Exercise 9.17.4 *(Solution on p. 384.)*

7. The middle 50% of the data are between _____ and _____.

8. Construct a boxplot of the data.

The next three questions refer to the following table: Seventy 5th and 6th graders were asked their favorite dinner.

	Pizza	**Hamburgers**	**Spaghetti**	**Fried shrimp**
5th grader	15	6	9	0
6th grader	15	7	10	8

Table 9.6

Exercise 9.17.5 *(Solution on p. 384.)*

9. Find the probability that one randomly chosen child is in the 6th grade and prefers fried shrimp.

---

[21]This content is available online at <http://cnx.org/content/m17013/1.9/>.

A. $\frac{32}{70}$
B. $\frac{8}{32}$
C. $\frac{8}{8}$
D. $\frac{8}{70}$

Exercise 9.17.6 *(Solution on p. 385.)*

10. Find the probability that a child does not prefer pizza.

A. $\frac{30}{70}$
B. $\frac{30}{40}$
C. $\frac{40}{70}$
D. 1

Exercise 9.17.7 *(Solution on p. 385.)*

11. Find the probability a child is in the 5th grade given that the child prefers spaghetti.

A. $\frac{9}{19}$
B. $\frac{9}{70}$
C. $\frac{9}{30}$
D. $\frac{19}{70}$

Exercise 9.17.8 *(Solution on p. 385.)*

12. A sample of convenience is a random sample.

A. true
B. false

Exercise 9.17.9 *(Solution on p. 385.)*

13. A statistic is a number that is a property of the population.

A. true
B. false

Exercise 9.17.10 *(Solution on p. 385.)*

14. You should always throw out any data that are outliers.

A. true
B. false

Exercise 9.17.11 *(Solution on p. 385.)*

15. Lee bakes pies for a little restaurant in Felton. She generally bakes 20 pies in a day, on the average.

a. Define the Random Variable X.
b. State the distribution for X.
c. Find the probability that Lee bakes more than 25 pies in any given day.

Exercise 9.17.12 *(Solution on p. 385.)*

16. Six different brands of Italian salad dressing were randomly selected at a supermarket. The grams of fat per serving are 7, 7, 9, 6, 8, 5. Assume that the underlying distribution is normal. Calculate a 95% confidence interval for the population average grams of fat per serving of Italian salad dressing sold in supermarkets.

Exercise 9.17.13 *(Solution on p. 385.)*

17. Given: uniform, exponential, normal distributions. Match each to a statement below.

a. mean = median $\neq$ mode
b. mean > median > mode
c. mean = median = mode

9.18 Lab: Hypothesis Testing of a Single Mean and Single Proportion[22]

Class Time:

Names:

9.18.1 Student Learning Outcomes:

- The student will select the appropriate distributions to use in each case.
- The student will conduct hypothesis tests and interpret the results.

9.18.2 Television Survey

In a recent survey, it was stated that Americans watch television on average four hours per day. Assume that $\sigma = 2$. Using your class as the sample, conduct a hypothesis test to determine if the average for students at your school is lower.

1. H_o:
2. H_a:
3. In words, define the random variable. __________ =
4. The distribution to use for the test is:
5. Determine the test statistic using your data.
6. Draw a graph and label it appropriately.Shade the actual level of significance.
   a. Graph:

[22]This content is available online at <http://cnx.org/content/m17007/1.9/>.

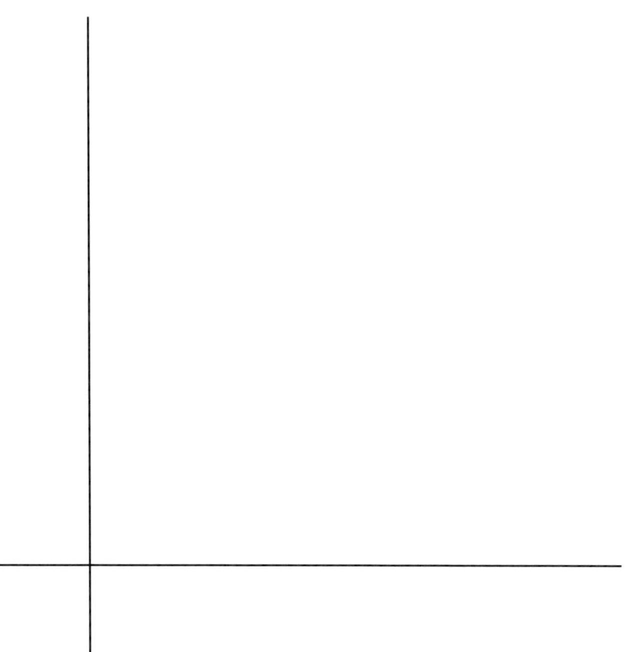

Figure 9.9

b. Determine the p-value:

7. Do you or do you not reject the null hypothesis? Why?
8. Write a clear conclusion using a complete sentence.

9.18.3 Language Survey

According to the 2000 Census, about 39.5% of Californians and 17.9% of all Americans speak a language other than English at home. Using your class as the sample, conduct a hypothesis test to determine if the percent of the students at your school that speak a language other than English at home is different from 39.5%.

1. H_o:
2. H_a:
3. In words, define the random variable. __________ =
4. The distribution to use for the test is:
5. Determine the test statistic using your data.
6. Draw a graph and label it appropriately. Shade the actual level of significance.

   a. Graph:

Figure 9.10

b. Determine the p-value:

7. Do you or do you not reject the null hypothesis? Why?
8. Write a clear conclusion using a complete sentence.

9.18.4 Jeans Survey

Suppose that young adults own an average of 3 pairs of jeans. Survey 8 people from your class to determine if the average is higher than 3.

1. H_o:
2. H_a:
3. In words, define the random variable. __________ =
4. The distribution to use for the test is:
5. Determine the test statistic using your data.
6. Draw a graph and label it appropriately. Shade the actual level of significance.

   a. Graph:

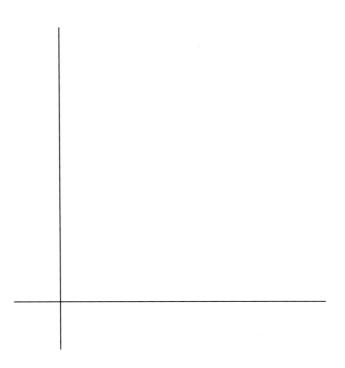

Figure 9.11

b. Determine the p-value:

7. Do you or do you not reject the null hypothesis? Why?
8. Write a clear conclusion using a complete sentence.

Solutions to Exercises in Chapter 9

Solution to Example 9.15, Problem 2 (p. 349)
$p' = \frac{x}{n}$ where x is the number of successes and n is the total number in the sample.

$x = 43, n = 150$

$p' = \frac{43}{150}$

Solution to Example 9.15, Problem 3 (p. 349)
A success is having three cell phones in a household.

Solution to Example 9.15, Problem 4 (p. 349)
The level of significance is the preset α. Since α is not given, assume that $\alpha = 0.5$.

Solution to Example 9.15, Problem 5 (p. 349)
p-value = 0.7216

Solution to Example 9.15, Problem 6 (p. 349)
Assuming that $\alpha = 0.5$, $\alpha < p\text{-value}$. The Decision is do not reject H_0 because there is not sufficient evidence to conclude that the proportion of households that have three cell phones is not 30%.

Solutions to Practice 1: Single Mean, Known Population Standard Deviation

Solution to Exercise 9.13.1 (p. 354)
Averages

Solution to Exercise 9.13.2 (p. 354)

- **a:** $H_o : \mu = 2.5$ (or, $H_o : \mu \leq 2.5$)
- **b:** $H_a : \mu > 2.5$

Solution to Exercise 9.13.3 (p. 354)
right-tailed

Solution to Exercise 9.13.4 (p. 354)
$\overline{X}$

Solution to Exercise 9.13.5 (p. 354)
The average time spent in jail for 26 first time convicted burglars

Solution to Exercise 9.13.6 (p. 354)
Yes, 1.5

Solution to Exercise 9.13.7 (p. 354)

- **a.** 3
- **b.** 1.5
- **c.** 1.8
- **d.** 26

Solution to Exercise 9.13.8 (p. 354)
σ

Solution to Exercise 9.13.9 (p. 354)
$\overline{X} \sim N\left(2.5, \frac{1.5}{\sqrt{26}}\right)$

Solution to Exercise 9.13.11 (p. 355)
0.0446

Solution to Exercise 9.13.12 (p. 355)

- **a.** Reject the null hypothesis

Solutions to Practice 2: Single Mean, Unknown Population Standard Deviation

Solution to Exercise 9.14.1 (p. 356)
averages
Solution to Exercise 9.14.2 (p. 356)

a. $H_o : \mu = 15$
b. $H_a : \mu \neq 15$

Solution to Exercise 9.14.3 (p. 356)
two-tailed
Solution to Exercise 9.14.4 (p. 356)
$\overline{X}$
Solution to Exercise 9.14.5 (p. 356)
the average time spent on death row
Solution to Exercise 9.14.6 (p. 356)
No
Solution to Exercise 9.14.7 (p. 356)

a. 17.4
b. s
c. 75

Solution to Exercise 9.14.8 (p. 356)
$t-$test
Solution to Exercise 9.14.9 (p. 356)
t_{74}
Solution to Exercise 9.14.11 (p. 357)
0.0015
Solution to Exercise 9.14.12 (p. 357)

a. Reject the null hypothesis

Solutions to Practice 3: Single Proportion

Solution to Exercise 9.15.1 (p. 358)
Proportions
Solution to Exercise 9.15.2 (p. 358)

a. $H_o : p = 0.095$
b. $H_a : P < 0.095$

Solution to Exercise 9.15.3 (p. 358)
left-tailed
Solution to Exercise 9.15.4 (p. 358)
P-hat
Solution to Exercise 9.15.5 (p. 358)
the proportion of people in that town suffering from depress. or a depr. illness
Solution to Exercise 9.15.6 (p. 358)

a. 7
b. 100
c. 0.07

Solution to Exercise 9.15.7 (p. 358)
2.93
Solution to Exercise 9.15.8 (p. 358)
Normal
Solution to Exercise 9.15.10 (p. 359)
0.1969
Solution to Exercise 9.15.11 (p. 359)

a. Do not reject the null hypothesis

Solutions to Homework

Solution to Exercise 9.16.1 (p. 360)

a. $H_o : \mu = 34$; $H_a : \mu \neq 34$
c. $H_o : \mu \geq 100,000$; $H_a : \mu < 100,000$
d. $H_o : p = 0.29$; $H_a : p \neq 0.29$
g. $H_o : p = 0.50$; $H_a : p \neq 0.50$
i. $H_o : p \geq 0.11$; $H_a : p < 0.11$

Solution to Exercise 9.16.2 (p. 360)

a. Type I error: We believe the average is not 34 years, when it really is 34 years. Type II error: We believe the average is 34 years, when it is not really 34 years.
c. Type I error: We believe the average is less than $100,000, when it really is at least $100,000. Type II error: We believe the average is at least $100,000, when it is really less than $100,000.
d. Type I error: We believe that the proportion of h.s. seniors who get drunk each month is not 29%, when it really is 29%. Type II error: We believe that 29% of h.s. seniors get drunk each month, when the proportion is really not 29%.
i. Type I error: We believe the proportion is less than 11%, when it is really at least 11%. Type II error: WE believe the proportion is at least 11%, when it really is less than 11%.

Solution to Exercise 9.16.5 (p. 360)

e. $z = -2.71$
f. 0.0034
h. Decision: Reject null; Conclusion: $\mu < 19$
i. $(17.449, 18.757)$

Solution to Exercise 9.16.7 (p. 361)

e. 3.5
f. 0.0005
h. Decision: Reject null; Conclusion: $\mu > 4.5$
i. $(4.7553, 5.4447)$

Solution to Exercise 9.16.9 (p. 361)

e. 2.7
f. 0.0042
h. Decision: Reject Null
i. $(80.789, 85.211)$

Solution to Exercise 9.16.11 (p. 361)

d. t_{11}

- **e.** 1.96
- **f.** 0.0380
- **h.** Decision: Reject null when $a = 0.05$; do not reject null when $a = 0.01$
- **i.** $(3.8865, 5.9468)$

Solution to Exercise 9.16.13 (p. 361)

- **e.** -1.64
- **f.** 0.1000
- **h.** Decision: Do not reject null
- **i.** $(0.3216, 0.4784)$

Solution to Exercise 9.16.15 (p. 362)

- **d.** t_9
- **e.** -1.33
- **f.** 0.1086
- **h.** Decision: Do not reject null
- **i.** $(51.886, 62.114)$

Solution to Exercise 9.16.19 (p. 363)

- **e.** 1.65
- **f.** 0.0984
- **h.** Decision: Do not reject null
- **i.** $(0.6836, 0.8533)$

Solution to Exercise 9.16.21 (p. 364)

- **e.** -2.39
- **f.** 0.0093
- **h.** Decision: Reject null
- **i.** $(91.854, 100.15)$

Solution to Exercise 9.16.23 (p. 365)

- **e.** -1.82
- **f.** 0.0345
- **h.** Decision: Do not reject null
- **i.** $(0.5331, 0.7685)$

Solution to Exercise 9.16.25 (p. 367)

- **e.** $z = -2.99$
- **f.** 0.0014
- **h.** Decision: Reject null; Conclusion: $p < .70$
- **i.** $(0.4529, 0.6582)$

Solution to Exercise 9.16.27 (p. 368)

- **e.** 0.57
- **f.** 0.7156
- **h.** Decision: Do not reject null
- **i.** $(0.3415, 0.6185)$

Solution to Exercise 9.16.29 (p. 369)

- **e.** 1.32

f. 0.1873
h. Decision: Do not reject null
i. $(0.65, 0.90)$

Solution to Exercise 9.16.31 (p. 369)

e. 9.98
f. 0.0000
h. Decision: Reject null
i. $(28.8, 30.0)$

Solution to Exercise 9.16.33 (p. 370)

e. -44.7
f. 0.0000
h. Decision: Reject null
i. $(0.60, 0.90)$ - in years

Solution to Exercise 9.16.34 (p. 370)
B
Solution to Exercise 9.16.35 (p. 370)
D
Solution to Exercise 9.16.36 (p. 370)
C
Solution to Exercise 9.16.37 (p. 370)
C
Solution to Exercise 9.16.38 (p. 371)
A
Solution to Exercise 9.16.39 (p. 371)
C
Solution to Exercise 9.16.40 (p. 371)
D
Solution to Exercise 9.16.41 (p. 371)
D
Solution to Exercise 9.16.42 (p. 371)
D
Solution to Exercise 9.16.43 (p. 372)
C
Solution to Exercise 9.16.44 (p. 372)
B

Solutions to Review

Solution to Exercise 9.17.1 (p. 373)
D
Solution to Exercise 9.17.2 (p. 373)

a. 2.8
b. 1.48

Solution to Exercise 9.17.3 (p. 373)
$M = 3$; $Q1 = 1$; $Q3 = 4$
Solution to Exercise 9.17.4 (p. 373)
1 and 4

Solution to Exercise 9.17.5 (p. 373)
D
Solution to Exercise 9.17.6 (p. 374)
C
Solution to Exercise 9.17.7 (p. 374)
A
Solution to Exercise 9.17.8 (p. 374)
B
Solution to Exercise 9.17.9 (p. 374)
B
Solution to Exercise 9.17.10 (p. 374)
B
Solution to Exercise 9.17.11 (p. 374)

- **b.** $P(20)$
- **c.** 0.1122

Solution to Exercise 9.17.12 (p. 374)
CI: $(5.52, 8.48)$
Solution to Exercise 9.17.13 (p. 374)

- **a.** uniform
- **b.** exponential
- **c.** normal

Chapter 10

Hypothesis Testing: Two Means, Paired Data, Two Proportions

10.1 Hypothesis Testing: Two Population Means and Two Population Proportions[1]

10.1.1 Student Learning Objectives

By the end of this chapter, the student should be able to:

- Classify hypothesis tests by type.
- Conduct and interpret hypothesis tests for two population means, population standard deviations known.
- Conduct and interpret hypothesis tests for two population means, population standard deviations unknown.
- Conduct and interpret hypothesis tests for two population proportions.
- Conduct and interpret hypothesis tests for matched or paired samples.

10.1.2 Introduction

Studies often compare two groups. For example, researchers are interested in the effect aspirin has in preventing heart attacks. Over the last few years, newspapers and magazines have reported about various aspirin studies involving two groups. Typically, one group is given aspirin and the other group is given a placebo. Then, the heart attack rate is studied over several years.

There are other situations that deal with the comparison of two groups. For example, studies compare various diet and exercise programs. Politicians compare the proportion of individuals from different income brackets who might vote for them. Students are interested in whether SAT or GRE preparatory courses really help raise their scores.

In the previous chapter, you learned to conduct hypothesis tests on single means and single proportions. You will expand upon that in this chapter. You will compare two averages or two proportions to each other. The general procedure is still the same, just expanded.

[1] This content is available online at <http://cnx.org/content/m17029/1.6/>.

To compare two averages or two proportions, you work with two groups. The groups are classified either as **independent** or **matched pairs**. **Independent groups** mean that the two samples taken are independent, that is, sample values selected from one population are not related in any way to sample values selected from the other population. **Matched pairs** consist of two samples that are dependent. The parameter tested using matched pairs is the population mean. The parameters tested using independent groups are either population means or population proportions.

> NOTE: This chapter relies on either a calculator or a computer to calculate the degrees of freedom, the test statistics, and p-values. TI-83+ and TI-84 instructions are included as well as the the test statistic formulas. Because of technology, we do not need to separate two population means, independent groups, population variances unknown into large and small sample sizes.

This chapter deals with the following hypothesis tests:

Independent groups (samples are independent)

- Test of two population means.
- Test of two population proportions.

Matched or paired samples (samples are dependent)

- Becomes a test of one population mean.

10.2 Comparing Two Independent Population Means with Unknown Population Standard Deviations[2]

1. The two independent samples are simple random samples from two distinct populations.
2. Both populations are normally distributed with the population means and standard deviations unknown unless the sample sizes are greater than 30. In that case, the populations need not be normally distributed.

The comparison of two population means is very common. A difference between the two samples depends on both the means and the standard deviations. Very different means can occur by chance if there is great variation among the individual samples. In order to account for the variation, we take the difference of the sample means, $\overline{X_1}$ - $\overline{X_2}$, and divide by the standard error (shown below) in order to standardize the difference. The result is a t-score test statistic (shown below).

Because we do not know the population standard deviations, we estimate them using the two sample standard deviations from our independent samples. For the hypothesis test, we calculate the estimated standard deviation, or **standard error**, of **the difference in sample means**, $\overline{X_1}$ - $\overline{X_2}$.

The standard error is:

$$\sqrt{\frac{(S_1)^2}{n_1}+\frac{(S_2)^2}{n_2}} \tag{10.1}$$

The test statistic (t-score) is calculated as follows:

T-score

$$\frac{(\overline{x_1}-\overline{x_2})-(\mu_1-\mu_2)}{\sqrt{\frac{(S_1)^2}{n_1}+\frac{(S_2)^2}{n_2}}} \tag{10.2}$$

where:

[2]This content is available online at <http://cnx.org/content/m17025/1.13/>.

- s_1 and s_2, the sample standard deviations, are estimates of σ_1 and σ_2, respectively.
- σ_1 and σ_2 are the unknown population standard deviations.
- $\overline{x_1}$ and $\overline{x_2}$ are the sample means. μ_1 and μ_2 are the population means.

The **degrees of freedom (df)** is a somewhat complicated calculation. However, a computer or calculator calculates it easily. The dfs are not always a whole number. The test statistic calculated above is approximated by the Student-t distribution with dfs as follows:

Degrees of freedom

$$df = \frac{\left[\frac{(s_1)^2}{n_1} + \frac{(s_2)^2}{n_2}\right]^2}{\frac{1}{n_1-1} \cdot \left[\frac{(s_1)^2}{n_1}\right]^2 + \frac{1}{n_2-1} \cdot \left[\frac{(s_2)^2}{n_2}\right]^2} \tag{10.3}$$

When both sample sizes n_1 and n_2 are five or larger, the Student-t approximation is very good. Notice that the sample variances ${s_1}^2$ and ${s_2}^2$ are not pooled. (If the question comes up, do not pool the variances.)

NOTE: It is not necessary to compute this by hand. A calculator or computer easily computes it.

Example 10.1: Independent groups

The average amount of time boys and girls ages 7 through 11 spend playing sports each day is believed to be the same. An experiment is done, data is collected, resulting in the table below:

	Sample Size	**Average Number of Hours Playing Sports Per Day**	**Sample Standard Deviation**
Girls	9	2 hours	$\sqrt{0.75}$
Boys	16	3.2 hours	1.00

Table 10.1

Problem

Is there a difference in the average amount of time boys and girls ages 7 through 11 play sports each day? Test at the 5% level of significance.

Solution

The population standard deviations are not known. Let g be the subscript for girls and b be the subscript for boys. Then, μ_g is the population mean for girls and μ_b is the population mean for boys. This is a test of two **independent groups**, two population **means.**

Random variable: $\overline{X_g} - \overline{X_b}$ = difference in the average amount of time girls and boys play sports each day.

H_o: $\mu_g = \mu_b \; (\mu_g - \mu_b = 0)$

H_a: $\mu_g \neq \mu_b \; (\mu_g - \mu_b \neq 0)$

The words **"the same"** tell you H_o has an "=". Since there are no other words to indicate H_a, then assume **"is different."** This is a two-tailed test.

Distribution for the test: Use t_{df} where df is calculated using the df formula for independent groups, two population means. Using a calculator, df is approximately 18.8462. **Do not pool the variances.**

Calculate the p-value using a Student-t distribution: p-value = 0.0054

Graph:

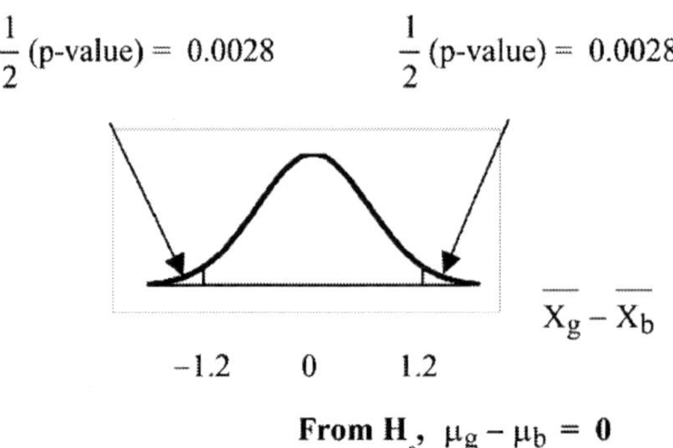

Figure 10.1

$s_g = \sqrt{0.75}$

$s_b = 1$

So, $\overline{x_g} - \overline{x_b} = 2 - 3.2 = -1.2$

Half the p-value is below -1.2 and half is above 1.2.

Make a decision: Since $\alpha >$ p-value, reject H_o.

This means you reject $\mu_g = \mu_b$. The means are different.

Conclusion: At the 5% level of significance, the sample data show there is sufficient evidence to conclude that the average number of hours that girls and boys aged 7 through 11 play sports per day is different.

NOTE: TI-83+ and TI-84: Press `STAT`. Arrow over to `TESTS` and press `4:2-SampTTest`. Arrow over to Stats and press `ENTER`. Arrow down and enter 2 for the first sample mean, $\sqrt{0.75}$ for Sx1, `9` for n1, `3.2` for the second sample mean, `1` for Sx2, and `16` for n2. Arrow down to μ1: and arrow to `does not equal` μ2. Press `ENTER`. Arrow down to Pooled: and `No`. Press `ENTER`. Arrow down to `Calculate` and press `ENTER`. The p-value is p = 0.0054, the dfs are approximately 18.8462, and the test statistic is -3.14. Do the procedure again but instead of Calculate do Draw.

Example 10.2

A study is done by a community group in two neighboring colleges to determine which one graduates students with more math classes. College A samples 11 graduates. Their average is 4 math classes with a standard deviation of 1.5 math classes. College B samples 9 graduates. Their average is 3.5 math classes with a standard deviation of 1 math class. The community group believes that a student who graduates from college A **has taken more math classes,** on the average. Test at a 1% significance level. Answer the following questions.

Problem 1 *(Solution on p. 424.)*
Is this a test of two means or two proportions?

Problem 2 *(Solution on p. 424.)*
Are the populations standard deviations known or unknown?

Problem 3 *(Solution on p. 424.)*
Which distribution do you use to perform the test?

Problem 4 *(Solution on p. 424.)*
What is the random variable?

Problem 5 *(Solution on p. 424.)*
What are the null and alternate hypothesis?

Problem 6 *(Solution on p. 424.)*
Is this test right, left, or two tailed?

Problem 7 *(Solution on p. 424.)*
What is the p-value?

Problem 8 *(Solution on p. 424.)*
Do you reject or not reject the null hypothesis?

Conclusion:
At the 1% level of significance, from the sample data, there is not sufficient evidence to conclude that a student who graduates from college A has taken more math classes, on the average, than a student who graduates from college B.

10.3 Comparing Two Independent Population Means with Known Population Standard Deviations[3]

Even though this situation is not likely (knowing the population standard deviations is not likely), the following example illustrates hypothesis testing for independent means, known population standard deviations. The distribution is Normal and is for the difference of sample means, $\overline{X_1} - \overline{X_2}$. The normal distribution has the following format:

Normal distribution

$$\overline{X_1} - \overline{X_2} \sim N\left[u_1 - u_2, \sqrt{\frac{(\sigma_1)^2}{n_1} + \frac{(\sigma_2)^2}{n_2}}\right] \tag{10.4}$$

The standard deviation is:

$$\sqrt{\frac{(\sigma_1)^2}{n_1} + \frac{(\sigma_2)^2}{n_2}} \tag{10.5}$$

The test statistic (z-score) is:

$$z = \frac{(\overline{x_1} - \overline{x_2}) - (\mu_1 - \mu_2)}{\sqrt{\frac{(\sigma_1)^2}{n_1} + \frac{(\sigma_2)^2}{n_2}}} \tag{10.6}$$

[3]This content is available online at <http://cnx.org/content/m17042/1.8/>.

Example 10.3
independent groups, population standard deviations known: The mean lasting time of 2 competing floor waxes is to be compared. **Twenty floors** are randomly assigned **to test each wax**. The following table is the result.

Wax	Sample Mean Number of Months Floor Wax Last	Population Standard Deviation
1	3	0.33
2	2.9	0.36

Table 10.2

Problem
Does the data indicate that **wax 1 is more effective than wax 2**? Test at a 5% level of significance.

Solution
This is a test of two independent groups, two population means, population standard deviations known.

Random Variable: $\overline{X_1} - \overline{X_2}$ = difference in the average number of months the competing floor waxes last.

$H_o : \mu_1 \leq \mu_2$

$H_a : \mu_1 > \mu_2$

The words **"is more effective"** says that **wax 1 lasts longer than wax 2**, on the average. "Longer" is a " $>$ " symbol and goes into H_a. Therefore, this is a right-tailed test.

Distribution for the test: The population standard deviations are known so the distribution is normal. Using the formula above, the distribution is:

$$\overline{X_1} - \overline{X_2} \sim N\left(0, \sqrt{\frac{0.33^2}{20} + \frac{0.36^2}{20}}\right)$$

Since $\mu_1 \leq \mu_2$ then $\mu_1 - \mu_2 \leq 0$ and the mean for the normal distribution is 0.

Calculate the p-value using the normal distribution: p-value = 0.1799

Graph:

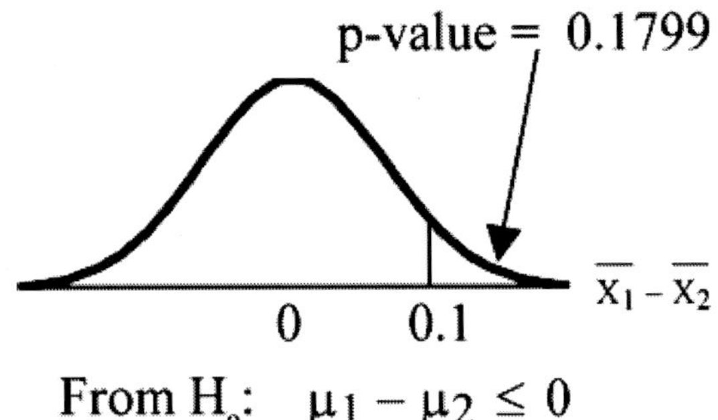

Figure 10.2

$\overline{x_1} - \overline{x_2} = 3 - 2.9 = 0.1$

Compare α and the p-value: $\alpha = 0.05$ and p-value = 0.1799. Therefore, $\alpha <$ p-value.

Make a decision: Since $\alpha <$ p-value, do not reject H_o.

Conclusion: At the 5% level of significance, from the sample data, there is not sufficient evidence to conclude that wax 1 lasts longer (wax 1 is more effective) than wax 2.

NOTE: TI-83+ and TI-84: Press `STAT`. Arrow over to `TESTS` and press `3:2-SampZTest`. Arrow over to `Stats` and press `ENTER`. Arrow down and enter `.33` for sigma1, `.36` for sigma2, `3` for the first sample mean, `20` for n1, `2.9` for the second sample mean, and `20` for n2. Arrow down to μ1: and arrow to $> \mu$2. Press `ENTER`. Arrow down to `Calculate` and press `ENTER`. The p-value is p = 0.1799 and the test statistic is 0.9157. Do the procedure again but instead of `Calculate` do `Draw`.

10.4 Comparing Two Independent Population Proportions[4]

1. The two independent samples are simple random samples that are independent.
2. The number of successes is at least five and the number of failures is at least five for each of the samples.

Comparing two proportions, like comparing two means, is common. If two estimated proportions are different, it may be due to a difference in the populations or it may be due to chance. A hypothesis test can help determine if a difference in the estimated proportions $(P'_A - P'_B)$ reflects a difference in the populations.

The difference of two proportions follows an approximate normal distribution. Generally, the null hypothesis states that the two proportions are the same. That is, $H_o : p_A = p_B$. To conduct the test, we use a pooled proportion, p_c.

[4] This content is available online at <http://cnx.org/content/m17043/1.8/>.

The pooled proportion is calculated as follows:

$$p_c = \frac{X_A + X_B}{n_A + n_B} \tag{10.7}$$

The distribution for the differences is:

$$P'_A - P'_B \sim N\left[0, \sqrt{p_c \cdot (1 - p_c) \cdot \left(\frac{1}{n_A} + \frac{1}{n_B}\right)}\right] \tag{10.8}$$

The test statistic (z-score) is:

$$z = \frac{(p'_A - p'_B) - (p_A - p_B)}{\sqrt{p_c \cdot (1 - p_c) \cdot \left(\frac{1}{n_A} + \frac{1}{n_B}\right)}} \tag{10.9}$$

Example 10.4: Two population proportions
Two types of medication for hives are being tested to determine if there is a **difference in the percentage of adult patient reactions. Twenty** out of a random **sample of 200** adults given medication A still had hives 30 minutes after taking the medication. **Twelve** out of another **random sample of 200 adults** given medication B still had hives 30 minutes after taking the medication. Test at a 1% level of significance.

10.4.1 Determining the solution

This is a test of 2 population proportions.

Problem ***(Solution on p. 424.)***

How do you know?

Let A and B be the subscripts for medication A and medication B. Then p_A and p_B are the desired population proportions.

Random Variable:
$P'_A - P'_B$ = difference in the percentages of adult patients who did not react after 30 minutes to medication A and medication B.

$H_o : p_A = p_B \qquad p_A - p_B = 0$

$H_a : p_A \neq p_B \qquad p_A - p_B \neq 0$

The words **"is a difference"** tell you the test is two-tailed.

Distribution for the test: Since this is a test of two binomial population proportions, the distribution is normal:

$p_c = \frac{X_A + X_B}{n_A + n_B} = \frac{20+12}{200+200} = 0.08 \quad 1 - p_c = 0.92$

Therefore, $P'_A - P'_B \sim N\left[0, \sqrt{(0.08) \cdot (0.92) \cdot \left(\frac{1}{200} + \frac{1}{200}\right)}\right]$

$P'_A - P'_B$ follows an approximate normal distribution.

Calculate the p-value using the normal distribution: p-value = 0.1404.

Estimated proportion for group A: $p'_A = \frac{x_A}{n_A} = \frac{20}{200} = 0.1$

Estimated proportion for group B: $p'_B = \frac{x_B}{n_B} = \frac{12}{200} = 0.06$

Graph:

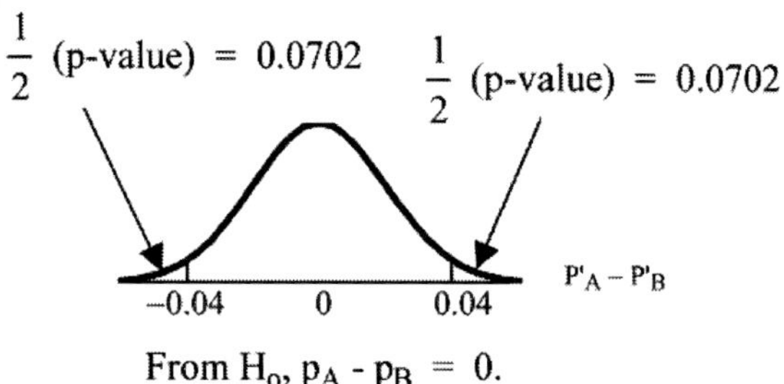

Figure 10.3

$P'_A - P'_B = 0.1 - 0.06 = 0.04.$

Half the p-value is below -0.04 and half is above 0.04.

Compare α and the p-value: $\alpha = 0.01$ and the *p-value* $= 0.1404$. $\alpha <$ p-value.

Make a decision: Since $\alpha <$ *p-value*, you cannot reject H_o.

Conclusion: At a 1% level of significance, from the sample data, there is not sufficient evidence to conclude that there is a difference in the percentages of adult patients who did not react after 30 minutes to medication A and medication B.

TI-83+ and TI-84: Press `STAT`. Arrow over to `TESTS` and press `6:2-PropZTest`. Arrow down and enter 20 for $x1$, 200 for $n1$, 12 for $x2$, and 200 for $n2$. Arrow down to `p1`: and arrow to `does not equal p2`. Press `ENTER`. Arrow down to `Calculate` and press `ENTER`. The p-value is $p = 0.1404$ and the test statistic is 1.47. Do the procedure again but instead of `Calculate` do `Draw`.

10.5 Matched or Paired Samples[5]

1. Simple random sampling is used.
2. Sample sizes are often small.
3. Two measurements (samples) are drawn from the same pair of individuals or objects.
4. Differences are calculated from the matched or paired samples.
5. The differences form the sample that is used for the hypothesis test.
6. The matched pairs have differences that either come from a population that is normal or the number of differences is greater than 30 or both.

[5]This content is available online at <http://cnx.org/content/m17033/1.11/>.

In a hypothesis test for matched or paired samples, subjects are matched in pairs and differences are calculated. The differences are the data. The population mean for the differences, μ_d, is then tested using a Student-t test for a single population mean with $n-1$ degrees of freedom where n is the number of differences.

The test statistic (t-score) is:

$$t = \frac{\overline{x_d} - \mu_d}{\left(\frac{s_d}{\sqrt{n}}\right)} \tag{10.10}$$

Example 10.5: Matched or paired samples

A study was conducted to investigate the effectiveness of hypnotism in reducing pain. Results for randomly selected subjects are shown in the table. The "before" value is matched to an "after" value.

Subject:	A	B	C	D	E	F	G	H
Before	6.6	6.5	9.0	10.3	11.3	8.1	6.3	11.6
After	6.8	2.5	7.4	8.5	8.1	6.1	3.4	2.0

Table 10.3

Problem

Are the sensory measurements, on average, lower after hypnotism? Test at a 5% significance level.

Solution

Corresponding "before" and "after" values form matched pairs.

After Data	**Before Data**	**Difference**
6.8	6.6	0.2
2.4	6.5	-4.1
7.4	9	-1.6
8.5	10.3	-1.8
8.1	11.3	-3.2
6.1	8.1	-2
3.4	6.3	-2.9
2	11.6	-9.6

Table 10.4

The data **for the test** are the differences: {0.2, -4.1, -1.6, -1.8, -3.2, -2, -2.9, -9.6}

The sample mean and sample standard deviation of the differences are: $\overline{x_d} = -3.13$ and $s_d = 2.91$ Verify these values.

Let μ_d be the population mean for the differences. We use the subscript d to denote "differences."

Random Variable: $\overline{X_d}$ = the average difference of the sensory measurements

$$H_o : \mu_d \geq 0 \tag{10.11}$$

There is no improvement. ($\mu_d$is the population mean of the differences.)

$$H_a : \mu_d < 0 \tag{10.12}$$

There is improvement. The score should be lower after hypnotism so the difference ought to be negative to indicate improvement.

Distribution for the test: The distribution is a student-t with $df = n - 1 = 8 - 1 = 7$. Use t_7. **(Notice that the test is for a single population mean.)**

Calculate the p-value using the Student-t distribution: $p\text{-}value = 0.0095$

Graph:

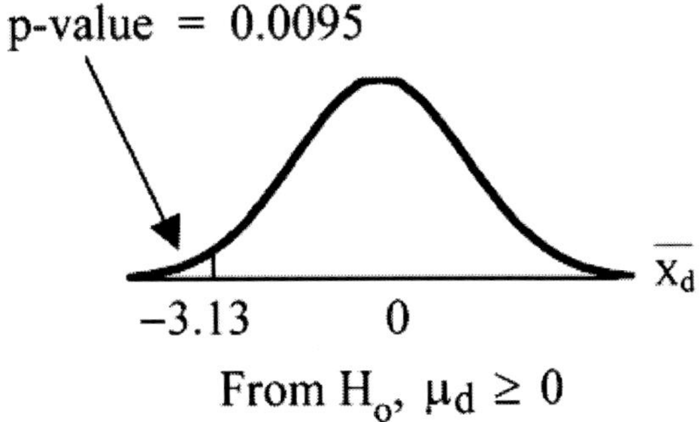

Figure 10.4

$\overline{X}_d$ is the random variable for the differences.

The sample mean and sample standard deviation of the differences are:

$\overline{x}_d = -3.13$

$\overline{s}_d = 2.91$

Compare α and the p-value: $\alpha = 0.05$ and $p\text{-}value = 0.0095$. $\alpha > p\text{-}value$.

Make a decision: Since $\alpha > p\text{-}value$, reject H_o.

This means that $\mu_d < 0$ and there is improvement.

Conclusion: At a 5% level of significance, from the sample data, there is sufficient evidence to conclude that the sensory measurements, on average, are lower after hypnotism. Hypnotism appears to be effective in reducing pain.

NOTE: For the TI-83+ and TI-84 calculators, you can either calculate the differences ahead of time (**after - before**) and put the differences into a list or you can put the **after** data into a first list and the **before** data into a second list. Then go to a third list and arrow up to the name. Enter 1st list name - 2nd list name. The calculator will do the subtraction and you will have the differences in the third list.

NOTE: TI-83+ and TI-84: Use your list of differences as the data. Press `STAT` and arrow over to `TESTS`. Press `2:T-Test`. Arrow over to `Data` and press `ENTER`. Arrow down and enter `0` for μ_0, the name of the list where you put the data, and `1` for Freq:. Arrow down to μ: and arrow over to $< \mu_0$. Press `ENTER`. Arrow down to `Calculate` and press `ENTER`. The p-value is 0.0094 and the test statistic is -3.04. Do these instructions again except arrow to `Draw` (instead of `Calculate`). Press `ENTER`.

Example 10.6

A college football coach was interested in whether the college's strength development class increased his players' maximum lift (in pounds) on the bench press exercise. He asked 4 of his players to participate in a study. The amount of weight they could each lift was recorded before they took the strength development class. After completing the class, the amount of weight they could each lift was again measured. The data are as follows:

Weight (in pounds)	Player 1	Player 2	Player 3	Player 4
Amount of weighted lifted prior to the class	205	241	338	368
Amount of weight lifted after the class	295	252	330	360

Table 10.5

The coach wants to know if the strength development class makes his players stronger, on average.

Problem *(Solution on p. 424.)*

Record the **differences** data. Calculate the differences by subtracting the amount of weight lifted prior to the class from the weight lifted after completing the class. The data for the differences are: {90, 11, -8, -8}

Using the differences data, calculate the sample mean and the sample standard deviation.

$\overline{x}_d = 21.3$ $\quad\quad$ $s_d = 46.7$

Using the difference data, this becomes a test of a single __________ (fill in the blank).

Define the random variable: $\overline{X}_d$ = average difference in the maximum lift per player.

The distribution for the hypothesis test is t_3.

$H_o : \mu d \leq 0$ $\quad\quad$ $H_a : \mu_d > 0$

Graph:

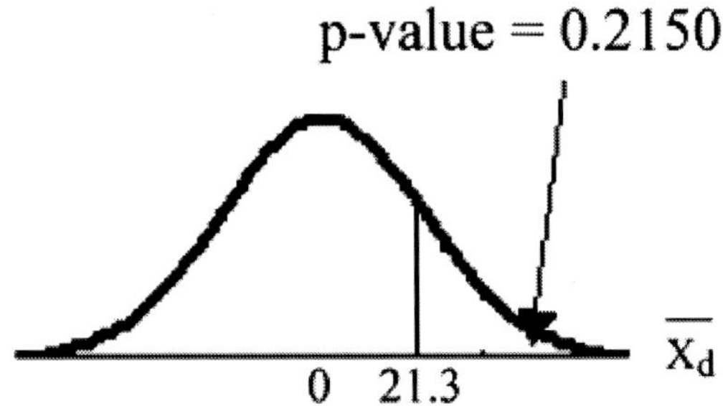

Figure 10.5

Calculate the p-value: The p-value is 0.2150

Decision: If the level of significance is 5%, the decision is to not reject the null hypothesis because $\alpha < p\text{-}value$.

What is the conclusion?

Example 10.7
Seven eighth graders at Kennedy Middle School measured how far they could push the shot-put with their dominant (writing) hand and their weaker (non-writing) hand. They thought that they could push equal distances with either hand. The following data was collected.

Distance (in feet) using	**Student 1**	**Student 2**	**Student 3**	**Student 4**	**Student 5**	**Student 6**	**Student 7**
Dominant Hand	30	26	34	17	19	26	20
Weaker Hand	28	14	27	18	17	26	16

Table 10.6

Problem *(Solution on p. 424.)*
Conduct a hypothesis test to determine whether the differences in distances between the children's dominant versus weaker hands is significant.

HINT: use a t-test on the difference data.

CHECK: The test statistic is 2.18 and the p-value is 0.0716.

What is your conclusion?

10.6 Summary of Types of Hypothesis Tests[6]

Two Population Means

- Populations are independent and population standard deviations are unknown.
- Populations are independent and population standard deviations are known (not likely).

Matched or Paired Samples

- Two samples are drawn from the same set of objects.
- Samples are dependent.

Two Population Proportions

- Populations are independent.

[6]This content is available online at <http://cnx.org/content/m17044/1.5/>.

10.7 Practice 1: Hypothesis Testing for Two Proportions[7]

10.7.1 Student Learning Outcomes

- The student will explore the properties of hypothesis testing with two proportions.

10.7.2 Given

In the 2000 Census, 2.4 percent of the U.S. population reported being two or more races. However, the percent varies tremendously from state to state. (http://www.census.gov/prod/2001pubs/c2kbr01-6.pdf) Suppose that two random surveys are conducted. In the first random survey, out of 1000 North Dakotans, only 9 people reported being of two or more races. In the second random survey, out of 500 Nevadans, 17 people reported being of two or more races. Conduct a hypothesis test to determine if the population percents are the same for the two states or if the percent for Nevada is statistically higher than for North Dakota.

10.7.3 Hypothesis Testing: Two Averages

Exercise 10.7.1 *(Solution on p. 424.)*
Is this a test of averages or proportions?

Exercise 10.7.2 *(Solution on p. 424.)*
State the null and alternative hypotheses.

a. H_0 :
b. H_a :

Exercise 10.7.3 *(Solution on p. 424.)*
Is this a right-tailed, left-tailed, or two-tailed test? How do you know?

Exercise 10.7.4
What is the Random Variable of interest for this test?

Exercise 10.7.5
In words, define the Random Variable for this test.

Exercise 10.7.6 *(Solution on p. 424.)*
Which distribution (Normal or student-t) would you use for this hypothesis test?

Exercise 10.7.7
Explain why you chose the distribution you did for the above question.

Exercise 10.7.8 *(Solution on p. 424.)*
Calculate the test statistic.

Exercise 10.7.9
Sketch a graph of the situation. Label the horizontal axis. Mark the hypothesized difference and the sample difference. Shade the area corresponding to the $p-$value.

---

[7]This content is available online at <http://cnx.org/content/m17027/1.9/>.

Figure 10.6

Exercise 10.7.10 *(Solution on p. 424.)*

Find the $p-$value:

Exercise 10.7.11 *(Solution on p. 424.)*

At a pre-conceived $\alpha = 0.05$, what is your:

a. Decision:
b. Reason for the decision:
c. Conclusion (write out in a complete sentence):

10.7.4 Discussion Question

Exercise 10.7.12

Does it appear that the proportion of Nevadans who are two or more races is higher than the proportion of North Dakotans? Why or why not?

10.8 Practice 2: Hypothesis Testing for Two Averages[8]

10.8.1 Student Learning Outcome

- The student will explore the properties of hypothesis testing with two averages.

10.8.2 Given

The U.S. Center for Disease Control reports that the average life expectancy for whites born in 1900 was 47.6 years and for nonwhites it was 33.0 years. (http://www.cdc.gov/nchs/data/dvs/nvsr53_06t12.pdf) Suppose that you randomly survey death records for people born in 1900 in a certain county. Of the 124 whites, the average life span was 45.3 years with a standard deviation of 12.7 years. Of the 82 nonwhites, the average life span was 34.1 years with a standard deviation of 15.6 years. Conduct a hypothesis test to see if the average life spans in the county were the same for whites and nonwhites.

10.8.3 Hypothesis Testing: Two Averages

Exercise 10.8.1 *(Solution on p. 425.)*
Is this a test of averages or proportions?

Exercise 10.8.2 *(Solution on p. 425.)*
State the null and alternative hypotheses.

a. H_0 :
b. H_a :

Exercise 10.8.3 *(Solution on p. 425.)*
Is this a right-tailed, left-tailed, or two-tailed test? How do you know?

Exercise 10.8.4 *(Solution on p. 425.)*
What is the Random Variable of interest for this test?

Exercise 10.8.5 *(Solution on p. 425.)*
In words, define the Random Variable for this test.

Exercise 10.8.6
Which distribution (Normal or student-t) would you use for this hypothesis test?

Exercise 10.8.7
Explain why you chose the distribution you did for the above question.

Exercise 10.8.8 *(Solution on p. 425.)*
Calculate the test statistic.

Exercise 10.8.9
Sketch a graph of the situation. Label the horizontal axis. Mark the hypothesized difference and the sample difference. Shade the area corresponding to the $p-$value.

---

[8]This content is available online at <http://cnx.org/content/m17039/1.7/>.

Figure 10.7

Exercise 10.8.10 *(Solution on p. 425.)*

Find the $p-$value:

Exercise 10.8.11 *(Solution on p. 425.)*

At a pre-conceived $\alpha = 0.05$, what is your:

- **a.** Decision:
- **b.** Reason for the decision:
- **c.** Conclusion (write out in a complete sentence):

10.8.4 Discussion Question

Exercise 10.8.12

Does it appear that the averages are the same? Why or why not?

10.9 Homework[9]

For questions Exercise 10.9.1 - Exercise 10.9.10, indicate which of the following choices best identifies the hypothesis test.

A. Independent group means, population standard deviations and/or variances known
B. Independent group means, population standard deviations and/or variances unknown
C. Matched or paired samples
D. Single mean
E. 2 proportions
F. Single proportion

Exercise 10.9.1 *(Solution on p. 425.)*
A powder diet is tested on 49 people and a liquid diet is tested on 36 different people. The population standard deviations are 2 pounds and 3 pounds, respectively. Of interest is whether the liquid diet yields a higher average weight loss than the powder diet.

Exercise 10.9.2
Two chocolate bars are taste-tested on consumers. Of interest is whether a larger percentage of consumers will prefer one bar over the other.

Exercise 10.9.3 *(Solution on p. 425.)*
The average number of English courses taken in a two–year time period by male and female college students is believed to be about the same. An experiment is conducted and data are collected from 9 males and 16 females.

Exercise 10.9.4
A football league reported that the average number of touchdowns per game was 5. A study is done to determine if the average number of touchdowns has decreased.

Exercise 10.9.5 *(Solution on p. 425.)*
A study is done to determine if students in the California state university system take longer to graduate than students enrolled in private universities. 100 students from both the California state university system and private universities are surveyed. From years of research, it is known that the population standard deviations are 1.5811 years and 1 year, respectively.

Exercise 10.9.6
According to a YWCA Rape Crisis Center newsletter, 75% of rape victims know their attackers. A study is done to verify this.

Exercise 10.9.7 *(Solution on p. 425.)*
According to a recent study, U.S. companies have an average maternity-leave of six weeks.

Exercise 10.9.8
A recent drug survey showed an increase in use of drugs and alcohol among local high school students as compared to the national percent. Suppose that a survey of 100 local youths and 100 national youths is conducted to see if the percentage of drug and alcohol use is higher locally than nationally.

Exercise 10.9.9 *(Solution on p. 425.)*
A new SAT study course is tested on 12 individuals. Pre-course and post-course scores are recorded. Of interest is the average increase in SAT scores.

Exercise 10.9.10
University of Michigan researchers reported in the *Journal of the National Cancer Institute* that quitting smoking is especially beneficial for those under age 49. In this American Cancer Society

[9]This content is available online at <http://cnx.org/content/m17023/1.10/>.

study, the risk (probability) of dying of lung cancer was about the same as for those who had never smoked.

10.9.1 For each problem below, fill in a hypothesis test solution sheet. The solution sheet is in the Appendix and can be copied. For the online version of the book, it is suggested that you copy the .doc or .pdf files.

NOTE: If you are using a student-t distribution for a homework problem below, including for paired data, you may assume that the underlying population is normally distributed. (In general, you must first prove that assumption, though.)

Exercise 10.9.11 ***(Solution on p. 425.)***

A powder diet is tested on 49 people and a liquid diet is tested on 36 different people. Of interest is whether the liquid diet yields a higher average weight loss than the powder diet. The powder diet group had an average weight loss of 42 pounds with a standard deviation of 12 pounds. The liquid diet group had an average weight loss of 45 pounds with a standard deviation of 14 pounds.

Exercise 10.9.12

The average number of English courses taken in a two–year time period by male and female college students is believed to be about the same. An experiment is conducted and data are collected from 29 males and 16 females. The males took an average of 3 English courses with a standard deviation of 0.8. The females took an average of 4 English courses with a standard deviation of 1.0. Are the averages statistically the same?

Exercise 10.9.13 ***(Solution on p. 425.)***

A study is done to determine if students in the California state university system take longer to graduate than students enrolled in private universities. 100 students from both the California state university system and private universities are surveyed. Suppose that from years of research, it is known that the population standard deviations are 1.5811 years and 1 year, respectively. The following data are collected. The California state university system students took on average 4.5 years with a standard deviation of 0.8. The private university students took on average 4.1 years with a standard deviation of 0.3.

Exercise 10.9.14

A new SAT study course is tested on 12 individuals. Pre-course and post-course scores are recorded. Of interest is the average increase in SAT scores. The following data is collected:

Pre-course score	Post-course score
1200	1300
960	920
1010	1100
840	880
1100	1070
1250	1320
860	860
1330	1370
790	770
990	1040
1110	1200
740	850

Table 10.7

Exercise 10.9.15 ***(Solution on p. 425.)***

A recent drug survey showed an increase in use of drugs and alcohol among local high school seniors as compared to the national percent. Suppose that a survey of 100 local seniors and 100 national seniors is conducted to see if the percentage of drug and alcohol use is higher locally than nationally. Locally, 65 seniors reported using drugs or alcohol within the past month, while 60 national seniors reported using them.

Exercise 10.9.16

A student at a four-year college claims that average enrollment at four–year colleges is higher than at two–year colleges in the United States. Two surveys are conducted. Of the 35 two–year colleges surveyed, the average enrollment was 5068 with a standard deviation of 4777. Of the 35 four-year colleges surveyed, the average enrollment was 5466 with a standard deviation of 8191. (Source: *Microsoft Bookshelf*)

Exercise 10.9.17 ***(Solution on p. 426.)***

A study was conducted by the U.S. Army to see if applying antiperspirant to soldiers' feet for a few days before a major hike would help cut down on the number of blisters soldiers had on their feet. In the experiment, for three nights before they went on a 13-mile hike, a group of 328 West Point cadets put an alcohol-based antiperspirant on their feet. A "control group" of 339 soldiers put on a similar, but inactive, preparation on their feet. On the day of the hike, the temperature reached 83 ° F. At the end of the hike, 21% of the soldiers who had used the antiperspirant and 48% of the control group had developed foot blisters. Conduct a hypothesis test to see if the percent of soldiers using the antiperspirant was significantly lower than the control group. (Source: U.S. Army study reported in *Journal of the American Academy of Dermatologists*)

Exercise 10.9.18

We are interested in whether the percents of female suicide victims for ages 15 to 24 are the same for the white and the black races in the United States. We randomly pick one year, 1992, to compare the races. The number of suicides estimated in the United States in 1992 for white females is 4930. 580 were aged 15 to 24. The estimate for black females is 330. 40 were aged 15 to 24. We will let female suicide victims be our population. (Source*: the National Center for Health Statistics, U.S. Dept. of Health and Human Services*)

Exercise 10.9.19 *(Solution on p. 426.)*

At Rachel's 11th birthday party, 8 girls were timed to see how long (in seconds) they could hold their breath in a relaxed position. After a two-minute rest, they timed themselves while jumping. The girls thought that the jumping would not affect their times, on average. Test their hypothesis.

Relaxed time (seconds)	Jumping time (seconds)
26	21
47	40
30	28
22	21
23	25
45	43
37	35
29	32

Table 10.8

Exercise 10.9.20

Elizabeth Mjelde, an art history professor, was interested in whether the value from the Golden Ratio formula, $\left(\frac{\textit{larger+smaller dimension}}{\textit{larger dimension}}\right)$was the same in the Whitney Exhibit for works from 1900 – 1919 as for works from 1920 – 1942. 37 early works were sampled. They averaged 1.74 with a standard deviation of 0.11. 65 of the later works were sampled. They averaged 1.746 with a standard deviation of 0.1064. Do you think that there is a significant difference in the Golden Ratio calculation? (Source: *data from Whitney Exhibit on loan to San Jose Museum of Art*)

Exercise 10.9.21 *(Solution on p. 426.)*

One of the questions in a study of marital satisfaction of dual–career couples was to rate the statement, "I'm pleased with the way we divide the responsibilities for childcare." The ratings went from 1 (strongly agree) to 5 (strongly disagree). Below are ten of the paired responses for husbands and wives. Conduct a hypothesis test to see if the average difference in the husband's versus the wife's satisfaction level is negative (meaning that, within the partnership, the husband is happier than the wife).

Wife's score	2	2	3	3	4	2	1	1	2	4
Husband's score	2	2	1	3	2	1	1	1	2	4

Table 10.9

Exercise 10.9.22

Ten individuals went on a low–fat diet for 12 weeks to lower their cholesterol. Evaluate the data below. Do you think that their cholesterol levels were significantly lowered?

Starting cholesterol level	Ending cholesterol level
140	140
220	230
110	120
240	220
200	190
180	150
190	200
360	300
280	300
260	240

Table 10.10

Exercise 10.9.23 ***(Solution on p. 426.)***

Average entry level salaries for college graduates with mechanical engineering degrees and electrical engineering degrees are believed to be approximately the same. (Source: *http:// www.graduatingengineer.com*[10]). A recruiting office thinks that the average mechanical engineering salary is actually lower than the average electrical engineering salary. The recruiting office randomly surveys 50 entry level mechanical engineers and 60 entry level electrical engineers. Their average salaries were $46,100 and $46,700, respectively. Their standard deviations were $3450 and $4210, respectively. Conduct a hypothesis test to determine if you agree that the average entry level mechanical engineering salary is lower than the average entry level electrical engineering salary.

Exercise 10.9.24

A recent year was randomly picked from 1985 to the present. In that year, there were 2051 Hispanic students at Cabrillo College out of a total of 12,328 students. At Lake Tahoe College, there were 321 Hispanic students out of a total of 2441 students. In general, do you think that the percent of Hispanic students at the two colleges is basically the same or different? (Source: *Chancellor's Office, California Community Colleges, November 1994*)

Exercise 10.9.25 ***(Solution on p. 426.)***

Eight runners were convinced that the average difference in their individual times for running one mile versus race walking one mile was at most 2 minutes. Below are their times. Do you agree that the average difference is at most 2 minutes?

[10] http://www.graduatingengineer.com/

Running time (minutes)	Race walking time (minutes)
5.1	7.3
5.6	9.2
6.2	10.4
4.8	6.9
7.1	8.9
4.2	9.5
6.1	9.4
4.4	7.9

Table 10.11

Exercise 10.9.26

Marketing companies have collected data implying that teenage girls use more ring tones on their cellular phones than teenage boys do. In one particular study of 40 randomly chosen teenage girls and boys (20 of each) with cellular phones, the average number of ring tones for the girls was 3.2 with a standard deviation of 1.5. The average for the boys was 1.7 with a standard deviation of 0.8. Conduct a hypothesis test to determine if the averages are approximately the same or if the girls' average is higher than the boys' average.

Exercise 10.9.27 *(Solution on p. 426.)*

While her husband spent 2½ hours picking out new speakers, a statistician decided to determine whether the percent of men who enjoy shopping for electronic equipment is higher than the percent of women who enjoy shopping for electronic equipment. The population was Saturday afternoon shoppers. Out of 67 men, 24 said they enjoyed the activity. 8 of the 24 women surveyed claimed to enjoy the activity. Interpret the results of the survey.

Exercise 10.9.28

We are interested in whether children's educational computer software costs less, on average, than children's entertainment software. 36 educational software titles were randomly picked from a catalog. The average cost was $31.14 with a standard deviation of $4.69. 35 entertainment software titles were randomly picked from the same catalog. The average cost was $33.86 with a standard deviation of $10.87. Decide whether children's educational software costs less, on average, than children's entertainment software. (Source: *Educational Resources*, December catalog)

Exercise 10.9.29 *(Solution on p. 426.)*

Parents of teenage boys often complain that auto insurance costs more, on average, for teenage boys than for teenage girls. A group of concerned parents examines a random sample of insurance bills. The average annual cost for 36 teenage boys was $679. For 23 teenage girls, it was $559. From past years, it is known that the population standard deviation for each group is $180. Determine whether or not you believe that the average cost for auto insurance for teenage boys is greater than that for teenage girls.

Exercise 10.9.30

A group of transfer bound students wondered if they will spend the same average amount on texts and supplies each year at their four-year university as they have at their community college. They conducted a random survey of 54 students at their community college and 66 students at their local four-year university. The sample means were $947 and $1011, respectively. The population standard deviations are known to be $254 and $87, respectively. Conduct a hypothesis test to determine if the averages are statistically the same.

Exercise 10.9.31 *(Solution on p. 426.)*

Joan Nguyen recently claimed that the proportion of college–age males with at least one pierced ear is as high as the proportion of college–age females. She conducted a survey in her classes. Out of 107 males, 20 had at least one pierced ear. Out of 92 females, 47 had at least one pierced ear. Do you believe that the proportion of males has reached the proportion of females?

Exercise 10.9.32

Some manufacturers claim that non-hybrid sedan cars have a lower average miles per gallon (mpg) than hybrid ones. Suppose that consumers test 21 hybrid sedans and get an average 31 mpg with a standard deviation of 7 mpg. Thirty-one non-hybrid sedans average 22 mpg with a standard deviation of 4 mpg. Suppose that the population standard deviations are known to be 6 and 3, respectively. Conduct a hypothesis test to the manufacturers claim.

Questions Exercise 10.9.33 – Exercise 10.9.37 refer to the Terri Vogel's data set (see Table of Contents).

Exercise 10.9.33 *(Solution on p. 426.)*

Using the data from Lap 1 only, conduct a hypothesis test to determine if the average time for completing a lap in races is the same as it is in practices.

Exercise 10.9.34

Repeat the test in Exercise 10.9.33, but use Lap 5 data this time.

Exercise 10.9.35 *(Solution on p. 427.)*

Repeat the test in Exercise 10.9.33, but this time combine the data from Laps 1 and 5.

Exercise 10.9.36

In 2 – 3 complete sentences, explain in detail how you might use Terri Vogel's data to answer the following question. "Does Terri Vogel drive faster in races than she does in practices?"

Exercise 10.9.37 *(Solution on p. 427.)*

Is the proportion of race laps Terri completes slower than 130 seconds less than the proportion of practice laps she completes slower than 135 seconds?

Exercise 10.9.38

"To Breakfast or Not to Breakfast?" by Richard Ayore

In the American society, birthdays are one of those days that everyone looks forward to. People of different ages and peer groups gather to mark the 18th, 20th, . . . birthdays. During this time, one looks back to see what he or she had achieved for the past year, and also focuses ahead for more to come.

If, by any chance, I am invited to one of these parties, my experience is always different. Instead of dancing around with my friends while the music is booming, I get carried away by memories of my family back home in Kenya. I remember the good times I had with my brothers and sister while we did our daily routine.

Every morning, I remember we went to the shamba (garden) to weed our crops. I remember one day arguing with my brother as to why he always remained behind just to join us an hour later. In his defense, he said that he preferred waiting for breakfast before he came to weed. He said, "This is why I always work more hours than you guys!"

And so, to prove his wrong or right, we decided to give it a try. One day we went to work as usual without breakfast, and recorded the time we could work before getting tired and stopping. On the next day, we all ate breakfast before going to work. We recorded how long we worked again before getting tired and stopping. Of interest was our average increase in work time. Though not sure, my brother insisted that it is more than two hours. Using the data below, solve our problem.

Work hours with breakfast	Work hours without breakfast
8	6
7	5
9	5
5	4
9	7
8	7
10	7
7	5
6	6
9	5

Table 10.12

10.9.2 Try these multiple choice questions.

For questions Exercise 10.9.39 – Exercise 10.9.40, use the following information.

A new AIDS prevention drugs was tried on a group of 224 HIV positive patients. Forty-five (45) patients developed AIDS after four years. In a control group of 224 HIV positive patients, 68 developed AIDS after four years. We want to test whether the method of treatment reduces the proportion of patients that develop AIDS after four years or if the proportions of the treated group and the untreated group stay the same.

Let the subscript t= treated patient and ut= untreated patient.

Exercise 10.9.39 *(Solution on p. 427.)*

The appropriate hypotheses are:

A. $H_o : p_t < p_{ut}$ and $H_a : p_t \geq p_{ut}$
B. $H_o : p_t \leq p_{ut}$ and $H_a : p_t > p_{ut}$
C. $H_o : p_t = p_{ut}$ and $H_a : p_t \neq p_{ut}$
D. $H_o : p_t = p_{ut}$ and $H_a : p_t < p_{ut}$

Exercise 10.9.40 *(Solution on p. 427.)*

If the p -value is 0.0062 what is the conclusion (use $\alpha = 5$)?

A. The method has no effect.
B. The method reduces the proportion of HIV positive patients that develop AIDS after four years.
C. The method increases the proportion of HIV positive patients that develop AIDS after four years.
D. The test does not determine whether the method helps or does not help.

Exercise 10.9.41 *(Solution on p. 427.)*

Lesley E. Tan investigated the relationship between left-handedness and right-handedness and motor competence in preschool children. Random samples of 41 left-handers and 41 right-handers were given several tests of motor skills to determine if there is evidence of a difference between the children based on this experiment. The experiment produced the means and standard deviations shown below. Determine the appropriate test and best distribution to use for that test.

	Left-handed	Right-handed
Sample size	41	41
Sample mean	97.5	98.1
Sample standard deviation	17.5	19.2

Table 10.13

A. Two independent means, normal distribution
B. Two independent means, student-t distribution
C. Matched or paired samples, student-t distribution
D. Two population proportions, normal distribution

For questions Exercise 10.9.42 – Exercise 10.9.43, use the following information.

An experiment is conducted to show that blood pressure can be consciously reduced in people trained in a "biofeedback exercise program." Six (6) subjects were randomly selected and the blood pressure measurements were recorded before and after the training. The difference between blood pressures was calculated $(after - before)$ producing the following results: $\overline{x}_d = -10.2$ $s_d = 8.4$. Using the data, test the hypothesis that the blood pressure has decreased after the training,

Exercise 10.9.42 *(Solution on p. 427.)*

The distribution for the test is

A. t_5
B. t_6
C. $N(-10.2, 8.4)$
D. $N\left(-10.2, \frac{8.4}{\sqrt{6}}\right)$

Exercise 10.9.43 *(Solution on p. 427.)*

If $\alpha = 0.05$, the p-value and the conclusion are

A. 0.0014; the blood pressure decreased after the training
B. 0.0014; the blood pressure increased after the training
C. 0.0155; the blood pressure decreased after the training
D. 0.0155; the blood pressure increased after the training

For questions Exercise 10.9.44– Exercise 10.9.45, use the following information.

The Eastern and Western Major League Soccer conferences have a new Reserve Division that allows new players to develop their skills. As of May 25, 2005, the Reserve Division teams scored the following number of goals for 2005.

Western	Eastern
Los Angeles 9	D.C. United 9
FC Dallas 3	Chicago 8
Chivas USA 4	Columbus 7
Real Salt Lake 3	New England 6
Colorado 4	MetroStars 5
San Jose 4	Kansas City 3

Table 10.14

Conduct a hypothesis test to determine if the Western Reserve Division teams score, on average, fewer goals than the Eastern Reserve Division teams. Subscripts: **1** Western Reserve Division (**W**); **2** Eastern Reserve Division (**E**)

Exercise 10.9.44 *(Solution on p. 427.)*

The **exact** distribution for the hypothesis test is:

A. The normal distribution.
B. The student-t distribution.
C. The uniform distribution.
D. The exponential distribution.

Exercise 10.9.45 *(Solution on p. 427.)*

If the level of significance is 0.05, the conclusion is:

A. The **W** Division teams score, on average, fewer goals than the **E** teams.
B. The **W** Division teams score, on average, more goals than the **E** teams.
C. The **W** teams score, on average, about the same number of goals as the **E** teams score.
D. Unable to determine.

Questions Exercise 10.9.46 – Exercise 10.9.48 refer to the following.

A researcher is interested in determining if a certain drug vaccine prevents West Nile disease. The vaccine with the drug is administered to 36 people and another 36 people are given a vaccine that does not contain the drug. Of the group that gets the vaccine with the drug, one (1) gets West Nile disease. Of the group that gets the vaccine without the drug, three (3) get West Nile disease. Conduct a hypothesis test to determine if the proportion of people that get the vaccine without the drug and get West Nile disease is more than the proportion of people that get the vaccine with the drug and get West Nile disease.

- "Drug" subscript: group who get the vaccine with the drug.
- "No Drug" subscript: group who get the vaccine without the drug

Exercise 10.9.46 *(Solution on p. 427.)*

This is a test of:

A. a test of two proportions
B. a test of two independent means
C. a test of a single mean
D. a test of matched pairs.

Exercise 10.9.47 *(Solution on p. 427.)*

An appropriate null hypothesis is:

A. $p_{No\ Drug} \leq p_{Drug}$
B. $p_{No\ Drug} \geq p_{Drug}$
C. $\mu_{No\ Drug} \leq \mu_{Drug}$
D. $p_{No\ Drug} > p_{Drug}$

Exercise 10.9.48 ***(Solution on p. 427.)***
The p-value is 0.1517. At a 1% level of significance, the appropriate conclusion is

A. the proportion of people that get the vaccine without the drug and get West Nile disease is less than the proportion of people that get the vaccine with the drug and get West Nile disease.
B. the proportion of people that get the vaccine without the drug and get West Nile disease is more than the proportion of people that get the vaccine with the drug and get West Nile disease.
C. the proportion of people that get the vaccine without the drug and get West Nile disease is more than or equal to the proportion of people that get the vaccine with the drug and get West Nile disease.
D. the proportion of people that get the vaccine without the drug and get West Nile disease is no more than the proportion of people that get the vaccine with the drug and get West Nile disease.

Questions Exercise 10.9.49 and Exercise 10.9.50 refer to the following:

A golf instructor is interested in determining if her new technique for improving players' golf scores is effective. She takes four (4) new students. She records their 18-holes scores before learning the technique and then after having taken her class. She conducts a hypothesis test. The data are as follows.

	Player 1	Player 2	Player 3	Player 4
Average score before class	83	78	93	87
Average score after class	80	80	86	86

Table 10.15

Exercise 10.9.49 ***(Solution on p. 427.)***
This is a test of:

A. a test of two independent means
B. a test of two proportions
C. a test of a single proportion
D. a test of matched pairs.

Exercise 10.9.50 ***(Solution on p. 427.)***
The correct decision is:

A. Reject H_o
B. Do not reject H_o
C. The test is inconclusive

Questions Exercise 10.9.51 and Exercise 10.9.52 refer to the following:

Suppose a statistics instructor believes that there is no significant difference between the average class scores of her two classes on Exam 2. The average and standard deviation for her 8:30 class of 35 students were 75.86 and 16.91. The average and standard deviation for her 11:30 class of 37 students were 75.41 and 19.73. "8:30" subscript refers to the 8:30 class. "11:30" subscript refers to the 11:30 class.

Exercise 10.9.51 *(Solution on p. 427.)*

An appropriate alternate hypothesis for the hypothesis test is:

A. $\mu_{8:30} > \mu_{11:30}$
B. $\mu_{8:30} < \mu_{11:30}$
C. $\mu_{8:30} = \mu_{11:30}$
D. $\mu_{8:30} \neq \mu_{11:30}$

Exercise 10.9.52 *(Solution on p. 427.)*

A concluding statement is:

A. The 11:30 class average is better than the 8:30 class average.
B. The 8:30 class average is better than the 11:30 class average.
C. There is no significant difference between the averages of the two classes.
D. There is a significant difference between the averages of the two classes.

10.10 Review[11]

The next three questions refer to the following information:

In a survey at Kirkwood Ski Resort the following information was recorded:

Sport Participation by Age

	0 – 10	11 - 20	21 - 40	40+
Ski	10	12	30	8
Snowboard	6	17	12	5

Table 10.16

Suppose that one person from of the above was randomly selected.

Exercise 10.10.1 *(Solution on p. 427.)*
Find the probability that the person was a skier or was age 11 – 20.

Exercise 10.10.2 *(Solution on p. 427.)*
Find the probability that the person was a snowboarder given he/she was age 21 – 40.

Exercise 10.10.3 *(Solution on p. 427.)*
Explain which of the following are true and which are false.

a. Sport and Age are independent events.
b. Ski and age 11 – 20 are mutually exclusive events.
c. $P(\textit{Ski and age } 21 - 40) < P(\textit{Ski} \mid \textit{age } 21 - 40)$
d. $P(\textit{Snowboard or age } 0 - 10) < P(\textit{Snowboard} \mid \textit{age } 0 - 10)$

Exercise 10.10.4 *(Solution on p. 428.)*
The average length of time a person with a broken leg wears a cast is approximately 6 weeks. The standard deviation is about 3 weeks. Thirty people who had recently healed from broken legs were interviewed. State the distribution that most accurately reflects total time to heal for the thirty people.

Exercise 10.10.5 *(Solution on p. 428.)*
The distribution for X is Uniform. What can we say for certain about the distribution for $\overline{X}$ when $n = 1$?

A. The distribution for $\overline{X}$ is still Uniform with the same mean and standard dev. as the distribution for X.
B. The distribution for $\overline{X}$is Normal with the different mean and a different standard deviation as the distribution for X.
C. The distribution for $\overline{X}$ is Normal with the same mean but a larger standard deviation than the distribution for X.
D. The distribution for $\overline{X}$ is Normal with the same mean but a smaller standard deviation than the distribution for X.

Exercise 10.10.6 *(Solution on p. 428.)*
The distribution for X is uniform. What can we say for certain about the distribution for $\sum X$ when $n = 50$?

[11]This content is available online at <http://cnx.org/content/m17021/1.8/>.

A. The distribution for $\sum X$is still uniform with the same mean and standard deviation as the distribution for X.
B. The distribution for $\sum X$ is Normal with the same mean but a larger standard deviation as the distribution for X.
C. The distribution for $\sum X$ is Normal with a larger mean and a larger standard deviation than the distribution for X.
D. The distribution for $\sum X$ is Normal with the same mean but a smaller standard deviation than the distribution for X.

The next three questions refer to the following information:

A group of students measured the lengths of all the carrots in a five-pound bag of baby carrots. They calculated the average length of baby carrots to be 2.0 inches with a standard deviation of 0.25 inches. Suppose we randomly survey 16 five-pound bags of baby carrots.

Exercise 10.10.7 ***(Solution on p. 428.)***

State the approximate distribution for $\overline{X}$, the distribution for the average lengths of baby carrots in 16 five-pound bags. $\overline{X}\sim$

Exercise 10.10.8

Explain why we cannot find the probability that one individual randomly chosen carrot is greater than 2.25 inches.

Exercise 10.10.9 ***(Solution on p. 428.)***

Find the probability that $\overline{X}$ is between 2 and 2.25 inches.

The next three questions refer to the following information:

At the beginning of the term, the amount of time a student waits in line at the campus store is normally distributed with a mean of 5 minutes and a standard deviation of 2 minutes.

Exercise 10.10.10 ***(Solution on p. 428.)***

Find the 90th percentile of waiting time in minutes.

Exercise 10.10.11 ***(Solution on p. 428.)***

Find the median waiting time for one student.

Exercise 10.10.12 ***(Solution on p. 428.)***

Find the probability that the average waiting time for 40 students is at least 4.5 minutes.

10.11 Lab: Hypothesis Testing for Two Means and Two Proportions[12]

Class Time:

Names:

10.11.1 Student Learning Outcomes:

- The student will select the appropriate distributions to use in each case.
- The student will conduct hypothesis tests and interpret the results.

10.11.2 Supplies:

- The business section from two consecutive days' newspapers
- 3 small packages of M&Ms®
- 5 small packages of Reeses Pieces®

10.11.3 Increasing Stocks Survey

Look at yesterday's newspaper business section. Conduct a hypothesis test to determine if the proportion of New York Stock Exchange (NYSE) stocks that increased is greater than the proportion of NASDAQ stocks that increased. As randomly as possible, choose 40 NYSE stocks and 32 NASDAQ stocks and complete the following statements.

1. H_o
2. H_a
3. In words, define the Random Variable. ____________=
4. The distribution to use for the test is:
5. Calculate the test statistic using your data.
6. Draw a graph and label it appropriately. Shade the actual level of significance.
   a. Graph:

[12]This content is available online at <http://cnx.org/content/m17022/1.10/>.

Figure 10.8

b. Calculate the p-value:

7. Do you reject or not reject the null hypothesis? Why?
8. Write a clear conclusion using a complete sentence.

10.11.4 Decreasing Stocks Survey

Randomly pick 8 stocks from the newspaper. Using two consecutive days' business sections, test whether the stocks went down, on average, for the second day.

1. H_o
2. H_a
3. In words, define the Random Variable. ____________=
4. The distribution to use for the test is:
5. Calculate the test statistic using your data.
6. Draw a graph and label it appropriately. Shade the actual level of significance.

    a. Graph:

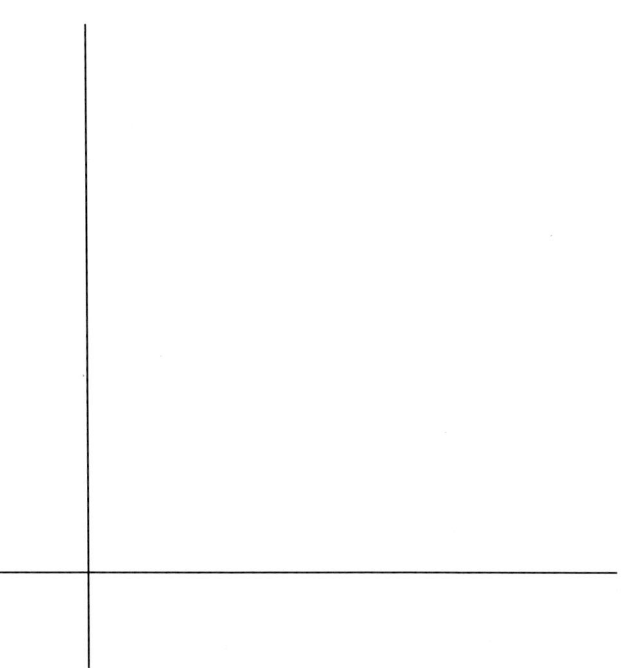

Figure 10.9

b. Calculate the p-value:

7. Do you reject or not reject the null hypothesis? Why?
8. Write a clear conclusion using a complete sentence.

10.11.5 Candy Survey

Buy three small packages of M&Ms and 5 small packages of Reeses Pieces (same net weight as the M&Ms). Test whether or not the average number of candy pieces per package is the same for the two brands.

1. H_o:
2. H_a:
3. In words, define the random variable. __________=
4. What distribution should be used for this test?
5. Calculate the test statistic using your data.
6. Draw a graph and label it appropriately. Shade the actual level of significance.

   a. Graph:

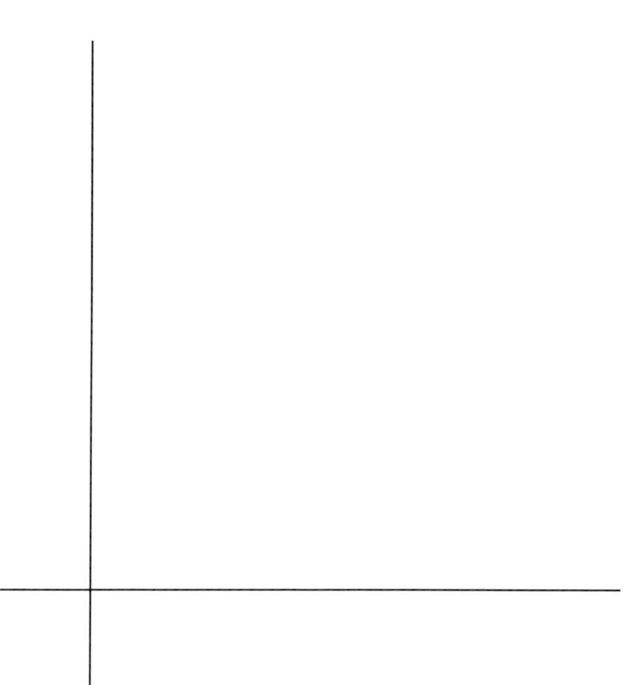

Figure 10.10

b. Calculate the p-value:

7. Do you reject or not reject the null hypothesis? Why?
8. Write a clear conclusion using a complete sentence.

10.11.6 Shoe Survey

Test whether women have, on average, more pairs of shoes than men. Include all forms of sneakers, shoes, sandals, and boots. Use your class as the sample.

1. H_o
2. H_a
3. In words, define the Random Variable. ____________=
4. The distribution to use for the test is:
5. Calculate the test statistic using your data.
6. Draw a graph and label it appropriately. Shade the actual level of significance.

   a. Graph:

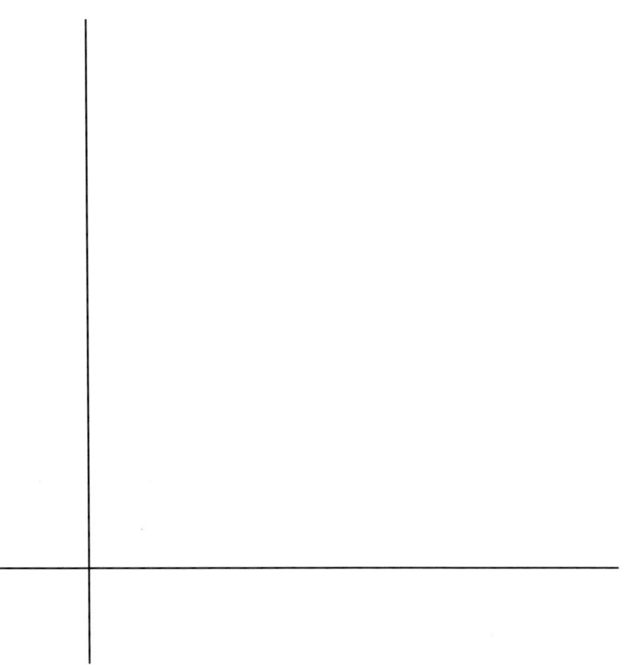

Figure 10.11

b. Calculate the p-value:

7. Do you reject or not reject the null hypothesis? Why?
8. Write a clear conclusion using a complete sentence.

Solutions to Exercises in Chapter 10

Solution to Example 10.2, Problem 1 (p. 391)
two means
Solution to Example 10.2, Problem 2 (p. 391)
unknown
Solution to Example 10.2, Problem 3 (p. 391)
Student-t
Solution to Example 10.2, Problem 4 (p. 391)
$\overline{X_A} - \overline{X_B}$
Solution to Example 10.2, Problem 5 (p. 391)

- $H_o : \mu_A \leq \mu_B$
- $H_a : \mu_A > \mu_B$

Solution to Example 10.2, Problem 6 (p. 391)
right
Solution to Example 10.2, Problem 7 (p. 391)
0.1928
Solution to Example 10.2, Problem 8 (p. 391)
Do not reject.
Solution to Example 10.4 (p. 394)
The problem asks for a difference in percentages.
Solution to Example 10.6 (p. 398)
means; At a 5% level of significance, from the sample data, there is not sufficient evidence to conclude that the strength development class helped to make the players stronger, on average.
Solution to Example 10.7 (p. 399)
H_0: μ_d equals 0; H_a: μ_d does not equal 0; Do not reject the null; At a 5% significance level, from the sample data, there is not sufficient evidence to conclude that the differences in distances between the children's dominant versus weaker hands is significant (there is not sufficient evidence to show that the children could push the shot-put further with their dominant hand). Alpha and the p-value are close so the test is not strong.

Solutions to Practice 1: Hypothesis Testing for Two Proportions

Solution to Exercise 10.7.1 (p. 401)
Proportions
Solution to Exercise 10.7.2 (p. 401)

a. H_0:PN=PND

a. H_a:PN> PND

Solution to Exercise 10.7.3 (p. 401)
right-tailed
Solution to Exercise 10.7.6 (p. 401)
Normal
Solution to Exercise 10.7.8 (p. 401)
3.50
Solution to Exercise 10.7.10 (p. 402)
0.0002
Solution to Exercise 10.7.11 (p. 402)

a. Reject the null hypothesis

Solutions to Practice 2: Hypothesis Testing for Two Averages

Solution to Exercise 10.8.1 (p. 403)
Averages
Solution to Exercise 10.8.2 (p. 403)

- **a.** $H_0 : \mu_W = \mu_{NW}$
- **b.** $H_a : \mu_W \neq \mu_{NW}$

Solution to Exercise 10.8.3 (p. 403)
two-tailed
Solution to Exercise 10.8.4 (p. 403)
$\overline{X}_W - \overline{X}_{NW}$
Solution to Exercise 10.8.5 (p. 403)
student-t
Solution to Exercise 10.8.8 (p. 403)
5.42
Solution to Exercise 10.8.10 (p. 404)
0.0000
Solution to Exercise 10.8.11 (p. 404)

- **a.** Reject the null hypothesis

Solutions to Homework

Solution to Exercise 10.9.1 (p. 405)
A
Solution to Exercise 10.9.3 (p. 405)
B
Solution to Exercise 10.9.5 (p. 405)
A
Solution to Exercise 10.9.7 (p. 405)
D
Solution to Exercise 10.9.9 (p. 405)
C
Solution to Exercise 10.9.11 (p. 406)

- **d.** $t_{68.44}$
- **e.** -1.04
- **f.** 0.1519
- **h.** Dec: do not reject null

Solution to Exercise 10.9.13 (p. 406)
Standard Normal

- **e.** $z = 2.14$
- **f.** 0.0163
- **h.** Decision: Reject null when $\alpha = 0.05$; Do not reject null when $\alpha = 0.01$

Solution to Exercise 10.9.15 (p. 407)

- **e.** 0.73
- **f.** 0.2326
- **h.** Decision: Do not reject null

Solution to Exercise 10.9.17 (p. 407)

- **e.** -7.33
- **f.** 0
- **h.** Decision: Reject null

Solution to Exercise 10.9.19 (p. 408)

- **d.** t_7
- **e.** -1.51
- **f.** 0.1755
- **h.** Decision: Do not reject null

Solution to Exercise 10.9.21 (p. 408)

- **d.** t_9
- **e.** $t = -1.86$
- **f.** 0.0479
- **h.** Decision: Reject null, but run another test

Solution to Exercise 10.9.23 (p. 409)

- **d.** t_{108}
- **e.** $t = -0.82$
- **f.** 0.2066
- **h.** Decision: Do not reject null

Solution to Exercise 10.9.25 (p. 409)

- **d.** t_7
- **e.** $t = 2.9850$
- **f.** 0.0103
- **h.** Decision: Reject null; The average difference is more than 2 minutes.

Solution to Exercise 10.9.27 (p. 410)

- **e.** 0.22
- **f.** 0.4133
- **h.** Decision: Do not reject null

Solution to Exercise 10.9.29 (p. 410)

- **e.** $z = 2.50$
- **f.** 0.0063
- **h.** Decision: Reject null

Solution to Exercise 10.9.31 (p. 411)

- **e.** -4.82
- **f.** 0
- **h.** Decision: Reject null

Solution to Exercise 10.9.33 (p. 411)

- **d.** $t_{20.32}$
- **e.** -4.70
- **f.** 0.0001
- **h.** Decision: Reject null

Solution to Exercise 10.9.35 (p. 411)

- **d.** $t_{40.94}$
- **e.** -5.08
- **f.** 0
- **h.** Decision: Reject null

Solution to Exercise 10.9.37 (p. 411)

- **e.** -0.95
- **f.** 0.1705
- **h.** Decision: Do not reject null

Solution to Exercise 10.9.39 (p. 412)
D
Solution to Exercise 10.9.40 (p. 412)
B
Solution to Exercise 10.9.41 (p. 412)
B
Solution to Exercise 10.9.42 (p. 413)
A
Solution to Exercise 10.9.43 (p. 413)
C
Solution to Exercise 10.9.44 (p. 414)
B
Solution to Exercise 10.9.45 (p. 414)
C
Solution to Exercise 10.9.46 (p. 414)
A
Solution to Exercise 10.9.47 (p. 414)
A
Solution to Exercise 10.9.48 (p. 415)
D
Solution to Exercise 10.9.49 (p. 415)
D
Solution to Exercise 10.9.50 (p. 415)
B
Solution to Exercise 10.9.51 (p. 416)
D
Solution to Exercise 10.9.52 (p. 416)
C

Solutions to Review

Solution to Exercise 10.10.1 (p. 417)
$\frac{77}{100}$
Solution to Exercise 10.10.2 (p. 417)
$\frac{12}{42}$
Solution to Exercise 10.10.3 (p. 417)

- **a.** False
- **b.** False
- **c.** True

d. False

Solution to Exercise 10.10.4 (p. 417)
$N\left(180, 16.43\right)$
Solution to Exercise 10.10.5 (p. 417)
A
Solution to Exercise 10.10.6 (p. 417)
C
Solution to Exercise 10.10.7 (p. 418)
$N\left(2, \frac{.25}{\sqrt{16}}\right)$
Solution to Exercise 10.10.9 (p. 418)
0.5000
Solution to Exercise 10.10.10 (p. 418)
7.6
Solution to Exercise 10.10.11 (p. 418)
5
Solution to Exercise 10.10.12 (p. 418)
0.9431

Chapter 11

The Chi-Square Distribution

11.1 The Chi-Square Distribution[1]

11.1.1 Student Learning Objectives

By the end of this chapter, the student should be able to:

- Interpret the chi-square probability distribution as the sample size changes.
- Conduct and interpret chi-square goodness-of-fit hypothesis tests.
- Conduct and interpret chi-square test of independence hypothesis tests.
- Conduct and interpret chi-square single variance hypothesis tests (optional).

11.1.2 Introduction

Have you ever wondered if lottery numbers were evenly distributed or if some numbers occurred with a greater frequency? How about if the types of movies people preferred were different across different age groups? What about if a coffee machine was dispensing approximately the same amount of coffee each time? You could answer these questions by conducting a hypothesis test.

You will now study a new distribution, one that is used to determine the answers to the above examples. This distribution is called the Chi-square distribution.

In this chapter, you will learn the three major applications of the Chi-square distribution:

- The goodness-of-fit test, which determines if data fit a particular distribution, such as with the lottery example
- The test of independence, which determines if events are independent, such as with the movie example
- The test of a single variance, which tests variability, such as with the coffee example

> NOTE: Though the Chi-square calculations depend on calculators or computers for most of the calculations, there is a table available (see the Table of Contents **15. Tables**). TI-83+ and TI-84 calculator instructions are included in the text.

[1]This content is available online at <http://cnx.org/content/m17048/1.7/>.

11.1.3 Optional Collaborative Classroom Activity

Look in the sports section of a newspaper or on the Internet for some sports data (baseball averages, basketball scores, golf tournament scores, football odds, swimming times, etc.). Plot a histogram and a boxplot using your data. See if you can determine a probability distribution that your data fits. Have a discussion with the class about your choice.

11.2 Notation[2]

The notation for the chi-square distribution is:

$\chi^2 \sim \chi^2_{df}$

where df = degrees of freedom depend on how chi-square is being used. (If you want to practice calculating chi-square probabilities then use $df = n - 1$. The degrees of freedom for the three major uses are each calculated differently.)

For the χ^2 distribution, the population mean is $\mu = df$ and the population standard deviation is $\sigma = \sqrt{2 \cdot df}$.

The random variable is shown as χ^2 but may be any upper case letter.

The random variable for a chi-square distribution with k degrees of freedom is the sum of k independent, squared standard normal variables.

$\chi^2 = (Z_1)^2 + (Z_2)^2 + ... + (Z_k)^2$

11.3 Facts About the Chi-Square Distribution[3]

1. The curve is nonsymmetrical and skewed to the right.
2. There is a different chi-square curve for each df.

[2]This content is available online at <http://cnx.org/content/m17052/1.5/>.
[3]This content is available online at <http://cnx.org/content/m17045/1.5/>.

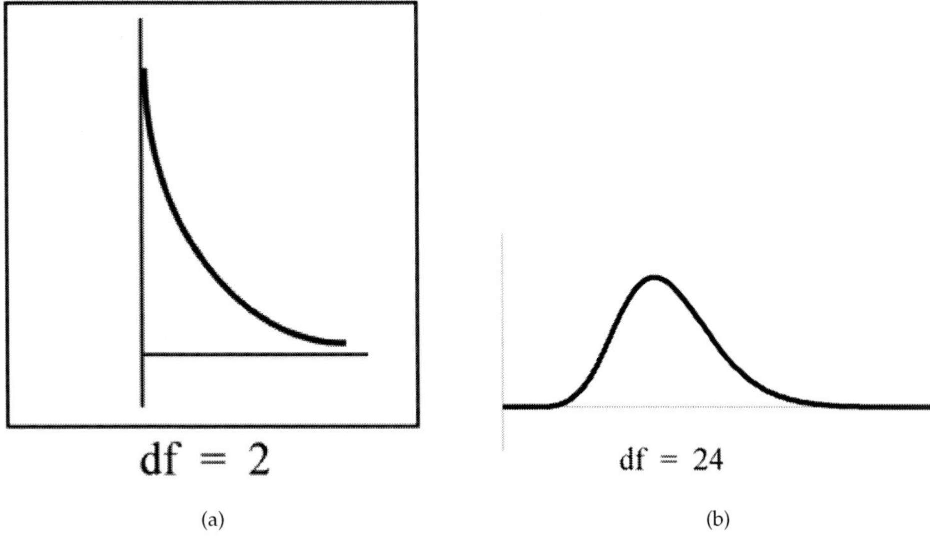

(a) (b)

Figure 11.1

3. The test statistic for any test is always greater than or equal to zero.
4. When $df > 90$, the chi-square curve approximates the normal. For $X \sim \chi^2_{1000}$ the mean, $\mu = df = 1000$ and the standard deviation, $\sigma = \sqrt{2 \cdot 1000} = 44.7$. Therefore, $X \sim N(1000, 44.7)$, approximately.
5. The mean, μ, is located just to the right of the peak.

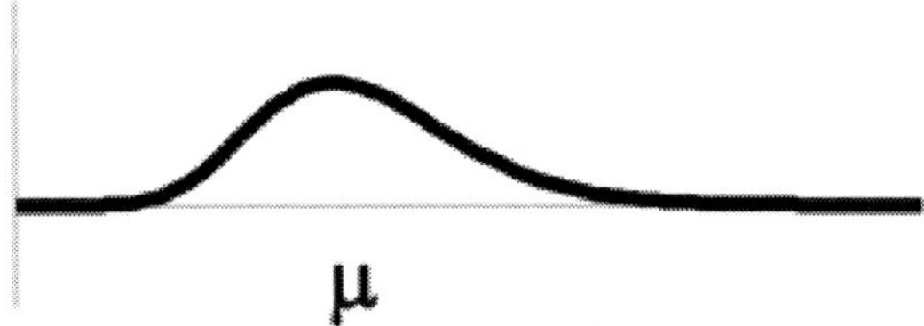

Figure 11.2

11.4 Goodness-of-Fit Test[4]

In this type of hypothesis test, you determine whether the data **"fit"** a particular distribution or not. For example, you may suspect your unknown data fit a binomial distribution. You use a chi-square test (meaning the distribution for the hypothesis test is chi-square) to determine if there is a fit or not. **The null and the alternate hypotheses for this test may be written in sentences or may be stated as equations or inequalities.**

[4]This content is available online at <http://cnx.org/content/m17192/1.7/>.

The test statistic for a goodness-of-fit test is:

$$\sum_{n} \frac{(O-E)^2}{E} \tag{11.1}$$

where:

- O = observed values (data)
- E = expected values (from theory)
- n = the number of different data cells or categories

The observed values are the data values and the expected values are the values you would expect to get if the null hypothesis were true. There are n terms of the form $\frac{(O-E)^2}{E}$.

The degrees of freedom are *df = (number of categories - 1).*

The goodness-of-fit test is almost always right tailed. If the observed values and the corresponding expected values are not close to each other, then the test statistic can get very large and will be way out in the right tail of the chi-square curve.

Example 11.1

Absenteeism of college students from math classes is a major concern to math instructors because missing class appears to increase the drop rate. Three statistics instructors wondered whether the absentee rate was the **same** for every day of the school week. They took a sample of absent students from three of their statistics classes during one week of the term. The results of the survey appear in the table.

	Monday	**Tuesday**	**Wednesday**	**Thursday**	**Friday**
# of students absent	28	22	18	20	32

Table 11.1

Determine the null and alternate hypotheses needed to run a goodness-of-fit test.

Since the instructors wonder whether the absentee rate is the same for every school day, we could say in the null hypothesis that the data **"fit"** a uniform distribution.

H_o**:** The rate at which college students are absent from their statistics class fits a uniform distribution.

The alternate hypothesis is the opposite of the null hypothesis.

H_a**:** The rate at which college students are absent from their statistics class does not fit a uniform distribution.

Problem 1

How many students do you **expect** to be absent on any given school day?

Solution

The total number of students in the sample is 120. **If the null hypothesis were true,** you would divide 120 by 5 to get 24 absences expected per day. **The expected number is based on a true null hypothesis.**

Problem 2
What are the degrees of freedom (df)?

Solution
There are 5 days of the week or 5 "cells" or categories.

$df = no.\ cells - 1 = 5 - 1 = 4$

Example 11.2
Employers particularly want to know which days of the week employees are absent in a five day work week. Most employers would like to believe that employees are absent equally during the week. That is, the average number of times an employee is absent is the same on Monday, Tuesday, Wednesday, Thursday, or Friday. Suppose a sample of 20 absent days was taken and the days absent were distributed as follows:

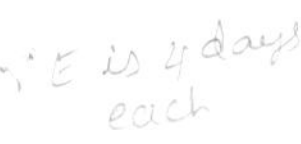

Day of the Week Absent

	Monday	**Tuesday**	**Wednesday**	**Thursday**	**Friday**
Number of Absences	5	4	2	3	6

Table 11.2

Problem
For the population of employees, do the absent days occur with equal frequencies during a five day work week? Test at a 5% significance level.

Solution
The null and alternate hypotheses are:

- H_o: The absent days occur with equal frequencies, that is, they fit a uniform distribution.
- H_a: The absent days occur with unequal frequencies, that is, they do not fit a uniform distribution.

If the absent days occur with equal frequencies, then, out of 20 absent days, there would be 4 absences on Monday, 4 on Tuesday, 4 on Wednesday, 4 on Thursday, and 4 on Friday. These numbers are the **expected** (E) values. The values in the table are the **observed** (O) values or data.

This time, calculate the χ^2 test statistic by hand. Make a chart with the following headings:

- Expected (E) values
- Observed (O) values
- $(O - E)$
- $(O - E)^2$
- $\frac{(O - E)^2}{E}$

Now add (sum) the last column. Verify that the sum is 2.5. This is the χ^2 test statistic.

To find the p-value, calculate $P\left(\chi^2 > 2.5\right)$. This test is right-tailed.

The dfs are the *number of cells* $- 1 = 4$.

Next, complete a graph like the one below with the proper labeling and shading. (You should shade the right tail. It will be a "large" right tail for this example because the p-value is "large.")

Use a computer or calculator to find the p-value. You should get *p-value* = 0.6446.

The decision is to not reject the null hypothesis.

Conclusion: At a 5% level of significance, from the sample data, there is not sufficient evidence to conclude that the absent days do not occur with equal frequencies.

TI-83+ and TI-84: Press `2nd DISTR`. Arrow down to χ^2`cdf`. Press `ENTER`. Enter `(2.5,1E99,4)`. Rounded to 4 places, you should see 0.6446 which is the p-value.

NOTE: TI-83+ and some TI-84 calculators do not have a special program for the test statistic for the goodness-of-fit test. The next example (Example 11-3) has the calculator instructions. The newer TI-84 calculators have in `STAT TESTS` the test `Chi2 GOF`. To run the test, put the observed values (the data) into a first list and the expected values (the values you expect if the null hypothesis is true) into a second list. Press `STAT TESTS` and `Chi2 GOF`. Enter the list names for the Observed list and the Expected list. Enter whatever else is asked and press `calculate` or `draw`. Make sure you clear any lists before you start. See below.

NOTE: **To Clear Lists in the calculators:** Go into `STAT EDIT` and arrow up to the list name area of the particular list. Press `CLEAR` and then arrow down. The list will be cleared. Or, you can press `STAT` and press 4 (for `ClrList`). Enter the list name and press `ENTER`.

Example 11.3

One study indicates that the number of televisions that American families have is distributed (this is the **given** distribution for the American population) as follows:

Number of Televisions	Percent
0	10
1	16
2	55
3	11
over 3	8

Table 11.3

The table contains expected (E) percents.

A random sample of 600 families in the far western United States resulted in the following data:

Number of Televisions	Frequency
0	66
1	119
2	340
3	60
over 3	15
	Total = 600

Table 11.4

The table contains observed (O) frequency values.

Problem

At the 1% significance level, does it appear that the distribution "number of televisions" of far western United States families is different from the distribution for the American population as a whole?

Solution

This problem asks you to test whether the far western United States families distribution fits the distribution of the American families. This test is always right-tailed.

The first table contains expected percentages. To get expected (E) frequencies, multiply the percentage by 600. The expected frequencies are:

Number of Televisions	Percent	Expected Frequency
0	10	$(0.10) \cdot (600) = 60$
1	16	$(0.16) \cdot (600) = 96$
2	55	$(0.55) \cdot (600) = 330$
3	11	$(0.11) \cdot (600) = 66$
over 3	8	$(0.08) \cdot (600) = 48$

Table 11.5

Therefore, the expected frequencies are 60, 96, 330, 66, and 48. In the TI calculators, you can let the calculator do the math. For example, instead of 60, enter .10*600.

H_o: The "number of televisions" distribution of far western United States families is the same as the "number of televisions" distribution of the American population.

H_a: The "number of televisions" distribution of far western United States families is different from the "number of televisions" distribution of the American population.

Distribution for the test: χ^2_4 where $df = \textit{(the number of cells)} - 1 = 5 - 1 = 4$.

NOTE: $df \neq 600 - 1$

Calculate the test statistic: $\chi^2 = 29.65$

Graph:

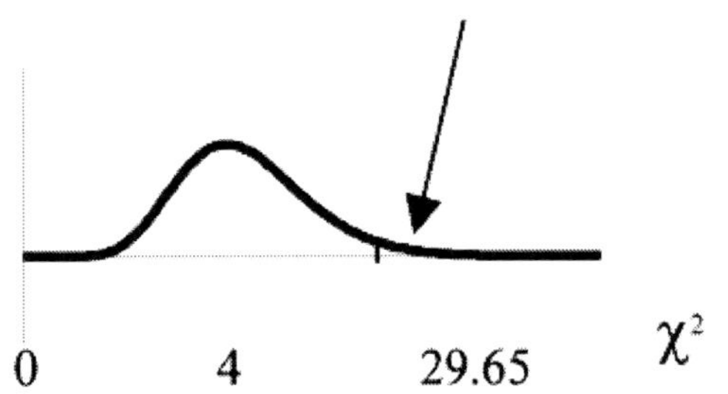

Probability statement: $p\text{-}value = P\left(\chi^2 > 29.65\right) = 0.000006$.

Compare α and the p-value:

- $\alpha = 0.01$
- $p\text{-}value = 0.000006$

So, $\alpha > p\text{-}value$.

Make a decision: Since $\alpha > p\text{-}value$, reject H_o.

This means you reject the belief that the distribution for the far western states is the same as that of the American population as a whole.

Conclusion: At the 1% significance level, from the data, there is sufficient evidence to conclude that the "number of televisions" distribution for the far western United States is different from the "number of televisions" distribution for the American population as a whole.

NOTE: TI 83+ and some TI-84 calculators: Press `STAT` and `ENTER`. Make sure to clear lists `L1`, `L2`, and `L3` if they have data in them (see the note at the end of Example 11-2). Into `L1`, put the observed frequencies `66, 119, 349, 60, 15`. Into `L2`, put the expected frequencies `.10*600, .16*600, .55*600, .11*600, .08*600`. Arrow over to list `L3` and up to the name area `"L3"`. Enter `(L1-L2)^2/L2` and `ENTER`. Press `2nd QUIT`. Press `2nd LIST` and arrow over to `MATH`. Press `5`. You should see `"sum" (Enter L3)`. Rounded to 2 decimal places, you should see `29.65`. Press `2nd DISTR`. Press `7` or Arrow down to `7:χ2cdf` and press `ENTER`. Enter `(29.65,1E99,4)`. Rounded to 4 places, you should see `5.77E-6 = .000006` (rounded to 6 decimal places) which is the p-value.

Example 11.4
Suppose you flip two coins 100 times. The results are 20 HH, 27 HT, 30 TH, and 23 TT. Are the coins fair? Test at a 5% significance level.

Solution
This problem can be set up as a goodness-of-fit problem. The sample space for flipping two fair coins is {HH, HT, TH, TT}. Out of 100 flips, you would expect 25 HH, 25 HT, 25 TH, and 25 TT. This is the expected distribution. The question, "Are the coins fair?" is the same as saying, "Does the distribution of the coins (20 HH, 27 HT, 30 TH, 23 TT) fit the expected distribution?"

Random Variable: Let X = the number of heads in one flip of the two coins. X takes on the value 0, 1, 2. (There are 0, 1, or 2 heads in the flip of 2 coins.) Therefore, the **number of cells is 3**. Since X = the number of heads, the observed frequencies are 20 (for 2 heads), 57 (for 1 head), and 23 (for 0 heads or both tails). The expected frequencies are 25 (for 2 heads), 50 (for 1 head), and 25 (for 0 heads or both tails). This test is right-tailed.

H_o: The coins are fair.

H_a: The coins are not fair.

Distribution for the test: χ^2_2 where $df = 3 - 1 = 2$.

Calculate the test statistic: $\chi^2 = 2.14$

Graph:

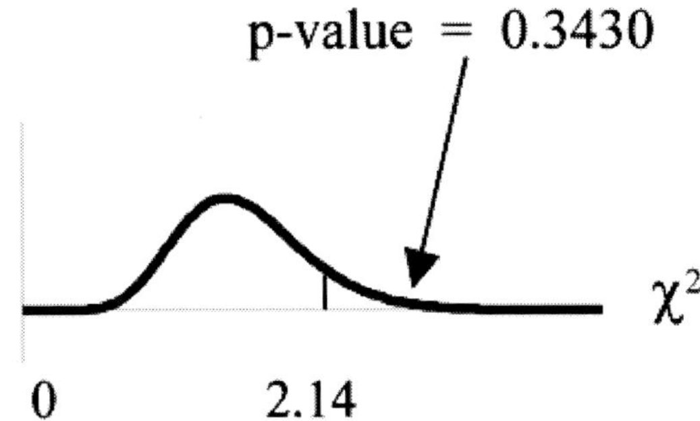

Probability statement: $p\text{-}value = P\left(\chi^2 > 2.14\right) = 0.3430$

Compare α **and the p-value:**

- $\alpha = 0.05$
- $p\text{-}value = 0.3430$

So, $\alpha < p\text{-}value$.

Make a decision: Since $\alpha < p\text{-}value$, do not reject H_o.

Conclusion: The coins are fair.

NOTE: TI-83+ and some TI- 84 calculators: Press `STAT` and `ENTER`. Make sure you clear lists `L1`, `L2`, and `L3` if they have data in them. Into `L1`, put the observed frequencies `20, 57, 23`. Into `L2`, put the expected frequencies `25, 50, 25`. Arrow over to list `L3` and up to the name area `"L3"`. Enter `(L1-L2)^2/L2` and `ENTER`. Press `2nd QUIT`. Press `2nd LIST` and arrow over to `MATH`. Press `5`. You should see `"sum"`.`Enter L3`. Rounded to 2 decimal places, you should see `2.14`. Press `2nd DISTR`. Arrow down to `7:χ2cdf` (or press `7`). Press `ENTER`. Enter `2.14,1E99,2)`. Rounded to 4 places, you should see `.3430` which is the p-value.

NOTE: For the newer TI-84 calculators, check `STAT TESTS` to see if you have `Chi2 GOF`. If you do, see the calculator instructions (a NOTE) before Example 11-3

11.5 Test of Independence[5]

Tests of independence involve using a **contingency table** of observed (data) values. You first saw a contingency table when you studied probability in the Probability Topics (Section 3.1) chapter.

The test statistic for a test of independence is similar to that of a goodness-of-fit test:

$$\sum_{(i \cdot j)} \frac{(O-E)^2}{E} \tag{11.2}$$

where:

- O = observed values
- E = expected values
- i = the number of rows in the table
- j = the number of columns in the table

There are $i \cdot j$ terms of the form $\frac{(O-E)^2}{E}$.

A test of independence determines whether two factors are independent or not. You first encountered the term independence in Chapter 3. As a review, consider the following example.

Example 11.5

Suppose A = a speeding violation in the last year and B = a car phone user. If A and B are independent then $P(A\ AND\ B) = P(A)\,P(B)$. $A\ AND\ B$ is the event that a driver received a speeding violation last year and is also a car phone user. Suppose, in a study of drivers who received speeding violations in the last year and who use car phones, that 755 people were surveyed. Out of the 755, 70 had a speeding violation and 685 did not; 305 were car phone users and 450 were not.

Let y = expected number of car phone users who received speeding violations.

If A and B are independent, then $P(A\ AND\ B) = P(A)\,P(B)$. By substitution,

$\frac{y}{755} = \frac{70}{755} \cdot \frac{305}{755}$

Solve for $y : y = \frac{70 \cdot 305}{755} = 28.3$

About 28 people from the sample are expected to be car phone users and to receive speeding violations.

In a test of independence, we state the null and alternate hypotheses in words. Since the contingency table consists of **two factors**, the null hypothesis states that the factors are **independent** and the alternate hypothesis states that they are **not independent (dependent)**. If we do a test of independence using the example above, then the null hypothesis is:

H_o: Being a car phone user and receiving a speeding violation are independent events.

If the null hypothesis were true, we would expect about 28 people to be car phone users and to receive a speeding violation.

The test of independence is always right-tailed because of the calculation of the test statistic. If the expected and observed values are not close together, then the test statistic is very large and way out in the right tail of the chi-square curve, like goodness-of-fit.

[5]This content is available online at <http://cnx.org/content/m17191/1.10/>.

The degrees of freedom for the test of independence are:

df = (*number of columns* - 1)(*number of rows* - 1)

The following formula calculates the **expected number** (E):

$$E = \frac{(row\ total)(column\ total)}{total\ number\ surveyed}$$

Example 11.6

In a volunteer group, adults 21 and older volunteer from one to nine hours each week to spend time with a disabled senior citizen. The program recruits among community college students, four-year college students, and nonstudents. The following table is a **sample** of the adult volunteers and the number of hours they volunteer per week.

Number of Hours Worked Per Week by Volunteer Type (Observed)

Type of Volunteer	1-3 Hours	4-6 Hours	7-9 Hours	Row Total
Community College Students	111	96	48	255
Four-Year College Students	96	133	61	290
Nonstudents	91	150	53	294
Column Total	298	379	162	839

Table 11.6: The table contains **observed (O)** values (data).

Problem

Are the number of hours volunteered **independent** of the type of volunteer?

Solution

The **observed table** and the question at the end of the problem, "Are the number of hours volunteered independent of the type of volunteer?" tell you this is a test of independence. The two factors are **number of hours volunteered** and **type of volunteer**. This test is always right-tailed.

H_0: The number of hours volunteered is **independent** of the type of volunteer.

H_a: The number of hours volunteered is **dependent** on the type of volunteer.

The expected table is:

Number of Hours Worked Per Week by Volunteer Type (Expected)

Type of Volunteer	1-3 Hours	4-6 Hours	7-9 Hours
Community College Students	90.57	115.19	49.24
Four-Year College Students	103.00	131.00	56.00
Nonstudents	104.42	132.81	56.77

Table 11.7: The table contains **expected** (E) values (data).

For example, the calculation for the expected frequency for the top left cell is

$$E = \frac{(row\ total)(column\ total)}{total\ number\ surveyed} = \frac{255 \cdot 298}{839} = 90.57$$

Calculate the test statistic: $\chi^2 = 12.99$ (calculator or computer)

Distribution for the test: χ^2_4

$df = (3\ columns - 1)\ (3\ rows - 1) = (2)\ (2) = 4$

Graph:

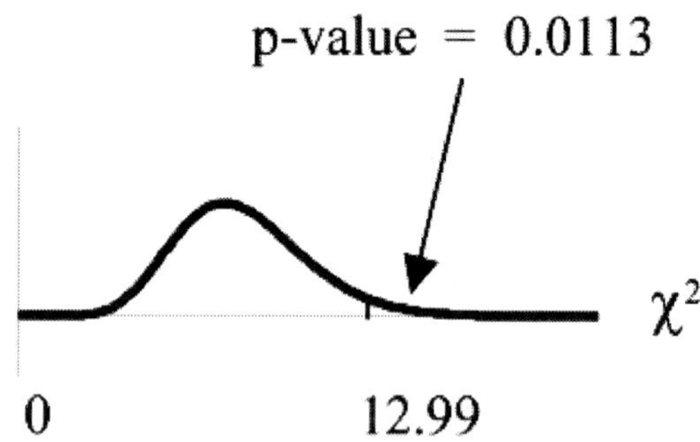

Probability statement: *p-value* $= P\left(\chi^2 > 12.99\right) = 0.0113$

Compare α **and the** *p-value*: Since no α is given, assume $\alpha = 0.05$. *p-value* $= 0.0113$. $\alpha >$ *p-value*.

Make a decision: Since $\alpha >$ *p-value*, reject H_o. This means that the factors are not independent.

Conclusion: At a 5% level of significance, from the data, there is sufficient evidence to conclude that the number of hours volunteered and the type of volunteer are dependent on one another.

For the above example, if there had been another type of volunteer, teenagers, what would the degrees of freedom be?

NOTE: Calculator instructions follow.

TI-83+ and TI-84 calculator: Press the `MATRX` key and arrow over to `EDIT`. Press `1:[A]`. Press `3 ENTER 3 ENTER`. Enter the table values by row from Example 11-6. Press `ENTER` after each. Press `2nd QUIT`. Press `STAT` and arrow over to `TESTS`. Arrow down to `C:`χ`2-TEST`. Press `ENTER`. You should see `Observed:[A] and Expected:[B]`. Arrow down to `Calculate`. Press `ENTER`. The test statistic is 12.9909 and the *p-value* $= 0.0113$. Do the procedure a second time but arrow down to `Draw` instead of `calculate`.

Example 11.7

De Anza College is interested in the relationship between anxiety level and the need to succeed in school. A random sample of 400 students took a test that measured anxiety level and need to succeed in school. The table shows the results. De Anza College wants to know if anxiety level and need to succeed in school are independent events.

Need to Succeed in School vs. Anxiety Level

Need to Succeed in School	High Anxiety	Med-high Anxiety	Medium Anxiety	Med-low Anxiety	Low Anxiety	Row Total
High Need	35	42	53	15	10	155
Medium Need	18	48	63	33	31	193
Low Need	4	5	11	15	17	52
Column Total	57	95	127	63	58	400

Table 11.8

Problem 1
How many high anxiety level students are expected to have a high need to succeed in school?

Solution
The column total for a high anxiety level is 57. The row total for high need to succeed in school is 155. The sample size or total surveyed is 400.

$E = \frac{(row\ total)(column\ total)}{total\ surveyed} = \frac{155 \cdot 57}{400} = 22.09$

The expected number of students who have a high anxiety level and a high need to succeed in school is about 22.

Problem 2
If the two variables are independent, how many students do you expect to have a low need to succeed in school and a med-low level of anxiety?

Solution
The column total for a med-low anxiety level is 63. The row total for a low need to succeed in school is 52. The sample size or total surveyed is 400.

Problem 3 *(Solution on p. 470.)*

a. $E = \frac{(row\ total)(column\ total)}{total\ surveyed} =$

b. The expected number of students who have a med-low anxiety level and a low need to succeed in school is about:

11.6 Test of a Single Variance (Optional)[6]

A test of a single variance assumes that the underlying distribution is **normal**. The null and alternate hypotheses are stated in terms of the **population variance** (or population standard deviation). The test

[6]This content is available online at <http://cnx.org/content/m17059/1.6/>.

statistic is:

$$\frac{(n-1)\cdot s^2}{\sigma^2} \tag{11.3}$$

where:

- n = the total number of data
- s^2 = sample variance
- σ^2 = population variance

You may think of s as the random variable in this test. The degrees of freedom are $df = n - 1$.

A test of a single variance may be right-tailed, left-tailed, or two-tailed.

The following example will show you how to set up the null and alternate hypotheses. The null and alternate hypotheses contain statements about the population variance.

Example 11.8
Math instructors are not only interested in how their students do on exams, on average, but how the exam scores vary. To many instructors, the variance (or standard deviation) may be more important than the average.

Suppose a math instructor believes that the standard deviation for his final exam is 5 points. One of his best students thinks otherwise. The student claims that the standard deviation is more than 5 points. If the student were to conduct a hypothesis test, what would the null and alternate hypotheses be?

Solution
Even though we are given the population standard deviation, we can set the test up using the population variance as follows.

- H_o: $\sigma^2 = 5^2$
- H_a: $\sigma^2 > 5^2$

Example 11.9
With individual lines at its various windows, a post office finds that the standard deviation for normally distributed waiting times for customers on Friday afternoon is 7.2 minutes. The post office experiments with a single main waiting line and finds that for a random sample of 25 customers, the waiting times for customers have a standard deviation of 3.5 minutes.

With a significance level of 5%, test the claim that **a single line causes lower variation among waiting times (shorter waiting times) for customers**.

Solution
Since the claim is that a single line causes lower variation, this is a test of a single variance. The parameter is the population variance, σ^2, or the population standard deviation, σ.

Random Variable: The sample standard deviation, s, is the random variable. Let s = standard deviation for the waiting times.

- H_o: $\sigma^2 = 7.2^2$
- H_a: $\sigma^2 < 7.2^2$

The word **"lower"** tells you this is a left-tailed test.

Distribution for the test: χ^2_{24}, where:

- n = the number of customers sampled
- $df = n - 1 = 25 - 1 = 24$

Calculate the test statistic:

$$\chi^2 = \frac{(n-1)\cdot s^2}{\sigma^2} = \frac{(25-1)\cdot 3.5^2}{7.2^2} = 5.67$$

where $n = 25$, $s = 3.5$, and $\sigma = 7.2$.

Graph:

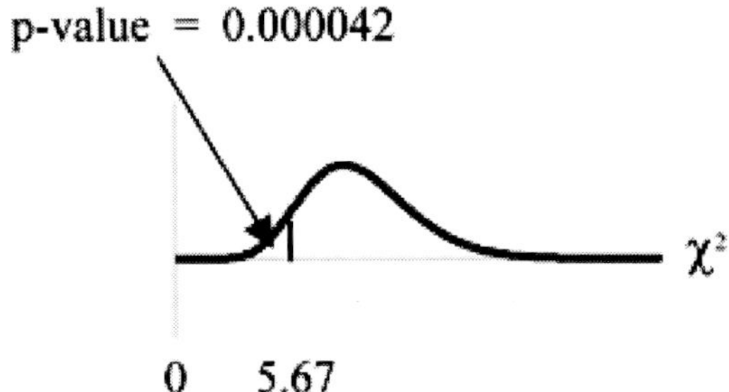

Probability statement: $p\text{-}value = P\left(\chi^2 < 5.67\right) = 0.000042$

Compare α **and the p-value:** $\alpha = 0.05$ $\quad p\text{-}value = 0.000042 \quad \alpha > p\text{-}value$

Make a decision: Since $\alpha > p\text{-}value$, reject H_o.

This means that you reject $\sigma^2 = 7.2^2$. In other words, you do not think the variation in waiting times is 7.2 minutes, but lower.

Conclusion: At a 5% level of significance, from the data, there is sufficient evidence to conclude that a single line causes a lower variation among the waiting times **or** with a single line, the customer waiting times vary less than 7.2 minutes.

TI-83+ and TI-84 calculators: In `2nd DISTR`, use `7:χ2cdf`. The syntax is `(lower, upper, df)` for the parameter list. For Example 11-9, `χ2cdf(-1E99,5.67,24)`. The *p-value* = 0.000042.

11.7 Summary of Formulas[7]

Formula 11.1: The Chi-square Probability Distribution
$\mu = df$ and $\sigma = \sqrt{2 \cdot df}$

Formula 11.2: Goodness-of-Fit Hypothesis Test

- Use goodness-of-fit to test whether a data set fits a particular probability distribution.
- The degrees of freedom are *number of cells or categories - 1.*
- The test statistic is $\sum_{n} \frac{(O-E)^2}{E}$, where O = observed values (data), E = expected values (from theory), and n = the number of different data cells or categories.
- The test is right-tailed.

Formula 11.3: Test of Independence

- Use the test of independence to test whether two factors are independent or not.
- The degrees of freedom are equal to *(number of columns - 1)(number of rows - 1).*
- The test statistic is $\sum_{(i \cdot j)} \frac{(O-E)^2}{E}$ where O = observed values, E = expected values, i = the number of rows in the table, and j = the number of columns in the table.
- The test is right-tailed.
- If the null hypothesis is true, the expected number $E = \frac{(row\ total)(column\ total)}{total\ surveyed}$.

Formula 11.4: Test of a Single Variance

- Use the test to determine variation.
- The degrees of freedom are the number of samples - 1.
- The test statistic is $\frac{(n-1) \cdot s^2}{\sigma^2}$, where n = the total number of data, s^2 = sample variance, and σ^2 = population variance.
- The test may be left, right, or two-tailed.

[7] This content is available online at <http://cnx.org/content/m17058/1.5/>.

11.8 Practice 1: Goodness-of-Fit Test[8]

11.8.1 Student Learning Outcomes

- The student will explore the properties of goodness-of-fit test data.

11.8.2 Given

The following data are real. The cumulative number of AIDS cases reported for Santa Clara County through December 31, 2003, is broken down by ethnicity as follows:

Ethnicity	**Number of Cases**
White	2032
Hispanic	897
African-American	372
Asian, Pacific Islander	168
Native American	20
	Total = 3489

Table 11.9

The percentage of each ethnic group in Santa Clara County is as follows:

Ethnicity	**Percentage of total county population**	**Number expected (round to 2 decimal places)**
White	47.79%	1667.39
Hispanic	24.15%	
African-American	3.55%	
Asian, Pacific Islander	24.21%	
Native American	0.29%	
	Total = 100%	

Table 11.10

11.8.3 Expected Results

If the ethnicity of AIDS victims followed the ethnicity of the total county population, fill in the expected number of cases per ethnic group.

[8]This content is available online at <http://cnx.org/content/m17054/1.8/>.

11.8.4 Goodness-of-Fit Test

Perform a goodness-of-fit test to determine whether the make-up of AIDS cases follows the ethnicity of the general population of Santa Clara County.

Exercise 11.8.1
H_o :

Exercise 11.8.2
H_a :

Exercise 11.8.3
Is this a right-tailed, left-tailed, or two-tailed test?

Exercise 11.8.4 *(Solution on p. 470.)*
degrees of freedom =

Exercise 11.8.5 *(Solution on p. 470.)*
Chi^2 test statistic =

Exercise 11.8.6 *(Solution on p. 470.)*
p-value =

Exercise 11.8.7
Graph the situation. Label and scale the horizontal axis. Mark the mean and test statistic. Shade in the region corresponding to the p-value.

Let $\alpha = 0.05$

Decision:

Reason for the Decision:

Conclusion (write out in complete sentences):

11.8.5 Discussion Question

Exercise 11.8.8
Does it appear that the pattern of AIDS cases in Santa Clara County corresponds to the distribution of ethnic groups in this county? Why or why not?

11.9 Practice 2: Contingency Tables[9]

11.9.1 Student Learning Outcomes

- The student will explore the properties of contingency tables.

Conduct a hypothesis test to determine if smoking level and ethnicity are independent.

11.9.2 Collect the Data

Copy the data provided in Probability Topics Practice 2: Calculating Probabilities into the table below.

Smoking Levels by Ethnicity (Observed)

Smoking Level Per Day	African American	Native Hawaiian	Latino	Japanese Americans	White	TOTALS
1-10						
11-20						
21-30						
31+						
TOTALS						

Table 11.11

11.9.3 Hypothesis

State the hypotheses.

- H_o :
- H_a :

11.9.4 Expected Values

Enter expected values in the above below. Round to two decimal places.

11.9.5 Analyze the Data

Calculate the following values:

Exercise 11.9.1 *(Solution on p. 470.)*
Degrees of freedom =

Exercise 11.9.2 *(Solution on p. 470.)*
Chi^2 test statistic =

Exercise 11.9.3 *(Solution on p. 470.)*
p-value =

Exercise 11.9.4 *(Solution on p. 470.)*
Is this a right-tailed, left-tailed, or two-tailed test? Explain why.

[9]This content is available online at <http://cnx.org/content/m17056/1.10/>.

11.9.6 Graph the Data

Exercise 11.9.5

Graph the situation. Label and scale the horizontal axis. Mark the mean and test statistic. Shade in the region corresponding to the p-value.

11.9.7 Conclusions

State the decision and conclusion (in a complete sentence) for the following preconceived levels of α .

Exercise 11.9.6 *(Solution on p. 470.)*

$\alpha = 0.05$

a. Decision:
b. Reason for the decision:
c. Conclusion (write out in a complete sentence):

Exercise 11.9.7

$\alpha = 0.01$

a. Decision:
b. Reason for the decision:
c. Conclusion (write out in a complete sentence):

11.10 Practice 3: Test of a Single Variance[10]

11.10.1 Student Learning Outcomes

- The student will explore the properties of data with a test of a single variance.

11.10.2 Given

Suppose an airline claims that its flights are consistently on time with an average delay of at most 15 minutes. It claims that the average delay is so consistent that the variance is no more than 150 minutes. Doubting the consistency part of the claim, a disgruntled traveler calculates the delays for his next 25 flights. The average delay for those 25 flights is 22 minutes with a standard deviation of 15 minutes.

11.10.3 Sample Variance

Exercise 11.10.1
Is the traveler disputing the claim about the average or about the variance?

Exercise 11.10.2 ***(Solution on p. 470.)***
A sample standard deviation of 15 minutes is the same as a sample variance of __________ minutes.

Exercise 11.10.3
Is this a right-tailed, left-tailed, or two-tailed test?

11.10.4 Hypothesis Test

Perform a hypothesis test on the consistency part of the claim.

Exercise 11.10.4
H_o :

Exercise 11.10.5
H_a :

Exercise 11.10.6 ***(Solution on p. 470.)***
Degrees of freedom =

Exercise 11.10.7 ***(Solution on p. 470.)***
Chi^2 test statistic =

Exercise 11.10.8 ***(Solution on p. 470.)***
p-value =

Exercise 11.10.9
Graph the situation. Label and scale the horizontal axis. Mark the mean and test statistic. Shade the p-value.

---

[10]This content is available online at <http://cnx.org/content/m17053/1.7/>.

Exercise 11.10.10
Let $\alpha = 0.05$

Decision:

Conclusion (write out in a complete sentence):

11.10.5 Discussion Questions

Exercise 11.10.11
How did you know to test the variance instead of the mean?

Exercise 11.10.12
If an additional test were done on the claim of the average delay, which distribution would you use?

Exercise 11.10.13
If an additional test was done on the claim of the average delay, but 45 flights were surveyed, which distribution would you use?

11.11 Homework[11]

Exercise 11.11.1

a. Explain why the "goodness of fit" test and the "test for independence" are generally right tailed tests.
b. If you did a left-tailed test, what would you be testing?

11.11.1 Word Problems

For each word problem, use a solution sheet to solve the hypothesis test problem. Go to The Table of Contents 14. Appendix for the solution sheet. Round expected frequency to two decimal places.

Exercise 11.11.2
A 6-sided die is rolled 120 times. Fill in the expected frequency column. Then, conduct a hypothesis test to determine if the die is fair. The data below are the result of the 120 rolls.

Face Value	Frequency	Expected Frequency
1	15	
2	29	
3	16	
4	15	
5	30	
6	15	

Table 11.12

Exercise 11.11.3 ***(Solution on p. 470.)***
The marital status distribution of the U.S. male population, age 15 and older, is as shown below. (*Source: U.S. Census Bureau, Current Population Reports*)

Marital Status	Percent	Expected Frequency
never married	31.3	
married	56.1	
widowed	2.5	
divorced/separated	10.1	

Table 11.13

Suppose that a random sample of 400 U.S. young adult males, 18 – 24 years old, yielded the following frequency distribution. We are interested in whether this age group of males fits the distribution of the U.S. adult population. Calculate the frequency one would expect when surveying 400 people. Fill in the above table, rounding to two decimal places.

[11]This content is available online at <http://cnx.org/content/m17028/1.10/>.

Marital Status	Frequency
never married	140
married	238
widowed	2
divorced/separated	20

Table 11.14

The next two questions refer to the following information: The real data below are from the California Reinvestment Committee and the California Economic Census. The data concern the percent of loans made by the Small Business Administration for Santa Clara County in recent years. (*Source: San Jose Mercury News*)

Ethnic Group	Percent of Loans	Percent of Population	Percent of Businesses Owned
Asian	22.48	16.79	12.17
Black	1.15	3.51	1.61
Latino	6.19	21.00	6.51
White	66.97	58.09	79.70

Table 11.15

Exercise 11.11.4

Perform a goodness-of-fit test to determine whether the percent of businesses owned in Santa Clara County fits the percent of the population, based on ethnicity.

Exercise 11.11.5 ***(Solution on p. 470.)***

Perform a goodness-of-fit test to determine whether the percent of loans fits the percent of the businesses owned in Santa Clara County, based on ethnicity.

Exercise 11.11.6

The City of South Lake Tahoe has an Asian population of 1419 people, out of a total population of 23,609 (*Source: U.S. Census Bureau, Census 2000*). Conduct a goodness of fit test to determine if the self-reported sub-groups of Asians are evenly distributed.

Race	Frequency	Expected Frequency
Asian Indian	131	
Chinese	118	
Filipino	1045	
Japanese	80	
Korean	12	
Vietnamese	9	
Other	24	

Table 11.16

Exercise 11.11.7 *(Solution on p. 471.)*

Long Beach is a city in Los Angeles County (L.A.C). The population of Long Beach is 461,522; the population of L.A.C. is 9,519,338 (*Source: U.S. Census Bureau, Census 2000*). Conduct a goodness of fit test to determine if the racial demographics of Long Beach fit that of L.A.C.

Race	Percent, L.A.C.	Expected #, L.B.	Actual #, L.B.
American Indian and Alaska Native	0.8	3692	3,881
Asian	11.9		55,591
Black or African American	9.8		68,618
Native Hawaiian and Other Pacific Islander	0.3		5,605
White, including Hispanic/Latino	48.7		208,410
Other	23.5		95,107
Two or more races	5.0		24,310

Table 11.17

Exercise 11.11.8

UCLA conducted a survey of more than 263,000 college freshmen from 385 colleges in fall 2005. The results of student expected majors by gender were reported in *The Chronicle of Higher Education (2/2/06).* Conduct a goodness of fit test to determine if the male distribution fits the female distribution.

Major	Women	Men
Arts & Humanities	14.0%	11.4%
Biological Sciences	8.4%	6.7%
Business	13.1%	22.7%
Education	13.0%	5.8%
Engineering	2.6%	15.6%
Physical Sciences	2.6%	3.6%
Professional	18.9%	9.3%
Social Sciences	13.0%	7.6%
Technical	0.4%	1.8%
Other	5.8%	8.2%
Undecided	8.0%	6.6%

Table 11.18

Exercise 11.11.9 *(Solution on p. 471.)*

A recent debate about where in the United States skiers believe the skiing is best prompted the following survey. Test to see if the best ski area is independent of the level of the skier.

U.S. Ski Area	Beginner	Intermediate	Advanced
Tahoe	20	30	40
Utah	10	30	60
Colorado	10	40	50

Table 11.19

Exercise 11.11.10

Car manufacturers are interested in whether there is a relationship between the size of car an individual drives and the number of people in the driver's family (that is, whether car size and family size are independent). To test this, suppose that 800 car owners were randomly surveyed with the following results. Conduct a test for independence.

Family Size	Sub & Compact	Mid-size	Full-size	Van & Truck
1	20	35	40	35
2	20	50	70	80
3 - 4	20	50	100	90
5+	20	30	70	70

Table 11.20

Exercise 11.11.11 *(Solution on p. 471.)*

College students may be interested in whether or not their majors have any effect on starting salaries after graduation. Suppose that 300 recent graduates were surveyed as to their majors in college and their starting salaries after graduation. Below are the data. Conduct a test for independence.

Major	$<$ $30,000	$30,000 - $39,999	$40,000 +
English	5	20	5
Engineering	10	30	60
Nursing	10	15	15
Business	10	20	30
Psychology	20	30	20

Table 11.21

Exercise 11.11.12

Some travel agents claim that honeymoon hot spots vary according to age of the bride and groom. Suppose that 280 East Coast recent brides were interviewed as to where they spent their honeymoons. The information is given below. Conduct a test for independence.

Location	20 - 29	30 - 39	40 - 49	50 and over
Niagara Falls	15	25	25	20
Poconos	15	25	25	10
Europe	10	25	15	5
Virgin Islands	20	25	15	5

Table 11.22

Exercise 11.11.13 *(Solution on p. 471.)*

A manager of a sports club keeps information concerning the main sport in which members participate and their ages. To test whether there is a relationship between the age of a member and his or her choice of sport, 643 members of the sports club are randomly selected. Conduct a test for independence.

Sport	18 - 25	26 - 30	31 - 40	41 and over
racquetball	42	58	30	46
tennis	58	76	38	65
swimming	72	60	65	33

Table 11.23

Exercise 11.11.14

A major food manufacturer is concerned that the sales for its skinny French fries have been decreasing. As a part of a feasibility study, the company conducts research into the types of fries sold across the country to determine if the type of fries sold is independent of the area of the country. The results of the study are below. Conduct a test for independence.

Type of Fries	Northeast	South	Central	West
skinny fries	70	50	20	25
curly fries	100	60	15	30
steak fries	20	40	10	10

Table 11.24

Exercise 11.11.15 *(Solution on p. 471.)*

According to Dan Lenard, an independent insurance agent in the Buffalo, N.Y. area, the following is a breakdown of the amount of life insurance purchased by males in the following age groups. He is interested in whether the age of the male and the amount of life insurance purchased are independent events. Conduct a test for independence.

Age of Males	None	$50,000 - $100,000	$100,001 - $150,000	$150,001 - $200,000	$200,000 +
20 - 29	40	15	40	0	5
30 - 39	35	5	20	20	10
40 - 49	20	0	30	0	30
50 +	40	30	15	15	10

Table 11.25

Exercise 11.11.16

Suppose that 600 thirty–year–olds were surveyed to determine whether or not there is a relationship between the level of education an individual has and salary. Conduct a test for independence.

Annual Salary	Not a high school grad.	High school graduate	College graduate	Masters or doctorate
< $30,000	15	25	10	5
$30,000 - $40,000	20	40	70	30
$40,000 - $50,000	10	20	40	55
$50,000 - $60,000	5	10	20	60
$60,000 +	0	5	10	150

Table 11.26

Exercise 11.11.17 *(Solution on p. 471.)*

A plant manager is concerned her equipment may need recalibrating. It seems that the actual weight of the 15 oz. cereal boxes it fills has been fluctuating. The standard deviation should be at most $\frac{1}{2}$ oz. In order to determine if the machine needs to be recalibrated, 84 randomly selected boxes of cereal from the next day's production were weighed. The standard deviation of the 84 boxes was 0.54. Does the machine need to be recalibrated?

Exercise 11.11.18

Consumers may be interested in whether the cost of a particular calculator varies from store to store. Based on surveying 43 stores, which yielded a sample mean of $84 and a sample standard deviation of $12, test the claim that the standard deviation is greater than $15.

Exercise 11.11.19 *(Solution on p. 471.)*

Isabella, an accomplished **Bay to Breakers** runner, claims that the standard deviation for her time to run the 7 ½ mile race is at most 3 minutes. To test her claim, Rupinder looks up 5 of her race times. They are 55 minutes, 61 minutes, 58 minutes, 63 minutes, and 57 minutes.

Exercise 11.11.20

Airline companies are interested in the consistency of the number of babies on each flight, so that they have adequate safety equipment. They are also interested in the variation of the number of babies. Suppose that an airline executive believes the average number of babies on flights is 6 with a variance of 9 at most. The airline conducts a survey. The results of the 18 flights surveyed give a sample average of 6.4 with a sample standard deviation of 3.9. Conduct a hypothesis test of the airline executive's belief.

Exercise 11.11.21 *(Solution on p. 472.)*

According to the *U.S. Bureau of the Census, United Nations,* in 1994 the number of births per woman in China was 1.8. This fertility rate has been attributed to the law passed in 1979 restricting births to one per woman. Suppose that a group of students studied whether or not the standard deviation of births per woman was greater than 0.75. They asked 50 women across China the number of births they had. Below are the results. Does the students' survey indicate that the standard deviation is greater than 0.75?

# of births	Frequency
0	5
1	30
2	10
3	5

Table 11.27

Exercise 11.11.22
According to an avid aquariest, the average number of fish in a 20–gallon tank is 10, with a standard deviation of 2. His friend, also an aquariest, does not believe that the standard deviation is 2. She counts the number of fish in 15 other 20–gallon tanks. Based on the results that follow, do you think that the standard deviation is different from 2? Data: 11; 10; 9; 10; 10; 11; 11; 10; 12; 9; 7; 9; 11; 10; 11

Exercise 11.11.23 ***(Solution on p. 472.)***
The manager of "Frenchies" is concerned that patrons are not consistently receiving the same amount of French fries with each order. The chef claims that the standard deviation for a 10–ounce order of fries is at most 1.5 oz., but the manager thinks that it may be higher. He randomly weighs 49 orders of fries, which yields: mean of 11 oz., standard deviation of 2 oz.

11.11.2 Try these true/false questions.

Exercise 11.11.24 ***(Solution on p. 472.)***
As the degrees of freedom increase, the graph of the chi-square distribution looks more and more symmetrical.

Exercise 11.11.25 ***(Solution on p. 472.)***
The standard deviation of the chi-square distribution is twice the mean.

Exercise 11.11.26 ***(Solution on p. 472.)***
The mean and the median of the chi-square distribution are the same if $df = 24$.

Exercise 11.11.27 ***(Solution on p. 472.)***
In a Goodness-of-Fit test, the expected values are the values we would expect if the null hypothesis were true.

Exercise 11.11.28 ***(Solution on p. 472.)***
In general, if the observed values and expected values of a Goodness-of-Fit test are not close together, then the test statistic can get very large and on a graph will be way out in the right tail.

Exercise 11.11.29 ***(Solution on p. 472.)***
The degrees of freedom for a Test for Independence are equal to the sample size minus 1.

Exercise 11.11.30 ***(Solution on p. 472.)***
Use a Goodness-of-Fit test to determine if high school principals believe that students are absent equally during the week or not.

Exercise 11.11.31 *(Solution on p. 472.)*

The Test for Independence uses tables of observed and expected data values.

Exercise 11.11.32 *(Solution on p. 472.)*

The test to use when determining if the college or university a student chooses to attend is related to his/her socioeconomic status is a Test for Independence.

Exercise 11.11.33 *(Solution on p. 472.)*

The test to use to determine if a coin is fair is a Goodness-of-Fit test.

Exercise 11.11.34 *(Solution on p. 472.)*

In a Test of Independence, the expected number is equal to the row total multiplied by the column total divided by the total surveyed.

Exercise 11.11.35 *(Solution on p. 472.)*

In a Goodness-of Fit test, if the p-value is 0.0113, in general, do not reject the null hypothesis.

Exercise 11.11.36 *(Solution on p. 472.)*

For a Chi-Square distribution with degrees of freedom of 17, the probability that a value is greater than 20 is 0.7258.

Exercise 11.11.37 *(Solution on p. 472.)*

If $df = 2$, the chi-square distribution has a shape that reminds us of the exponential.

11.12 Review[12]

The next two questions refer to the following real study:

A recent survey of U.S. teenage pregnancy was answered by 720 girls, age 12 - 19. 6% of the girls surveyed said they have been pregnant. (*Parade Magazine*) We are interested in the true proportion of U.S. girls, age 12 - 19, who have been pregnant.

Exercise 11.12.1 *(Solution on p. 472.)*

Find the 95% confidence interval for the true proportion of U.S. girls, age 12 - 19, who have been pregnant.

Exercise 11.12.2 *(Solution on p. 472.)*

The report also stated that the results of the survey are accurate to within $\pm$ 3.7% at the 95% confidence level. Suppose that a new study is to be done. It is desired to be accurate to within 2% of the 95% confidence level. What will happen to the minimum number that should be surveyed?

Exercise 11.12.3

Given: $X \sim Exp\left(\frac{1}{3}\right)$. Sketch the graph that depicts: $P(X > 1)$.

The next four questions refer to the following information:

Suppose that the time that owners keep their cars (purchased new) is normally distributed with a mean of 7 years and a standard deviation of 2 years. We are interested in how long an individual keeps his car (purchased new). Our population is people who buy their cars new.

Exercise 11.12.4 *(Solution on p. 472.)*

60% of individuals keep their cars **at most** how many years?

Exercise 11.12.5 *(Solution on p. 473.)*

Suppose that we randomly survey one person. Find the probability that person keeps his/her car **less than** 2.5 years.

Exercise 11.12.6 *(Solution on p. 473.)*

If we are to pick individuals 10 at a time, find the distribution for the **average** car length ownership.

Exercise 11.12.7 *(Solution on p. 473.)*

If we are to pick 10 individuals, find the probability that the **sum** of their ownership time is more than 55 years.

Exercise 11.12.8 *(Solution on p. 473.)*

For which distribution is the median not equal to the mean?

A. Uniform
B. Exponential
C. Normal
D. Student-t

Exercise 11.12.9 *(Solution on p. 473.)*

Compare the standard normal distribution to the student-t distribution, centered at 0. Explain which of the following are true and which are false.

a. As the number surveyed increases, the area to the left of -1 for the student-t distribution approaches the area for the standard normal distribution.
b. As the number surveyed increases, the area to the left of -1 for the standard normal distribution approaches the area for the student-t distribution.

[12]This content is available online at <http://cnx.org/content/m17057/1.8/>.

c. As the degrees of freedom decrease, the graph of the student-t distribution looks more like the graph of the standard normal distribution.
d. If the number surveyed is less than 30, the normal distribution should never be used.

The next five questions refer to the following information:

We are interested in the checking account balance of a twenty-year-old college student. We randomly survey 16 twenty-year-old college students. We obtain a sample mean of $640 and a sample standard deviation of $150. Let X = checking account balance of an individual twenty year old college student.

Exercise 11.12.10
Explain why we cannot determine the distribution of X.

Exercise 11.12.11 *(Solution on p. 473.)*
If you were to create a confidence interval or perform a hypothesis test for the population average checking account balance of 20-year old college students, what distribution would you use?

Exercise 11.12.12 *(Solution on p. 473.)*
Find the 95% confidence interval for the true average checking account balance of a twenty-year-old college student.

Exercise 11.12.13 *(Solution on p. 473.)*
What type of data is the balance of the checking account considered to be?

Exercise 11.12.14 *(Solution on p. 473.)*
What type of data is the number of 20 year olds considered to be?

Exercise 11.12.15 *(Solution on p. 473.)*
On average, a busy emergency room gets a patient with a shotgun wound about once per week. We are interested in the number of patients with a shotgun wound the emergency room gets per 28 days.

a. Define the random variable X.
b. State the distribution for X.
c. Find the probability that the emergency room gets no patients with shotgun wounds in the next 28 days.

The next two questions refer to the following information:

The probability that a certain slot machine will pay back money when a quarter is inserted is 0.30 . Assume that each play of the slot machine is independent from each other. A person puts in 15 quarters for 15 plays.

Exercise 11.12.16 *(Solution on p. 473.)*
Is the expected number of plays of the slot machine that will pay back money greater than, less than or the same as the median? Explain your answer.

Exercise 11.12.17 *(Solution on p. 473.)*
Is it likely that exactly 8 of the 15 plays would pay back money? Justify your answer numerically.

Exercise 11.12.18 *(Solution on p. 473.)*
A game is played with the following rules:

- it costs $10 to enter
- a fair coin is tossed 4 times
- if you do not get 4 heads or 4 tails, you lose your $10
- if you get 4 heads or 4 tails, you get back your $10, plus $30 more

Over the long run of playing this game, what are your expected earnings?

Exercise 11.12.19 *(Solution on p. 473.)*

- The average grade on a math exam in Rachel's class was 74, with a standard deviation of 5. Rachel earned an 80.
- The average grade on a math exam in Becca's class was 47, with a standard deviation of 2. Becca earned a 51.
- The average grade on a math exam in Matt's class was 70, with a standard deviation of 8. Matt earned an 83.

Find whose score was the best, compared to his or her own class. Justify your answer numerically.

The next two questions refer to the following information:

70 compulsive gamblers were asked the number of days they go to casinos per week. The results are given in the following graph:

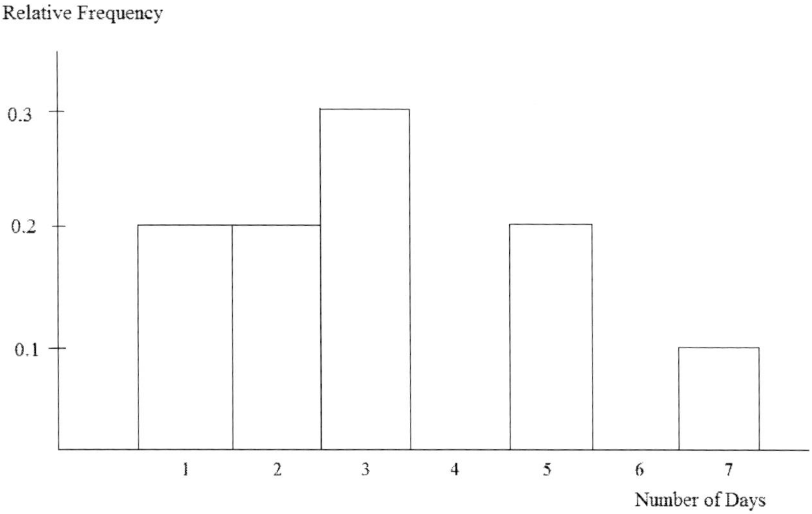

Figure 11.3

Exercise 11.12.20 *(Solution on p. 473.)*

Find the number of responses that were "5".

Exercise 11.12.21 *(Solution on p. 473.)*

Find the mean, standard deviation, all four quartiles and IQR.

Exercise 11.12.22 *(Solution on p. 473.)*

Based upon research at De Anza College, it is believed that about 19% of the student population speaks a language other than English at home.

Suppose that a study was done this year to see if that percent has decreased. Ninety-eight students were randomly surveyed with the following results. Fourteen said that they speak a language other than English at home.

a. State an appropriate **null** hypothesis.
b. State an appropriate **alternate** hypothesis.

c. Define the Random Variable, P'.
d. Calculate the test statistic.
e. Calculate the p-value.
f. At the 5% level of decision, what is your decision about the null hypothesis?
g. What is the Type I error?
h. What is the Type II error?

Exercise 11.12.23

Assume that you are an emergency paramedic called in to rescue victims of an accident. You need to help a patient who is bleeding profusely. The patient is also considered to be a high risk for contracting AIDS. Assume that the null hypothesis is that the patient does **not** have the HIV virus. What is a Type I error?

Exercise 11.12.24 *(Solution on p. 473.)*

It is often said that Californians are more casual than the rest of Americans. Suppose that a survey was done to see if the proportion of Californian professionals that wear jeans to work is greater than the proportion of non-Californian professionals. Fifty of each was surveyed with the following results. 10 Californians wear jeans to work and 4 non-Californians wear jeans to work.

- C = Californian professional
- NC = non-Californian professional

a. State appropriate **null** and **alternate** hypotheses.
b. Define the Random Variable.
c. Calculate the test statistic and p-value.
d. At the 5% level of decision, do you accept or reject the null hypothesis?
e. What is the Type I error?
f. What is the Type II error?

The next two questions refer to the following information:

A group of Statistics students have developed a technique that they feel will lower their anxiety level on statistics exams. They measured their anxiety level at the start of the quarter and again at the end of the quarter. Recorded is the paired data in that order: (1000, 900); (1200, 1050); (600, 700); (1300, 1100); (1000, 900); (900, 900).

Exercise 11.12.25 *(Solution on p. 474.)*

This is a test of (pick the best answer):

A. large samples, independent means
B. small samples, independent means
C. dependent means

Exercise 11.12.26 *(Solution on p. 474.)*

State the distribution to use for the test.

11.13 Lab 1: Chi-Square Goodness-of-Fit[13]

Class Time:

Names:

11.13.1 Student Learning Outcome:

- The student will evaluate data collected to determine if they fit either the uniform or exponential distributions.

11.13.2 Collect the Data

Go to your local supermarket. Ask 30 people as they leave for the total amount on their grocery receipts. (Or, ask 3 cashiers for the last 10 amounts. Be sure to include the express lane, if it is open.)

1. Record the values.

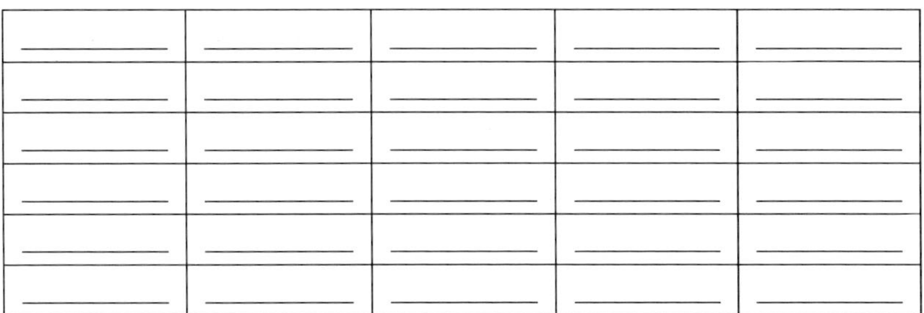

Table 11.28

2. Construct a histogram of the data. Make 5 - 6 intervals. Sketch the graph using a ruler and pencil. Scale the axes.

[13]This content is available online at <http://cnx.org/content/m17049/1.8/>.

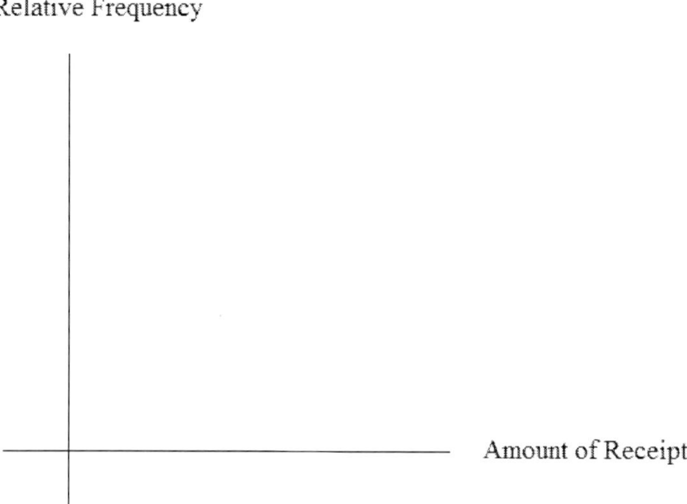

Figure 11.4

3. Calculate the following:
   a. $\overline{x} =$
   b. $s =$
   c. $s^2 =$

11.13.3 Uniform Distribution

Test to see if grocery receipts follow the uniform distribution.

1. Using your lowest and highest values, $X \sim U$ (_______,_______)
2. Divide the distribution above into fifths.
3. Calculate the following:
   a. Lowest value =
   b. 20th percentile =
   c. 40th percentile =
   d. 60th percentile =
   e. 80th percentile =
   f. Highest value =
4. For each fifth, count the observed number of receipts and record it. Then determine the expected number of receipts and record that.

Fifth	Observed	Expected
1st		
2nd		
3rd		
4th		
5th		

Table 11.29

5. H_o:
6. H_a:
7. What distribution should you use for a hypothesis test?
8. Why did you choose this distribution?
9. Calculate the test statistic.
10. Find the p-value.
11. Sketch a graph of the situation. Label and scale the x-axis. Shade the area corresponding to the p-value.

Figure 11.5

12. State your decision.
13. State your conclusion in a complete sentence.

11.13.4 Exponential Distribution

Test to see if grocery receipts follow the exponential distribution with decay parameter $\frac{1}{\bar{x}}$.

1. Using $\frac{1}{\bar{x}}$ as the decay parameter, $X \sim Exp$ (_______).
2. Calculate the following:
   - **a.** Lowest value =
   - **b.** First quartile =
   - **c.** 37th percentile =
   - **d.** Median =
   - **e.** 63rd percentile =
   - **f.** 3rd quartile =
   - **g.** Highest value =
3. For each cell, count the observed number of receipts and record it. Then determine the expected number of receipts and record that.

Cell	Observed	Expected
1st		
2nd		
3rd		
4th		
5th		
6th		

Table 11.30

4. H_o
5. H_a
6. What distribution should you use for a hypothesis test?
7. Why did you choose this distribution?
8. Calculate the test statistic.
9. Find the p-value.
10. Sketch a graph of the situation. Label and scale the x-axis. Shade the area corresponding to the p-value.

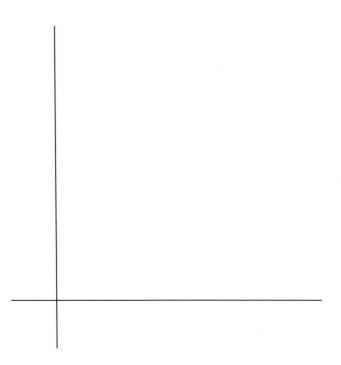

Figure 11.6

11. State your decision.
12. State your conclusion in a complete sentence.

11.13.5 Discussion Questions

1. Did your data fit either distribution? If so, which?
2. In general, do you think it's likely that data could fit more than one distribution? In complete sentences, explain why or why not.

11.14 Lab 2: Chi-Square Test for Independence[14]

Class Time:

Names:

11.14.1 Student Learning Outcome:

- The student will evaluate if there is a significant relationship between favorite type of snack and gender.

11.14.2 Collect the Data

1. Using your class as a sample, complete the following chart.

Favorite type of snack

	sweets (candy & baked goods)	ice cream	chips & pretzels	fruits & vegetables	Total
male					
female					
Total					

Table 11.31

2. Looking at the above chart, does it appear to you that there is dependence between gender and favorite type of snack food? Why or why not?

11.14.3 Hypothesis Test

Conduct a hypothesis test to determine if the factors are independent

1. H_0:
2. H_a:
3. What distribution should you use for a hypothesis test?
4. Why did you choose this distribution?
5. Calculate the test statistic.
6. Find the p-value.
7. Sketch a graph of the situation. Label and scale the x-axis. Shade the area corresponding to the p-value.

[14]This content is available online at <http://cnx.org/content/m17050/1.9/>.

Figure 11.7

8. State your decision.
9. State your conclusion in a complete sentence.

11.14.4 Discussion Questions

1. Is the conclusion of your study the same as or different from your answer to (I2) above?
2. Why do you think that occurred?

Solutions to Exercises in Chapter 11

Solution to Example 11.7, Problem 3 (p. 441)

- **a.** $E = \frac{\text{(row total)(column total)}}{\text{total surveyed}} = 8.19$
- **b.** 8

Solutions to Practice 1: Goodness-of-Fit Test

Solution to Exercise 11.8.4 (p. 446)
degrees of freedom = 4
Solution to Exercise 11.8.5 (p. 446)
951.69
Solution to Exercise 11.8.6 (p. 446)
0

Solutions to Practice 2: Contingency Tables

Solution to Exercise 11.9.1 (p. 447)
12
Solution to Exercise 11.9.2 (p. 447)
10301.8
Solution to Exercise 11.9.3 (p. 447)
0
Solution to Exercise 11.9.4 (p. 447)
right
Solution to Exercise 11.9.6 (p. 448)

- **a.** Reject the null hypothesis

Solutions to Practice 3: Test of a Single Variance

Solution to Exercise 11.10.2 (p. 449)
225
Solution to Exercise 11.10.6 (p. 449)
24
Solution to Exercise 11.10.7 (p. 449)
36
Solution to Exercise 11.10.8 (p. 449)
0.0549

Solutions to Homework

Solution to Exercise 11.11.3 (p. 451)

- **a.** The data fits the distribution
- **b.** The data does not fit the distribution
- **c.** 3
- **e.** 19.27
- **f.** 0.0002
- **h.** Decision: Reject Null; Conclusion: Data does not fit the distribution.

Solution to Exercise 11.11.5 (p. 452)

c. 3
e. 10.91
f. 0.0122
g. Decision: Reject null when $a = 0.05$; Conclusion: Percent of loans does not fit the distribution.
Decision: Do not reject null when $a = 0.01$; Conclusion Percent of loans fits the distribution.

Solution to Exercise 11.11.7 (p. 453)

c. 6
e. 27,876
f. 0
h. Decision: Reject null; Conclusion: L.B. does not fit L.A.C.

Solution to Exercise 11.11.9 (p. 453)

c. 4
e. 10.53
f. 0.0324
h. Decision: Reject null; Conclusion: Best ski area and level of skier are not independent.

Solution to Exercise 11.11.11 (p. 454)

c. 8
e. 33.55
f. 0
h. Decision: Reject null; Conclusion: Major and starting salary are not independent events.

Solution to Exercise 11.11.13 (p. 455)

c. 6
e. 25.21
f. 0.0003
h. Decision: Reject null

Solution to Exercise 11.11.15 (p. 455)

c. 12
e. 125.74
f. 0
h. Decision: Reject null

Solution to Exercise 11.11.17 (p. 456)

c. 83
d. 96.81
e. 0.1426
g. Decision: Do not reject null; Conclusion: The standard deviation is at most 0.5 oz.
h. It does not need to be calibrated

Solution to Exercise 11.11.19 (p. 456)

c. 4
d. 4.52
e. 0.3402
g. Decision: Do not reject null.
h. No

Solution to Exercise 11.11.21 (p. 456)

- **c.** 49
- **d.** 54.37
- **e.** 0.2774
- **g.** Decision: Do not reject null; Conclusion: The standard deviation is at most 0.75.
- **h.** No

Solution to Exercise 11.11.23 (p. 457)

- **a.** $\sigma^2 \leq (1.5)^2$
- **c.** 48
- **d.** 85.33
- **e.** 0.0007
- **g.** Decision: Reject null.
- **h.** Yes

Solution to Exercise 11.11.24 (p. 457)
True
Solution to Exercise 11.11.25 (p. 457)
False
Solution to Exercise 11.11.26 (p. 457)
False
Solution to Exercise 11.11.27 (p. 457)
True
Solution to Exercise 11.11.28 (p. 457)
True
Solution to Exercise 11.11.29 (p. 457)
False
Solution to Exercise 11.11.30 (p. 457)
True
Solution to Exercise 11.11.31 (p. 458)
True
Solution to Exercise 11.11.32 (p. 458)
True
Solution to Exercise 11.11.33 (p. 458)
True
Solution to Exercise 11.11.34 (p. 458)
True
Solution to Exercise 11.11.35 (p. 458)
False
Solution to Exercise 11.11.36 (p. 458)
False
Solution to Exercise 11.11.37 (p. 458)
True

Solutions to Review

Solution to Exercise 11.12.1 (p. 459)
$(0.0424, 0.0770)$
Solution to Exercise 11.12.2 (p. 459)
2401

Solution to Exercise 11.12.4 (p. 459)
7.5
Solution to Exercise 11.12.5 (p. 459)
0.0122
Solution to Exercise 11.12.6 (p. 459)
$N(7, 0.63)$
Solution to Exercise 11.12.7 (p. 459)
0.9911
Solution to Exercise 11.12.8 (p. 459)
B
Solution to Exercise 11.12.9 (p. 459)

- **a.** True
- **b.** False
- **c.** False
- **d.** False

Solution to Exercise 11.12.11 (p. 460)
student-t with $df = 15$
Solution to Exercise 11.12.12 (p. 460)
$(560.07, 719.93)$
Solution to Exercise 11.12.13 (p. 460)
quantitative - continuous
Solution to Exercise 11.12.14 (p. 460)
quantitative - discrete
Solution to Exercise 11.12.15 (p. 460)

- **b.** $P(4)$
- **c.** 0.0183

Solution to Exercise 11.12.16 (p. 460)
greater than
Solution to Exercise 11.12.17 (p. 460)
No; $P(X = 8) = 0.0348$
Solution to Exercise 11.12.18 (p. 460)
You will lose $5
Solution to Exercise 11.12.19 (p. 461)
Becca
Solution to Exercise 11.12.20 (p. 461)
14
Solution to Exercise 11.12.21 (p. 461)

- Mean = 3.2
- Quartiles = 1.85, 2, 3, and 5
- IQR = 3

Solution to Exercise 11.12.22 (p. 461)

- **d.** $z = -1.19$
- **e.** 0.1171
- **f.** Do not reject the null

Solution to Exercise 11.12.24 (p. 462)

- **c.** $z = 1.73$; $p = 0.0419$

d. Reject the null

Solution to Exercise 11.12.25 (p. 462)
C
Solution to Exercise 11.12.26 (p. 462)
t_5

Chapter 12

Linear Regression and Correlation

12.1 Linear Regression and Correlation[1]

12.1.1 Student Learning Objectives

By the end of this chapter, the student should be able to:

- Discuss basic ideas of linear regression and correlation.
- Create and interpret a line of best fit.
- Calculate and interpret the correlation coefficient.
- Calculate and interpret outliers.

12.1.2 Introduction

Professionals often want to know how two or more variables are related. For example, is there a relationship between the grade on the second math exam a student takes and the grade on the final exam? If there is a relationship, what is it and how strong is the relationship?

In another example, your income may be determined by your education, your profession, your years of experience, and your ability. The amount you pay a repair person for labor is often determined by an initial amount plus an hourly fee. These are all examples in which regression can be used.

The type of data described in the examples is **bivariate** data - "bi" for two variables. In reality, statisticians use **multivariate** data, meaning many variables.

In this chapter, you will be studying the simplest form of regression, "linear regression" with one independent variable (x). This involves data that fits a line in two dimensions. You will also study correlation which measures how strong the relationship is.

12.2 Linear Equations[2]

Linear regression for two variables is based on a linear equation with one independent variable. It has the form:

$$y = a + bx \tag{12.1}$$

[1]This content is available online at <http://cnx.org/content/m17089/1.5/>.
[2]This content is available online at <http://cnx.org/content/m17086/1.4/>.

where a and b are constant numbers.

x is the independent variable, and y is the dependent variable. Typically, you choose a value to substitute for the independent variable and then solve for the dependent variable.

Example 12.1
The following examples are linear equations.

$$y = 3 + 2x \tag{12.2}$$

$$y = -0.01 + 1.2x \tag{12.3}$$

The graph of a linear equation of the form $y = a + bx$ is a **straight line**. Any line that is not vertical can be described by this equation.

Example 12.2

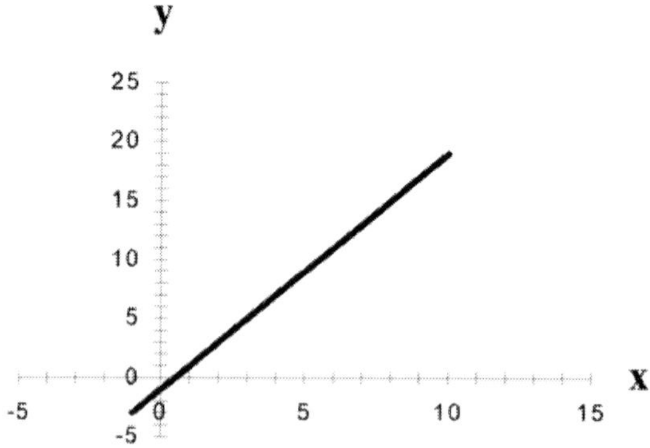

Figure 12.1: Graph of the equation $y = -1 + 2x$.

Linear equations of this form occur in applications of life sciences, social sciences, psychology, business, economics, physical sciences, mathematics, and other areas.

Example 12.3
Aaron's Word Processing Service (AWPS) does word processing. Its rate is \$32 per hour plus a \$31.50 one-time charge. The total cost to a customer depends on the number of hours it takes to do the word processing job.

Problem
Find the equation that expresses the **total cost** in terms of the **number of hours** required to finish the word processing job.

Solution
Let x = the number of hours it takes to get the job done.

Let y = the total cost to the customer.

The \$31.50 is a fixed cost. If it takes x hours to complete the job, then $(32)(x)$ is the cost of the word processing only. The total cost is:

$y = 31.50 + 32x$

12.3 Slope and Y-Intercept of a Linear Equation[3]

For the linear equation $y = a + bx$, b = slope and a = y-intercept.

From algebra recall that the slope is a number that describes the steepness of a line and the y-intercept is the y coordinate of the point $(0, a)$ where the line crosses the y-axis.

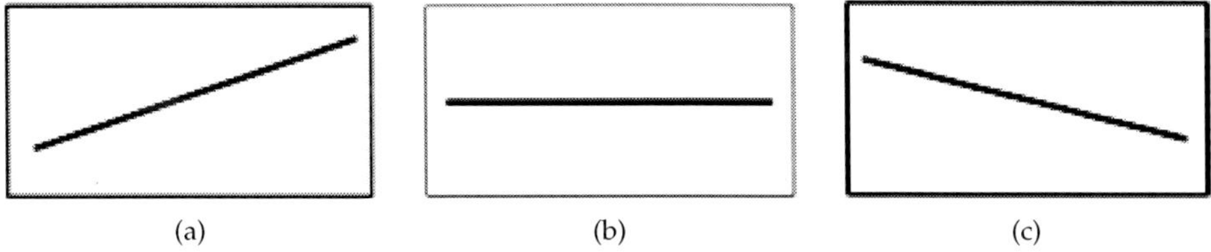

Figure 12.2: Three possible graphs of $y = a + bx$. (a) If $b > 0$, the line slopes upward to the right. (b) If $b = 0$, the line is horizontal. (c) If $b < 0$, the line slopes downward to the right.

Example 12.4
Svetlana tutors to make extra money for college. For each tutoring session, she charges a one time fee of $25 plus $15 per hour of tutoring. A linear equation that expresses the total amount of money Svetlana earns for each session she tutors is $y = 25 + 15x$.

Problem
What are the independent and dependent variables? What is the y-intercept and what is the slope? Interpret them using complete sentences.

Solution
The independent variable (x) is the number of hours Svetlana tutors each session. The dependent variable (y) is the amount, in dollars, Svetlana earns for each session.

The y-intercept is 25 (a = 25). At the start of the tutoring session, Svetlana charges a one-time fee of $25 (this is when x = 0). The slope is 15 (b = 15). For each session, Svetlana earns $15 for each hour she tutors.

12.4 Scatter Plots[4]

Before we take up the discussion of linear regression and correlation, we need to examine a way to display the relation between two variables x and y. The most common and easiest way is a **scatter plot**. The following example illustrates a scatter plot.

[3]This content is available online at <http://cnx.org/content/m17083/1.5/>.
[4]This content is available online at <http://cnx.org/content/m17082/1.6/>.

Example 12.5
From an article in the *Wall Street Journal*: In Europe and Asia, m-commerce is becoming more popular. M-commerce users have special mobile phones that work like electronic wallets as well as provide phone and Internet services. Users can do everything from paying for parking to buying a TV set or soda from a machine to banking to checking sports scores on the Internet. In the next few years, will there be a relationship between the year and the number of m-commerce users? Construct a scatter plot. Let x = the year and let y = the number of m-commerce users, in millions.

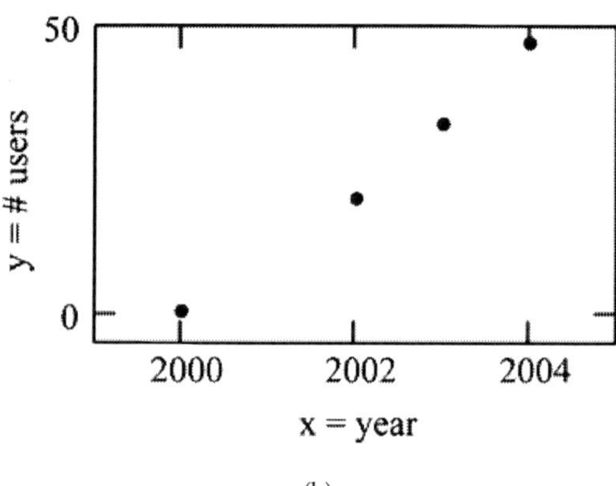

(b)

x (year)	y (# of users)
2000	0.5
2002	20.0
2003	33.0
2004	47.0

(a)

Figure 12.3: (a) Table showing the number of m-commerce users (in millions) by year. (b) Scatter plot showing the number of m-commerce users (in millions) by year.

A scatter plot shows the **direction** and **strength** of a relationship between the variables. A clear direction happens when there is either:

- High values of one variable occurring with high values of the other variable or low values of one variable occurring with low values of the other variable.
- High values of one variable occurring with low values of the other variable.

You can determine the strength of the relationship by looking at the scatter plot and seeing how close the points are to a line, a power function, an exponential function, or to some other type of function.

When you look at a scatterplot, you want to notice the **overall pattern** and any **deviations** from the pattern. The following scatterplot examples illustrate these concepts.

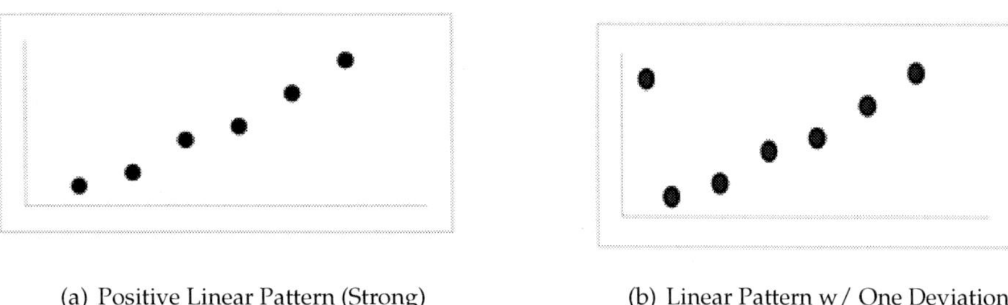

(a) Positive Linear Pattern (Strong) (b) Linear Pattern w/ One Deviation

Figure 12.4

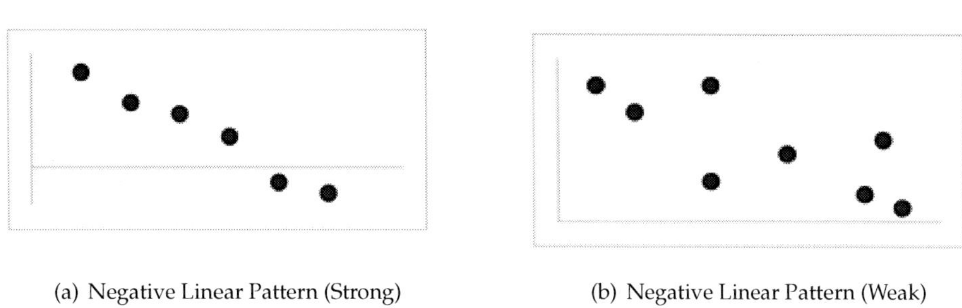

(a) Negative Linear Pattern (Strong) (b) Negative Linear Pattern (Weak)

Figure 12.5

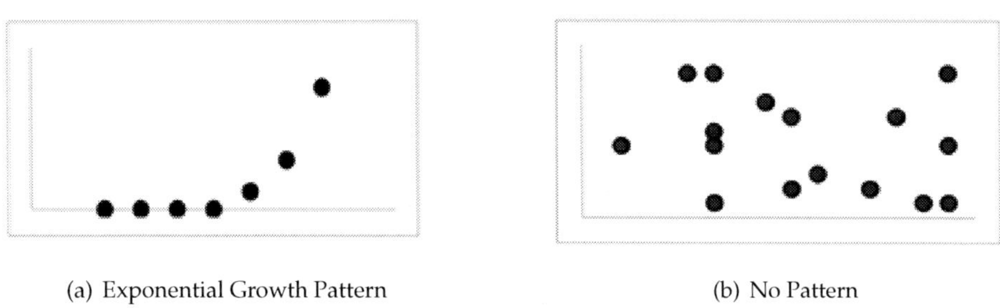

(a) Exponential Growth Pattern (b) No Pattern

Figure 12.6

In this chapter, we are interested in scatter plots that show a linear pattern. Linear patterns are quite common. The linear relationship is strong if the points are close to a straight line. If we think that the points show a linear relationship, we would like to draw a line on the scatter plot. This line can be calculated through a process called **linear regression**. However, we only calculate a regression line if one of the variables helps to explain or predict the other variable. If x is the independent variable and y the dependent variable, then we can use a regression line to predict y for a given value of x.

12.5 The Regression Equation[5]

Data rarely fit a straight line exactly. Usually, you must be satisfied with rough predictions. Typically, you have a set of data whose scatter plot appears to **"fit"** a straight line. This is called a **Line of Best Fit or Least Squares Line**.

12.5.1 Optional Collaborative Classroom Activity

If you know a person's pinky (smallest) finger length, do you think you could predict that person's height? Collect data from your class (pinky finger length, in inches). The independent variable, x, is pinky finger length and the dependent variable, y, is height.

For each set of data, plot the points on graph paper. Make your graph big enough and **use a ruler**. Then "by eye" draw a line that appears to "fit" the data. For your line, pick two convenient points and use them to find the slope of the line. Find the y-intercept of the line by extending your lines so they cross the y-axis. Using the slopes and the y-intercepts, write your equation of "best fit". Do you think everyone will have the same equation? Why or why not?

Using your equation, what is the predicted height for a pinky length of 2.5 inches?

Example 12.6

A random sample of 11 statistics students produced the following data where x is the third exam score, out of 80, and y is the final exam score, out of 200. Can you predict the final exam score of a random student if you know the third exam score?

[5]This content is available online at <http://cnx.org/content/m17090/1.8/>.

x (third exam score)	y (final exam score)
65	175
67	133
71	185
71	163
66	126
75	198
67	153
70	163
71	159
69	151
69	159

(a)

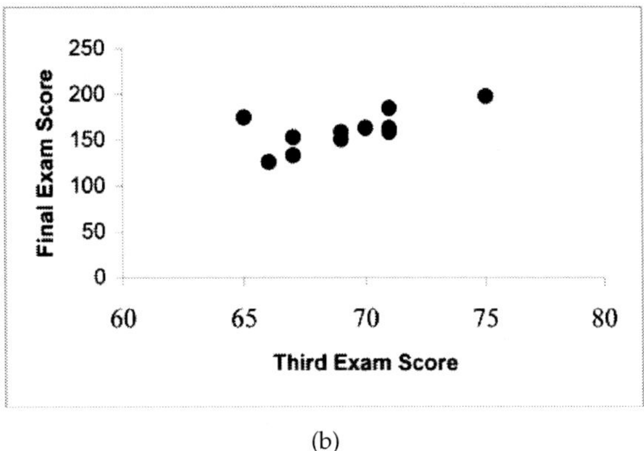

(b)

Figure 12.7: (a) Table showing the scores on the final exam based on scores from the third exam. (b) Scatter plot showing the scores on the final exam based on scores from the third exam.

The third exam score, x, is the independent variable and the final exam score, y, is the dependent variable. We will plot a regression line that best "fits" the data. If each of you were to fit a line "by eye", you would draw different lines. We can use what is called a **least-squares regression line** to obtain the best fit line.

Consider the following diagram. Each point of data is of the the form $(x, y)$and each point of the line of best fit using least-squares linear regression has the form $\left(x, \hat{y}\right)$.

The $\hat{y}$ is read **"y hat"** and is the **estimated value of** y. It is the value of y obtained using the regression line. It is not generally equal to y from data.

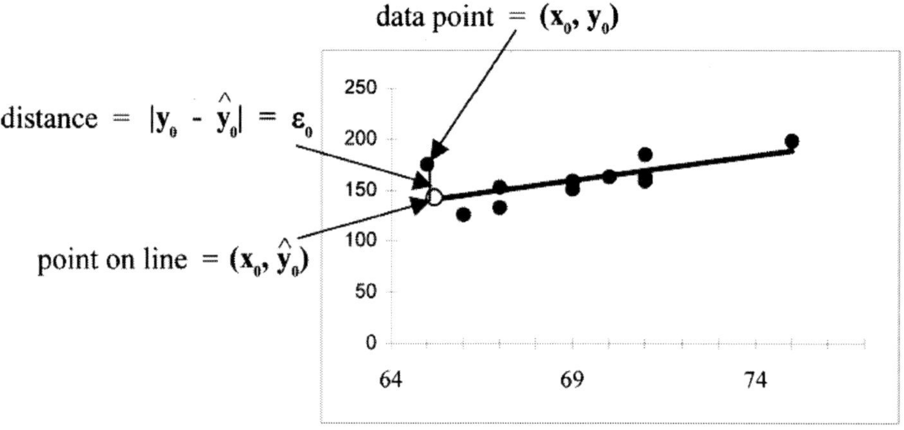

Figure 12.8

The term $|y_0 - \hat{y}_0| = \epsilon_0$ is called the **"error" or residual**. It is not an error in the sense of a mistake, but measures the vertical distance between the actual value of y and the estimated value of y.

ϵ = the Greek letter **epsilon**

For each data point, you can calculate, $|y_i - \hat{y}_i| = \epsilon_i$ for $i = 1, 2, 3, ..., 11$.

Each ϵ is a vertical distance.

For the example about the third exam scores and the final exam scores for the 11 statistics students, there are 11 data points. Therefore, there are 11 ϵ values. If you square each ϵ and add, you get

$$(\epsilon_1)^2 + (\epsilon_2)^2 + ... + (\epsilon_{11})^2 = \sum_{i=1}^{11} \epsilon^2$$

This is called the **Sum of Squared Errors (SSE)**.

Using calculus, you can determine the values of a and b that make the **SSE** a minimum. When you make the **SSE** a minimum, you have determined the points that are on the line of best fit. It turns out that the line of best fit has the equation:

$$\hat{y} = a + bx \tag{12.4}$$

where $a = \overline{y} - b \cdot \overline{x}$ and $b = \frac{\Sigma(x-\overline{x})\cdot(y-\overline{y})}{\Sigma(x-\overline{x})}^2$.

$\overline{x}$ and $\overline{y}$ are the averages of the x values and the y values, respectively. The best fit line always passes through the point $(\overline{x}, \overline{y})$.

The slope b can be written as $b = r \cdot \left(\frac{s_y}{s_x}\right)$ where s_y = the standard deviation of the y values and s_x = the standard deviation of the x values. r is the correlation coefficient which is discussed in the next section.

NOTE: Many calculators or any linear regression and correlation computer program can calculate the best fit line. The calculations tend to be tedious if done by hand. **In the technology section, there are instructions for calculating the best fit line.**

The graph of the line of best fit for the third exam/final exam example is shown below:

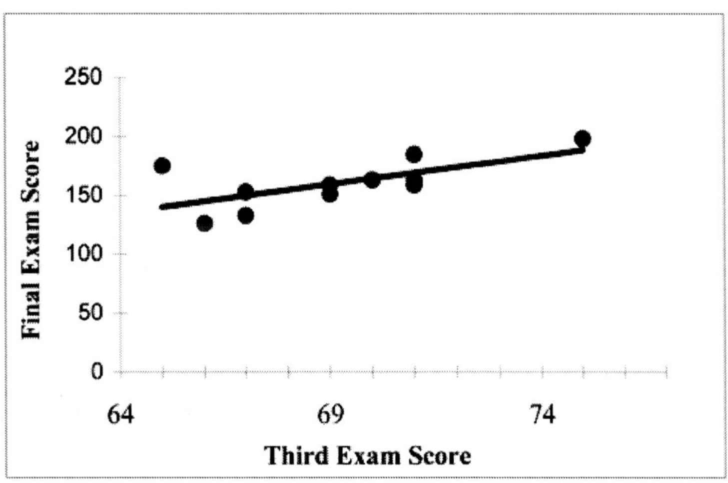

Figure 12.9

Remember, the best fit line is called the **least squares regression line** (it is sometimes referred to as the **LSL** which is an acronym for least squares line). The best fit line for the third exam/final exam example has the equation:

$$\hat{y} = -173.51 + 4.83x \quad (12.5)$$

The idea behind finding the best fit line is based on the assumption that the data are actually scattered about a straight line. Remember, it is always important to plot a scatter diagram first (which many calculators and computer programs can do) to see if it is worth calculating the line of best fit.

The slope of the line is 4.83 (b = 4.83). We can interpret the slope as follows: As the third exam score increases by one point, the final exam score increases by 4.83 points.

NOTE: If the scatter plot indicates that there is a linear relationship between the variables, then it is reasonable to use a best fit line to make predictions for y given x within the domain of x-values in the sample data, **but not necessarily for x-values outside that domain.**

12.6 The Correlation Coefficient[6]

Besides looking at the scatter plot and seeing that a line seems reasonable, how can you tell if the line is a good predictor? Use the correlation coefficient as another indicator (besides the scatterplot) of the strength of the relationship between x and y. The correlation coefficient, r, is defined as:

[6]This content is available online at <http://cnx.org/content/m17092/1.6/>.

$$r = \frac{n \cdot \Sigma x \cdot y - (\Sigma x) \cdot (\Sigma y)}{\sqrt{\left[n \cdot \Sigma x^2 - (\Sigma x)^2\right] \cdot \left[n \cdot \Sigma y^2 - (\Sigma y)^2\right]}}$$

where n = the number of data points.

If you suspect a linear relationship between x and y, then r can measure how strong the linear relationship is.

One property of r is that $-1 \leq r \leq 1$. If $r = 1$, there is perfect positive correlation. If $r = -1$, there is perfect negative correlation. In both these cases, the original data points lie on a straight line. Of course, in the real world, this will not generally happen.

The formula for r looks formidable. However, many calculators and any regression and correlation computer program can calculate r. The sign of r is the same as the slope, b, of the best fit line.

12.7 Facts About the Correlation Coefficient for Linear Regression[7]

- A positive r means that when x increases, y increases and when x decreases, y decreases **(positive correlation)**.
- A negative r means that when x increases, y decreases and when x decreases, y increases **(negative correlation)**.
- An r of zero means there is absolutely no linear relationship between x and y **(no correlation)**.
- High correlation does not suggest that x causes y or y causes x. We say **"correlation does not imply causation."** For example, every person who learned math in the 17th century is dead. However, learning math does not necessarily cause death!

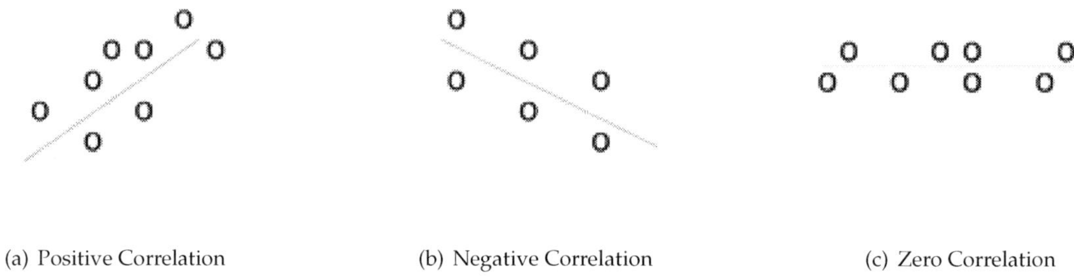

(a) Positive Correlation (b) Negative Correlation (c) Zero Correlation

Figure 12.10: (a) A scatter plot showing data with a positive correlation. (b) A scatter plot showing data with a negative correlation. (c) A scatter plot showing data with zero correlation.

The 95% Critical Values of the Sample Correlation Coefficient Table (Section 12.10) at the end of this chapter (before the Summary (Section 12.11)) may be used to give you a good idea of whether the computed value of r **is significant or not**. Compare r to the appropriate critical value in the table. If r is significant, then you may want to use the line for prediction.

Example 12.7
Suppose you computed $r = 0.801$ using $n = 10$ data points. $df = n - 2 = 10 - 2 = 8$. The critical values associated with $df = 8$ are -0.632 and + 0.632. If $r <$ *negative critical value* or $r >$

[7]This content is available online at <http://cnx.org/content/m17077/1.7/>.

positive critical value, then r is significant. Since $r = 0.801$ and $0.801 > 0.632$, r is significant and the line may be used for prediction. If you view this example on a number line, it will help you.

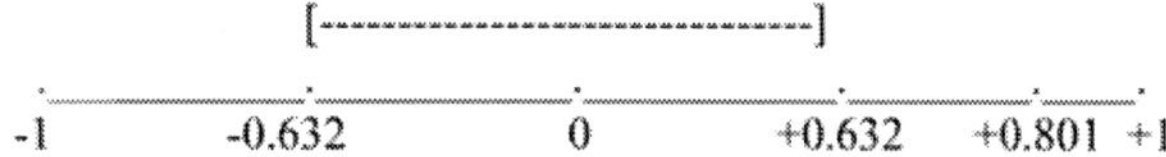

Figure 12.11: r is not significant between -0.632 and +0.632. $r = 0.801 > +0.632$. Therefore, r is significant.

Example 12.8

Suppose you computed $r = -0.624$ with 14 data points. $df = 14 - 2 = 12$. The critical values are -0.532 and 0.532. Since $-0.624 < -0.532$, r is significant and the line may be used for prediction

-0.624 -0.532 +0.532

Figure 12.12: $r = -0.624 < -0.532$. Therefore, r is significant.

Example 12.9

Suppose you computed $r = 0.776$ and $n = 6$. $df = 6 - 2 = 4$. The critical values are -0.811 and 0.811. Since $-0.811 < 0.776 < 0.811$, r is not significant and the line should not be used for prediction.

-0.811 0.776 0.811

Figure 12.13: $-0.811 < r = 0.776 < 0.811$. Therefore, r is not significant.

If $r = -1$ or $r = +1$, then all the data points lie exactly on a straight line.

If the line is significant, then **within the range of the x-values,** the line can be used to predict a y value.

As an illustration, consider the third exam/final exam example. The line of best fit is: $\hat{y} = -173.51 + 4.83x$ with $r = 0.6631$

Can the line be used for prediction? **Given a third exam score (x value), can we successfully predict the final exam score (predicted y value).** Test $r = 0.6631$ with its appropriate critical value.

Using the table with $df = 11 - 2 = 9$, the critical values are -0.602 and +0.602. Since $0.6631 > 0.602$, r is significant. **Because r is significant and the scatter plot shows a reasonable linear trend, the line can be used to predict final exam scores.**

Example 12.10
Suppose you computed the following correlation coefficients. Using the table at the end of the chapter, determine if r is significant and the line of best fit associated with each r can be used to predict a y value. If it helps, draw a number line.

- $r = -0.567$ and the sample size, n, is 19. The $df = n - 2 = 17$. The critical value is -0.456. $-0.567 < -0.456$ so r is significant.
- $r = 0.708$ and the sample size, n, is 9. The $df = n - 2 = 7$. The critical value is 0.666. $0.708 > 0.666$ so r is significant.
- $r = 0.134$ and the sample size, n, is 14. The $df = 14 - 2 = 12$. The critical value is 0.532. 0.134 is between -0.532 and 0.532 so r is not significant.
- $r = 0$ and the sample size, n, is 5. No matter what the dfs are, $r = 0$ is between the two critical values so r is not significant.

12.8 Prediction[8]

The exam scores **(x-values)** range from 65 to 75. Suppose you want to know the final exam score of statistics students who received 73 on the third exam. **Since 73 is between the x-values 65 and 75,** substitute $x = 73$ into the equation. Then:

$$\hat{y} = -173.51 + 4.83\,(73) = 179.08 \quad (12.7)$$

We predict that a statistics student who receives a 73 on the third exam will receive 179.08 on the final exam. **Remember, do not use the regression equation to predict values outside the domain of x.**

Example 12.11
Recall the third exam/final exam example.

Problem 1
What would you predict the final exam score to be for a student who scored a 66 on the third exam?

Solution
145.27

Problem 2 *(Solution on p. 520.)*
What would you predict the final exam score to be for a student who scored a 78 on the third exam?

12.9 Outliers[9]

In some data sets, there are values **(points)** called **outliers. Outliers are points that are far from the least squares line.** They have large "errors." Outliers need to be examined closely. Sometimes, for some reason or another, they should not be included in the analysis of the data. It is possible that an outlier is a result of erroneous data. Other times, an outlier may hold valuable information about the population under study. The key is to carefully examine what causes a data point to be an outlier.

[8]This content is available online at <http://cnx.org/content/m17095/1.6/>.
[9]This content is available online at <http://cnx.org/content/m17094/1.7/>.

Example 12.12
In the third exam/final exam example, you can determine if there is an outlier or not. If there is one, as an exercise, delete it and fit the remaining data to a new line. For this example, the new line ought to fit the remaining data better. This means the **SSE** should be smaller and the correlation coefficient ought to be closer to 1 or -1.

Solution
Computers and many calculators can determine outliers from the data. However, as an exercise, we will go through the steps that are needed to calculate an outlier. In the table below, the first two columns are the third exam and the final exam data. The third column shows the y-hat values calculated from the line of best fit.

x	y	$\hat{y}$
65	175	140
67	133	150
71	185	169
71	163	169
66	126	145
75	198	189
67	153	150
70	163	164
71	159	169
69	151	160
69	159	160

Table 12.1

A **Residual** is the $Actual\ y\ value - predicted\ y\ value = y - \hat{y}$

Calculate the absolute value of each residual.

Calculate each $|y - \hat{y}|$**:**

x	y	$\hat{y}$	$\lvert y - \hat{y} \rvert$
65	175	140	$\lvert 175-140 \rvert = 35$
67	133	150	$\lvert 133-150 \rvert = 17$
71	185	169	$\lvert 185-169 \rvert = 16$
71	163	169	$\lvert 163-169 \rvert = 6$
66	126	145	$\lvert 126-145 \rvert = 19$
75	198	189	$\lvert 198-189 \rvert = 9$
67	153	150	$\lvert 153-150 \rvert = 3$
70	163	164	$\lvert 163-164 \rvert = 1$
71	159	169	$\lvert 159-169 \rvert = 10$
69	151	160	$\lvert 151-160 \rvert = 9$
69	159	160	$\lvert 159-160 \rvert = 1$

Table 12.2

Square each $\lvert y - \hat{y} \rvert$:

$35^2; 17^2; 16^2; 6^2; 19^2; 9^2; 3^2; 1^2; 10^2; 9^2; 1^2$

Then, add (sum) all the $\lvert y - \hat{y} \rvert$ squared terms:

$$\sum_{i=1}^{11} \left(\lvert y - \hat{y} \rvert \right)^2 = \sum_{i=1}^{11} \epsilon^2 \qquad \text{(Recall that } \lvert y_i - \hat{y}_i \rvert = \epsilon_i \text{.)}$$

$$= 35^2 + 17^2 + 16^2 + 6^2 + 19^2 + 9^2 + 3^2 + 1^2 + 10^2 + 9^2 + 1^2$$

$= 2440 =$ **SSE**

Next, calculate s, the standard deviation of all the $\lvert y - \hat{y} \rvert = \epsilon$ values where n = the total number of data points. (Calculate the standard deviation of 35; 17; 16; 6; 19; 9; 3; 1; 10; 9; 1.)

$$s = \sqrt{\frac{SSE}{n-2}}$$

For the third exam/final exam problem, $s = \sqrt{\frac{2440}{11-2}} = 16.47$

Next, multiply s by 1.9 and get $(1.9) \cdot (16.47) = 31.29$ (the value 31.29 is almost 2 standard deviations away from the mean of the $\lvert y - \hat{y} \rvert$ values.)

NOTE: The number $1.9s$ is equal to **1.9 standard deviations**. It is a measure that is almost 2 standard deviations. If we were to measure the vertical distance from any data point to the corresponding point on the line of best fit and that distance was equal to $1.9s$ or greater, then we would consider the data point to be "too far" from the line of best fit. We would call that point a **potential outlier**.

For the example, if any of the $|y - \hat{y}|$ values are **at least** 31.29, the corresponding (x, y) point (data point) is a potential outlier.

Mathematically, we say that if $|y - \hat{y}| \geq (1.9) \cdot (s)$, then the corresponding point is an outlier.

For the third exam/final exam problem, all the $|y - \hat{y}|$'s are less than 31.29 except for the first one which is 35.

$35 > 31.29$ That is, $|y - \hat{y}| \geq (1.9) \cdot (s)$

The point which corresponds to $|y - \hat{y}| = 35$ is $(65, 175)$. **Therefore, the point** $(65, 175)$ **is an outlier.** For this example, we will delete it. (Remember, we do not always delete an outlier.) The next step is to compute a new best-fit line using the 10 remaining points. The new line of best fit and the correlation coefficient are:

$\hat{y} = -355.19 + 7.39x$ and $r = 0.9121$

If you compare $r = 0.9121$ to its critical value 0.632, $0.9121 > 0.632$. Therefore, r is significant. In fact, $r = 0.9121$ is a better r than the original (0.6631) because $r = 0.9121$ is closer to 1. This means that the 10 points fit the line better. The line can better predict the final exam score given the third exam score.

Example 12.13
Using the new line of best fit (calculated with 10 points), what would a student who receives a 73 on the third exam expect to receive on the final exam?

Example 12.14
(*From The Consumer Price Indexes Web site*) The Consumer Price Index (CPI) measures the average change over time in the prices paid by urban consumers for consumer goods and services. The CPI affects nearly all Americans because of the many ways it is used. One of its biggest uses is as a measure of inflation. By providing information about price changes in the Nation's economy to government, business, and labor, the CPI helps them to make economic decisions. The President, Congress, and the Federal Reserve Board use the CPI's trends to formulate monetary and fiscal policies. In the following table, x is the year and y is the CPI.

Data:

x	y
1915	10.1
1926	17.7
1935	13.7
1940	14.7
1947	24.1
1952	26.5
1964	31.0
1969	36.7
1975	49.3
1979	72.6
1980	82.4
1986	109.6
1991	130.7
1999	166.6

Table 12.3

Problem

- Make a scatterplot of the data.
- Calculate the least squares line. Write the equation in the form $\hat{y} = a + bx$.
- Draw the line on the scatterplot.
- Find the correlation coefficient. Is it significant?
- What is the average CPI for the year 1990?

Solution

- Scatter plot and line of best fit.
- $\hat{y} = -3204 + 1.662x$ is the equation of the line of best fit.
- $r = 0.8694$
- The number of data points is $n = 14$. Use the 95% Critical Values of the Sample Correlation Coefficient table at the end of Chapter 12. $n - 2 = 12$. The corresponding critical value is 0.532. Since $0.8694 > 0.532$, r is significant.
- $\hat{y} = -3204 + 1.662\,(1990) = 103.4$ CPI

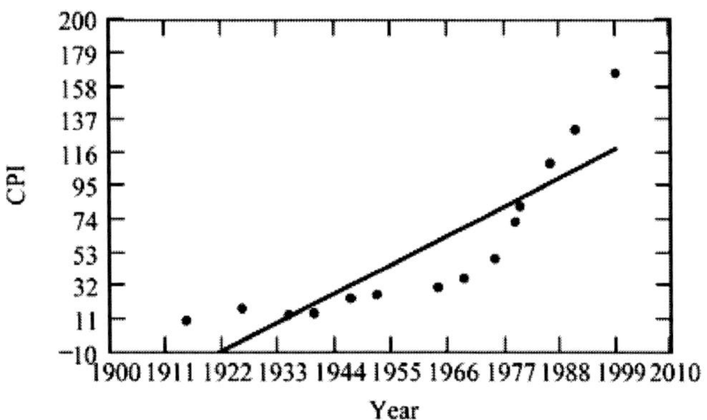

Figure 12.14

12.10 95% Critical Values of the Sample Correlation Coefficient Table[10]

Degrees of Freedom: $n-2$	Critical Values: (+ and −)
1	0.997
2	0.950
3	0.878
4	0.811
5	0.754
6	0.707
7	0.666
8	0.632
9	0.602
10	0.576
11	0.555
12	0.532
	continued on next page

[10]This content is available online at <http://cnx.org/content/m17098/1.5/>.

13	0.514
14	0.497
15	0.482
16	0.468
17	0.456
18	0.444
19	0.433
20	0.423
21	0.413
22	0.404
23	0.396
24	0.388
25	0.381
26	0.374
27	0.367
28	0.361
29	0.355
30	0.349
40	0.304
50	0.273
60	0.250
70	0.232
80	0.217
90	0.205
100 and over	0.195

Table 12.4

12.11 Summary[11]

Bivariate Data: Each data point has two values. The form is (x, y).

Line of Best Fit or Least Squares Line (LSL): $\hat{y} = a + bx$

x = independent variable; y = dependent variable

Residual: $Actual\ y\ value - predicted\ y\ value = y - \hat{y}$

Correlation Coefficient r:

1. Used to determine whether a line of best fit is good for prediction.
2. Between -1 and 1 inclusive. The closer r is to 1 or -1, the closer the original points are to a straight line.
3. If r is negative, the slope is negative. If r is positive, the slope is positive.
4. If $r = 0$, then the line is horizontal.

Sum of Squared Errors (SSE): The smaller the **SSE**, the better the original set of points fits the line of best fit.

Outlier: A point that does not seem to fit the rest of the data.

[11]This content is available online at <http://cnx.org/content/m17081/1.4/>.

12.12 Practice: Linear Regression[12]

12.12.1 Student Learning Outcomes

- The student will explore the properties of linear regression.

12.12.2 Given

The data below are real. Keep in mind that these are only reported figures. (*Source: Centers for Disease Control and Prevention, National Center for HIV, STD, and TB Prevention, October 24, 2003*)

Adults and Adolescents only, United States

Year	# AIDS cases diagnosed	# AIDS deaths
Pre-1981	91	29
1981	319	121
1982	1,170	453
1983	3,076	1,482
1984	6,240	3,466
1985	11,776	6,878
1986	19,032	11,987
1987	28,564	16,162
1988	35,447	20,868
1989	42,674	27,591
1990	48,634	31,335
1991	59,660	36,560
1992	78,530	41,055
1993	78,834	44,730
1994	71,874	49,095
1995	68,505	49,456
1996	59,347	38,510
1997	47,149	20,736
1998	38,393	19,005
1999	25,174	18,454
2000	25,522	17,347
2001	25,643	17,402
2002	26,464	16,371
Total	**802,118**	**489,093**

Table 12.5

[12]This content is available online at <http://cnx.org/content/m17088/1.8/>.

NOTE: We will use the columns "year" and "# AIDS cases diagnosed" for all questions unless otherwise stated.

12.12.3 Graphing

Graph "year" vs. "# AIDS cases diagnosed." **Plot the points on the graph located below in the section titled "Plot"**. Do not include pre-1981. Label both axes with words. Scale both axes.

12.12.4 Data

Exercise 12.12.1
Enter your data into your calculator or computer. The pre-1981 data should not be included. Why is that so?

12.12.5 Linear Equation

Write the linear equation below, rounding to 4 decimal places:

Exercise 12.12.2 *(Solution on p. 520.)*
Calculate the following:

a. $a =$
b. $b =$
c. $corr. =$
d. $n =$(# of pairs)

Exercise 12.12.3 *(Solution on p. 520.)*
equation: $\hat{y}=$

12.12.6 Solve

Exercise 12.12.4 *(Solution on p. 520.)*
Solve.

a. When $x = 1985, \hat{y}=$
b. When $x = 1990, \hat{y}=$

12.12.7 Plot

Plot the 2 above points on the graph below. Then, connect the 2 points to form the regression line.

Obtain the graph on your calculator or computer.

12.12.8 Discussion Questions

Look at the graph above.

Exercise 12.12.5
Does the line seem to fit the data? Why or why not?

Exercise 12.12.6
Do you think a linear fit is best? Why or why not?

Exercise 12.12.7
Hand draw a smooth curve on the graph above that shows the flow of the data.

Exercise 12.12.8
What does the correlation imply about the relationship between time (years) and the number of diagnosed AIDS cases reported in the U.S.?

Exercise 12.12.9
Why is "year" the independent variable and "# AIDS cases diagnosed." the dependent variable (instead of the reverse)?

Exercise 12.12.10 ***(Solution on p. 520.)***
Solve.

- **a.** When $x = 1970$, $\hat{y}=$:
- **b.** Why doesn't this answer make sense?

12.13 Homework[13]

Exercise 12.13.1 *(Solution on p. 520.)*

For each situation below, state the independent variable and the dependent variable.

a. A study is done to determine if elderly drivers are involved in more motor vehicle fatalities than all other drivers. The number of fatalities per 100,000 drivers is compared to the age of drivers.
b. A study is done to determine if the weekly grocery bill changes based on the number of family members.
c. Insurance companies base life insurance premiums partially on the age of the applicant.
d. Utility bills vary according to power consumption.
e. A study is done to determine if a higher education reduces the crime rate in a population.

Exercise 12.13.2

In 1990 the number of driver deaths per 100,000 for the different age groups was as follows (Source: *The National Highway Traffic Safety Administration's National Center for Statistics and Analysis*):

Age	Number of Driver Deaths per 100,000
15-24	28
25-39	15
40-69	10
70-79	15
80+	25

Table 12.6

a. For each age group, pick the midpoint of the interval for the x value. (For the 80+ group, use 85.)
b. Using "ages" as the independent variable and "Number of driver deaths per 100,000" as the dependent variable, make a scatter plot of the data.
c. Calculate the least squares (best–fit) line. Put the equation in the form of: $\hat{y} = a + bx$
d. Find the correlation coefficient. Is it significant?
e. Pick two ages and find the estimated fatality rates.
f. Use the two points in (e) to plot the least squares line on your graph from (b).
g. Based on the above data, is there a linear relationship between age of a driver and driver fatality rate?
h. What is the slope of the least squares (best-fit) line? Interpret the slope.

Exercise 12.13.3 *(Solution on p. 520.)*

The average number of people in a family that received welfare for various years is given below. (Source: *House Ways and Means Committee, Health and Human Services Department*)

[13]This content is available online at <http://cnx.org/content/m17085/1.8/>.

Year	Welfare family size
1969	4.0
1973	3.6
1975	3.2
1979	3.0
1983	3.0
1988	3.0
1991	2.9

Table 12.7

a. Using "year" as the independent variable and "welfare family size" as the dependent variable, make a scatter plot of the data.
b. Calculate the least squares line. Put the equation in the form of: $\hat{y} = a + bx$
c. Find the correlation coefficient. Is it significant?
d. Pick two years between 1969 and 1991 and find the estimated welfare family sizes.
e. Use the two points in (d) to plot the least squares line on your graph from (b).
f. Based on the above data, is there a linear relationship between the year and the average number of people in a welfare family?
g. Using the least squares line, estimate the welfare family sizes for 1960 and 1995. Does the least squares line give an accurate estimate for those years? Explain why or why not.
h. Are there any outliers in the above data?
i. What is the estimated average welfare family size for 1986? Does the least squares line give an accurate estimate for that year? Explain why or why not.
j. What is the slope of the least squares (best-fit) line? Interpret the slope.

Exercise 12.13.4

Use the AIDS data from the practice for this section (Section 12.12.2: Given), but this time use the columns "year #" and "# new AIDS deaths in U.S." Answer all of the questions from the practice again, using the new columns.

Exercise 12.13.5 ***(Solution on p. 520.)***

The height (sidewalk to roof) of notable tall buildings in America is compared to the number of stories of the building (beginning at street level). (Source: *Microsoft Bookshelf*)

Height (in feet)	Stories
1050	57
428	28
362	26
529	40
790	60
401	22
380	38
1454	110
1127	100
700	46

Table 12.8

a. Using "stories" as the independent variable and "height" as the dependent variable, make a scatter plot of the data.
b. Does it appear from inspection that there is a relationship between the variables?
c. Calculate the least squares line. Put the equation in the form of: $\hat{y} = a + bx$
d. Find the correlation coefficient. Is it significant?
e. Find the estimated heights for 32 stories and for 94 stories.
f. Use the two points in (e) to plot the least squares line on your graph from (b).
g. Based on the above data, is there a linear relationship between the number of stories in tall buildings and the height of the buildings?
h. Are there any outliers in the above data? If so, which point(s)?
i. What is the estimated height of a building with 6 stories? Does the least squares line give an accurate estimate of height? Explain why or why not.
j. Based on the least squares line, adding an extra story adds about how many feet to a building?
k. What is the slope of the least squares (best-fit) line? Interpret the slope.

Exercise 12.13.6

Below is the life expectancy for an individual born in the United States in certain years. (Source: *National Center for Health Statistics*)

Year of Birth	Life Expectancy
1930	59.7
1940	62.9
1950	70.2
1965	69.7
1973	71.4
1982	74.5
1987	75
1992	75.7

Table 12.9

a. Decide which variable should be the independent variable and which should be the dependent variable.
b. Draw a scatter plot of the ordered pairs.
c. Calculate the least squares line. Put the equation in the form of: $\hat{y} = a + bx$
d. Find the correlation coefficient. Is it significant?
e. Find the estimated life expectancy for an individual born in 1950 and for one born in 1982.
f. Why aren't the answers to part (e) the values on the above chart that correspond to those years?
g. Use the two points in (e) to plot the least squares line on your graph from (b).
h. Based on the above data, is there a linear relationship between the year of birth and life expectancy?
i. Are there any outliers in the above data?
j. Using the least squares line, find the estimated life expectancy for an individual born in 1850. Does the least squares line give an accurate estimate for that year? Explain why or why not.
k. What is the slope of the least squares (best-fit) line? Interpret the slope.

Exercise 12.13.7 *(Solution on p. 521.)*

The percent of female wage and salary workers who are paid hourly rates is given below for the years 1979 - 1992. (Source: *Bureau of Labor Statistics, U.S. Dept. of Labor*)

Year	Percent of workers paid hourly rates
1979	61.2
1980	60.7
1981	61.3
1982	61.3
1983	61.8
1984	61.7
1985	61.8
1986	62.0
1987	62.7
1990	62.8
1992	62.9

Table 12.10

a. Using "year" as the independent variable and "percent" as the dependent variable, make a scatter plot of the data.
b. Does it appear from inspection that there is a relationship between the variables? Why or why not?
c. Calculate the least squares line. Put the equation in the form of: $\hat{y} = a + bx$
d. Find the correlation coefficient. Is it significant?
e. Find the estimated percents for 1991 and 1988.
f. Use the two points in (e) to plot the least squares line on your graph from (b).
g. Based on the above data, is there a linear relationship between the year and the percent of female wage and salary earners who are paid hourly rates?
h. Are there any outliers in the above data?
i. What is the estimated percent for the year 2050? Does the least squares line give an accurate estimate for that year? Explain why or why not?
j. What is the slope of the least squares (best-fit) line? Interpret the slope.

Exercise 12.13.8

The maximum discount value of the Entertainment® card for the "Fine Dining" section, Edition 10, for various pages is given below.

Page number	Maximum value ($)
4	16
14	19
25	15
32	17
43	19
57	15
72	16
85	15
90	17

Table 12.11

a. Decide which variable should be the independent variable and which should be the dependent variable.
b. Draw a scatter plot of the ordered pairs.
c. Calculate the least squares line. Put the equation in the form of: $\hat{y} = a + bx$
d. Find the correlation coefficient. Is it significant?
e. Find the estimated maximum values for the restaurants on page 10 and on page 70.
f. Use the two points in (e) to plot the least squares line on your graph from (b).
g. Does it appear that the restaurants giving the maximum value are placed in the beginning of the "Fine Dining" section? How did you arrive at your answer?
h. Suppose that there were 200 pages of restaurants. What do you estimate to be the maximum value for a restaurant listed on page 200?
i. Is the least squares line valid for page 200? Why or why not?
j. What is the slope of the least squares (best-fit) line? Interpret the slope.

The next two questions refer to the following data: The cost of a leading liquid laundry detergent in different sizes is given below.

Size (ounces)	Cost ($)	Cost per ounce
16	3.99	
32	4.99	
64	5.99	
200	10.99	

Table 12.12

Exercise 12.13.9 *(Solution on p. 521.)*

a. Using "size" as the independent variable and "cost" as the dependent variable, make a scatter plot.
b. Does it appear from inspection that there is a relationship between the variables? Why or why not?

c. Calculate the least squares line. Put the equation in the form of: $\hat{y} = a + bx$
d. Find the correlation coefficient. Is it significant?
e. If the laundry detergent were sold in a 40 ounce size, find the estimated cost.
f. If the laundry detergent were sold in a 90 ounce size, find the estimated cost.
g. Use the two points in (e) and (f) to plot the least squares line on your graph from (a).
h. Does it appear that a line is the best way to fit the data? Why or why not?
i. Are there any outliers in the above data?
j. Is the least squares line valid for predicting what a 300 ounce size of the laundry detergent would cost? Why or why not?
k. What is the slope of the least squares (best-fit) line? Interpret the slope.

Exercise 12.13.10

a. Complete the above table for the cost per ounce of the different sizes.
b. Using "Size" as the independent variable and "Cost per ounce" as the dependent variable, make a scatter plot of the data.
c. Does it appear from inspection that there is a relationship between the variables? Why or why not?
d. Calculate the least squares line. Put the equation in the form of: $\hat{y} = a + bx$
e. Find the correlation coefficient. Is it significant?
f. If the laundry detergent were sold in a 40 ounce size, find the estimated cost per ounce.
g. If the laundry detergent were sold in a 90 ounce size, find the estimated cost per ounce.
h. Use the two points in (f) and (g) to plot the least squares line on your graph from (b).
i. Does it appear that a line is the best way to fit the data? Why or why not?
j. Are there any outliers in the above data?
k. Is the least squares line valid for predicting what a 300 ounce size of the laundry detergent would cost per ounce? Why or why not?
l. What is the slope of the least squares (best-fit) line? Interpret the slope.

Exercise 12.13.11 *(Solution on p. 521.)*

According to flyer by a Prudential Insurance Company representative, the costs of approximate probate fees and taxes for selected net taxable estates are as follows:

Net Taxable Estate ($)	Approximate Probate Fees and Taxes ($)
600,000	30,000
750,000	92,500
1,000,000	203,000
1,500,000	438,000
2,000,000	688,000
2,500,000	1,037,000
3,000,000	1,350,000

Table 12.13

a. Decide which variable should be the independent variable and which should be the dependent variable.
b. Make a scatter plot of the data.

c. Does it appear from inspection that there is a relationship between the variables? Why or why not?
d. Calculate the least squares line. Put the equation in the form of: $\hat{y} = a + bx$
e. Find the correlation coefficient. Is it significant?
f. Find the estimated total cost for a net taxable estate of $1,000,000. Find the cost for $2,500,000.
g. Use the two points in (f) to plot the least squares line on your graph from (b).
h. Does it appear that a line is the best way to fit the data? Why or why not?
i. Are there any outliers in the above data?
j. Based on the above, what would be the probate fees and taxes for an estate that does not have any assets?
k. What is the slope of the least squares (best-fit) line? Interpret the slope.

Exercise 12.13.12
The following are advertised sale prices of color televisions at Anderson's.

Size (inches)	Sale Price ($)
9	147
20	197
27	297
31	447
35	1177
40	2177
60	2497

Table 12.14

a. Decide which variable should be the independent variable and which should be the dependent variable.
b. Make a scatter plot of the data.
c. Does it appear from inspection that there is a relationship between the variables? Why or why not?
d. Calculate the least squares line. Put the equation in the form of: $\hat{y} = a + bx$
e. Find the correlation coefficient. Is it significant?
f. Find the estimated sale price for a 32 inch television. Find the cost for a 50 inch television.
g. Use the two points in (f) to plot the least squares line on your graph from (b).
h. Does it appear that a line is the best way to fit the data? Why or why not?
i. Are there any outliers in the above data?
j. What is the slope of the least squares (best-fit) line? Interpret the slope.

Exercise 12.13.13 ***(Solution on p. 521.)***
Below are the average heights for American boys. (Source: *Physician's Handbook, 1990*)

Age (years)	Height (cm)
birth	50.8
2	83.8
3	91.4
5	106.6
7	119.3
10	137.1
14	157.5

Table 12.15

a. Decide which variable should be the independent variable and which should be the dependent variable.
b. Make a scatter plot of the data.
c. Does it appear from inspection that there is a relationship between the variables? Why or why not?
d. Calculate the least squares line. Put the equation in the form of: $\hat{y} = a + bx$
e. Find the correlation coefficient. Is it significant?
f. Find the estimated average height for a one year–old. Find the estimated average height for an eleven year–old.
g. Use the two points in (f) to plot the least squares line on your graph from (b).
h. Does it appear that a line is the best way to fit the data? Why or why not?
i. Are there any outliers in the above data?
j. Use the least squares line to estimate the average height for a sixty–two year–old man. Do you think that your answer is reasonable? Why or why not?
k. What is the slope of the least squares (best-fit) line? Interpret the slope.

Exercise 12.13.14
The following chart gives the gold medal times for every other Summer Olympics for the women's 100 meter freestyle (swimming).

Year	Time (seconds)
1912	82.2
1924	72.4
1932	66.8
1952	66.8
1960	61.2
1968	60.0
1976	55.65
1984	55.92
1992	54.64

Table 12.16

a. Decide which variable should be the independent variable and which should be the dependent variable.
b. Make a scatter plot of the data.
c. Does it appear from inspection that there is a relationship between the variables? Why or why not?
d. Calculate the least squares line. Put the equation in the form of: $\hat{y} = a + bx$
e. Find the correlation coefficient. Is the decrease in times significant?
f. Find the estimated gold medal time for 1932. Find the estimated time for 1984.
g. Why are the answers from (f) different from the chart values?
h. Use the two points in (f) to plot the least squares line on your graph from (b).
i. Does it appear that a line is the best way to fit the data? Why or why not?
j. Use the least squares line to estimate the gold medal time for the next Summer Olympics. Do you think that your answer is reasonable? Why or why not?

The next three questions use the following state information.

State	**# letters in name**	**Year entered the Union**	**Rank for entering the Union**	**Area (square miles)**
Alabama	7	1819	22	52,423
Colorado		1876	38	104,100
Hawaii		1959	50	10,932
Iowa		1846	29	56,276
Maryland		1788	7	12,407
Missouri		1821	24	69,709
New Jersey		1787	3	8,722
Ohio		1803	17	44,828
South Carolina	13	1788	8	32,008
Utah		1896	45	84,904
Wisconsin		1848	30	65,499

Table 12.17

Exercise 12.13.15 *(Solution on p. 521.)*

We are interested in whether or not the number of letters in a state name depends upon the year the state entered the Union.

a. Decide which variable should be the independent variable and which should be the dependent variable.
b. Make a scatter plot of the data.
c. Does it appear from inspection that there is a relationship between the variables? Why or why not?
d. Calculate the least squares line. Put the equation in the form of: $\hat{y} = a + bx$
e. Find the correlation coefficient. What does it imply about the significance of the relationship?
f. Find the estimated number of letters (to the nearest integer) a state would have if it entered the Union in 1900. Find the estimated number of letters a state would have if it entered the Union in 1940.

g. Use the two points in (f) to plot the least squares line on your graph from (b).
h. Does it appear that a line is the best way to fit the data? Why or why not?
i. Use the least squares line to estimate the number of letters a new state that enters the Union this year would have. Can the least squares line be used to predict it? Why or why not?

Exercise 12.13.16

We are interested in whether there is a relationship between the ranking of a state and the area of the state.

a. Let rank be the independent variable and area be the dependent variable.
b. What do you think the scatter plot will look like? Make a scatter plot of the data.
c. Does it appear from inspection that there is a relationship between the variables? Why or why not?
d. Calculate the least squares line. Put the equation in the form of: $\hat{y} = a + bx$
e. Find the correlation coefficient. What does it imply about the significance of the relationship?
f. Find the estimated areas for Alabama and for Colorado. Are they close to the actual areas?
g. Use the two points in (f) to plot the least squares line on your graph from (b).
h. Does it appear that a line is the best way to fit the data? Why or why not?
i. Are there any outliers?
j. Use the least squares line to estimate the area of a new state that enters the Union. Can the least squares line be used to predict it? Why or why not?
k. Delete "Hawaii" and substitute "Alaska" for it. Alaska is the fortieth state with an area of 656,424 square miles.
l. Calculate the new least squares line.
m. Find the estimated area for Alabama. Is it closer to the actual area with this new least squares line or with the previous one that included Hawaii? Why do you think that's the case?
n. Do you think that, in general, newer states are larger than the original states?

Exercise 12.13.17 *(Solution on p. 522.)*

We are interested in whether there is a relationship between the rank of a state and the year it entered the Union.

a. Let year be the independent variable and rank be the dependent variable.
b. What do you think the scatter plot will look like? Make a scatter plot of the data.
c. Why must the relationship be positive between the variables?
d. Calculate the least squares line. Put the equation in the form of: $\hat{y} = a + bx$
e. Find the correlation coefficient. What does it imply about the significance of the relationship?
f. Let's say a fifty-first state entered the union. Based upon the least squares line, when should that have occurred?
g. Using the least squares line, how many states do we currently have?
h. Why isn't the least squares line a good estimator for this year?

Exercise 12.13.18

Below are the percents of the U.S. labor force (excluding self-employed and unemployed) that are members of a union. We are interested in whether the decrease is significant. (Source: *Bureau of Labor Statistics, U.S. Dept. of Labor*)

Year	Percent
1945	35.5
1950	31.5
1960	31.4
1970	27.3
1980	21.9
1986	17.5
1993	15.8

Table 12.18

a. Let year be the independent variable and percent be the dependent variable.
b. What do you think the scatter plot will look like? Make a scatter plot of the data.
c. Why will the relationship between the variables be negative?
d. Calculate the least squares line. Put the equation in the form of: $\hat{y} = a + bx$
e. Find the correlation coefficient. What does it imply about the significance of the relationship?
f. Based on your answer to (e), do you think that the relationship can be said to be decreasing?
g. If the trend continues, when will there no longer be any union members? Do you think that will happen?

The next two questions refer to the following information: The data below reflects the 1991-92 Reunion Class Giving. (Source: *SUNY Albany alumni magazine*)

Class Year	Average Gift	Total Giving
1922	41.67	125
1927	60.75	1,215
1932	83.82	3,772
1937	87.84	5,710
1947	88.27	6,003
1952	76.14	5,254
1957	52.29	4,393
1962	57.80	4,451
1972	42.68	18,093
1976	49.39	22,473
1981	46.87	20,997
1986	37.03	12,590

Table 12.19

Exercise 12.13.19 ***(Solution on p. 522.)***

We will use the columns "class year" and "total giving" for all questions, unless otherwise stated.

a. What do you think the scatter plot will look like? Make a scatter plot of the data.
b. Calculate the least squares line. Put the equation in the form of: $\hat{y} = a + bx$
c. Find the correlation coefficient. What does it imply about the significance of the relationship?
d. For the class of 1930, predict the total class gift.
e. For the class of 1964, predict the total class gift.
f. For the class of 1850, predict the total class gift. Why doesn't this value make any sense?

Exercise 12.13.20
We will use the columns "class year" and "average gift" for all questions, unless otherwise stated.

a. What do you think the scatter plot will look like? Make a scatter plot of the data.
b. Calculate the least squares line. Put the equation in the form of: $\hat{y} = a + bx$
c. Find the correlation coefficient. What does it imply about the significance of the relationship?
d. For the class of 1930, predict the average class gift.
e. For the class of 1964, predict the average class gift.
f. For the class of 2010, predict the average class gift. Why doesn't this value make any sense?

12.13.1 Try these multiple choice questions

Exercise 12.13.21 *(Solution on p. 522.)*
A correlation coefficient of -0.95 means there is a ____________ between the two variables.

A. Strong positive correlation
B. Weak negative correlation
C. Strong negative correlation
D. No Correlation

Exercise 12.13.22 *(Solution on p. 522.)*
According to the data reported by the New York State Department of Health regarding West Nile Virus for the years 2000-2004, the least squares line equation for the number of reported dead birds (x) versus the number of human West Nile virus cases (y) is $\hat{y} = -10.2638 + 0.0491x$. If the number of dead birds reported in a year is 732, how many human cases of West Nile virus can be expected?

A. 25.7
B. 46.2
C. -25.7
D. 7513

The next three questions refer to the following data: (showing the number of hurricanes by category to directly strike the mainland U.S. each decade) obtained from *www.nhc.noaa.gov/gifs/table6.gif*[14] A major hurricane is one with a strength rating of 3, 4 or 5.

[14] http://www.nhc.noaa.gov/gifs/table6.gif

Decade	Total Number of Hurricanes	Number of Major Hurricanes
1941-1950	24	10
1951-1960	17	8
1961-1970	14	6
1971-1980	12	4
1981-1990	15	5
1991-2000	14	5
2001 – 2004	9	3

Table 12.20

Exercise 12.13.23 *(Solution on p. 522.)*

Using only completed decades (1941 – 2000), calculate the least squares line for the number of major hurricanes expected based upon the total number of hurricanes.

A. $\hat{y} = -1.67x + 0.5$
B. $\hat{y} = 0.5x - 1.67$
C. $\hat{y} = 0.94x - 1.67$
D. $\hat{y} = -2x + 1$

Exercise 12.13.24 *(Solution on p. 522.)*

The correlation coefficient is 0.942. Is this considered significant? Why or why not?

A. No, because 0.942 is greater than the critical value of 0.707
B. Yes, because 0.942 is greater than the critical value of 0.707
C. No, because 0942 is greater than the critical value of 0.811
D. Yes, because 0.942 is greater than the critical value of 0.811

Exercise 12.13.25 *(Solution on p. 522.)*

The data for 2001-2004 show 9 hurricanes have hit the mainland United States. The line of best fit predicts 2.83 major hurricanes to hit mainland U.S. Can the least squares line be used to make this prediction?

A. No, because 9 lies outside the independent variable values
B. Yes, because, in fact, there have been 3 major hurricanes this decade
C. No, because 2.83 lies outside the dependent variable values
D. Yes, because how else could we predict what is going to happen this decade.

12.14 Lab 1: Regression (Distance from School)[15]

Class Time:

Names:

12.14.1 Student Learning Outcomes:

- The student will calculate and construct the line of best fit between two variables.
- The student will evaluate the relationship between two variables to determine if that relationship is significant.

12.14.2 Collect the Data

Use 8 members of your class for the sample. Collect bivariate data (distance an individual lives from school, the cost of supplies for the current term).

1. Complete the table.

Distance from school	Cost of supplies this term

Table 12.21

2. Which variable should be the dependent variable and which should be the independent variable? Why?
3. Graph "distance" vs. "cost." Plot the points on the graph. Label both axes with words. Scale both axes.

[15]This content is available online at <http://cnx.org/content/m17080/1.10/>.

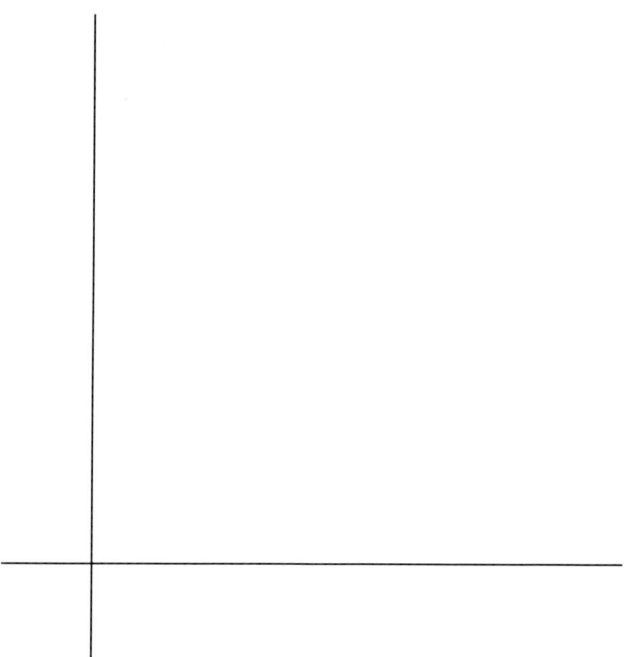

Figure 12.15

12.14.3 Analyze the Data

Enter your data into your calculator or computer. Write the linear equation below, rounding to 4 decimal places.

1. Calculate the following:
   a. $a =$
   b. $b=$
   c. correlation =
   d. $n =$
   e. equation: $\hat{y} =$
   f. Is the correlation significant? Why or why not? (Answer in 1-3 complete sentences.)
2. Supply an answer for the following senarios:
   a. For a person who lives 8 miles from campus, predict the total cost of supplies this term:
   b. For a person who lives 80 miles from campus, predict the total cost of supplies this term:
3. Obtain the graph on your calculator or computer. Sketch the regression line below.

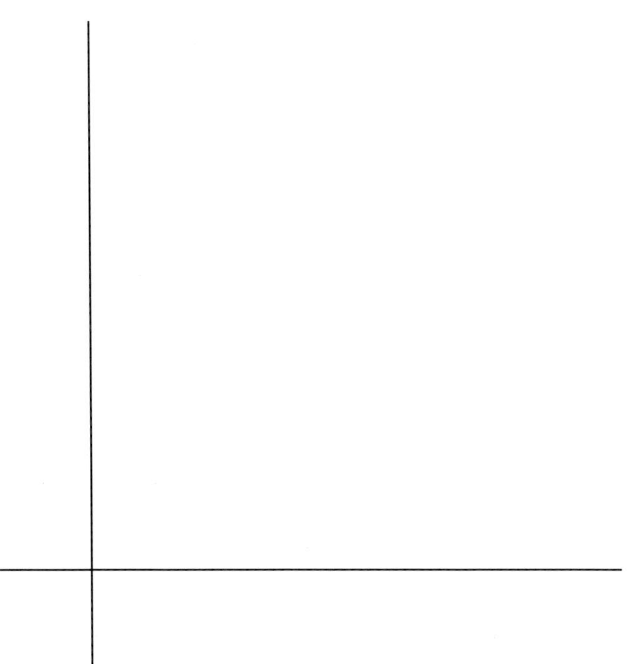

Figure 12.16

12.14.4 Discussion Questions

1. Answer each with 1-3 complete sentences.
   a. Does the line seem to fit the data? Why?
   b. What does the correlation imply about the relationship between the distance and the cost?
2. Are there any outliers? If so, which point is an outlier?
3. Should the outlier, if it exists, be removed? Why or why not?

12.15 Lab 2: Regression (Textbook Cost)[16]

Class Time:

Names:

12.15.1 Student Learning Outcomes:

- The student will calculate and construct the line of best fit between two variables.
- The student will evaluate the relationship between two variables to determine if that relationship is significant.

12.15.2 Collect the Data

Survey 10 textbooks. Collect bivariate data (number of pages in a textbook, the cost of the textbook).

1. Complete the table.

Number of pages	Cost of textbook

Table 12.22

2. Which variable should be the dependent variable and which should be the independent variable? Why?
3. Graph "distance" vs. "cost." Plot the points on the graph in "Analyze the Data". Label both axes with words. Scale both axes.

12.15.3 Analyze the Data

Enter your data into your calculator or computer. Write the linear equation below, rounding to 4 decimal places.

1. Calculate the following:
   - **a.** $a =$
   - **b.** $b =$
   - **c.** correlation =
   - **d.** $n =$

[16]This content is available online at <http://cnx.org/content/m17087/1.9/>.

e. equation: $y =$

f. Is the correlation significant? Why or why not? (Answer in 1-3 complete sentences.)

2. Supply an answer for the following senarios:
   a. For a textbook with 400 pages, predict the cost:
   b. For a textbook with 600 pages, predict the cost:
3. Obtain the graph on your calculator or computer. Sketch the regression line below.

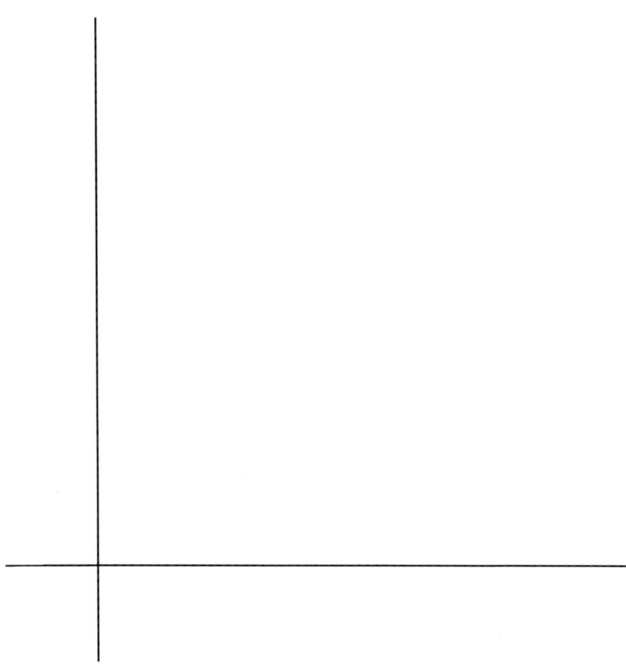

Figure 12.17

12.15.4 Discussion Questions

1. Answer each with 1-3 complete sentences.
   a. Does the line seem to fit the data? Why?
   b. What does the correlation imply about the relationship between the number of pages and the cost?
2. Are there any outliers? If so, which point(s) is an outlier?
3. Should the outlier, if it exists, be removed? Why or why not?

12.16 Lab 3: Regression (Fuel Efficiency)[17]

Class Time:

Names:

12.16.1 Student Learning Outcomes:

- The student will calculate and construct the line of best fit between two variables.
- The student will evaluate the relationship between two variables to determine if that relationship is significant.

12.16.2 Collect the Data

Use the most recent April issue of Consumer Reports. It will give the total fuel efficiency (in miles per gallon) and weight (in pounds) of new model cars with automatic transmissions. We will use this data to determine the relationship, if any, between the fuel efficiency of a car and its weight.

1. Which variable should be the independent variable and which should be the dependent variable? Explain your answer in one or two complete sentences.
2. Using your random number generator, randomly select 20 cars from the list and record their weights and fuel efficiency into the table below.

[17] This content is available online at <http://cnx.org/content/m17079/1.8/>.

Weight	Fuel Efficiency

Table 12.23

3. Which variable should be the dependent variable and which should be the independent variable? Why?
4. By hand, do a scatterplot of "weight" vs. "fuel efficiency". Plot the points on graph paper. Label both axes with words. Scale both axes accurately.

Figure 12.18

12.16.3 Analyze the Data

Enter your data into your calculator or computer. Write the linear equation below, rounding to 4 decimal places.

1. Calculate the following:
   - **a.** $a =$
   - **b.** $b =$
   - **c.** correlation =
   - **d.** $n =$
   - **e.** equation: $\hat{y} =$
2. Obtain the graph of the regression line on your calculator. Sketch the regression line on the same axes as your scatterplot.

12.16.4 Discussion Questions

1. Is the correlation significant? Explain how you determined this in complete sentences.

2. Is the relationship a positive one or a negative one? Explain how you can tell and what this means in terms of weight and fuel efficiency.
3. In one or two complete sentences, what is the practical interpretation of the slope of the least squares line in terms of fuel efficiency and weight?
4. For a car that weighs 4000 pounds, predict its fuel efficiency. Include units.
5. Can we predict the fuel efficiency of a car that weighs 10000 pounds using the least squares line? Explain why or why not.
6. Questions. Answer each in 1 to 3 complete sentences.
   - **a.** Does the line seem to fit the data? Why or why not?
   - **b.** What does the correlation imply about the relationship between fuel efficiency and weight of a car? Is this what you expected?
7. Are there any outliers? If so, which point is an outlier?

** This lab was designed and contributed by Diane Mathios.

Solutions to Exercises in Chapter 12

Solution to Example 12.11, Problem 2 (p. 486)
78 is outside of the domain of x values (independent variables), so you cannot reliably predict the final exam score for this student.
Solution to Example 12.13 (p. 489)
184.28

Solutions to Practice: Linear Regression

Solution to Exercise 12.12.2 (p. 495)

- **a.** $a =$ -3,448,225
- **b.** $b = 1750$
- **c.** $corr. = 0.4526$
- **d.** $n = 22$

Solution to Exercise 12.12.3 (p. 495)
$\hat{y}$= -3,448,225 +1750x
Solution to Exercise 12.12.4 (p. 495)

- **a.** 25082
- **b.** 33,831

Solution to Exercise 12.12.10 (p. 496)

- **a.** -1164

Solutions to Homework

Solution to Exercise 12.13.1 (p. 497)

- **a.** Independent: Age; Dependent: Fatalities
- **d.** Independent: Power Consumption; Dependent: Utility

Solution to Exercise 12.13.3 (p. 497)

- **b.** $\hat{y} = 88.7206 - 0.0432x$
- **c.** -0.8533, Yes
- **g.** No
- **h.** No.
- **i.** 2.97, Yes
- **j.** slope = -0.0432. As the year increases by one, the welfare family size decreases by 0.0432 people.

Solution to Exercise 12.13.5 (p. 498)

- **b.** Yes
- **c.** $\hat{y} = 102.4287 + 11.7585x$
- **d.** 0.9436; yes
- **e.** 478.70 feet; 1207.73 feet
- **g.** Yes
- **h.** Yes; $(57, 1050)$
- **i.** 172.98; No
- **j.** 11.7585 feet

k. slope = 11.7585. As the number of stories increases by one, the height of the building increases by 11.7585 feet.

Solution to Exercise 12.13.7 (p. 500)

b. Yes
c. $\hat{y} = -266.8863 + 0.1656x$
d. 0.9448; Yes
e. 62.9206; 62.4237
h. No
i. 72.639; No
j. slope = 0.1656. As the year increases by one, the percent of workers paid hourly rates increases by 0.1565.

Solution to Exercise 12.13.9 (p. 502)

b. Yes
c. $\hat{y} = 3.5984 + 0.0371x$
d. 0.9986; Yes
e. $5.08
f. $6.93
i. No
j. Not valid
k. slope = 0.0371. As the number of ounces increases by one, the cost of the liquid detergent increases by $0.0371 (or about 4 cents).

Solution to Exercise 12.13.11 (p. 503)

c. Yes
d. $\hat{y} = -337,424.6478 + 0.5463x$
e. 0.9964; Yes
f. $208,872.49; $1,028,318.20
h. Yes
i. No
k. slope = 0.5463. As the net taxable estate increases by one dollar, the approximate probate fees and taxes increases by 0.5463 dollars (about 55 cents).

Solution to Exercise 12.13.13 (p. 504)

c. Yes
d. $\hat{y} = 65.0876 + 7.0948x$
e. 0.9761; yes
f. 72.2 cm; 143.13 cm
h. Yes
i. No
j. 505.0 cm; No
k. slope = 7.0948. As the age of an American boy increases by one year, the average height increases by 7.0948 cm.

Solution to Exercise 12.13.15 (p. 506)

c. No
d. $\hat{y} = 47.03 - 0.216x$

e. -0.4280
f. 6; 5

Solution to Exercise 12.13.17 (p. 507)

d. $\hat{y} = -480.5845 + 0.2748x$
e. 0.9553
f. 1934

Solution to Exercise 12.13.19 (p. 508)

b. $\hat{y} = -569,770.2796 + 296.0351$
c. 0.8302
d. $1577.48
e. $11,642.68
f. -$22,105.33

Solution to Exercise 12.13.21 (p. 509)
C
Solution to Exercise 12.13.22 (p. 509)
A
Solution to Exercise 12.13.23 (p. 510)
A
Solution to Exercise 12.13.24 (p. 510)
D
Solution to Exercise 12.13.25 (p. 510)
A

Chapter 13

F Distribution and ANOVA

13.1 F Distribution and ANOVA[1]

13.1.1 Student Learning Objectives

By the end of this chapter, the student should be able to:

- Interpret the F probability distribution as the number of groups and the sample size change.
- Discuss two uses for the F distribution, ANOVA and the test of two variances.
- Conduct and interpret ANOVA.
- Conduct and interpret hypothesis tests of two variances (optional).

13.1.2 Introduction

Many statistical applications in psychology, social science, business administration, and the natural sciences involve several groups. For example, an environmentalist is interested in knowing if the average amount of pollution varies in several bodies of water. A sociologist is interested in knowing if the amount of income a person earns varies according to his or her upbringing. A consumer looking for a new car might compare the average gas mileage of several models.

For hypothesis tests involving more than two averages, statisticians have developed a method called Analysis of Variance" (abbreviated ANOVA). In this chapter, you will study the simplest form of ANOVA called single factor or one-way ANOVA. You will also study the F distribution, used for ANOVA, and the test of two variances. This is just a very brief overview of ANOVA. You will study this topic in much greater detail in future statistics courses.

- ANOVA, as it is presented here, relies heavily on a calculator or computer.
- For further information about ANOVA, use the online link ANOVA[2] . Use the back button to return here. (The url is http://en.wikipedia.org/wiki/Analysis_of_variance.)

[1]This content is available online at <http://cnx.org/content/m17065/1.7/>.

[2]http://en.wikipedia.org/wiki/Analysis_of_variance

13.2 ANOVA[3]

13.2.1 F Distribution and ANOVA: Purpose and Basic Assumption of ANOVA

The purpose of an **ANOVA** test is to determine the existence of a statistically significant difference among several group means. The test actually uses **variances** to help determine if the means are equal or not.

In order to perform an ANOVA test, there are three basic **assumptions** to be fulfilled:

- Each population from which a sample is taken is assumed to be normal.
- Each sample is randomly selected and independent.
- The populations are assumed to have **equal standard deviations (or variances).**

13.2.2 The Null and Alternate Hypotheses

The null hypothesis is simply that all the group population means are the same. The alternate hypothesis is that at least one pair of means is different. For example, if there are k groups:

$H_o : \mu_1 = \mu_2 = \mu_3 = ... = \mu_k$

H_a : At least two of the group means $\mu_1, \mu_2, \mu_3, ..., \mu_k$ are not equal.

13.3 The F Distribution and the F Ratio[4]

The distribution used for the hypothesis test is a new one. It is called the F distribution, named after Sir Ronald Fisher, an English statistician. The F statistic is a ratio (a fraction). There are two sets of degrees of freedom; one for the numerator and one for the denominator.

For example, if F follows an F distribution and the degrees of freedom for the numerator are 4 and the degrees of freedom for the denominator are 10, then $F \sim F_{4,10}$.

To calculate the F ratio, two estimates of the variance are made.

1. **Variance between samples:** An estimate of σ^2 that is the variance of the sample means. If the samples are different sizes, the variance between samples is weighted to account for the different sample sizes. The variance is also called **variation due to treatment or explained variation.**
2. **Variance within samples:** An estimate of σ^2 that is the average of the sample variances (also known as a pooled variance). When the sample sizes are different, the variance within samples is weighted. The variance is also called the **variation due to error or unexplained variation.**

- $SS_{between}$ = the sum of squares that represents the variation among the different samples.
- SS_{within} = the sum of squares that represents the variation within samples that is due to chance.

To find a "sum of squares" means to add together squared quantities which, in some cases, may be weighted. We used sum of squares to calculate the sample variance and the sample standard deviation in **Descriptive Statistics**.

MS means "mean square." $MS_{between}$ is the variance between groups and MS_{within} is the variance within groups.

[3] This content is available online at <http://cnx.org/content/m17068/1.5/>.
[4] This content is available online at <http://cnx.org/content/m17076/1.7/>.

Calculation of Sum of Squares and Mean Square

- k = the number of different groups
- n_j = the size of the jth group
- s_j= the sum of the values in the jth group
- N = total number of all the values combined. (total sample size: $\sum n_j$)
- x = one value: $\sum x = \sum s_j$
- Sum of squares of all values from every group combined: $\sum x^2$
- Between group variability: $SS_{total} = \sum x^2 - \frac{(\sum x)^2}{N}$
- Total sum of squares: $\sum x^2 - \frac{(\sum x)^2}{N}$
- Explained variation- sum of squares representing variation among the different samples $SS_{between} = \sum \left[\frac{(sj)^2}{n_j}\right] - \frac{(\sum s_j)^2}{N}$
- Unexplained variation- sum of squares representing variation within samples due to chance: $SS_{within} = SS_{total} - SS_{between}$
- df's for different groups (df's for the numerator): $df_{between} = k - 1$
- Equation for errors within samples (df's for the denominator): $df_{within} = N - k$
- Mean square (variance estimate) explained by the different groups: $MS_{between} = \frac{SS_{between}}{df_{between}}$
- Mean square (variance estimate) that is due to chance (unexplained): $MS_{within} = \frac{SS_{within}}{df_{within}}$

$MS_{between}$ and MS_{within} can be written as follows:

- $MS_{between} = \frac{SS_{between}}{df_{between}} = \frac{SS_{between}}{k-1}$
- $MS_{within} = \frac{SS_{within}}{df_{within}} = \frac{SS_{within}}{N-k}$

The ANOVA test depends on the fact that $MS_{between}$ can be influenced by population differences among means of the several groups. Since MS_{within} compares values of each group to its own group mean, the fact that group means might be different does not affect MS_{within}.

The null hypothesis says that all groups are samples from populations having the same normal distribution. The alternate hypothesis says that at least two of the sample groups come from populations with different normal distributions. If the null hypothesis is true, $MS_{between}$ and MS_{within} should both estimate the same value.

> NOTE: The null hypothesis says that all the group population means are equal. The hypothesis of equal means implies that the populations have the same normal distribution because it is assumed that the populations are normal and that they have equal variances.

F-Ratio or F Statistic

$$F = \frac{MS_{between}}{MS_{within}} \tag{13.1}$$

If $MS_{between}$ and MS_{within} estimate the same value (following the belief that H_o is true), then the F-ratio should be approximately equal to 1. Only sampling errors would contribute to variations away from 1. As it turns out, $MS_{between}$ consists of the population variance plus a variance produced from the differences between the samples. MS_{within} is an estimate of the population variance. Since variances are always positive, if the null hypothesis is false, $MS_{between}$ will be larger than MS_{within}. The F-ratio will be larger than 1.

The above calculations were done with groups of different sizes. If the groups are the same size, the calculations simplify somewhat and the F ratio can be written as:

F-Ratio Formula when the groups are the same size

$$F = \frac{n \cdot (s_{-x})^{2^2}}{(s_{pooled})} \tag{13.2}$$

where ...

- $(s_{-x})^2$ =the variance of the sample means
- n =the sample size of each group
- $(s_{pooled})^2$ =the mean of the sample variances (pooled variance)
- $df_{numerator} = k - 1$
- $df_{denominator} = k(n-1) = N - k$

The ANOVA hypothesis test is always right-tailed because larger F-values are way out in the right tail of the F-distribution curve and tend to make us reject H_o.

13.3.1 Notation

The notation for the F distribution is $F \sim F_{df(num),df(denom)}$

where $df(num) = df_{between}$ and $df(denom) = df_{within}$

The mean for the F distribution is $\mu = \frac{df(num)}{df(denom)-1}$

13.4 Facts About the F Distribution[5]

1. The curve is not symmetrical but skewed to the right.
2. There is a different curve for each set of *dfs*.
3. The F statistic is greater than or equal to zero.
4. As the degrees of freedom for the numerator and for the denominator get larger, the curve approximates the normal.
5. Other uses for the F distribution include comparing two variances and Two-Way Analysis of Variance. Comparing two variances is discussed at the end of the chapter. Two-Way Analysis is mentioned for your information only.

[5]This content is available online at <http://cnx.org/content/m17062/1.9/>.

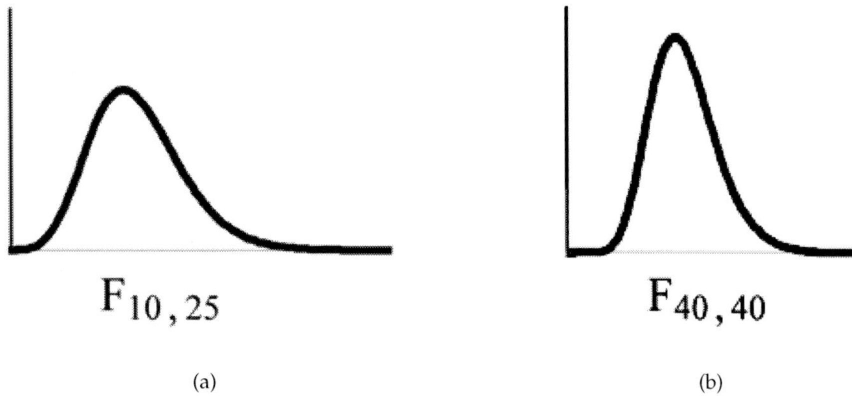

(a) (b)

Figure 13.1

Example 13.1

One-Way ANOVA: Four sororities took a random sample of sisters regarding their grade averages for the past term. The results are shown below:

GRADE AVERAGES FOR FOUR SORORITIES			
Sorority 1	**Sorority 2**	**Sorority 3**	**Sorority 4**
2.17	2.63	2.63	3.79
1.85	1.77	3.78	3.45
2.83	3.25	4.00	3.08
1.69	1.86	2.55	2.26
3.33	2.21	2.45	3.18

Table 13.1

Problem

Using a significance level of 1%, is there a difference in grade averages among the sororities?

Solution

Let μ_1, μ_2, μ_3, μ_4 be the population means of the sororities. Remember that the null hypothesis claims that the sorority groups are from the same normal distribution. The alternate hypothesis says that at least two of the sorority groups come from populations with different normal distributions. Notice that the four sample sizes are each size 5.

$H_o : \mu_1 = \mu_2 = \mu_3 = \mu_4$

H_a: Not all of the means $\mu_1, \mu_2, \mu_3, \mu_4$ are equal.

Distribution for the test: $F_{3,16}$

where $k = 4\ groups$ and $N = 20\ samples\ in\ total$

$df\ (num) = k - 1 = 4 - 1 = 3$

$df\ (denom) = N - k = 20 - 4 = 16$

Calculate the test statistic: $F = 2.23$

Graph:

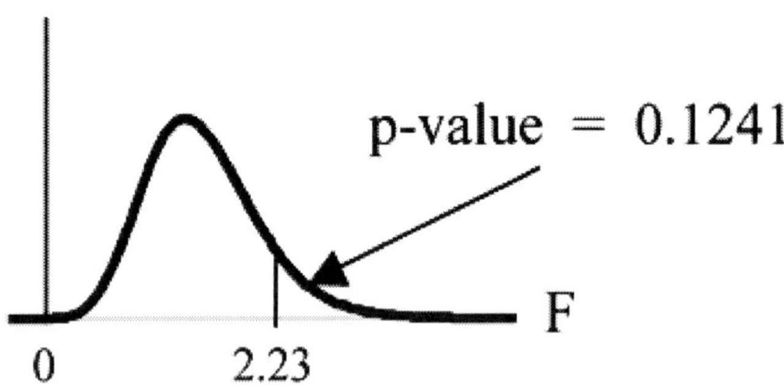

Figure 13.2

Probability statement: $p\text{-}value = P(F > 2.23) = 0.1241$

Compare α **and the** $p - value$**:** $\alpha = 0.01$ $p\text{-}value = 0.1242$ $\alpha < p\text{-}value$

Make a decision: Since $\alpha < p\text{-}value$, you cannot reject H_0.

This means that the population averages appear to be the same.

Conclusion: There is not sufficient evidence to conclude that there is a difference among the grade averages for the sororities.

TI-83+ or TI 84: Put the data into lists L1, L2, L3, and L4. Press `STAT` and arrow over to `TESTS`. Arrow down to `F:ANOVA`. Press `ENTER` and Enter `(L1,L2,L3,L4)`. The F statistic is 2.2303 and the *p-value* is 0.1241. *df(numerator) = 3* (under `"Factor"`) and *df(denominator) = 16* (under `Error`).

Example 13.2
A fourth grade class is studying the environment. One of the assignments is to grow bean plants in different soils. Tommy chose to grow his bean plants in soil found outside his classroom mixed with dryer lint. Tara chose to grow her bean plants in potting soil bought at the local nursery. Nick chose to grow his bean plants in soil from his mother's garden. No chemicals were used on the plants, only water. They were grown inside the classroom next to a large window. Each child grew 5 plants. At the end of the growing period, each plant was measured, producing the following data (in inches):

Tommy's Plants	Tara's Plants	Nick's Plants
24	25	23
21	31	27
23	23	22
30	20	30
23	28	20

Table 13.2

Problem 1

Does it appear that the three media in which the bean plants were grown produce the same average height? Test at a 3% level of significance.

Solution

This time, we will perform the calculations that lead to the F′ statistic. Notice that each group has the same number of plants so we will use the formula $F' = \frac{n \cdot (s_{-x})^{2\,2}}{(s_{pooled})}$.

First, calculate the sample mean and sample variance of each group.

	Tommy's Plants	Tara's Plants	Nick's Plants
Sample Mean	24.2	25.4	24.4
Sample Variance	11.7	18.3	16.3

Table 13.3

Next, calculate the variance of the three group means (Calculate the variance of 24.2, 25.4, and 24.4). **Variance of the group means = 0.413** = $(s_{-x})^2$

Then $MS_{between} = n\,(s_{-x})^2 = (5)\,(0.413)$ where $n = 5$ is the sample size (number of plants each child grew).

Calculate the average of the three sample variances (Calculate the average of 11.7, 18.3, and 16.3). **Average of the sample variances = 15.433** = $(s_{pooled})^2$

Then $MS_{within} = (s_{pooled})^2 = 15.433$.

The F statistic (or F ratio) is $F = \frac{MS_{between}}{MS_{within}} = \frac{n \cdot (s_{-x})^{2\,2}}{(s_{pooled})} = \frac{(5)\cdot(0.413)}{15.433} = 0.134$

The dfs for the numerator = *the number of groups* $- 1 = 3 - 1 = 2$

The dfs for the denominator = *the total number of samples* – *the number of groups* $= 15 - 3 = 12$

The distribution for the test is $F_{2,12}$ and the F statistic is $F = 0.134$

The p-value is $P\,(F > 0.134) = 0.8759$.

Decision: Since $\alpha = 0.03$ and the *p-value* $= 0.8759$, do not reject H_o. (Why?)

Conclusion: With a 3% the level of significance, from the sample data, the evidence is not sufficient to conclude that the average heights of the bean plants are not different. Of the three media tested, it appears that it does not matter which one the bean plants are grown in.

(This experiment was actually done by three classmates of the son of one of the authors.)

Another fourth grader also grew bean plants but this time in a jelly-like mass. The heights were (in inches) 24, 28, 25, 30, and 32.

Problem 2 ***(Solution on p. 544.)***

Do an ANOVA test on the 4 groups. You may use your calculator or computer to perform the test. Are the heights of the bean plants different? Use a solution sheet (Section 14.5.4).

13.4.1 Optional Classroom Activity

Randomly divide the class into four groups of the same size. Have each member of each group record the number of states in the United States he or she has visited. Run an ANOVA test to determine if the average number of states visited in the four groups are the same. Test at a 1% level of significance. Use one of the solution sheets (Section 14.5.4) at the end of the chapter (after the homework).

13.5 Test of Two Variances[6]

Another of the uses of the F distribution is testing two variances. It is often desirable to compare two variances rather than two averages. For instance, college administrators would like two college professors grading exams to have the same variation in their grading. In order for a lid to fit a container, the variation in the lid and the container should be the same. A supermarket might be interested in the variability of check-out times for two checkers.

In order to perform a F test of two variances, it is important that the following are true:

1. The populations from which the two samples are drawn are normally distributed.
2. The two populations are independent of each other.

Suppose we sample randomly from two independent normal populations. Let σ_1^2 and σ_2^2 be the population variances and s_1^2 and s_2^2 be the sample variances. Let the sample sizes be n_1 and n_2. Since we are interested in comparing the two sample variances, we use the F ratio

$$F = \frac{\left[\frac{(s_1)^2}{(\sigma_1)^2}\right]}{\left[\frac{(s_2)^2}{(\sigma_2)^2}\right]}$$

F has the distribution $F \sim F(n_1 - 1, n_2 - 1)$

where $n_1 - 1$ are the degrees of freedom for the numerator and $n_2 - 1$ are the degrees of freedom for the denominator.

[6] This content is available online at <http://cnx.org/content/m17075/1.6/>.

If the null hypothesis is $\sigma_1^2 = \sigma_2^2$, then the F-Ratio becomes $F = \frac{\left[\frac{(s_1)^2}{(\sigma_1)^2}\right]}{\left[\frac{(s_2)^2}{(\sigma_2)^2}\right]} = \frac{(s_1)^2}{(s_2)^2}$.

If the two populations have equal variances, then s_1^2 and s_2^2 are close in value and $F = \frac{(s_1)^2}{(s_2)^2}$ is close to 1. But if the two population variances are very different, s_1^2 and s_2^2 tend to be very different, too. Choosing s_1^2 as the larger sample variance causes the ratio $\frac{(s_1)^2}{(s_2)^2}$ to be greater than 1. If s_1^2 and s_2^2 are far apart, then $F = \frac{(s_1)^2}{(s_2)^2}$ is a large number.

Therefore, if F is close to 1, the evidence favors the null hypothesis (the two population variances are equal). But if F is much larger than 1, then the evidence is against the null hypothesis.

A test of two variances may be left, right, or two-tailed.

Example 13.3
Two college instructors are interested in whether or not there is any variation in the way they grade math exams. They each grade the same set of 30 exams. The first instructor's grades have a variance of 52.3. The second instructor's grades have a variance of 89.9.

Problem
Test the claim that the first instructor's variance is smaller. (In most colleges, it is desirable for the variances of exam grades to be nearly the same among instructors.) The level of significance is 10%.

Solution
Let 1 and 2 be the subscripts that indicate the first and second instructor, respectively.

$n_1 = n_2 = 30.$

H_o: $\sigma_1^2 = \sigma_2^2$ and H_a: $\sigma_1^2 < \sigma_2^2$

Calculate the test statistic: By the null hypothesis $(\sigma_1^2 = \sigma_2^2)$, the F statistic is

$$F = \frac{\left[\frac{(s_1)^2}{(\sigma_1)^2}\right]}{\left[\frac{(s_2)^2}{(\sigma_2)^2}\right]} = \frac{(s_1)^2}{(s_2)^2} = \frac{52.3}{89.9} = 0.6$$

Distribution for the test: $F_{29,29}$ where $n_1 - 1 = 29$ and $n_2 - 1 = 29$.

Graph: **This test is left tailed.**

Draw the graph labeling and shading appropriately.

Figure 13.3

Probability statement: $p\text{-}value = P(F < 0.582) = 0.0755$

Compare α and the p-value: $\alpha = 0.10 \qquad \alpha > p\text{-}value.$

Make a decision: Since $\alpha > p\text{-}value$, reject H_o.

Conclusion: With a 10% level of significance, from the data, there is sufficient evidence to conclude that the variance in grades for the first instructor is smaller.

TI-83+ and TI-84: Press `STAT` and arrow over to `TESTS`. Arrow down to `D:2-SampFTest`. Press `ENTER`. Arrow to `Stats` and press `ENTER`. For `Sx1`, `n1`, `Sx2`, and `n2`, enter $\sqrt{(52.3)}$, `30`, $\sqrt{(89.9)}$, and `30`. Press `ENTER` after each. Arrow to σ`1:` and $<\sigma 2$. Press `ENTER`. Arrow down to `Calculate` and press `ENTER`. $F = 0.5818$ and $p\text{-}value = 0.0753$. Do the procedure again and try `Draw` instead of `Calculate`.

13.6 Summary[7]

- An **ANOVA** hypothesis test determines if several population means are equal. The distribution for the test is the F distribution with 2 different degrees of freedom.

  Assumptions:

  1. Each population from which a sample is taken is assumed to be normal.
  2. Each sample is randomly selected and independent.
  3. The populations are assumed to have equal standard deviations (or variances)

- A **Test of Two Variances** hypothesis test determines if two variances are the same. The distribution for the hypothesis test is the F distribution with 2 different degrees of freedom.

  Assumptions:

  1. The populations from which the two samples are drawn are normally distributed.
  2. The two populations are independent of each other.

[7]This content is available online at <http://cnx.org/content/m17072/1.3/>.

13.7 Practice: ANOVA[8]

13.7.1 Student Learning Outcome

- The student will explore the properties of ANOVA.

13.7.2 Given

Suppose a group is interested in determining whether teenagers obtain their drivers licenses at approximately the same average age across the country. Suppose that the following data are randomly collected from five teenagers in each region of the country. The numbers represent the age at which teenagers obtained their drivers licenses.

	Northeast	South	West	Central	East
	16.3	16.9	16.4	16.2	17.1
	16.1	16.5	16.5	16.6	17.2
	16.4	16.4	16.6	16.5	16.6
	16.5	16.2	16.1	16.4	16.8
$\overline{x} =$	________	________	________	________	________
$s^2 =$	________	________	________	________	________

Table 13.4

13.7.3 Hypothesis

Exercise 13.7.1
State the hypotheses.

H_o:

H_a:

13.7.4 Data Entry

Enter the data into your calculator or computer.

Exercise 13.7.2 ***(Solution on p. 544.)***
degrees of freedom - numerator: $df(n) =$

Exercise 13.7.3 ***(Solution on p. 544.)***
degrees of freedom - denominator: $df(d) =$

Exercise 13.7.4 ***(Solution on p. 544.)***
F test statistic =

Exercise 13.7.5 ***(Solution on p. 544.)***
p-value =

[8]This content is available online at <http://cnx.org/content/m17067/1.8/>.

13.7.5 Decisions and Conclusions

State the decisions and conclusions (in complete sentences) for the following preconceived levels of α .

Exercise 13.7.6
$\alpha = 0.05$

Decision:

Conclusion:

Exercise 13.7.7
$\alpha = 0.01$

Decision:

Conclusion:

13.8 Homework[9]

DIRECTIONS: Use a solution sheet to conduct the following hypothesis tests. The solution sheet can be found in the Table of Contents 14. Appendix.

Exercise 13.8.1 *(Solution on p. 544.)*

Three students, Linda, Tuan, and Javier, are given 5 laboratory rats each for a nutritional experiment. Each rat's weight is recorded in grams. Linda feeds her rats Formula A, Tuan feeds his rats Formula B, and Javier feeds his rats Formula C. At the end of a specified time period, each rat is weighed again and the net gain in grams is recorded. Using a significance level of 10%, test the hypothesis that the three formulas produce the same average weight gain.

Weights of Student Lab Rats

Linda's rats	Tuan's rats	Javier's rats
43.5	47.0	51.2
39.4	40.5	40.9
41.3	38.9	37.9
46.0	46.3	45.0
38.2	44.2	48.6

Table 13.5

Exercise 13.8.2

A grassroots group opposed to a proposed increase in the gas tax claimed that the increase would hurt working-class people the most, since they commute the farthest to work. Suppose that the group randomly surveyed 24 individuals and asked them their daily one-way commuting mileage. The results are below:

working-class	professional (middle incomes)	professional (wealthy)
17.8	16.5	8.5
26.7	17.4	6.3
49.4	22.0	4.6
9.4	7.4	12.6
65.4	9.4	11.0
47.1	2.1	28.6
19.5	6.4	15.4
51.2	13.9	9.3

Table 13.6

Exercise 13.8.3 *(Solution on p. 544.)*

Refer to Exercise 13.8.1. Determine whether or not the variance in weight gain is statistically the same among Javier's and Linda's rats.

[9]This content is available online at <http://cnx.org/content/m17063/1.9/>.

Exercise 13.8.4
Refer to Exercise 13.8.2 above (Exercise 13.8.2). Determine whether or not the variance in mileage driven is statistically the same among the working class and professional (middle income) groups.

For the next two problems, refer to the data from Terri Vogel's Log Book [link pending].

Exercise 13.8.5 *(Solution on p. 544.)*
Examine the 7 practice laps. Determine whether the average lap time is statistically the same for the 7 practice laps, or if there is at least one lap that has a different average time from the others.

Exercise 13.8.6
Examine practice laps 3 and 4. Determine whether or not the variance in lap time is statistically the same for those practice laps.

For the next four problems, refer to the following data.

The following table lists the number of pages in four different types of magazines.

home decorating	news	health	computer
172	87	82	104
286	94	153	136
163	123	87	98
205	106	103	207
197	101	96	146

Table 13.7

Exercise 13.8.7 *(Solution on p. 544.)*
Using a significance level of 5%, test the hypothesis that the four magazine types have the same average length.

Exercise 13.8.8
Eliminate one magazine type that you now feel has an average length different than the others. Redo the hypothesis test, testing that the remaining three averages are statistically the same. Use a new solution sheet. Based on this test, are the average lengths for the remaining three magazines statistically the same?

Exercise 13.8.9
Which two magazine types do you think have the same variance in length?

Exercise 13.8.10
Which two magazine types do you think have different variances in length?

13.9 Review[10]

The next two questions refer to the following situation:

Suppose that the probability of a drought in any independent year is 20%. Out of those years in which a drought occurs, the probability of water rationing is 10%. However, in any year, the probability of water rationing is 5%.

Exercise 13.9.1 *(Solution on p. 545.)*

What is the probability of both a drought **and** water rationing occurring?

Exercise 13.9.2 *(Solution on p. 545.)*

Out of the years with water rationing, find the probability that there is a drought.

The next three questions refer to the following survey:

Favorite Type of Pie by Gender

	apple	pumpkin	pecan
female	40	10	30
male	20	30	10

Table 13.8

Exercise 13.9.3 *(Solution on p. 545.)*

Suppose that one individual is randomly chosen. Find the probability that the person's favorite pie is apple **or** the person is male.

Exercise 13.9.4 *(Solution on p. 545.)*

Suppose that one male is randomly chosen. Find the probability his favorite pie is pecan.

Exercise 13.9.5 *(Solution on p. 545.)*

Conduct a hypothesis test to determine if favorite pie type and gender are independent.

The next two questions refer to the following situation:

Let's say that the probability that an adult watches the news at least once per week is 0.60.

Exercise 13.9.6 *(Solution on p. 545.)*

We randomly survey 14 people. On average, how many people do we expect to watch the news at least once per week?

Exercise 13.9.7 *(Solution on p. 545.)*

We randomly survey 14 people. Of interest is the number that watch the news at least once per week. State the distribution of X. $X \sim$

Exercise 13.9.8 *(Solution on p. 545.)*

The following histogram is most likely to be a result of sampling from which distribution?

[10]This content is available online at <http://cnx.org/content/m17070/1.8/>.

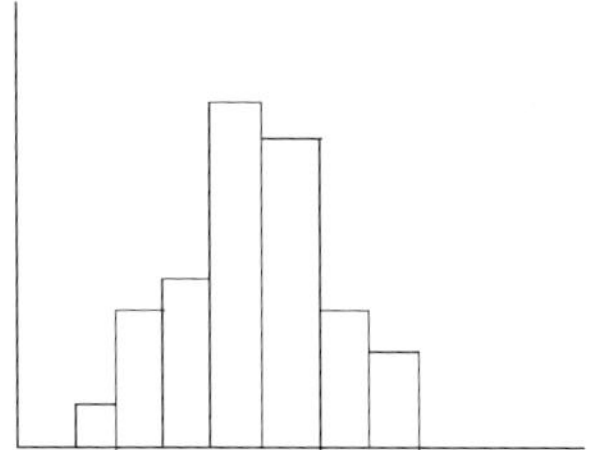

Figure 13.4

A. Chi-Square
B. Geometric
C. Uniform
D. Binomial

Exercise 13.9.9

The ages of De Anza evening students is known to be normally distributed. A sample of 6 De Anza evening students reported their ages (in years) as: 28; 35; 47; 45; 30; 50. Find the probability that the average of 6 ages of randomly chosen students is less than 35 years.

The next three questions refer to the following situation:

The amount of money a customer spends in one trip to the supermarket is known to have an exponential distribution. Suppose the average amount of money a customer spends in one trip to the supermarket is \$72.

Exercise 13.9.10 *(Solution on p. 545.)*

Find the probability that one customer spends less than \$72 in one trip to the supermarket?

Exercise 13.9.11 *(Solution on p. 545.)*

Suppose 5 customers pool their money. (They are poor college students.) How much money altogether would you expect the 5 customers to spend in one trip to the supermarket (in dollars)?

Exercise 13.9.12 *(Solution on p. 545.)*

State the distribution to use is if you want to find the probability that the **average** amount spent by 5 customers in one trip to the supermarket is less than \$60.

Exercise 13.9.13 *(Solution on p. 545.)*

A math exam was given to all the fifth grade children attending Country School. Two random samples of scores were taken. The null hypothesis is that the average math scores for boys and girls in fifth grade are the same. Conduct a hypothesis test.

	n	$\overline{x}$	s^2
Boys	55	82	29
Girls	60	86	46

Table 13.9

Exercise 13.9.14 *(Solution on p. 545.)*

In a survey of 80 males, 55 had played an organized sport growing up. Of the 70 females surveyed, 25 had played an organized sport growing up. We are interested in whether the proportion for males is higher than the proportion for females. Conduct a hypothesis test.

Exercise 13.9.15 *(Solution on p. 545.)*

Which of the following is preferable when designing a hypothesis test?

A. Maximize α and minimize β
B. Minimize α and maximize β
C. Maximize α and β
D. Minimize α and β

The next three questions refer to the following situation:

120 people were surveyed as to their favorite beverage (non-alcoholic). The results are below.

Preferred Beverage by Age

	0 – 9	**10 – 19**	**20 – 29**	**30 +**	**Totals**
Milk	14	10	6	0	30
Soda	3	8	26	15	52
Juice	7	12	12	7	38
Totals	24	30	44	22	120

Table 13.10

Exercise 13.9.16 *(Solution on p. 545.)*

Are the events of **milk** and **30+**:

a. Independent events? Justify your answer.
b. Mutually exclusive events? Justify your answer.

Exercise 13.9.17 *(Solution on p. 545.)*

Suppose that one person is randomly chosen. Find the probability that person is **10 – 19** given that he/she **prefers juice**.

Exercise 13.9.18 *(Solution on p. 545.)*

Are **Preferred Beverage** and **Age** independent events? Conduct a hypothesis test.

Exercise 13.9.19 *(Solution on p. 545.)*

Given the following histogram, which distribution is the data most likely to come from?

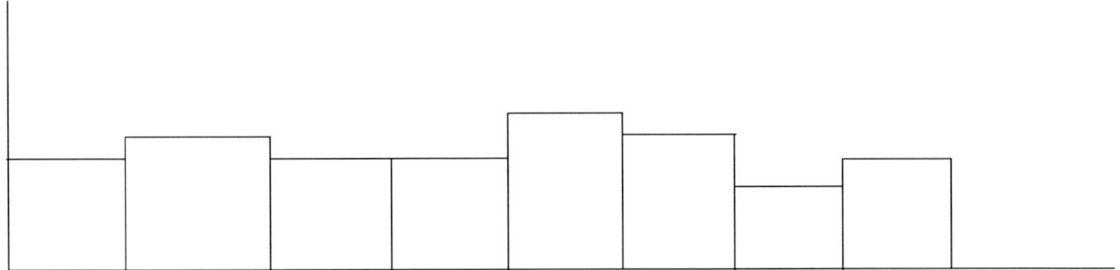

Figure 13.5

A. uniform
B. exponential
C. normal
D. chi-square

13.10 Lab: ANOVA[11]

Class Time:

Names:

13.10.1 Student Learning Outcome:

- The student will conduct a simple ANOVA test involving three variables.

13.10.2 Collect the Data

1. Record the price per pound of 8 fruits, 8 vegetables, and 8 breads in your local supermarket.

Fruits	Vegetables	Breads

Table 13.11

2. Explain how you could try to collect the data randomly.

13.10.3 Analyze the Data and Conduct a Hypothesis Test

1. Compute the following:
   a. Fruit:
      i. $\overline{x}$ =
      ii. s_x=
      iii. n =
   a. Vegetables:
      i. $\overline{x}$ =
      ii. s_x =
      iii. n=
   a. Bread:
      i. $\overline{x}$ =
      ii. s_x=
      iii. n =

[11]This content is available online at <http://cnx.org/content/m17061/1.8/>.

2. Find the following:
   a. $df\ (num) =$
   b. $df\ (denom) =$
3. State the approximate distribution for the test.
4. Test statistic: $F =$
5. Sketch a graph of this situation. CLEARLY, label and scale the horizontal axis and shade the region(s) corresponding to the p-value.
6. p-value =
7. Test at $\alpha = 0.05$. State your decision and conclusion.
8. **a.** Decision: Why did you make this decision?
   b. Conclusion (write a complete sentence).
   c. Based on the results of your study, is there a need to further investigate any of the food groups' prices? Why or why not?

Solutions to Exercises in Chapter 13

Solution to Example 13.2, Problem 2 (p. 530)

- $F = 0.9496$
- $p - value = 0.4401$

The heights of the bean plants are the same.

Solutions to Practice: ANOVA

Solution to Exercise 13.7.2 (p. 534)
$df(1) = 4$
Solution to Exercise 13.7.3 (p. 534)
$df(2) = 15$
Solution to Exercise 13.7.4 (p. 534)
Test statistic $= F = 4.22$
Solution to Exercise 13.7.5 (p. 534)
0.017

Solutions to Homework

Solution to Exercise 13.8.1 (p. 536)

a. H_o: $\mu_L = \mu_T = \mu_J$
c. $df(n) = 2; df(d) = 12$
e. 0.67
f. 0.5305
h. Decision: Do not reject null; Conclusion: Means are same

Solution to Exercise 13.8.3 (p. 536)

c. $df(n) = 4; df(d) = 4$
e. 3.00
f. $2(0.1563) = 0.3126$
h. Decision: Do not reject null; Conclusion: Variances are same

Solution to Exercise 13.8.5 (p. 537)

c. $df(n) = 6; df(d) = 98$
e. 1.69
f. 0.1319
h. Decision: Do not reject null; Conclusion: Average lap times are the same

Solution to Exercise 13.8.7 (p. 537)

a. H_o: $\mu_d = \mu_n = \mu_h = \mu_c$
b. At least one average is different
c. $df(n) = 3; df(d) = 16$
e. 8.69
f. 0.0012
h. Decision: Reject null; Conclusion: At least one average is different

Solutions to Review

Solution to Exercise 13.9.1 (p. 538)
0.02
Solution to Exercise 13.9.2 (p. 538)
0.40
Solution to Exercise 13.9.3 (p. 538)
$\frac{100}{140}$
Solution to Exercise 13.9.4 (p. 538)
$\frac{10}{60}$
Solution to Exercise 13.9.5 (p. 538)
$p\text{-}value = 0$; Reject null; Conclude dependent events
Solution to Exercise 13.9.6 (p. 538)
8.4
Solution to Exercise 13.9.7 (p. 538)
$B\,(14, 0.60)$
Solution to Exercise 13.9.8 (p. 538)
D
Solution to Exercise 13.9.10 (p. 539)
0.6321
Solution to Exercise 13.9.11 (p. 539)
$360
Solution to Exercise 13.9.12 (p. 539)
$N\left(72, \frac{72}{\sqrt{5}}\right)$
Solution to Exercise 13.9.13 (p. 539)
$p\text{-}value = 0.0006$; Reject null; Conclude averages are not equal
Solution to Exercise 13.9.14 (p. 540)
$p\text{-}value = 0$; Reject null; Conclude proportion of males is higher
Solution to Exercise 13.9.15 (p. 540)
D
Solution to Exercise 13.9.16 (p. 540)

- **a.** No
- **b.** Yes, $P\,(M \text{ and } 30+) = 0$

Solution to Exercise 13.9.17 (p. 540)
$\frac{12}{38}$
Solution to Exercise 13.9.18 (p. 540)
No; $p\text{-}value = 0$
Solution to Exercise 13.9.19 (p. 540)
A

Appendix

14.1 Practice Final Exam 1[12]

Questions 1-2 refer to the following:

An experiment consists of tossing two 12-sided dice (the numbers 1-12 are printed on the sides of each dice).

- Let Event A = both dice show an even number
- Let Event B = both dice show a number more than 8

Exercise 14.1.1 *(Solution on p. 602.)*

Events A and B are:

A. Mutually exclusive.
B. Independent.
C. Mutually exclusive and independent.
D. Neither mutually exclusive nor independent.

Exercise 14.1.2 *(Solution on p. 602.)*

Find $P(A|B)$

A. $\frac{2}{4}$
B. $\frac{16}{144}$
C. $\frac{4}{16}$
C. $\frac{2}{144}$

Exercise 14.1.3 *(Solution on p. 602.)*

Which of the following are TRUE when we perform a hypothesis test on matched or paired samples?

A. Sample sizes are almost never small.
B. Two measurements are drawn from the same pair of individuals or objects.
C. Two sample averages are compared to each other.
D. Answer choices B and C are both true.

Questions 4 - 5 refer to the following:

118 students were asked what type of color their bedrooms were painted: light colors, dark colors or vibrant colors. The results were tabulated according to gender.

[12] This content is available online at <http://cnx.org/content/m16304/1.15/>.

	Light colors	Dark colors	Vibrant colors
Female	20	22	28
Male	10	30	8

Table 14.1

Exercise 14.1.4 *(Solution on p. 602.)*

Find the probability that a randomly chosen student is male or has a bedroom painted with light colors.

A. $\frac{10}{118}$
B. $\frac{68}{118}$
C. $\frac{48}{118}$
D. $\frac{10}{48}$

Exercise 14.1.5 *(Solution on p. 602.)*

Find the probability that a randomly chosen student is male given the student's bedroom is painted with dark colors.

A. $\frac{30}{118}$
B. $\frac{30}{48}$
C. $\frac{22}{118}$
D. $\frac{30}{52}$

Questions 6 – 7 refer to the following:

We are interested in the number of times a teenager must be reminded to do his/her chores each week. A survey of 40 mothers was conducted. The table below shows the results of the survey.

X	$P(x)$
0	$\frac{2}{40}$
1	$\frac{5}{40}$
2	
3	$\frac{14}{40}$
4	$\frac{7}{40}$
5	$\frac{4}{40}$

Table 14.2

Exercise 14.1.6 *(Solution on p. 602.)*

Find the probability that a teenager is reminded 2 times.

A. 8
B. $\frac{8}{40}$
C. $\frac{6}{40}$
D. 2

Exercise 14.1.7 *(Solution on p. 602.)*

Find the expected number of times a teenager is reminded to do his/her chores.

A. 15
B. 2.78
C. 1.0
D. 3.13

Questions 8 – 9 refer to the following:

On any given day, approximately 37.5% of the cars parked in the De Anza parking structure are parked crookedly. (Survey done by Kathy Plum.) We randomly survey 22 cars. We are interested in the number of cars that are parked crookedly.

Exercise 14.1.8 *(Solution on p. 602.)*

For every 22 cars, how many would you expect to be parked crookedly, on average?

A. 8.25
B. 11
C. 18
D. 7.5

Exercise 14.1.9 *(Solution on p. 602.)*

What is the probability that at least 10 of the 22 cars are parked crookedly.

A. 0.1263
B. 0.1607
C. 0.2870
D. 0.8393

Exercise 14.1.10 *(Solution on p. 602.)*

Using a sample of 15 Stanford-Binet IQ scores, we wish to conduct a hypothesis test. Our claim is that the average IQ score on the Stanford-Binet IQ test is more than 100. It is known that the standard deviation of all Stanford-Binet IQ scores is 15 points. The correct distribution to use for the hypothesis test is:

A. Binomial
B. Student-t
C. Normal
D. Uniform

Questions 11 – 13 refer to the following:

De Anza College keeps statistics on the pass rate of students who enroll in math classes. According to the statistics kept from Fall 1997 through Fall 1999, 1795 students enrolled in Math 1A (1st quarter calculus) and 1428 passed the course. In the same time period, of the 856 students enrolled in Math 1B (2nd quarter calculus), 662 passed. In general, are the pass rates of Math 1A and Math 1B statistically the same? Let A = the subscript for Math 1A and B = the subscript for Math 1B.

Exercise 14.1.11 *(Solution on p. 602.)*

If you were to conduct an appropriate hypothesis test, the alternate hypothesis would be:

A. H_a: $p_A = p_B$
B. H_a: $p_A > p_B$
C. H_o: $p_A = p_B$
D. H_a: $p_A \neq p_B$

Exercise 14.1.12 *(Solution on p. 602.)*

The Type I error is to:

A. believe that the pass rate for Math 1A is the same as the pass rate for Math 1B when, in fact, the pass rates are different.
B. believe that the pass rate for Math 1A is different than the pass rate for Math 1B when, in fact, the pass rates are the same.
C. believe that the pass rate for Math 1A is greater than the pass rate for Math 1B when, in fact, the pass rate for Math 1A is less than the pass rate for Math 1B.
D. believe that the pass rate for Math 1A is the same as the pass rate for Math 1B when, in fact, they are the same.

Exercise 14.1.13 *(Solution on p. 602.)*

The correct decision is to:

A. reject H_0
B. not reject H_0
C. not make a decision because of lack of information

Kia, Alejandra, and Iris are runners on the track teams at three different schools. Their running times, in minutes, and the statistics for the track teams at their respective schools, for a one mile run, are given in the table below:

	Running Time	School Average Running Time	School Standard Deviation
Kia	4.9	5.2	.15
Alejandra	4.2	4.6	.25
Iris	4.5	4.9	.12

Table 14.3

Exercise 14.1.14 *(Solution on p. 602.)*

Which student is the BEST when compared to the other runners at her school?

A. Kia
B. Alejandra
C. Iris
D. Impossible to determine

Questions 15 – 16 refer to the following:

The following adult ski sweater prices are from the Gorsuch Ltd. Winter catalog:

$\{\$212, \$292, \$278, \$199\$280, \$236\}$

Assume the underlying sweater price population is approximately normal. The null hypothesis is that the average price of adult ski sweaters from Gorsuch Ltd. is at least $275.

Exercise 14.1.15 *(Solution on p. 602.)*

The correct distribution to use for the hypothesis test is:

A. Normal
B. Binomial

C. Student-t
D. Exponential

Exercise 14.1.16 *(Solution on p. 602.)*
The hypothesis test:

A. is two-tailed
B. is left-tailed
C. is right-tailed
D. has no tails

Exercise 14.1.17 *(Solution on p. 602.)*
Sara, a statistics student, wanted to determine the average number of books that college professors have in their office. She randomly selected 2 buildings on campus and asked each professor in the selected buildings how many books are in his/her office. Sara surveyed 25 professors. The type of sampling selected is a:

A. simple random sampling
B. systematic sampling
C. cluster sampling
D. stratified sampling

Exercise 14.1.18 *(Solution on p. 602.)*
A clothing store would use which measure of the center of data when placing orders?

A. Mean
B. Median
C. Mode
D. IQR

Exercise 14.1.19 *(Solution on p. 602.)*
In a hypothesis test, the p-value is

A. the probability that an outcome of the data will happen purely by chance when the null hypothesis is true.
B. called the preconceived alpha.
C. compared to beta to decide whether to reject or not reject the null hypothesis.
D. Answer choices A and B are both true.

Questions 20 - 22 refer to the following:

A community college offers classes 6 days a week: Monday through Saturday. Maria conducted a study of the students in her classes to determine how many days per week the students who are in her classes come to campus for classes. In each of her 5 classes she randomly selected 10 students and asked them how many days they come to campus for classes. The results of her survey are summarized in the table below.

Number of Days on Campus	Frequency	Relative Frequency	Cumulative Relative Frequency
1	2		
2	12	.24	
3	10	.20	
4			.98
5	0		
6	1	.02	1.00

Table 14.4

Exercise 14.1.20 *(Solution on p. 602.)*
Combined with convenience sampling, what other sampling technique did Maria use?

A. simple random
B. systematic
C. cluster
D. stratified

Exercise 14.1.21 *(Solution on p. 602.)*
How many students come to campus for classes 4 days a week?

A. 49
B. 25
C. 30
D. 13

Exercise 14.1.22 *(Solution on p. 602.)*
What is the 60th percentile for the this data?

A. 2
B. 3
C. 4
D. 5

The next two questions refer to the following:

The following data are the results of a random survey of 110 Reservists called to active duty to increase security at California airports.

Number of Dependents	Frequency
0	11
1	27
2	33
3	20
4	19

Table 14.5

Exercise 14.1.23 *(Solution on p. 602.)*

Construct a 95% Confidence Interval for the true population average number of dependents of Reservists called to active duty to increase security at California airports.

A. (1.85, 2.32)
B. (1.80, 2.36)
C. (1.97, 2.46)
D. (1.92, 2.50)

Exercise 14.1.24 *(Solution on p. 602.)*

The 95% confidence Interval above means:

A. 5% of Confidence Intervals constructed this way will not contain the true population aveage number of dependents.
B. We are 95% confident the true population average number of dependents falls in the interval.
C. Both of the above answer choices are correct.
D. None of the above.

Exercise 14.1.25 *(Solution on p. 603.)*

$X \sim U(4, 10)$. Find the 30th percentile.

A. 0.3000
B. 3
C. 5.8
D. 6.1

Exercise 14.1.26 *(Solution on p. 603.)*

If $X \sim Exp(0.8)$, then $P(X < \mu) =$

A. 0.3679
B. 0.4727
C. 0.6321
D. cannot be determined

Exercise 14.1.27 *(Solution on p. 603.)*

The lifetime of a computer circuit board is normally distributed with a mean of 2500 hours and a standard deviation of 60 hours. What is the probability that a randomly chosen board will last at most 2560 hours?

A. 0.8413
B. 0.1587
C. 0.3461
D. 0.6539

Exercise 14.1.28 *(Solution on p. 603.)*

A survey of 123 Reservists called to active duty as a result of the September 11, 2001, attacks was conducted to determine the proportion that were married. Eighty-six reported being married. Construct a 98% confidence interval for the true population proportion of reservists called to active duty that are married.

A. (0.6030, 0.7954)
B. (0.6181, 0.7802)

C. (0.5927, 0.8057)
D. (0.6312, 0.7672)

Exercise 14.1.29 *(Solution on p. 603.)*
Winning times in 26 mile marathons run by world class runners average 145 minutes with a standard deviation of 14 minutes. A sample of the last 10 marathon winning times is collected.

Let $\overline{x}$ = average winning times for 10 marathons.

The distribution for $\overline{x}$ is:

A. $N\left(145, \frac{14}{\sqrt{10}}\right)$
B. $N\,(145, 14)$
C. t_9
D. t_{10}

Exercise 14.1.30 *(Solution on p. 603.)*
Suppose that Phi Beta Kappa honors the top 1% of college and university seniors. Assume that grade point averages (G.P.A.) at a certain college are normally distributed with a 2.5 average and a standard deviation of 0.5. What would be the minimum G.P.A. needed to become a member of Phi Beta Kappa at that college?

A. 3.99
B. 1.34
C. 3.00
D. 3.66

The number of people living on American farms has declined steadily during this century. Here are data on the farm population (in millions of persons) from 1935 to 1980.

Year	1935	1940	1945	1950	1955	1960	1965	1970	1975	1980
Population	32.1	30.5	24.4	23.0	19.1	15.6	12.4	9.7	8.9	7.2

Table 14.6

The linear regression equation is y-hat = 1166.93 – 0.5868x

Exercise 14.1.31 *(Solution on p. 603.)*
What was the expected farm population (in millions of persons) for 1980?

A. 7.2
B. 5.1
C. 6.0
D. 8.0

Exercise 14.1.32 *(Solution on p. 603.)*
In linear regression, which is the best possible SSE?

A. 13.46
B. 18.22
C. 24.05
D. 16.33

Exercise 14.1.33 *(Solution on p. 603.)*

In regression analysis, if the correlation coefficient is close to 1 what can be said about the best fit line?

A. It is a horizontal line. Therefore, we can not use it.
B. There is a strong linear pattern. Therefore, it is most likely a good model to be used.
C. The coefficient correlation is close to the limit. Therefore, it is hard to make a decision.
D. We do not have the equation. Therefore, we can not say anything about it.

Question 34-36 refer to the following:

A study of the career plans of young women and men sent questionnaires to all 722 members of the senior class in the College of Business Administration at the University of Illinois. One question asked which major within the business program the student had chosen. Here are the data from the students who responded.

	Female	**Male**
Accounting	68	56
Administration	91	40
Ecomonics	5	6
Finance	61	59

Table 14.7: Does the data suggest that there is a relationship between the gender of students and their choice of major?

Exercise 14.1.34 *(Solution on p. 603.)*

The distribution for the test is:

A. $Chi^2{}_8$
B. $Chi^2{}_3$
C. t_{722}
D. $N(0,1)$

Exercise 14.1.35 *(Solution on p. 603.)*

The expected number of female who choose Finance is :

A. 37
B. 61
C. 60
D. 70

Exercise 14.1.36 *(Solution on p. 603.)*

The p-value is 0.0127. The conclusion to the test is:

A. The choice of major and the gender of the student are independent of each other.
B. The choice of major and the gender of the student are not independent of each other.
C. Students find Economics very hard.
D. More females prefer Administration than males.

Exercise 14.1.37 **(Solution on p. 603.)**

An agency reported that the work force nationwide is composed of 10% professional, 10% clerical, 30% skilled, 15% service, and 35% semiskilled laborers. A random sample of 100 San Jose residents indicated 15 professional, 15 clerical, 40 skilled, 10 service, and 20 semiskilled laborers. At $\alpha = .10$ does the work force in San Jose appear to be consistent with the agency report for the nation? Which kind of test is it?

A. *Chi*^{2} goodness of fit
B. *Chi*^{2} test of independence
C. Independent groups proportions
D. Unable to determine

14.2 Practice Final Exam 2[13]

Exercise 14.2.1 *(Solution on p. 603.)*

A study was done to determine the proportion of teenagers that own a car. The true proportion of teenagers that own a car is the:

A. statistic
B. parameter
C. population
D. variable

The next two questions refer to the following data:

value	frequency
0	1
1	4
2	7
3	9
6	4

Table 14.8

Exercise 14.2.2 *(Solution on p. 603.)*

The box plot for the data is:

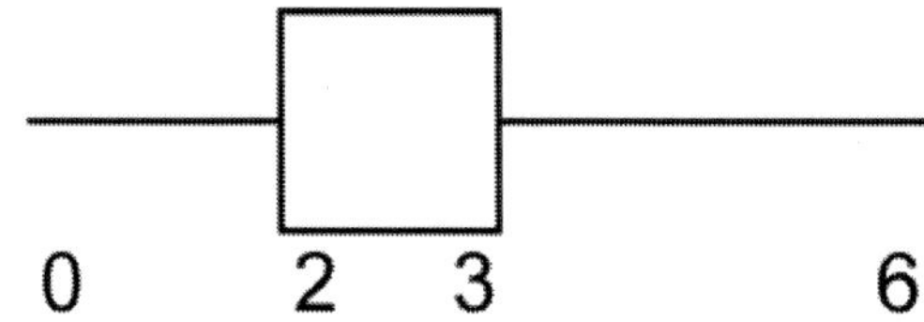

A.

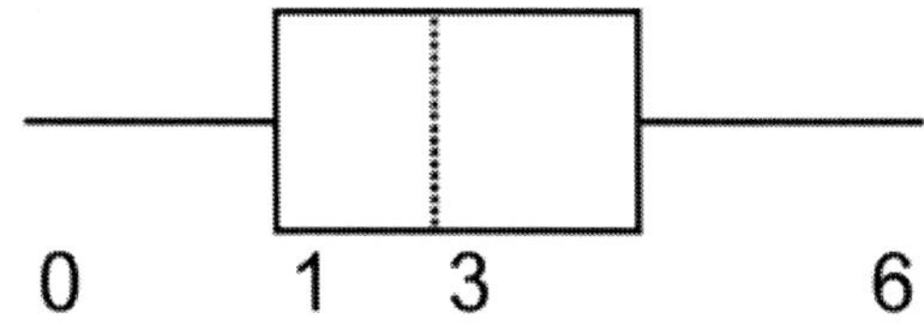

B.

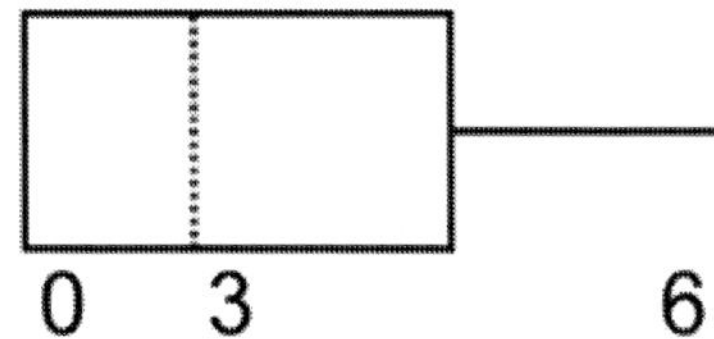

C.

[13]This content is available online at <http://cnx.org/content/m16303/1.13/>.

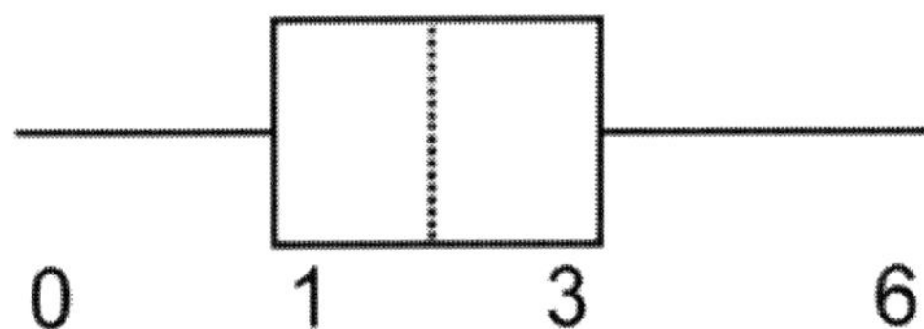

D.

Exercise 14.2.3 *(Solution on p. 603.)*

If 6 were added to each value of the data in the table, the 15th percentile of the new list of values is:

A. 6
B. 1
C. 7
D. 8

The next two questions refer to the following situation:

Suppose that the probability of a drought in any independent year is 20%. Out of those years in which a drought occurs, the probability of water rationing is 10%. However, in any year, the probability of water rationing is 5%.

Exercise 14.2.4 *(Solution on p. 603.)*

What is the probability of both a drought and water rationing occurring?

A. 0.05
B. 0.01
C. 0.02
D. 0.30

Exercise 14.2.5 *(Solution on p. 603.)*

Which of the following is true?

A. drought and water rationing are independent events
B. drought and water rationing are mutually exclusive events
C. none of the above

The next two questions refer to the following situation:

Suppose that a survey yielded the following data:

Favorite Pie Type

gender	apple	pumpkin	pecan
female	40	10	30
male	20	30	10

Table 14.9

Exercise 14.2.6 *(Solution on p. 603.)*

Suppose that one individual is randomly chosen. The probability that the person's favorite pie is apple or the person is male is:

A. $\frac{40}{60}$
B. $\frac{60}{140}$
C. $\frac{120}{140}$
D. $\frac{100}{140}$

Exercise 14.2.7 *(Solution on p. 603.)*

Suppose H_o is: Favorite pie type and gender are independent.

The *p-value* is:

A. ≈ 0
B. 1
C. 0.05
D. cannot be determined

The next two questions refer to the following situation:

Let's say that the probability that an adult watches the news at least once per week is 0.60. We randomly survey 14 people. Of interest is the number that watch the news at least once per week.

Exercise 14.2.8 *(Solution on p. 603.)*

Which of the following statements is FALSE?

A. $X \sim B(14, 0.60)$
B. The values for x are: $\{1, 2, 3, ..., 14\}$
C. $\mu = 8.4$
D. $P(X = 5) = 0.0408$

Exercise 14.2.9 *(Solution on p. 603.)*

Find the probability that at least 6 adults watch the news.

A. $\frac{6}{14}$
B. 0.8499
C. 0.9417
D. 0.6429

Exercise 14.2.10 *(Solution on p. 603.)*

The following histogram is most likely to be a result of sampling from which distribution?

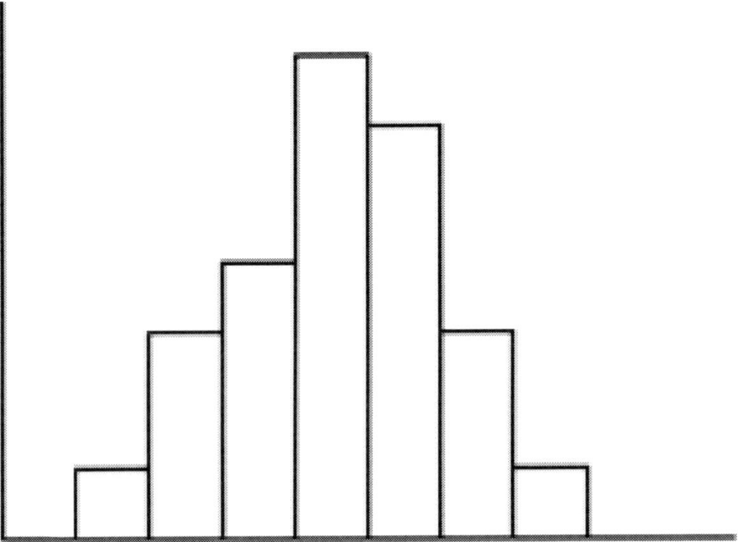

A. Chi-Square
B. Exponential
C. Uniform
D. Binomial

The ages of campus day and evening students is known to be normally distributed. A sample of 6 campus day and evening students reported their ages (in years) as: $\{18, 35, 27, 45, 20, 20\}$

Exercise 14.2.11 *(Solution on p. 603.)*

What is the error bound for the 90% confidence interval of the true average age?

A. 11.2
B. 22.3
C. 17.5
D. 8.7

Exercise 14.2.12 *(Solution on p. 604.)*

If a normally distributed random variable has $\mu = 0$ and $\sigma = 1$, then 97.5% of the population values lie above:

A. -1.96
B. 1.96
C. 1
D. -1

The next three questions refer to the following situation:

The amount of money a customer spends in one trip to the supermarket is known to have an exponential distribution. Suppose the average amount of money a customer spends in one trip to the supermarket is $72.

Exercise 14.2.13 *(Solution on p. 604.)*

What is the probability that one customer spends less than \$72 in one trip to the supermarket?

A. 0.6321
B. 0.5000
C. 0.3714
D. 1

Exercise 14.2.14 *(Solution on p. 604.)*

How much money altogether would you expect next 5 customers to spend in one trip to the supermarket (in dollars)?

A. 72
B. $\frac{72^2}{5}$
C. 5184
D. 360

Exercise 14.2.15 *(Solution on p. 604.)*

If you want to find the probability that the average of 5 customers is less than \$60, the distribution to use is:

A. $N(72, 72)$
B. $N\left(72, 72\sqrt{5}\right)$
C. $Exp(72)$
D. $Exp\left(\frac{1}{72}\right)$

The next three questions refer to the following situation:

The amount of time it takes a fourth grader to carry out the trash is uniformly distributed in the interval from 1 to 10 minutes.

Exercise 14.2.16 *(Solution on p. 604.)*

What is the probability that a randomly chosen fourth grader takes more than 7 minutes to take out the trash?

A. $\frac{3}{9}$
B. $\frac{7}{9}$
C. $\frac{3}{10}$
D. $\frac{7}{10}$

Exercise 14.2.17 *(Solution on p. 604.)*

Which graph best shows the probability that a randomly chosen fourth grader takes more than 6 minutes to take out the trash given that he/she has already taken more than 3 minutes?

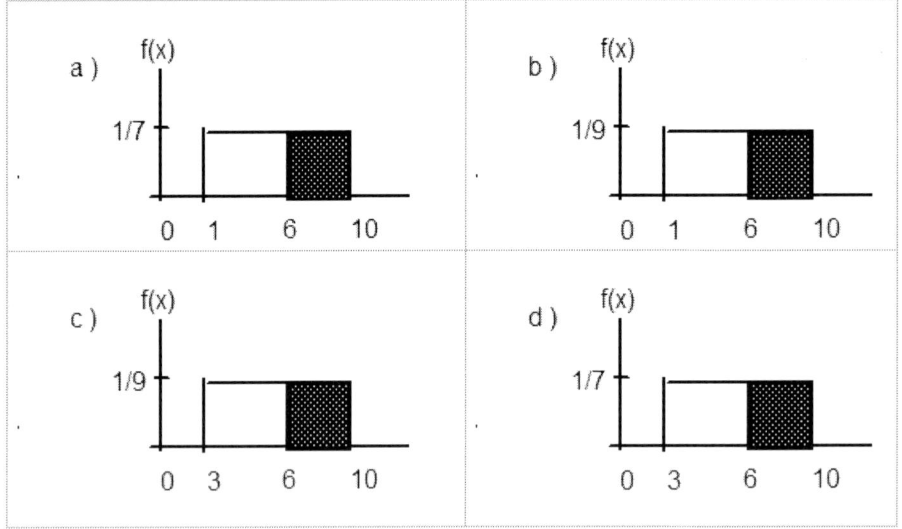

Exercise 14.2.18 *(Solution on p. 604.)*

We should expect a fourth grader to take how many minutes to take out the trash?

A. 4.5
B. 5.5
C. 5
D. 10

The next three questions refer to the following situation:

At the beginning of the quarter, the amount of time a student waits in line at the campus cafeteria is normally distributed with a mean of 5 minutes and a standard deviation of 2 minutes.

Exercise 14.2.19 *(Solution on p. 604.)*

What is the 90th percentile of waiting times (in minutes)?

A. 1.28
B. 90
C. 8.29
D. 7.56

Exercise 14.2.20 *(Solution on p. 604.)*

The median waiting time (in minutes) for one student is:

A. 5
B. 50
C. 2.5
D. 2

Exercise 14.2.21 *(Solution on p. 604.)*

A sample of 10 students has an average waiting time of 5. 5 minutes. The 95% confidence interval for the true population mean is:

A. (4.46 , 6.04)
B. (4.26 , 6.74)

C. (2.4 , 8.6)
D. (1.58 , 9.42)

Exercise 14.2.22 *(Solution on p. 604.)*

A sample of 80 software engineers in Silicon Valley is taken and it is found that 20% of them earn approximately $50,000 per year. A point estimate for the true proportion of engineers in Silicon Valley who earn $50,000 per year is:

A. 16
B. 0.2
C. 1
D. 0.95

Exercise 14.2.23 *(Solution on p. 604.)*

If $P(Z < z_\alpha) = 0.1587$ where $Z \sim N(0,1)$, then α is equal to:

A. -1
B. 0.1587
C. 0.8413
D. 1

Exercise 14.2.24 *(Solution on p. 604.)*

A professor tested 35 students to determine their entering skills. At the end of the term, after completing the course, the same test was administered to the same 35 students to study their improvement. This would be a test of:

A. independent groups
B. 2 proportions
C. dependent groups
D. exclusive groups

Exercise 14.2.25 *(Solution on p. 604.)*

A math exam was given to all the third grade children attending ABC School. Two random samples of scores were taken.

	n	$\overline{x}$	s
Boys	55	82	5
Girls	60	86	7

Table 14.10

Which of the following correctly describes the results of a hypothesis test of the claim, "There is a difference between the mean scores obtained by third grade girls and boys at the 5 % level of significance"?

A. Do not reject H_o. There is no difference in the mean scores.
B. Do not reject H_o. There is a difference in the mean scores.
C. Reject H_o. There is no difference in the mean scores.
D. Reject H_o. There is a difference in the mean scores.

Exercise 14.2.26 *(Solution on p. 604.)*

In a survey of 80 males, 45 had played an organized sport growing up. Of the 70 females surveyed, 25 had played an organized sport growing up. We are interested in whether the proportion for males is higher than the proportion for females. The correct conclusion is:

A. The proportion for males is the same as the proportion for females.
B. The proportion for males is not the same as the proportion for females.
C. The proportion for males is higher than the proportion for females.
D. Not enough information to determine.

Exercise 14.2.27 *(Solution on p. 604.)*

From past experience, a statistics teacher has found that the average score on a midterm is 81 with a standard deviation of 5.2. This term, a class of 49 students had a standard deviation of 5 on the midterm. Do the data indicate that we should reject the teacher's claim that the standard deviation is 5.2? Use $\alpha = 0.05$.

A. Yes
B. No
C. Not enough information given to solve the problem

Exercise 14.2.28 *(Solution on p. 604.)*

Three loading machines are being compared. Machine I took 31 minutes to load packages. Machine II took 28 minutes to load packages. Machine III took 29 minutes to load packages. The expected time for any machine to load packages is 29 minutes. Find the *p-value* when testing that the loading times are the same.

A. the *p–value* is close to 0
B. *p–value* is close to 1
C. Not enough information given to solve the problem

The next three questions refer to the following situation:

A corporation has offices in different parts of the country. It has gathered the following information concerning the number of bathrooms and the number of employees at seven sites:

Number of employees x	650	730	810	900	102	107	1150
Number of bathrooms y	40	50	54	61	82	110	121

Table 14.11

Exercise 14.2.29 *(Solution on p. 604.)*

Is there a correlation between the number of employees and the number of bathrooms significant?

A. Yes
B. No
C. Not enough information to answer question

Exercise 14.2.30 *(Solution on p. 604.)*

The linear regression equation is:

A. $\hat{y} = 0.0094 - 79.96x$
B. $\hat{y} = -79.96 + 0.0094x$

C. $\hat{y} = -79.96 - 0.0094x$
D. $\hat{y} = -0.0094 + 79.96x$

Exercise 14.2.31 *(Solution on p. 604.)*

If a site has 1150 employees, approximately how many bathrooms should it have?

A. 69
B. 121
C. 101
D. 86

Exercise 14.2.32 *(Solution on p. 604.)*

Suppose that a sample of size 10 was collected, with $\overline{x} = 4.4$ and $s = 1.4$.

$H_o : \sigma^2 = 1.6$ vs. $H_a : \sigma^2 \neq 1.6$. Which graph best describes the results of the test?

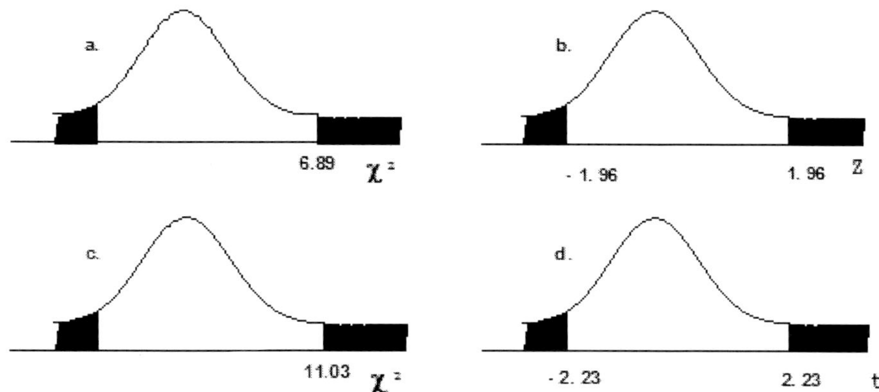

Exercise 14.2.33 *(Solution on p. 604.)*

64 backpackers were asked the number of days their latest backpacking trip was. The number of days is given in the table below:

# of days	1	2	3	4	5	6	7	8
Frequency	5	9	6	12	7	10	5	10

Table 14.12

Conduct an appropriate test to determine if the distribution is uniform.

A. The *p–value* is > 0.10 , the distribution is uniform.
B. The *p–value* is < 0.01 , the distribution is uniform.
C. The *p–value* is between 0.01 and 0.10, but without α there is not enough information
D. There is no such test that can be conducted.

Exercise 14.2.34 *(Solution on p. 604.)*

Which of the following statements is true when using one-way ANOVA?

A. The populations from which the samples are selected have different distributions.
B. The sample sizes are large.
C. The test is to determine if the different groups have the same averages.
D. There is a correlation between the factors of the experiment.

14.3 Data Sets[14]

14.3.1 Lap Times

The following tables provide lap times from Terri Vogel's Log Book. Times are recorded in seconds for 2.5-mile laps completed in a series of races and practice runs.

Race Lap Times (in Seconds)

	Lap 1	Lap 2	Lap 3	Lap 4	Lap 5	Lap 6	Lap 7
Race 1	135	130	131	132	130	131	133
Race 2	134	131	131	129	128	128	129
Race 3	129	128	127	127	130	127	129
Race 4	125	125	126	125	124	125	125
Race 5	133	132	132	132	131	130	132
Race 6	130	130	130	129	129	130	129
Race 7	132	131	133	131	134	134	131
Race 8	127	128	127	130	128	126	128
Race 9	132	130	127	128	126	127	124
Race 10	135	131	131	132	130	131	130
Race 11	132	131	132	131	130	129	129
Race 12	134	130	130	130	131	130	130
Race 13	128	127	128	128	128	129	128
Race 14	132	131	131	131	132	130	130
Race 15	136	129	129	129	129	129	129
Race 16	129	129	129	128	128	129	129
Race 17	134	131	132	131	132	132	132
Race 18	129	129	130	130	133	133	127
Race 19	130	129	129	129	129	129	128
Race 20	131	128	130	128	129	130	130

Table 14.13

[14]This content is available online at <http://cnx.org/content/m17132/1.4/>.

Practice Lap Times (in Seconds)

	Lap 1	Lap 2	Lap 3	Lap 4	Lap 5	Lap 6	Lap 7
Practice 1	142	143	180	137	134	134	172
Practice 2	140	135	134	133	128	128	131
Practice 3	130	133	130	128	135	133	133
Practice 4	141	136	137	136	136	136	145
Practice 5	140	138	136	137	135	134	134
Practice 6	142	142	139	138	129	129	127
Practice 7	139	137	135	135	137	134	135
Practice 8	143	136	134	133	134	133	132
Practice 9	135	134	133	133	132	132	133
Practice 10	131	130	128	129	127	128	127
Practice 11	143	139	139	138	138	137	138
Practice 12	132	133	131	129	128	127	126
Practice 13	149	144	144	139	138	138	137
Practice 14	133	132	137	133	134	130	131
Practice 15	138	136	133	133	132	131	131

Table 14.14

14.3.2 Stock Prices

The following table lists initial public offering (IPO) stock prices for all 1999 stocks that at least doubled in value during the first day of trading.

IPO Offer Prices

$17.00	$23.00	$14.00	$16.00	$12.00	$26.00
$20.00	$22.00	$14.00	$15.00	$22.00	$18.00
$18.00	$21.00	$21.00	$19.00	$15.00	$21.00
$18.00	$17.00	$15.00	$25.00	$14.00	$30.00
$16.00	$10.00	$20.00	$12.00	$16.00	$17.44
$16.00	$14.00	$15.00	$20.00	$20.00	$16.00
$17.00	$16.00	$15.00	$15.00	$19.00	$48.00
$16.00	$18.00	$9.00	$18.00	$18.00	$20.00
$8.00	$20.00	$17.00	$14.00	$11.00	$16.00
$19.00	$15.00	$21.00	$12.00	$8.00	$16.00
$13.00	$14.00	$15.00	$14.00	$13.41	$28.00
$21.00	$17.00	$28.00	$17.00	$19.00	$16.00
$17.00	$19.00	$18.00	$17.00	$15.00	
$14.00	$21.00	$12.00	$18.00	$24.00	
$15.00	$23.00	$14.00	$16.00	$12.00	
$24.00	$20.00	$14.00	$14.00	$15.00	
$14.00	$19.00	$16.00	$38.00	$20.00	
$24.00	$16.00	$8.00	$18.00	$17.00	
$16.00	$15.00	$7.00	$19.00	$12.00	
$8.00	$23.00	$12.00	$18.00	$20.00	
$21.00	$34.00	$16.00	$26.00	$14.00	

Table 14.15

NOTE: *Data compiled by Jay R. Ritter of Univ. of Florida using data from Securities Data Co. and Bloomberg.*

14.4 Group Projects

14.4.1 Group Project: Univariate Data[15]

14.4.1.1 Student Learning Objectives

- The student will design and carry out a survey.
- The student will analyze and graphically display the results of the survey.

14.4.1.2 Instructions

As you complete each task below, check it off. Answer all questions in your summary.

____ Decide what data you are going to study.

EXAMPLES: Here are two examples, but you may **NOT** use them: number of M&M's per small bag, number of pencils students have in their backpacks.

____ Is your data discrete or continuous? How do you know?
____ Decide how you are going to collect the data (for instance, buy 30 bags of M&M's; collect data from the World Wide Web).
____ Describe your sampling technique in detail. Use cluster, stratified, systematic, or simple random (using a random number generator) sampling. Do not use convenience sampling. What method did you use? Why did you pick that method?
____ Conduct your survey. **Your data size must be at least 30.**
____ Summarize your data in a chart with columns showing **data value, frequency, relative frequency and cumulative relative frequency.**
____ Answer the following (rounded to 2 decimal places):

1. $\overline{x}$ =
2. s =
3. First quartile =
4. Median =
5. 70th percentile =

____ What value is 2 standard deviations above the mean?
____ What value is 1.5 standard deviations below the mean?
____ Construct a histogram displaying your data.
____ In complete sentences, describe the shape of your graph.
____ Do you notice any potential outliers? If so, what values are they? Show your work in how you used the potential outlier formula in Chapter 2 (since you have univariate data) to determine whether or not the values might be outliers.
____ Construct a box plot displaying your data.
____ Does the middle 50% of the data appear to be concentrated together or spread apart? Explain how you determined this.
____ Looking at both the histogram and the box plot, discuss the distribution of your data.

14.4.1.3 Assignment Checklist

You need to turn in the following typed and stapled packet, with pages in the following order:

____ **Cover sheet**: name, class time, and name of your study

[15]This content is available online at <http://cnx.org/content/m17142/1.7/>.

____ **Summary page**: This should contain paragraphs written with complete sentences. It should include answers to all the questions above. It should also include statements describing the population under study, the sample, a parameter or parameters being studied, and the statistic or statistics produced.

____ **URL** for data, if your data are from the World Wide Web.

____ **Chart of data, frequency, relative frequency and cumulative relative frequency.**

____ **Page(s) of graphs:** histogram and box plot.

14.4.2 Group Project: Continuous Distributions and Central Limit Theorem[16]

14.4.2.1 Student Learning Objectives

- The student will collect a sample of continuous data.
- The student will attempt to fit the data sample to various distribution models.
- The student will validate the Central Limit Theorem.

14.4.2.2 Instructions

As you complete each task below, check it off. Answer all questions in your summary.

14.4.2.3 Part I: Sampling

____ Decide what **continuous** data you are going to study. (Here are two examples, but you may NOT use them: the amount of money a student spends on college supplies this term or the length of a long distance telephone call.)

____ Describe your sampling technique in detail. Use cluster, stratified, systematic, or simple random (using a random number generator) sampling. Do not use convenience sampling. What method did you use? Why did you pick that method?

____ Conduct your survey. Gather **at least 150 pieces of continuous quantitative data**.

____ Define (in words) the random variable for your data. $X =$ _______

____ Create 2 lists of your data: (1) unordered data, (2) in order of smallest to largest.

____ Find the sample mean and the sample standard deviation (rounded to 2 decimal places).

1. $\overline{x} =$
2. $s =$

____ Construct a histogram of your data containing 5 - 10 intervals of equal width. The histogram should be a representative display of your data. Label and scale it.

14.4.2.4 Part II: Possible Distributions

____ Suppose that X followed the theoretical distributions below. Set up each distribution using the appropriate information from your data.

____ Uniform: $X \sim U$ ____________ Use the lowest and highest values as a and b.

____ Exponential: $X \sim Exp$ ____________Use $\overline{x}$ to estimate μ .

____ Normal: $X \sim N$ ____________ Use $\overline{x}$ to estimate for μ and s to estimate for σ.

____ **Must** your data fit one of the above distributions? Explain why or why not.

____ **Could** the data fit 2 or 3 of the above distributions (at the same time)? Explain.

____ Calculate the value k(an X value) that is 1.75 standard deviations above the sample mean. $k =$ _________ (rounded to 2 decimal places) Note: $k = \overline{x} + (1.75) * s$

____ Determine the relative frequencies (RF) rounded to 4 decimal places.

1. $RF = \frac{frequency}{total\ number\ surveyed}$
2. $RF\left(X < k\right) =$
3. $RF\left(X > k\right) =$
4. $RF\left(X = k\right) =$

Use a separate piece of paper for EACH distribution (uniform, exponential, normal) to respond to the following questions.

---

[16] This content is available online at <http://cnx.org/content/m17141/1.9/>.

NOTE: You should have one page for the uniform, one page for the exponential, and one page for the normal

____ State the distribution: $X \sim$ ________
____ Draw a graph for each of the three theoretical distributions. Label the axes and mark them appropriately.
____ Find the following theoretical probabilities (rounded to 4 decimal places).

1. $P(X < k) =$
2. $P(X > k) =$
3. $P(X = k) =$

____ Compare the relative frequencies to the corresponding probabilities. Are the values close?
____ Does it appear that the data fit the distribution well? Justify your answer by comparing the probabilities to the relative frequencies, and the histograms to the theoretical graphs.

14.4.2.5 Part III: CLT Experiments

______ From your original data (before ordering), use a random number generator to pick 40 samples of size 5. For each sample, calculate the average.
______ On a separate page, attached to the summary, include the 40 samples of size 5, along with the 40 sample averages.
______ List the 40 averages in order from smallest to largest.
______ Define the random variable, $\overline{X}$, in words. $\overline{X} =$
______ State the approximate theoretical distribution of $\overline{X}$. $\overline{X} \sim$
______ Base this on the mean and standard deviation from your original data.
______ Construct a histogram displaying your data. Use 5 to 6 intervals of equal width. Label and scale it.

Calculate the value $\overline{k}$ (an $\overline{X}$ value) that is 1.75 standard deviations above the sample mean. $\overline{k}=$ _____ (rounded to 2 decimal places)

Determine the relative frequencies (RF) rounded to 4 decimal places.

1. $RF(\overline{X} < \overline{k}) =$
2. $RF(\overline{X} > \overline{k}) =$
3. $RF(\overline{X} = \overline{k}) =$

Find the following theoretical probabilities (rounded to 4 decimal places).

- $P(\overline{X} < \overline{k}) =$
- $P(\overline{X} > \overline{k}) =$
- $P(\overline{X} = \overline{k}) =$

______ Draw the graph of the theoretical distribution of X.
______ Answer the questions below.
______ Compare the relative frequencies to the probabilities. Are the values close?
______ Does it appear that the data of averages fit the distribution of $\overline{X}$ well? Justify your answer by comparing the probabilities to the relative frequencies, and the histogram to the theoretical graph.
______ In 3 - 5 complete sentences for each, answer the following questions. Give thoughtful explanations.
______ In summary, do your original data seem to fit the uniform, exponential, or normal distributions? Answer why or why not for each distribution. If the data do not fit any of those distributions, explain why.

______ What happened to the shape and distribution when you averaged your data? **In theory,** what should have happened? In theory, would "it" always happen? Why or why not?

______ Were the relative frequencies compared to the theoretical probabilities closer when comparing the X or $\bar{X}$ distributions? Explain your answer.

14.4.2.6 Assignment Checklist

You need to turn in the following typed and stapled packet, with pages in the following order:

____ **Cover sheet**: name, class time, and name of your study

____ **Summary pages**: These should contain several paragraphs written with complete sentences that describe the experiment, including what you studied and your sampling technique, as well as answers to all of the questions above.

____ **URL** for data, if your data are from the World Wide Web.

____ **Pages, one for each theoretical distribution**, with the distribution stated, the graph, and the probability questions answered

____ **Pages of the data requested**

____ **All graphs required**

14.4.3 Partner Project: Hypothesis Testing - Article[17]

14.4.3.1 Student Learning Objectives

- The student will identify a hypothesis testing problem in print.
- The student will conduct a survey to verify or dispute the results of the hypothesis test.
- The student will summarize the article, analysis, and conclusions in a report.

14.4.3.2 Instructions

As you complete teach task below, check it off. Answer all questions in your summary.

____ **Find an article** in a newspaper, magazine or on the internet which makes a claim about **ONE** population mean or **ONE** population proportion. The claim may be based upon a survey that the article was reporting on. Decide whether this claim is the null or alternate hypothesis.
____ **Copy or print out the article** and include a copy in your project, along with the source.
____ **State how you will collect your data.** (Convenience sampling is not acceptable.)
____ **Conduct your survey. You must have more than 50 responses in your sample.** When you hand in your final project, attach the tally sheet or the packet of questionnaires that you used to collect data. Your data must be real.
____ **State the statistics** that are a result of your data collection: sample size, sample mean, and sample standard deviation, OR sample size and number of successes.
____ **Make 2 copies of the appropriate solution sheet.**
____ **Record the hypothesis test** on the solution sheet, based on your experiment. **Do a DRAFT solution** first on one of the solution sheets and check it over carefully. Have a classmate check your solution to see if it is done correctly. Make your decision using a 5% level of significance. Include the 95% confidence interval on the solution sheet.
____ **Create a graph that illustrates your data.** This may be a pie or bar chart or may be a histogram or box plot, depending on the nature of your data. Produce a graph that makes sense for your data and gives useful visual information about your data. You may need to look at several types of graphs before you decide which is the most appropriate for the type of data in your project.
____ **Write your summary** (in complete sentences and paragraphs, with proper grammar and correct spelling) that describes the project. The summary **MUST** include:

1. Brief discussion of the article, including the source.
2. Statement of the claim made in the article (one of the hypotheses).
3. Detailed description of how, where, and when you collected the data, including the sampling technique. Did you use cluster, stratified, systematic, or simple random sampling (using a random number generator)? As stated above, convenience sampling is not acceptable.
4. Conclusion about the article claim in light of your hypothesis test. This is the conclusion of your hypothesis test, stated in words, in the context of the situation in your project in sentence form, as if you were writing this conclusion for a non-statistician.
5. Sentence interpreting your confidence interval in the context of the situation in your project.

14.4.3.3 Assignment Checklist

Turn in the following typed (12 point) and stapled packet for your final project:

____ **Cover sheet** containing your name(s), class time, and the name of your study.
____ **Summary**, which includes all items listed on summary checklist.
____ **Solution sheet** neatly and completely filled out. The solution sheet does not need to be typed.

[17]This content is available online at <http://cnx.org/content/m17140/1.7/>.

____ **Graphic representation of your data**, created following the guidelines discussed above. Include only graphs which are appropriate and useful.

____ **Raw data collected AND a table summarizing the sample data** (n, xbar and s; or x, n, and p′, as appropriate for your hypotheses). The raw data does not need to be typed, but the summary does. Hand in the data as you collected it. (Either attach your tally sheet or an envelope containing your questionnaires.)

14.4.4 Partner Project: Hypothesis Testing - Word Problem[18]

14.4.4.1 Student Learning Objectives

- The student will write, edit, and solve a hypothesis testing word problem.

14.4.4.2 Instructions

Write an original hypothesis testing problem for either **ONE** population mean or **ONE** population proportion. As you complete each task, check it off. Answer all questions in your summary. Look at the homework for the Hypothesis Testing: Single Mean and Single Proportion chapter for examples (poems, two acts of a play, a work related problem). The problems with names attached to them are problems written by students in past quarters. Some other examples that are not in the homework include: a soccer hypothesis testing poster, a cartoon, a news reports, a children's story, a song.

____ Your problem must be original and creative. It also must be in proper English. If English is difficult for you, have someone edit your problem.
____ Your problem must be at least ½ page, typed and singled spaced. This **DOES NOT** include the data. Data will make the problem longer and that is fine. For this problem, the data and story may be real or fictional.
____ In the narrative of the problem, make it very clear what the null and alternative hypotheses are.
____ Your sample size must be **LARGER THAN 50** (even if it is fictional).
____ State in your problem how you will collect your data.
____ Include your data with your word problem.
____ State the statistics that are a result of your data collection: sample size, sample mean, and sample standard deviation, OR sample size and number of successes.
____ Create a graph that illustrates your problem. This may be a pie or bar chart or may be a histogram or box plot, depending on the nature of your data. Produce a graph that makes sense for your data and gives useful visual information about your data. You may need to look at several types of graphs before you decide which is the most appropriate for your problem.
____ Make 2 copies of the appropriate solution sheet.
____ Record the hypothesis test on the solution sheet, based on your problem. Do a **DRAFT** solution first on one of the solution sheets and check it over carefully. Make your decision using a 5% level of significance. Include the 95% confidence interval on the solution

14.4.4.3 Assignment Checklist

You need to turn in the following typed (12 point) and stapled packet for your final project:

____ **Cover sheet** containing your name, the name of your problem, and the date
____ **The problem**
____ **Data for the problem**
____ **Solution sheet** neatly and completely filled out. The solution sheet does not need to be typed.
____ **Graphic representation of the data**, created following the guidelines discussed above. Include only graphs that are appropriate and useful.
____ **Sentences interpreting the results of the hypothesis test and the confidence interval** in the context of the situation in the project.

[18]This content is available online at <http://cnx.org/content/m17144/1.7/>.

14.4.5 Group Project: Bivariate Data, Linear Regression, and Univariate Data[19]

14.4.5.1 Student Learning Objectives

- The students will collect a bivariate data sample through the use of appropriate sampling techniques.
- The student will attempt to fit the data to a linear model.
- The student will determine the appropriateness of linear fit of the model.
- The student will analyze and graph univariate data.

14.4.5.2 Instructions

1. As you complete each task below, check it off. Answer all questions in your introduction or summary.
2. Check your course calendar for intermediate and final due dates.
3. Graphs may be constructed by hand or by computer, unless your instructor informs you otherwise. All graphs must be neat and accurate.
4. All other responses must be done on the computer.
5. Neatness and quality of explanations are used to determine your final grade.

14.4.5.3 Part I: Bivariate Data

Introduction

____ State the bivariate data your group is going to study.

EXAMPLES: Here are two examples, but you may **NOT** use them: height vs. weight and age vs. running distance.

____ Describe how your group is going to collect the data (for instance, collect data from the web, survey students on campus).

____ Describe your sampling technique in detail. Use cluster, stratified, systematic, or simple random sampling (using a random number generator) sampling. Convenience sampling is **NOT** acceptable.

____ Conduct your survey. Your number of pairs must be at least 30.

____ Print out a copy of your data.

Analysis

____ On a separate sheet of paper construct a scatter plot of the data. Label and scale both axes.

____ State the least squares line and the correlation coefficient.

____ On your scatter plot, in a different color, construct the least squares line.

____ Is the correlation coefficient significant? Explain and show how you determined this.

____ Interpret the slope of the linear regression line in the context of the data in your project. Relate the explanation to your data, and quantify what the slope tells you.

____ Does the regression line seem to fit the data? Why or why not? If the data does not seem to be linear, explain if any other model seems to fit the data better.

____ Are there any outliers? If so, what are they? Show your work in how you used the potential outlier formula in the Linear Regression and Correlation chapter (since you have bivariate data) to determine whether or not any pairs might be outliers.

[19]This content is available online at <http://cnx.org/content/m17143/1.6/>.

14.4.5.4 Part II: Univariate Data

In this section, you will use the data for **ONE** variable only. Pick the variable that is more interesting to analyze. For example: if your independent variable is sequential data such as year with 30 years and one piece of data per year, your x-values might be 1971, 1972, 1973, 1974, . . ., 2000. This would not be interesting to analyze. In that case, choose to use the dependent variable to analyze for this part of the project.

_____ Summarize your data in a chart with columns showing data value, frequency, relative frequency, and cumulative relative frequency.

_____ Answer the following, rounded to 2 decimal places:

1. Sample mean =
2. Sample standard deviation =
3. First quartile =
4. Third quartile =
5. Median =
6. 70th percentile =
7. Value that is 2 standard deviations above the mean =
8. Value that is 1.5 standard deviations below the mean =

_____ Construct a histogram displaying your data. Group your data into 6 – 10 intervals of equal width. Pick regularly spaced intervals that make sense in relation to your data. For example, do NOT group data by age as 20-26,27-33,34-40,41-47,48-54,55-61 . . . Instead, maybe use age groups 19.5-24.5, 24.5-29.5, . . . or 19.5-29.5, 29.5-39.5, 39.5-49.5, . . .

_____ In complete sentences, describe the shape of your histogram.

_____ Are there any potential outliers? Which values are they? Show your work and calculations as to how you used the potential outlier formula in chapter 2 (since you are now using univariate data) to determine which values might be outliers.

_____ Construct a box plot of your data.

_____ Does the middle 50% of your data appear to be concentrated together or spread out? Explain how you determined this.

_____ Looking at both the histogram AND the box plot, discuss the distribution of your data. For example: how does the spread of the middle 50% of your data compare to the spread of the rest of the data represented in the box plot; how does this correspond to your description of the shape of the histogram; how does the graphical display show any outliers you may have found; does the histogram show any gaps in the data that are not visible in the box plot; are there any interesting features of your data that you should point out.

14.4.5.5 Due Dates

- Part I, Intro: __________ (keep a copy for your records)
- Part I, Analysis: __________ (keep a copy for your records)
- Entire Project, typed and stapled: __________

  ____ Cover sheet: names, class time, and name of your study.

  ____ Part I: label the sections "Intro" and "Analysis."

  ____ Part II:

  ____ Summary page containing several paragraphs written in complete sentences describing the experiment, including what you studied and how you collected your data. The summary page should also include answers to ALL the questions asked above.

  ____ All graphs requested in the project.

  ____ All calculations requested to support questions in data.

  ____ Description: what you learned by doing this project, what challenges you had, how you overcame the challenges.

NOTE: **Include answers to ALL questions asked, even if not explicitly repeated in the items above.**

14.5 Solution Sheets

14.5.1 Solution Sheet: Hypothesis Testing for Single Mean and Single Proportion[20]

Class Time:

Name:

a. H_o:
b. H_a:
c. In words, **CLEARLY** state what your random variable $\overline{X}$ or P' represents.
d. State the distribution to use for the test.
e. What is the test statistic?
f. What is the p-value? In 1 – 2 complete sentences, explain what the p-value means for this problem.
g. Use the previous information to sketch a picture of this situation. CLEARLY, label and scale the horizontal axis and shade the region(s) corresponding to the p-value.

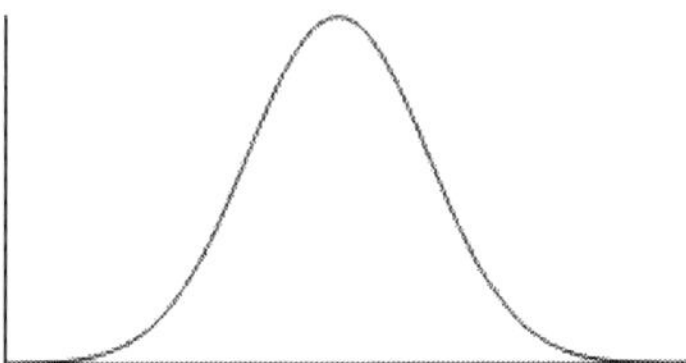

Figure 14.1

h. Indicate the correct decision ("reject" or "do not reject" the null hypothesis), the reason for it, and write an appropriate conclusion, using **complete sentences**.

i. Alpha:
ii. Decision:
iii. Reason for decision:
iv. Conclusion:

i. Construct a 95% Confidence Interval for the true mean or proportion. Include a sketch of the graph of the situation. Label the point estimate and the lower and upper bounds of the Confidence Interval.

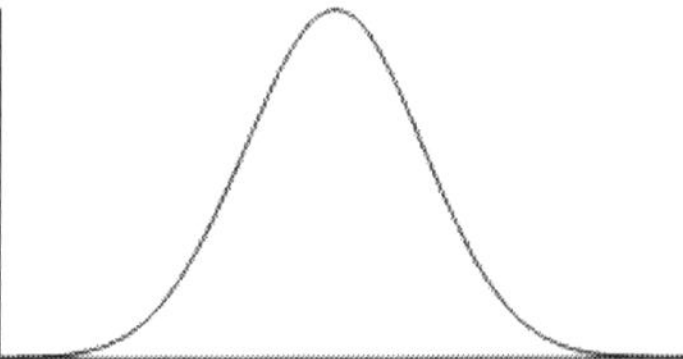

Figure 14.2

[20]This content is available online at <http://cnx.org/content/m17134/1.6/>.

14.5.2 Solution Sheet: Hypothesis Testing for Two Means, Paired Data, and Two Proportions[21]

Class Time:

Name:

- **a.** H_o: _______
- **b.** H_a: _______
- **c.** In words, **clearly** state what your random variable $\overline{X}_1 - \overline{X}_2$, $P_1' - P_2'$- or $\overline{X}_d$ represents.
- **d.** State the distribution to use for the test.
- **e.** What is the test statistic?
- **f.** What is the p-value? In 1 – 2 complete sentences, explain what the p-value means for this problem.
- **g.** Use the previous information to sketch a picture of this situation. **CLEARLY** label and scale the horizontal axis and shade the region(s) corresponding to the p-value.

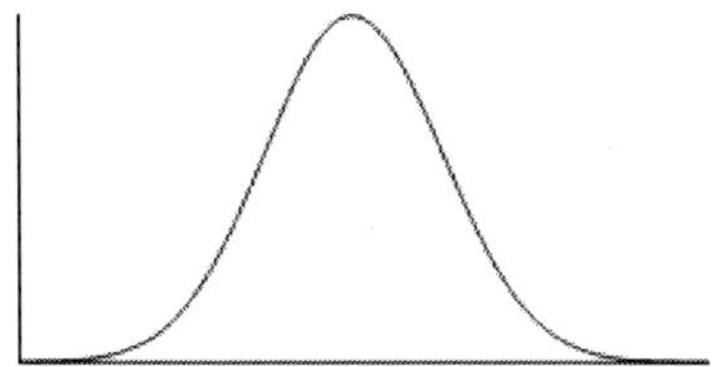

Figure 14.3

- **h.** Indicate the correct decision ("reject" or "do not reject" the null hypothesis), the reason for it, and write an appropriate conclusion, using **complete sentences**.
  - **i.** Alpha:
  - **ii.** Decision:
  - **iii.** Reason for decision:
  - **iv.** Conclusion:
- **i.** In complete sentences, explain how you determined which distribution to use.

[21]This content is available online at <http://cnx.org/content/m17133/1.6/>.

14.5.3 Solution Sheet: The Chi-Square Distribution[22]

Class Time:

Name:

- **a.** H_o: _______
- **b.** H_a:
- **c.** What are the degrees of freedom?
- **d.** State the distribution to use for the test.
- **e.** What is the test statistic?
- **f.** What is the p-value? In 1 – 2 complete sentences, explain what the p-value means for this problem.
- **g.** Use the previous information to sketch a picture of this situation. **Clearly** label and scale the horizontal axis and shade the region(s) corresponding to the p-value.

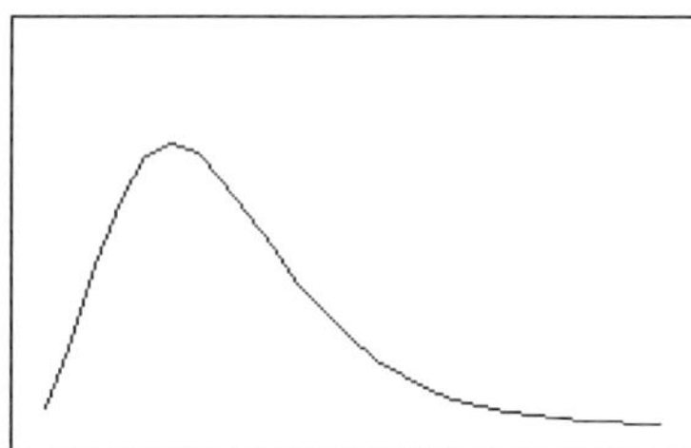

Figure 14.4

- **h.** Indicate the correct decision ("reject" or "do not reject" the null hypothesis) and write appropriate conclusions, using **complete sentences.**
  - **i.** Alpha:
  - **ii.** Decision:
  - **iii.** Reason for decision:
  - **iv.** Conclusion:

[22]This content is available online at <http://cnx.org/content/m17136/1.5/>.

14.5.4 Solution Sheet: F Distribution and ANOVA[23]

Class Time:

Name:

a. H_o:
b. H_a:
c. $df(n) =$
d. $df(d) =$
e. State the distribution to use for the test.
f. What is the test statistic?
g. What is the p-value? In $1 - 2$ complete sentences, explain what the p-value means for this problem.
h. Use the previous information to sketch a picture of this situation. **Clearly** label and scale the horizontal axis and shade the region(s) corresponding to the p-value.

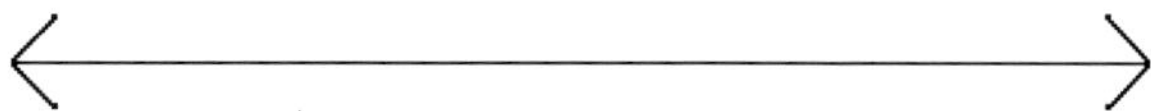

Figure 14.5

i. Indicate the correct decision ("reject" or "do not reject" the null hypothesis) and write appropriate conclusions, using **complete sentences**.

i. Alpha:
ii. Decision:
iii. Reason for decision:
iv. Conclusion:

[23]This content is available online at <http://cnx.org/content/m17135/1.5/>.

14.6 English Phrases Written Mathematically[24]

14.6.1 English Phrases Written Mathematically

When the English says:	Interpret this as:
$X$is at least 4.	$X \geq 4$
XThe minimum is 4.	$X \geq 4$
X is no less than 4.	$X \geq 4$
X is greater than or equal to 4.	$X \geq 4$
X is at most 4.	$X \leq 4$
XThe maximum is 4.	$X \leq 4$
$X$is no more than 4.	$X \leq 4$
X is less than or equal to 4.	$X \leq 4$
Xdoes not exceed 4.	$X \leq 4$
$X$is greater than 4.	$X > 4$
XThere are more than 4.	$X > 4$
Xexceeds 4.	$X > 4$
$X$is less than 4.	$X < 4$
XThere are fewer than 4.	$X < 4$
$X$is 4.	$X = 4$
$X$is equal to 4.	$X = 4$
$X$is the same as 4.	$X = 4$
$X$is not 4.	$X \neq 4$
$X$is not equal to 4.	$X \neq 4$
$X$is not the same as 4.	$X \neq 4$
$X$is different than 4.	$X \neq 4$

Table 14.16

[24]This content is available online at <http://cnx.org/content/m16307/1.5/>.

14.7 Symbols and their Meanings[25]

Symbols and their Meanings

Chapter (1st used)	Symbol	Spoken	Meaning
Sampling and Data	$\sqrt{}$	The square root of	same
Sampling and Data	π	Pi	3.14159... (a specific number)
Descriptive Statistics	$Q1$	Quartile one	the first quartile
Descriptive Statistics	$Q2$	Quartile two	the second quartile
Descriptive Statistics	$Q3$	Quartile three	the third quartile
Descriptive Statistics	IQR	inter-quartile range	Q3-Q1=IQR
Descriptive Statistics	$\overline{x}$	x-bar	sample mean
Descriptive Statistics	μ	mu	population mean
Descriptive Statistics	$s\ s_x\ sx$	s	sample standard deviation
Descriptive Statistics	$s^2\ s_x^2$	s-sqaured	sample variance
Descriptive Statistics	$\sigma\ \sigma_x\ \sigma x$	sigma	population standard deviation
Descriptive Statistics	$\sigma^2\ \sigma_x^2$	sigma-squared	population variance
Descriptive Statistics	Σ	capital sigma	sum
Probability Topics	$\{\}$	brackets	set notation
Probability Topics	S	S	sample space
Probability Topics	A	Event A	event A
Probability Topics	$P(A)$	probability of A	probability of A occurring
Probability Topics	$P(A \mid B)$	probability of A given B	prob. of A occurring given B has occurred
Probability Topics	$P(AorB)$	prob. of A or B	prob. of A or B or both occurring
			continued on next page

[25]This content is available online at <http://cnx.org/content/m16302/1.8/>.

Probability Topics	$P(AandB)$	prob. of A and B	prob. of both A and B occurring (same time)
Probability Topics	A'	A-prime, complement of A	complement of A, not A
Probability Topics	$P(A')$	prob. of complement of A	same
Probability Topics	G_1	green on first pick	same
Probability Topics	$P(G_1)$	prob. of green on first pick	same
Discrete Random Variables	PDF	prob. distribution function	same
Discrete Random Variables	X	X	the random variable X
Discrete Random Variables	$X \sim$	the distribution of X	same
Discrete Random Variables	B	binomial distribution	same
Discrete Random Variables	G	geometric distribution	same
Discrete Random Variables	H	hypergeometric dist.	same
Discrete Random Variables	P	Poisson dist.	same
Discrete Random Variables	λ	Lambda	average of Poisson distribution
Discrete Random Variables	$\geq$	greater than or equal to	same
Discrete Random Variables	$\leq$	less than or equal to	same
Discrete Random Variables	$=$	equal to	same
Discrete Random Variables	$\neq$	not equal to	same

continued on next page

Continuous Random Variables	$f(x)$	f of x	function of x
Continuous Random Variables	pdf	prob. density function	same
Continuous Random Variables	U	uniform distribution	same
Continuous Random Variables	Exp	exponential distribution	same
Continuous Random Variables	k	k	critical value
Continuous Random Variables	$f(x) =$	f of x equals	same
Continuous Random Variables	m	m	decay rate (for exp. dist.)
The Normal Distribution	N	normal distribution	same
The Normal Distribution	z	z-score	same
The Normal Distribution	Z	standard normal dist.	same
The Central Limit Theorem	CLT	Central Limit Theorem	same
The Central Limit Theorem	$\overline{X}$	X-bar	the random variable X-bar
The Central Limit Theorem	μ_x	mean of X	the average of X
The Central Limit Theorem	$\mu_{\overline{x}}$	mean of X-bar	the average of X-bar
The Central Limit Theorem	σ_x	standard deviation of X	same
The Central Limit Theorem	$\sigma_{\overline{x}}$	standard deviation of X-bar	same
The Central Limit Theorem	ΣX	sum of X	same

continued on next page

The Central Limit Theorem	Σx	sum of x	same
Confidence Intervals	CL	confidence level	same
Confidence Intervals	CI	confidence interval	same
Confidence Intervals	EBM	error bound for a mean	same
Confidence Intervals	EBP	error bound for a proportion	same
Confidence Intervals	t	student-t distribution	same
Confidence Intervals	df	degrees of freedom	same
Confidence Intervals	$t_{\frac{\alpha}{2}}$	student-t with a/2 area in right tail	same
Confidence Intervals	p' $\hat{p}$	p-hat	sample proportion of success
Confidence Intervals	q' $\hat{q}$	q-hat	sample proportion of failure
Hypothesis Testing	H_0	H-naught, H-sub 0	null hypothesis
Hypothesis Testing	H_a	H-a, H-sub a	alternate hypothesis
Hypothesis Testing	H_1	H-1, H-sub 1	alternate hypothesis
Hypothesis Testing	α	alpha	probability of Type I error
Hypothesis Testing	β	beta	probability of Type II error
Hypothesis Testing	$\overline{X1} - \overline{X2}$	X1-bar minus X2-bar	difference in sample means
	$\mu_1 - \mu_2$	mu-1 minus mu-2	difference in population means
	$P'_1 - P'_2$	P1-hat minus P2-hat	difference in sample proportions
	$p_1 - p_2$	p1 minus p2	difference in population proportions
Chi-Square Distribution	X^2	Ky-square	Chi-square
			continued on next page

	O	Observed	Observed frequency
	E	Expected	Expected frequency
Linear Regression and Correlation	$y = a + bx$	y equals a plus b-x	equation of a line
	$\hat{y}$	y-hat	estimated value of y
	r	correlation coefficient	same
	ϵ	error	same
	SSE	Sum of Squared Errors	same
	$1.9s$	1.9 times s	cut-off value for outliers
F-Distribution and ANOVA	F	F-ratio	F ratio

Table 14.17

14.8 Formulas[26]

Formula 14.1: Factorial
$n! = n(n-1)(n-2)\ldots(1)$

$0! = 1$

Formula 14.2: Combinations
$\binom{n}{r} = \frac{n!}{(n-r)!r!}$

Formula 14.3: Binomial Distribution
$X \sim B(n, p)$

$P(X = x) = \binom{n}{x} p^x q^{n-x}$, for $x = 0, 1, 2, \ldots, n$

Formula 14.4: Geometric Distribution
$X \sim G(p)$

$P(X = x) = q^{x-1} p$, for $x = 1, 2, 3, \ldots$

Formula 14.5: Hypergeometric Distribution
$X \sim H(r, b, n)$

$$P(X = x) = \left(\frac{\binom{r}{x}\binom{b}{n-x}}{\binom{r+b}{n}} \right)$$

Formula 14.6: Poisson Distribution
$X \sim P(\mu)$

$P(X = x) = \frac{\mu^x e^{-\mu}}{x!}$

Formula 14.7: Uniform Distribution
$X \sim U(a, b)$

$f(X) = \frac{1}{b-a}, a < x < b$

Formula 14.8: Exponential Distribution
$X \sim Exp(m)$

$f(x) = me^{-mx}$, $m > 0, x \geq 0$

Formula 14.9: Normal Distribution
$X \sim N(\mu, \sigma^2)$

$$f(x) = \frac{1}{\sigma\sqrt{2\pi}} e^{\frac{-(x-\mu)^2}{2\sigma^2}}, \quad -\infty < x < \infty$$

Formula 14.10: Gamma Function
$\Gamma(z) = \int_0^\infty x^{z-1} e^{-x} dx \; z > 0$

$\Gamma\left(\frac{1}{2}\right) = \sqrt{\pi}$

$\Gamma(m+1) = m!$ for m, a nonnegative integer

otherwise: $\Gamma(a+1) = a\Gamma(a)$

Formula 14.11: Student-t Distribution
$X \sim t_{df}$

[26] This content is available online at <http://cnx.org/content/m16301/1.6/>.

$$f(x) = \frac{\left(1+\frac{x^2}{n}\right)^{\frac{-(n+1)}{2}} \Gamma\left(\frac{n+1}{2}\right)}{\sqrt{n\pi}\Gamma\left(\frac{n}{2}\right)}$$

$$X = \frac{Z}{\sqrt{\frac{Y}{n}}}$$

$Z \sim N(0,1)$, $Y \sim X^2_{df}$, n = degrees of freedom

Formula 14.12: Chi-Square Distribution
$X \sim X^2_{df}$

$f(x) = \frac{x^{\frac{n-2}{2}} e^{\frac{-x}{2}}}{2^{\frac{n}{2}} \Gamma\left(\frac{n}{2}\right)}$, $x > 0$, n = positive integer and degrees of freedom

Formula 14.13: F Distribution
$X \sim F_{df(n),df(d)}$

$df(n)$ =degrees of freedom for the numerator

$df(d)$ =degrees of freedom for the denominator

$$f(x) = \frac{\Gamma\left(\frac{u+v}{2}\right)}{\Gamma\left(\frac{u}{2}\right)\Gamma\left(\frac{v}{2}\right)} \left(\frac{u}{v}\right)^{\frac{u}{2}} x^{\left(\frac{u}{2}-1\right)} \left[1 + \left(\frac{u}{v}\right) x^{-.5(u+v)}\right]$$

$X = \frac{Y_u}{W_v}$, Y, W are chi-square

14.9 Notes for the TI-83, 83+, 84 Calculator[27]

14.9.1 Quick Tips

Legend

- ( ) represents a button press
- [ ] represents yellow command or green letter behind a key
- < > represents items on the scren

To adjust the contrast

Press 2nd, then hold ▲ to increase the contrast or ▼ to decrease the contrast.

To capitalize letters and words

Press ALPHA to get one capital letter, or press 2nd, then ALPHA to set all button presses to capital letters. You can return to the top-level button values by pressing ALPHA again.

To correct a mistake

If you hit a wrong button, just hit CLEAR and start again.

To write in scientific notation

Numbers in scientific notation are expressed on the TI-83, 83+, and 84 using E notation, such that...

- 4.321 E 4 = 4.321×10^4
- 4.321 E -4 = 4.321×10^{-4}

To transfer programs or equations from one calculator to another:

Both calculators: Insert your respective end of the link cable cable and press 2nd, then [LINK].

Calculator receiving information:

1. Use the arrows to navigate to and select <RECEIVE>
2. Press ENTER

Calculator sending information:

1. Press appropriate number or letter.
2. Use up and down arrows to access the appropriate item.
3. Press ENTER to select item to transfer.
4. Press right arrow to navigate to and select <TRANSMIT>.
5. Press ENTER

NOTE: ERROR 35 LINK generally means that the cables have not been inserted far enough.

Both calculators: Insert your respective end of the link cable cable Both calculators: press 2nd, then [QUIT] To exit when done.

[27]This content is available online at <http://cnx.org/content/m19710/1.3/>.

14.9.2 Manipulating One-Variable Statistics

NOTE: These directions are for entering data with the built-in statistical program.

Sample Data

Data	Frequency
-2	10
-1	3
0	4
1	5
3	8

Table 14.18: We are manipulating 1-variable statistics.

To begin:

1. Turn on the calculator.
   ON
2. Access statistics mode.
   STAT
3. Select <4:ClrList> to clear data from lists, if desired.
   4 , ENTER
4. Enter list [L1] to be cleared.
   2nd , [L1] , ENTER
5. Display last instruction.
   2nd , [ENTRY]
6. Continue clearing remaining lists in the same fashion, if desired.
   ◀ , 2nd , [L2] , ENTER
7. Access statistics mode.
   STAT
8. Select <1:Edit . . .>
   ENTER
9. Enter data. Data values go into [L1]. (You may need to arrow over to [L1])
   - Type in a data value and enter it. (For negative numbers, use the negate (-) key at the bottom of the keypad)
     (−) , 9 , ENTER
   - Continue in the same manner until all data values are entered.
10. In [L2], enter the frequencies for each data value in [L1].
    - Type in a frequency and enter it. (If a data value appears only once, the frequency is "1")
      4 , ENTER
    - Continue in the same manner until all data values are entered.
11. Access statistics mode.
    STAT
12. Navigate to <CALC>

13. Access <1:1-var Stats>
    ENTER
14. Indicate that the data is in [L1]...
    2nd, [L1] , ,
15. ...and indicate that the frequencies are in [L2].
    2nd, [L2] , ENTER
16. The statistics should be displayed. You may arrow down to get remaining statistics. Repeat as necessary.

14.9.3 Drawing Histograms

NOTE: We will assume that the data is already entered

We will construct 2 histograms with the built-in STATPLOT application. The first way will use the default ZOOM. The second way will involve customizing a new graph.

1. Access graphing mode.
   2nd, [STAT PLOT]
2. Select <1:plot 1> To access plotting - first graph.
   ENTER
3. Use the arrows navigate go to <ON> to turn on Plot 1.
   <ON> , ENTER
4. Use the arrows to go to the histogram picture and select the histogram.
   ENTER
5. Use the arrows to navigate to <Xlist>
6. If "L1" is not selected, select it.
   2nd, [L1] , ENTER
7. Use the arrows to navigate to <Freq>.
8. Assign the frequencies to [L2].
   2nd, [L2] , ENTER
9. Go back to access other graphs.
   2nd, [STAT PLOT]
10. Use the arrows to turn off the remaining plots.
11. **Be sure to deselect or clear all equations before graphing.**

To deselect equations:

1. Access the list of equations.

2. Select each equal sign (=).

3. Continue, until all equations are deselected.

To clear equations:

1. Access the list of equations.

2. Use the arrow keys to navigate to the right of each equal sign (=) and clear them.
   ▼ ▶ CLEAR
3. Repeat until all equations are deleted.

To draw default histogram:

1. Access the ZOOM menu.
   ZOOM
2. Select <`9:ZoomStat`>
   9
3. The histogram will show with a window automatically set.

To draw custom histogram:

1. Access WINDOW to set the graph parameters.
2.
   - $X_{min} = -2.5$
   - $X_{max} = 3.5$
   - $X_{scl} = 1$ (width of bars)
   - $Y_{min} = 0$
   - $Y_{max} = 10$
   - $Y_{scl} = 1$ (spacing of tick marks on y-axis)
   - $X_{res} = 1$
3. Access GRAPH to see the histogram.

To draw box plots:

1. Access graphing mode.
   2nd, `[STAT PLOT]`
2. Select <`1:Plot 1`> to access the first graph.
   ENTER
3. Use the arrows to select <`ON`> and turn on Plot 1.
   ENTER
4. Use the arrows to select the box plot picture and enable it.
   ENTER
5. Use the arrows to navigate to <`Xlist`>
6. If "L1" is not selected, select it.
   2nd, `[L1]`, ENTER
7. Use the arrows to navigate to <`Freq`>.
8. Indicate that the frequencies are in `[L2]`.
   2nd, `[L2]`, ENTER
9. Go back to access other graphs.
   2nd, `[STAT PLOT]`
10. **Be sure to deselect or clear all equations before graphing** using the method mentioned above.
11. View the box plot.
    GRAPH, `[STAT PLOT]`

14.9.4 Linear Regression

14.9.4.1 Sample Data

The following data is real. The percent of declared ethnic minority students at De Anza College for selected years from 1970 - 1995 was:

Year	Student Ethnic Minority Percentage
1970	14.13
1973	12.27
1976	14.08
1979	18.16
1982	27.64
1983	28.72
1986	31.86
1989	33.14
1992	45.37
1995	53.1

Table 14.19: The independent variable is "Year," while the independent variable is "Student Ethnic Minority Percent."

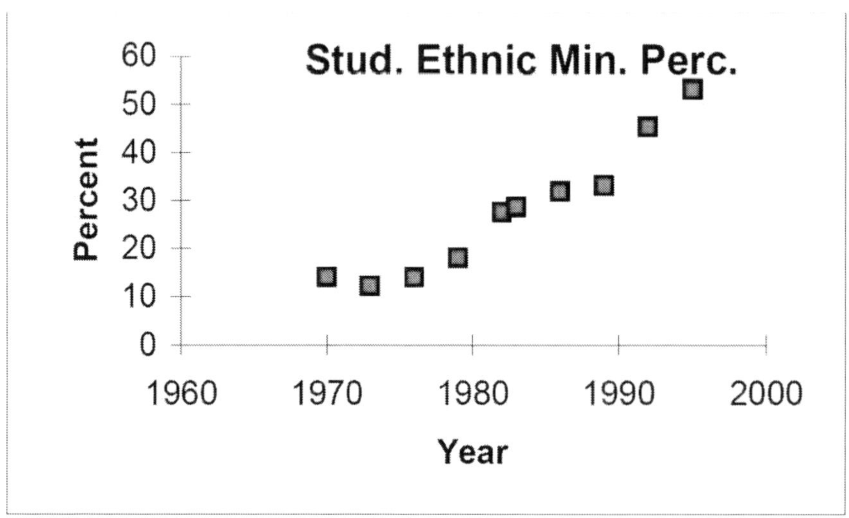

Figure 14.6: By hand, verify the scatterplot above.

NOTE: The TI-83 has a built-in linear regression feature, which allows the data to be edited.The x-values will be in [L1]; the y-values in [L2].

To enter data and do linear regression:

1. ON Turns calculator on

2. Before accessing this program, be sure to turn off all plots.
   - Access graphing mode.
     2nd, [STAT PLOT]
   - Turn off all plots.
     4, ENTER
3. Round to 3 decimal places. To do so:
   - Access the mode menu.
     MODE, [STAT PLOT]
   - Navigate to <Float> and then to the right to <3>.

   - All numbers will be rounded to 3 decimal places until changed.

4. Enter statistics mode and clear lists [L1] and [L2], as describe above.

5. Enter editing mode to insert values for x and y.
   STAT, ENTER
6. Enter each value. Press ENTER to continue.

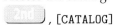

To display the correlation coefficient:

1. Access the catalog.
   2nd, [CATALOG]
2. Arrow down and select <DiagnosticOn>

3. r and r^2 will be displayed during regression calculations.
4. Access linear regression.

5. Select the form of $y = a + bx$

The display will show:

LinReg

- $y = a + bx$
- $a = -3176.909$
- $b = 1.617$
- $r^2 = 0.924$
- $r = 0.961$

This means the Line of Best Fit (Least Squares Line) is:

- $y = -3176.909 + 1.617x$
- $Percent = -3176.909 + 1.617(year\ \#)$

The correlation coefficient $r = 0.961$

To see the scatter plot:

1. Access graphing mode.
   2nd, [STAT PLOT]
2. Select <1:plot 1> To access plotting - first graph.
   ENTER
3. Navigate and select <ON> to turn on Plot 1.
   <ON> ENTER
4. Navigate to the first picture.
5. Select the scatter plot.
   ENTER
6. Navigate to <Xlist>
7. If [L1] is not selected, press 2nd, [L1] to select it.
8. Confirm that the data values are in [L1].
   <ON> ENTER
9. Navigate to <Ylist>
10. Select that the frequencies are in [L2].
    2nd, [L2], ENTER
11. Go back to access other graphs.
    2nd, [STAT PLOT]
12. Use the arrows to turn off the remaining plots.
13. Access WINDOW to set the graph parameters.
    - $X_{min} = 1970$
    - $X_{max} = 2000$
    - $X_{scl} = 10$ (spacing of tick marks on x-axis)
    - $Y_{min} = -0.05$
    - $Y_{max} = 60$
    - $Y_{scl} = 10$ (spacing of tick marks on y-axis)
    - $X_{res} = 1$
14. Be sure to deselect or clear all equations before graphing, using the instructions above.
15. Press GRAPH to see the scatter plot.

To see the regression graph:

1. Access the equation menu. The regression equation will be put into Y1.

2. Access the vars menu and navigate to <5: Statistics>
   VARS, 5
3. Navigate to <EQ>.
4. <1: RegEQ> contains the regression equation which will be entered in Y1.
   ENTER
5. Press GRAPH. The regression line will be superimposed over scatter plot.

To see the residuals and use them to calculate the critical point for an outlier:

1. Access the list. RESID will be an item on the menu. Navigate to it.
   2nd, [LIST], <RESID>
2. Confirm twice to view the list of residuals. Use the arrows to select them.
   ENTER, ENTER
3. The critical point for an outlier is: $1.9\sqrt{\frac{SSE}{n-2}}$ where:
   - n = number of pairs of data
   - SSE = sum of the squared errors
   - $\sum\left(residual^2\right)$
4. Store the residuals in [L3].

   STO▶, 2nd, [L3], ENTER
5. Calculate the $\frac{(residual)^2}{n-2}$. Note that $n-2=8$
   2nd, [L3], x², ÷, 8
6. Store this value in [L4].
   STO▶, 2nd, [L4], ENTER
7. Calculate the critical value using the equation above.
   1, ., 9, X, 2nd, [V], 2nd, [LIST] ▶, ▶, 5, 2nd, [L4],),), ENTER
8. Verify that the calculator displays: 7.642669563. This is the critical value.
9. Compare the absolute value of each residual value in [L3] to 7.64 . If the absolute value is greater than 7.64, then the (x, y) corresponding point is an outlier. In this case, none of the points is an outlier.

To obtain estimates of y for various x-values:
There are various ways to determine estimates for "y". One way is to substitute values for "x" in the equation. Another way is to use the TRACE on the graph of the regression line.

14.9.5 TI-83, 83+, 84 instructions for distributions and tests

14.9.5.1 Distributions

Access DISTR (for "Distributions").

For technical assistance, visit the Texas Instruments website at http://www.ti.com[28] and enter your calculator model into the "search" box.

Binomial Distribution

- `binompdf(n,p,x)` corresponds to P(X = x)
- `binomcdf(n,p,x)` corresponds to P(X ≤ x)
- To see a list of all probabilities for x: 0, 1, . . . , n, leave off the "x" parameter.

Poisson Distribution

- `poissonpdf(λ,x)` corresponds to P(X = x)
- `poissoncdf(λ,x)` corresponds to P(X ≤ x)

[28]http://www.ti.com

Continuous Distributions (general)

- $-\infty$ uses the value -1EE99 for left bound
- $+\infty$ uses the value 1EE99 for right bound

Normal Distribution

- `normalpdf(x,μ,σ)` yields a probability density function value (only useful to plot the normal curve, in which case "x" is the variable)
- `normalcdf(left bound, right bound, μ,σ)` corresponds to P(left bound $<$ X $<$ right bound)
- `normalcdf(left bound, right bound)` corresponds to P(left bound $<$ Z $<$ right bound) - standard normal
- `invNorm(p,μ,σ)` yields the critical value, k: P(X $<$ k) = p
- `invNorm(p)` yields the critical value, k: P(Z $<$ k) = p for the standard normal

Student-t Distribution

- `tpdf(x,df)` yields the probability density function value (only useful to plot the student-t curve, in which case "x" is the variable)
- `tcdf(left bound, right bound, df)` corresponds to P(left bound $<$ t $<$ right bound)

Chi-square Distribution

- X^2`pdf(x,df)` yields the probability density function value (only useful to plot the chi^2 curve, in which case "x" is the variable)
- X^2`cdf(left bound, right bound, df)` corresponds to P(left bound $< X^2 <$ right bound)

F Distribution

- `Fpdf(x,dfnum,dfdenom)` yields the probability density function value (only useful to plot the F curve, in which case "x" is the variable)
- `Fcdf(left bound,right bound,dfnum,dfdenom)` corresponds to P(left bound $<$ F $<$ right bound)

14.9.5.2 Tests and Confidence Intervals

Access `STAT` and `TESTS`.

For the Confidence Intervals and Hypothesis Tests, you may enter the data into the appropriate lists and press `DATA` to have the calculator find the sample means and standard deviations. Or, you may enter the sample means and sample standard deviations directly by pressing `STAT` once in the appropriate tests.

Confidence Intervals

- `ZInterval` is the confidence interval for mean when σ is known
- `TInterval` is the confidence interval for mean when σ is unknown; s estimates σ.
- `Z-PropZInt` is the confidence interval for proportion

NOTE: The confidence levels should be given as percents (ex. enter "95" for a 95% confidence level).

Hypothesis Tests

- `Z-Test` is the hypothesis test for single mean when σ is known
- `T-Test` is the hypothesis test for single mean when σ is unknown; s estimates σ.
- `2-SampZTest` is the hypothesis test for 2 independent means when both σ's are known

- `2-SampTTest` is the hypothesis test for 2 independent means when both σ's are unknown
- `1-PropZTest` is the hypothesis test for single proportion.
- `2-PropZTest` is the hypothesis test for 2 proportions.
- X^2`-Test` is the hypothesis test for independence.

NOTE: Input the null hypothesis value in the row below "`Inpt`." For a test of a single mean, "$\mu\varnothing$" represents the null hypothesis. For a test of a single proportion, "p$\varnothing$" represents the null hypothesis. Enter the alternate hypothesis on the bottom row.

Solutions to Exercises in Chapter 14

Solutions to Practice Final Exam 1

Solution to Exercise 14.1.1 (p. 547)
B: Independent.
Solution to Exercise 14.1.2 (p. 547)
C: $\frac{4}{16}$
Solution to Exercise 14.1.3 (p. 547)
B: Two measurements are drawn from the same pair of individuals or objects.
Solution to Exercise 14.1.4 (p. 548)
B: $\frac{68}{118}$
Solution to Exercise 14.1.5 (p. 548)
D: $\frac{30}{52}$
Solution to Exercise 14.1.6 (p. 548)
B: $\frac{8}{40}$
Solution to Exercise 14.1.7 (p. 548)
B: 2.78
Solution to Exercise 14.1.8 (p. 549)
A: 8.25
Solution to Exercise 14.1.9 (p. 549)
C: 0.2870
Solution to Exercise 14.1.10 (p. 549)
C: Normal
Solution to Exercise 14.1.11 (p. 549)
D: H_a: $p_A \neq p_B$
Solution to Exercise 14.1.12 (p. 550)
B: believe that the pass rate for Math 1A is different than the pass rate for Math 1B when, in fact, the pass rates are the same.
Solution to Exercise 14.1.13 (p. 550)
B: not reject H_0
Solution to Exercise 14.1.14 (p. 550)
C: Iris
Solution to Exercise 14.1.15 (p. 550)
C: Student-t
Solution to Exercise 14.1.16 (p. 551)
B: is left-tailed
Solution to Exercise 14.1.17 (p. 551)
C: cluster sampling
Solution to Exercise 14.1.18 (p. 551)
C: Mode
Solution to Exercise 14.1.19 (p. 551)
A: the probability that an outcome of the data will happen purely by chance when the null hypothesis is true.
Solution to Exercise 14.1.20 (p. 552)
D: stratified
Solution to Exercise 14.1.21 (p. 552)
B: 25
Solution to Exercise 14.1.22 (p. 552)
C: 4
Solution to Exercise 14.1.23 (p. 553)
A: (1.85, 2.32)

Solution to Exercise 14.1.24 (p. 553)
C: Both above are correct.
Solution to Exercise 14.1.25 (p. 553)
C: 5.8
Solution to Exercise 14.1.26 (p. 553)
C: 0.6321
Solution to Exercise 14.1.27 (p. 553)
A: 0.8413
Solution to Exercise 14.1.28 (p. 553)
A: (0.6030, 0.7954)
Solution to Exercise 14.1.29 (p. 554)
A: $N\left(145, \frac{14}{\sqrt{10}}\right)$
Solution to Exercise 14.1.30 (p. 554)
D: 3.66
Solution to Exercise 14.1.31 (p. 554)
B: 5.1
Solution to Exercise 14.1.32 (p. 554)
A: 13.46
Solution to Exercise 14.1.33 (p. 555)
B: There is a strong linear pattern. Therefore, it is most likely a good model to be used.
Solution to Exercise 14.1.34 (p. 555)
B: ${Chi^2}_3$
Solution to Exercise 14.1.35 (p. 555)
D: 70
Solution to Exercise 14.1.36 (p. 555)
B: The choice of major and the gender of the student are not independent of each other.
Solution to Exercise 14.1.37 (p. 556)
A: Chi^2 goodness of fit

Solutions to Practice Final Exam 2

Solution to Exercise 14.2.1 (p. 557)
B: parameter
Solution to Exercise 14.2.2 (p. 557)
A
Solution to Exercise 14.2.3 (p. 558)
A: 6
Solution to Exercise 14.2.4 (p. 558)
C: 0.02
Solution to Exercise 14.2.5 (p. 558)
C: none of the above
Solution to Exercise 14.2.6 (p. 558)
D: $\frac{100}{140}$
Solution to Exercise 14.2.7 (p. 559)
A: ≈ 0
Solution to Exercise 14.2.8 (p. 559)
B: The values for x are: $\{1, 2, 3, ..., 14\}$
Solution to Exercise 14.2.9 (p. 559)
C: 0.9417
Solution to Exercise 14.2.10 (p. 560)
D: Binomial

Solution to Exercise 14.2.11 (p. 560)
D: 8.7
Solution to Exercise 14.2.12 (p. 560)
A: -1.96
Solution to Exercise 14.2.13 (p. 561)
A: 0.6321
Solution to Exercise 14.2.14 (p. 561)
D: 360
Solution to Exercise 14.2.15 (p. 561)
B: $N\left(72, 72\sqrt{5}\right)$
Solution to Exercise 14.2.16 (p. 561)
A: $\frac{3}{9}$
Solution to Exercise 14.2.17 (p. 561)
(D)
Solution to Exercise 14.2.18 (p. 562)
B: 5.5
Solution to Exercise 14.2.19 (p. 562)
D: 7.56
Solution to Exercise 14.2.20 (p. 562)
A: 5
Solution to Exercise 14.2.21 (p. 562)
B: (4.26, 6.74)
Solution to Exercise 14.2.22 (p. 563)
B: 0.2
Solution to Exercise 14.2.23 (p. 563)
A: -1
Solution to Exercise 14.2.24 (p. 563)
C: dependent groups
Solution to Exercise 14.2.25 (p. 563)
D: Reject H_o. There is a difference in the mean scores.
Solution to Exercise 14.2.26 (p. 564)
C: The proportion for males is higher than the proportion for females.
Solution to Exercise 14.2.27 (p. 564)
B: No
Solution to Exercise 14.2.28 (p. 564)
C: Not enough information given to solve the problem
Solution to Exercise 14.2.29 (p. 564)
B: No
Solution to Exercise 14.2.30 (p. 564)
C: $\hat{y} = -79.96x - 0.0094$
Solution to Exercise 14.2.31 (p. 565)
A: 69
Solution to Exercise 14.2.32 (p. 565)
C
Solution to Exercise 14.2.33 (p. 565)
B: The *p–value* is < 0.01 , the distribution is uniform.
Solution to Exercise 14.2.34 (p. 565)
C: The test is to determine if the different groups have the same averages.

Tables[29]

NOTE: When you are finished with the table link, use the back button on your browser to return here.

Tables (NIST/SEMATECH e-Handbook of Statistical Methods, http://www.itl.nist.gov/div898/handbook/, January 3, 2009)

- Student-t table[30]
- Normal table[31]
- Chi-Square table[32]
- F-table[33]
- All four tables can be accessed by going to http://www.itl.nist.gov/div898/handbook/eda/section3/eda367.htm[34]

95% Critical Values of the Sample Correlation Coefficient Table

- 95% Critical Values of the Sample Correlation Coefficient[35]

  NOTE: The url for this table is http://cnx.org/content/m17098/latest/

[29]This content is available online at <http://cnx.org/content/m19138/1.3/>.
[30]http://www.itl.nist.gov/div898/handbook/eda/section3/eda3672.htm
[31]http://www.itl.nist.gov/div898/handbook/eda/section3/eda3671.htm
[32]http://www.itl.nist.gov/div898/handbook/eda/section3/eda3674.htm
[33]http://www.itl.nist.gov/div898/handbook/eda/section3/eda3673.htm
[34]http://www.itl.nist.gov/div898/handbook/eda/section3/eda367.htm
[35]http://cnx.org/content/m17098/latest/

Glossary

A

Addition Rule

For any events A and B in the sample space $P(A \, or \, B) = P(A) + P(B) - P(A and B)$.

Analysis of Variance

Also referred to as ANOVA. A method of testing whether or not the means of three or more populations are equal. The method is applicable if:

- All populations of interest are normally distributed.
- The populations have equal standard deviations.
- Samples (not necessarily of the same size) are randomly and independently selected from each population.

The test statistic for analysis of variance is the F-ratio.

AND

Logical operation over the subsets of a set. In statistics, if A and B are any two events (subsets in the sample space), then the event " A and B" consists of all possible outcomes that are common to both A and B.

Arithmetic Mean

The sum of the values divided by the number of values. The notation for the mean of a sample is $\overline{x}$. The notation for the mean of a population is μ.

Average

A number that describes the central tendency of the data. There are a number of specialized averages, including the arithmetic mean, weighted mean, median, mode, and geometric mean.

B

Bayes' Theorem

Developed by Reverend Bayes in the 1700s. A rule designed to find the probability of one event, A, occurring, given that a finite set of other events, $\{B_i, i = 1, 2, ..., l\}$, has occurred.

Bernoulli Trials

An experiment with the following characteristics:

- There are only 2 possible outcomes called "success" and "failure" for each trial.
- The probabilities p of success and $q = 1 - p$ of failure are the same for any trial.

Bias

A possible consequence if certain members of the population are denied the chance to be selected for the sample.

Binomial Distribution

A discrete random variable (RV) which arises from the Bernoulli trials. There are a fixed number, n, of independent trials. "Independent" means that the result of any trial (for example, trial 1) does not affect the results of all the following trials, and all trials are conducted under the same conditions. Under these circumstances the binomial RV X is defined as the number of successes in n trials. The notation is: $X \sim B(n, p)$. The mean is $\mu = np$ and the standard deviation is $\sigma = \sqrt{npq}$. The probability of having exactly x successes in n trials is $P(X = x) = \binom{n}{x} p^x q^{n-x}$.

C

Central Limit Theorem

Given a random variable (RV) with known mean μ and known standard deviation σ. We are sampling with size n and we are interested in two new RVs - the sample mean, $\bar{X}$, and the sample sum, ΣX. If the size n of the sample is sufficiently large, then $\bar{X} \sim N\left(\mu, \frac{\sigma}{\sqrt{n}}\right)$ and $\Sigma X \sim N\left(n\mu, \sqrt{n}\sigma\right)$. If the size n of the sample is sufficiently large, then the distribution of the sample means and the distribution of the sample sums will approximate a normal distribution regardless of the shape of the population. The mean of the sample means will equal the population mean and the mean of the sample sums will equal n times the population mean. The standard deviation of the distribution of the sample means, $\frac{\sigma}{\sqrt{n}}$, is called the standard error of the mean.

Charts

Special graphical formats used to visualize a frequency distribution. They include, but are not limited to: **histograms, frequency polygons, cumulative frequency polygons, box plots, stemplots, bar charts, Venn and tree diagrams, and pie charts**.

Chi-square Distribution

A continuous distribution with the following characteristics:

- The random variable (RV) is continuous and takes on only nonnegative values (in fact, it is the sum of squares of k independent normal distributions).
- There is a "family" of Chi-square distributions. Each representative of the family is completely defined by the number of degrees of freedom, $k-1$, where k is the number of categories (not the size of sample).
- The pdf is positively skewed (skewed right). However, as k increases (k>90), the distribution approximates the normal distribution.

The notation is: $\chi^2 \sim \chi^2_{df}$. For the χ^2 distribution, the population mean is $\mu = df$ and the population standard deviation is $\sigma = \sqrt{2 \cdot df}$. The Chi-square distribution is used to calculate the test statistic for the **Goodness-of-fit Test** (to determine if a population follows a specified distribution), for the **Test of Independence** (to determine if two factors are related or not), and for the **Test of a Single Variance.**

Class Mark

Midpoint of the class.

Classes

Intervals in which the data are grouped. It is convenient to group outcomes into classes when working with large amounts of data. For example, every bar in a histogram corresponds to one class (one interval) and the midpoint of the interval can be chosen as a representative of all outcomes in the class. The Midpoint of the class is often called the **class mark**.

Cluster Sampling

A procedure that is used if the population is dispersed over a wide geographic area. The population is divided into units or groups (counties, precincts, blocks, etc.) called primary units. Then some of the primary units are randomly chosen, and all members of those primary units are the sample.

Coefficient of Correlation

A measure developed by Karl Pearson (early 1900s) that gives the strength of association between the independent variable and the dependent variable. The formula is:

$$r = \frac{n\sum XY - (\sum X)(\sum Y)}{\sqrt{\left[n\sum X^2 - (\sum X)^2\right]\left[n\sum Y^2 - (\sum Y)^2\right]}}, \qquad ()$$

where n is the number of data points. The coefficient r is not more then 1 nor less then -1. The closer the coefficient is to ± 1, the stronger the evidence of a significant linear relationship between X and Y.

Complement Event

The event consisting of all outcomes that are in the sample space but are not in the given event.

Conditional Probability

The likelihood that an event will occur given that another event has already occurred.

Confidence Interval (CI)

An interval estimate for an unknown population parameter. This depends on:

- The desired confidence level.
- Information that is known about the distribution (for example, known standard deviation).
- The sample and its size.

Confidence Level (CL)

The percent expression for the probability that the confidence interval contains the true population parameter. For example, if the $CL = 90\%$, then in 90 out of 100 samples the interval estimate will enclose the true population parameter.

Contingency Table

The method of displaying a frequency distribution as a table with rows and columns to show how two variables may be dependent (contingent) upon each other. The table provides an easy way to calculate conditional probabilities.

Continuous Random Variable (RV)

A random variable (RV) whose outcomes are measured.

Example: The height of trees in the forest is a continuous RV.

Correlation Analysis

A group of statistical procedures used to measure the strength of the relationship between two variables.

Counting Principal

If there are m ways of doing one thing and n ways of doing another, then there are $m \times n$ ways of doing both.

Example: A cafe offers $m = 5$ kinds of coffee and $n = 7$ kinds of cake. There are 35 ways to serve coffee with cake.

Critical Value

The dividing point between the region where the null hypothesis is not rejected and the region where it is rejected. For a one-tailed hypothesis test, there is only one critical value. For a two-tailed hypothesis test, there are two critical values—one in each tail— with the same absolute value and opposite signs.

Cumulative Distribution Function (CDF)

Given a quantitative random variable (RV) X, the function $P(X \leq x)$ is called the Cumulative Distribution Function (CDF). The CDF is the sum of the probabilities of all values of X that are less than or equal to a particular x.

Cumulative Relative Frequency

The term applies to an ordered set of observations from smallest to largest. The Cumulative Relative Frequency is the sum of the relative frequencies for all values that are less than or equal to the given value.

D Data

A set of observations (a set of possible outcomes). Most data can be put into two groups: **qualitative** (hair color, ethnic groups and other **attributes** of the population) and **quantitative** (distance traveled to college, number of children in a family, etc.). Quantitative data can be separated into two subgroups: **discrete** and **continuous**. Data is discrete if it is the result of counting (the number of students of a given ethnic group in a class, the number of books on a shelf, etc.). Data is continuous if it is the result of measuring (distance traveled, weight of luggage, etc.)

Degrees of Freedom (df)

The number of sample values that are free to vary.

Dependant Samples

Samples chosen in such a way that they are not independent of each other. Paired samples are dependent because two measurements are taken from the same individual or item.

Example: If the test scores of 13 individuals were recorded before a new teaching method was introduced, and then after using the new method, the paired samples are dependent.

Descriptive Statistics

The numerical and graphical ways used to describe and display the important characteristics of data; for example, charts, frequency distributions, measures of central tendency and measures of spread and skewness.

Discrete Random Variable

A random variable (RV) whose outcomes are counted.

Domain

The set of possible values for the independent variable.

Example:

- We are interested in the longevity of human life in years. The domain is $\{0, 1, 2, 3..., 120\}$.
- We are interested in the suit of a regular 52-card deck. The domain is $\{\heartsuit; \diamondsuit; \clubsuit; \spadesuit\}$.

E Equally Likely

Each outcome of an experiment has the same probability of occurring.

Error Bound for a Population Mean (EBM)

The margin of error. It depends on the confidence level, sample size, and the known or estimated population standard deviation.

Error Bound for a Proportion (EBP)

The margin of error. It depends on the confidence level, sample size, and the estimated (from the sample) proportion of successes.

Event

A subset in the set of all outcomes of an experiment. The set of all outcomes of an experiment is called a **sample space** and denoted usually by S. An event is any arbitrary subset in **S**. It can contain one outcome, two outcomes, no outcomes (empty subset), the entire sample space, etc. Standard notations for events are capital letters such as A, B, C, etc.

Expected Value

The arithmetic average when an experiment is repeated many times. The Expected Value is called the long-term mean or average. Notation: $E(x), \mu$. For a discrete random variable (RV) with probability distribution function $P(X = x)$, the definition also can be written in the form $E(X) = \mu = \sum xP(x)$.

Experiment

A planned activity carried out under controlled conditions.

Exponential Distribution

A continuous random variable (RV) that appears when we are interested in the intervals of time between some random events, for example, the length of time between emergency arrivals at a hospital. Notation: $X{\sim}Exp(m)$. The mean is $\mu = \frac{1}{m}$ and the standard deviation is $\sigma = \frac{1}{m}$. The probability density function is $f(x) = me^{-mx}$, $x \geq 0$ and the cumulative distribution function is $P(X \leq x) = 1 - e^{-mx}$.

F

F Distribution

Developed by Sir Ronald Fisher. The F Distribution has the following characteristics:

- The random variable (RV) is a ratio (called the F-ratio) of two sums of weighted squares. It is continuous and takes on only nonnegative value.
- The pdf is positively skewed (skewed to the right).
- There is a "family" of F distributions.

Every representative of the family is defined by 2 parameters: the number of degrees of freedom for the numerator in the F-ratio and the number of degrees of freedom in the denominator in the F-ratio. The F Distribution is used to test of 2 population variances and in ANOVA hypothesis tests.

Frequency Distribution

A grouping of data into mutually exclusive classes showing the number of outcomes in each class.

Frequency

The number of times a value of the data occurs.

G

Geometric Distribution

A discrete random variable (RV) which arises from the Bernoulli trials. The trials are repeated until the first success. The geometric variable X is defined as the number of trials until the first success. Notation: $X{\sim}\ G(p)$. The mean is $\mu = \frac{1}{p}$ and the standard deviation is $\sigma = \sqrt{\frac{1}{p} \cdot \left(\frac{1}{p} - 1\right)}$ The probability of exactly x failures before the first success is given by the formula: $P(X = x) = p(1 - p)^{x-1}$.

H

Hypergeometric Distribution

A discrete random variable (RV) that is characterized by

- A fixed number of trials.
- The probability of success is not the same from trial to trial.

We sample from two groups of items when we are interested in only one group. X is defined as the number of successes out of the total number chosen. Notation: $X{\sim}H(r, b, n)$., where r = the number of items in the group of interest, b = the number of items in the group not of interest, and n = the number of items chosen.

Hypothesis

A statement about the value of a population parameter. In case of two hypotheses, the statement assumed to be true is called the null hypothesis (notation H_0) and the contradictory statement is called the alternate hypothesis (notation H_a).

Hypothesis Testing

Based on sample evidence, hypothesis testing is a procedure that determines whether the null hypothesis is a reasonable statement and cannot be rejected, or is unreasonable and should be rejected.

I

Independent Events

The occurrence of one event has no effect on the probability of the occurrence of any other event. Events A and B are independent if any of the following is true:

- $P(A|B) = P(A)$
- $P(B|A) = P(B)$
- $P(AandB) = P(A)P(B)$

.

Independent Samples

Samples that are not related in any way.

Inferential Statistics

Also called statistical inference or inductive statistics. This facet of statistics deals with estimating a population parameter based on a sample statistic. For example, if 4 out of the 100 calculators sampled are defective we might infer that 4 percent of the production is defective.

Interquartile Range (IRQ)

The distance between the third quartile (Q3) and the first quartile (Q1). IQR = Q3 - Q1.

Interval Estimate

Based on sample information, an Interval Estimate is an interval of numbers that may contain a population parameter.

L

Level of Significance of the Test

Probability of a Type I error (reject the null hypothesis when it is true). Notation: α. In hypothesis testing, the Level of Significance is called the preconceived α or the preset α.

Linear Regression Equation

A linear equation in the form $\hat{y} = a + bx$, that defines the relationship between two variables. It is used to predict the dependent variable y based on a selected value of independent variable x.

M

Mean

A number that measures the central tendency. A common name for mean is 'average.' The term 'mean' is a shortened form of 'arithmetic mean.' By definition, the mean for a sample (denoted by $\overline{x}$) is $\overline{x} = \frac{\textit{Sum of all values in the sample}}{\textit{Number of values in the sample}}$, and the mean for a population (denoted by μ) is $\mu = \frac{\textit{Sum of all values in the population}}{\textit{Number of values in the population}}$.

Median

A number that separates ordered data into halves. Half the values are the same number or smaller than the median and half the values are the same number or larger than the median. The median may or may not be part of the data.

Mode

The value that appears most frequently in a set of data.

Multiplication Rule

For any events A and B in the sample space, $P(AandB) = P(A \mid B) \cdot P(B) = P(B \mid A) \cdot P(A)$.

Mutually Exclusive

An observation cannot fall into more than one class (category). Being in one category prevents being in a mutually exclusive category.

N

Normal Distribution

A continuous random variable (RV) with pdf $f(x) = \frac{1}{\sigma\sqrt{2\pi}}e^{\frac{-(x-\mu)^2}{2\sigma^2}}$ where μ is the mean of the distribution and σ is the standard deviation. Notation: $X \sim N(\mu, \sigma)$. If $\mu = 0$ and $\sigma = 1$, the RV is called **the standard normal distribution**.

O

One-Tailed Test

Used when the alternate hypothesis states a direction. The rejection region is in one tail. Example: H_a:$\mu > 40$ with the rejection region in the right tail.

OR

Logical operation over the subsets of a set. In statistics, if A and B are any two events (subsets in the sample space), then the event "A **or** B" consists of all outcomes that are in A, or in B, or in both A and B.

Outcome (observation)

A particular result of an experiment.

Outlier

An observation that does not fit the rest of the data.

P

p-value

The probability that an event will happen purely by chance assuming the null hypothesis is true. The smaller the p-value, the stronger the evidence is against the null hypothesis.

Parameter

A numerical characteristic of the population.

Example: The mean price to rent a 1-bedroom apartment in California.

pdf

see **Probability Density Function**

PDF

see **Probability Distribution Function**

Percentile

A number that divides ordered data into hundredths.

Example: Let a data set contain 200 ordered observations starting with $\{2.3, 2.7, 2.8, 2.9, 2.9, 3.0...\}$. Then the first percentile is $\frac{(2.7+2.8)}{2} = 2.75$, because 1% of the data is to the left of this point on the number line and 99% of the data is on its right. The second percentile is $\frac{(2.9+2.9)}{2} = 2.9$. Percentiles may or may not be part of the data. In this example, the first percentile is not in the data, but the second percentile is. The median of the data is the second quartile and the 50th percentile. The first and third quartiles are the 25th and the 75th percentiles, respectively.

Point Estimate

A single number computed from a sample and used to estimate a population parameter.

Poisson Distribution

A discrete random variable (RV) is the number of times a certain event will occur in a specific interval. Characteristics of the variable:

- The probability that the event occurs in a given interval is the same for all intervals.
- The events occur with a known mean and independently of the time since the last event.

The distribution is defined by the mean μ of the event in the interval. Notation: $X \sim P(\mu)$. The mean is $\mu = np$. The standard deviation is $\sigma = \mu$. The probability of having exactly x successes in r trials is $P(X = x) = e^{-\mu}\frac{\mu^x}{x!}$. The Poisson distribution is often used to approximate the binomial distribution when n is "large" and p is "small" (a general rule is that n should be greater than or equal to 20 and p should be less than or equal to .05).

Population

The collection, or set, of all individuals, objects, or measurements whose properties are being studied.

Preconceived α

The probability of rejecting the null hypothesis when the null hypothesis is true (α is equal to the probability of a Type I error). α is called the **level of significance of the test**. Also called the preset α.

Probability

A number between 0 and 1, inclusive, that gives the likelihood that a specific event will occur. The foundation of statistics is given by the following 3 axioms (by A. N. Kolmogorov, 1930's): Let S denote the sample space and A and B are two events in S . Then:

- $0 \leq P(A) \leq 1$.
- If A and B are any two mutually exclusive events, then $P(AorB) = P(A) + P(B)$.
- $P(S) = 1$.

Probability Density Function (pdf)

A mathematical description of a continuous random variable (RV). For any specific value x, $P(X = x) = 0$. By definition, the pdf is any positive function $f(x)$ over the real numbers such that the area bounded above by $f(x)$, below by the $x - axis$ and from the right by a vertical line $X = x$ is equal to the probability $P(X \leq x)$.

Probability Distribution Function (PDF)

A mathematical description of a discrete random variable (RV), given either in the form of an equation (by formula) or in the form of a table listing all the possible outcomes of an experiment and the probability associated with each outcome.

Example: A biased coin with probability 0.7 of heads is tossed 5 times. We are interested in the number of heads (X = the number of heads). X is Binomial: $X \sim B(5, .7)$. $P(X = x) = \begin{pmatrix} 5 \\ x \end{pmatrix} .7^x .3^{5-x}$ or in the form of the table.

x	$P(X=x)$
0	0.0024
1	0.0284
2	0.1323
3	0.3087
4	0.3602
5	0.1681

Table

Probability Distribution

The common name for **Probability Density Function (pdf)** and **Probability Distribution Function (PDF)**.

Proportion

- As a number: A proportion is the number of successes divided by the total number in the sample.
- As a probability distribution: Given a binomial random variable (RV), $X \sim B(n, p)$, consider the ratio of the number X of successes in n Bernouli trials to the number n of trials. $P' = \frac{X}{n}$. This new RV is called a proportion, and if the number of trials, n, is large enough, $P' \sim N\left(p, \frac{pq}{n}\right)$.

Q **Qualitative Data**

see **Data**.

Quantitative Data

Quartiles

The numbers that separate the data into quarters. Quartiles may or may not be part of the data. The second quartile is the median of the data.

R **Random Variable (RV)**

see **Variable**

Range

Difference between the highest and lowest values: Range = Highest value – Lowest value.

Relative Frequency

The ratio of the number of times a value of the data occurs in the set of all outcomes to the number of all outcomes.

S **Sample**

A portion of the population under study. A sample is representative if it characterizes the population being studied.

Sample Error

The difference between a sample statistic and the corresponding population parameter that can be attributed to sampling (to chance).

Sample Space

The set of all possible outcomes of an experiment.

Sampling

A procedure for gathering information about the entire population by selecting only a portion of the population. The more popular random procedures are systematic sampling, simple random sampling, stratified sampling, and cluster sampling.

Scatter Diagram

A chart that visually depicts the relationship between two variables.

Simple Random Sampling

A sampling scheme in which every member of the population has the same chance of being selected.

Special Rule for Addition

For this rule to apply the events must be mutually exclusive: $P(AorB) = P(A) + P(B)$.

Special Rule for Multiplication

For this rule to apply the events must be independent: $P(AandB) = P(A)P(B)$.

Standard Deviation

A number that is equal to the square root of the variance and measures how far data values are from their mean. Notation: s for sample standard deviation and $\sigma$for population standard deviation.

Standard Error of the Mean

The standard deviation of the distribution of the sample means, $\frac{\sigma}{\sqrt{n}}$.

Standard Normal Distribution

A continuous random variable (RV) $X \sim N(0,1)$. When X follows the standard normal distribution, it is often noted as $Z \sim N(0,1)$.

Statistic

A numerical characteristic of the sample. A statistic estimates the corresponding population parameter. For example, the average number of full-time students in a 7:30 a.m. class for this term (statistic) is an estimate for the average number of full-time students in any class this term (parameter).

Statistics

The science of collecting, organizing, analyzing, and interpreting numerical data.

Stratified Random Sampling

A population is divided into groups (called strata) and then a random sample is selected from each stratum.

Student-t Distribution

Investigated and reported by William S. Gossett in 1908 and published under the pseudonym Student. The major characteristics of the random variable (RV) are:

- The Student-t is continuous and assumes any real values.
- The pdf is symmetrical about its mean of zero. However, it is more spread out and flatter at the apex than the normal distribution.
- The Student-t approaches the standard normal distribution as n gets larger.
- There is a "family" of t distributions: every representative of the family is completely defined by the number of degrees of freedom (one less than the number, n, of data).

Notation: t_{df} where df is the degrees of freedom. $df = n - 1$.

Systematic Sampling

A population is arranged in some standard list (for example, alphabetically) and then every m-th (for example, every fifth) representative of the list is taken in the sample starting from a random initial representative.

T t statistic

Calculated from the data according to the Student-t distribution statistic that is used to conduct a hypothesis test and to make the statistical inference about the whole population. If data contains n observations, then the number of degrees of freedom for the Student-t distribution is $n - 1$. The t statistic is used, for example, when the population standard deviation is unknown, when n is small, and when samples are dependent (matched pairs hypothesis test). The t statistic formula is $t = \frac{\overline{x}-\mu}{\frac{s}{\sqrt{n}}}$.

Test Statistic

Calculated from the sample value that is used to conduct the hypothesis test and that makes the statistical inference about the whole population. The calculation depends on the choice of the appropriate distribution, which often is reflected in the name of statistic: z-score, **t** statistic, **F** statistic (**F** Ratio), etc.

Tree Diagram

The useful visual representation of a sample space and events in the form of a "tree" with branches marked by possible outcomes simultaneously with associated probabilities (frequencies, relative frequencies).

Type 1 Error

The decision is to reject the Null hypothesis, when, in fact, the Null hypothesis is true.

Type 2 Error

The decision is not to reject the Null hypothesis when, in fact, the Null hypothesis is false.

U Uniform Distribution

A continuous random variable (RV) that has equally likely outcomes over the domain, $a < x < b$. Often referred as the **Rectangular distribution** because the graph of the pdf has the form of a rectangle. Notation: $X \sim U(a, b)$. The mean is $\mu = \frac{a+b}{2}$ and the standard deviation is $\sigma = \sqrt{\frac{(b-a)^2}{12}}$ The probability density function is $f(X) = \frac{1}{b-a}$ for $a < X < b$ or $a \leq X \leq b$. The cumulative distribution is $P(X \leq x) = \frac{x-a}{b-a}$.

V Variable (Random Variable)

A characteristic of interest in a population being studied. The common notation for variables are upper case Latin letters X, Y, Z,.... The common notation for a specific value of a variable) are lower case Latin letters x, y, z,.... The variable in statistics differs from the variable in intermediate algebra in two following ways:

- The domain of a random variable (RV) is not necessarily a numerical set but it may be words. For example, if X = hair color then the domain is {black, blond, gray, red, brown}.
- We can tell a specific value of x that the variable X takes on only after performing the experiment.

Variance

Mean of the squared deviations from the mean. Square of the standard deviation. For a set of data, a deviation can be represented as $x - \overline{x}$ where x is a value of the data and $\overline{x}$ is the sample mean. The sample variance is equal to the sum of the squares of the deviations divided by the difference of the sample size and 1.

Venn Diagram

The useful visual representation of a sample space and events in the form of circles or ovals showing their intersections.

Z

z-score

The linear transformation of the form $z = \frac{x-\mu}{\sigma}$. If this transformation is applied to any normal distribution $X \sim N(\mu, \sigma)$, the result is the standard normal distribution $Z \sim N(0,1)$. If this transformation is applied to any specific value x of the RV with mean μ and standard deviation σ, the result is called the z-score of x. Z-scores allow us to compare data that are normally distributed but scaled differently.

Index of Keywords and Terms

Keywords are listed by the section with that keyword (page numbers are in parentheses). Keywords do not necessarily appear in the text of the page. They are merely associated with that section. *Ex.* apples, § 1.1 (1) **Terms** are referenced by the page they appear on. *Ex.* apples, 1